Mobile Antenna Systems Handbook

The Artech House Mobile Communications Series

John Walker, *Series Editor*

Cellular Radio: Analog and Digital Systems, Asha Mehrotra

The Evolution of Mobile Communications in the U.S. and Europe: Regulation, Technology, and Markets, Michael Paetsch

Land-Mobile Radio System Engineering, Garry C. Hess

Mobile Antenna Systems Handbook, K. Fujimoto, J. R. James

Mobile Information Systems, John Walker, editor

Narrowband Land-Mobile Radio Networks, Jean-Paul Linnartz

For a complete listing of *The Artech House Telecommunications Library,* turn to the back of this book

Mobile Antenna Systems Handbook

K. Fujimoto
J. R. James

Editors

Artech House
Boston • London

Library of Congress Cataloging-in-Publication Data
Mobile antenna systems handbook/edited by K. Fujimoto, J. R. James.
Includes bibliographical references and index.
ISBN 0-89006-539-X
1. Mobile communication systems. 2. Antennas (Electronics)
I. Fujimoto, K (Kyohei), 1929– . II. James, James R.
TK6570.M6M57 1994 94-7672
621.3845–dc20 CIP

A catalogue record for this book is available from the British Library

685 Canton Street
Norwood, MA 02062

International Standard Book Number: 0-89006-539-X
Library of Congress Catalog Card Number: 94-7672

10 9 8 7 6 5 4 3 2 1

Contents

Authorship by Chapter

Chapter 1 J.R. James and K. Fujimoto

Chapter 2
K. Kagoshima: 2.1.1, 2.1.2
W.C.Y. Lee: 2.1.3, 2.2, 2.3.3
K. Fujimoto: 2.3.1
K. Hirasawa: 2.3.2
T. Taga: 2.4

Chapter 3
K. Kagoshima: 3.1, 3.2, 3.3, 3.5
T. Taga: 3.4

Chapter 4
R. Mumford: 4.1.1, 4.1.2
Q. Balzano: 4.1.3, 4.1.4, 4.1.5, 4.3, 4.5
T. Taga: 4.2, 4.4

Chapter 5
H.K. Lindenmeier, L. Reiter, and J. Hopf: 5.1
K. Fujimoto: 5.2, 5.4
K. Hirasawa: 5.3

Chapter 6
T. Shiokawa and S. Ohmori: 6.1, 6.2, 6.3, 6.4
T. Teshirogi: 6.5, 6.6

Chapter 7 Y. Suzuki

Appendix K. Fujimoto and J.R. James

Preface

People worldwide are traveling more and more, and at the same time they wish to keep in contact. It is therefore no surprise that mobile communications is one of the most vibrant, expanding, challenging fields of present-day electronics, and it has been our pleasure and privilege to contribute this book on mobile antenna systems. Our mission has been all the more stimulating because antennas are a fascinating topic for many reasons.

Few topics can claim to be part of everyday life as mobile communications does: with people's hands-on daily experience as they adjust their portable radio and telephone whip antennas, most people have some knowledge of antennas and perhaps a widespread notion that there is nothing much left to be invented. Wire dipole and Yagi antennas are of course commonplace, but each application is different and conceals subtle differences in design and behavior. New antenna types continue to evolve, and printed antennas are a particular example of the impact of new materials and modern computational techniques. The new printed radiators have in turn made possible many innovative communication systems. There seems to be no end to the new scenarios and applications, and mobile communication is the predominant driving force worldwide in antenna creation. The unseen factor and perhaps the most difficult thing to appreciate about antennas is the often incommensurate amount of design work necessary to adapt a conventional antenna type to a new system application. Such design work has to be resourced and is an expensive commitment for industrial research and development laboratories. The quest is to make antennas satisfy more critical specifications and outperform their generic forms while satisfying new, eagerly sought system objectives. Compactness of size, bandwidth, robustness, and ease of operation are some of the many system parameters to be optimized, together with the all-important lowering of manufacturing costs. Learned journals and conferences reflect the intensity of effort and the demand for innovative, skillful antenna system designers. Such is the fascinating world of antenna system design.

Our book is, we believe, a timely account of the state of affairs, and we aim to present in one volume a variety of antenna techniques relating to communication, radar, navigation, and so forth that are not covered in other books, the distinction being the emphasis throughout on systems. Therefore, we have not set out to fill the text with fundamental antenna theory, but rather to illustrate by applications what is currently being achieved, what the system constraints are, and what scope exists for further advances or adaptation. We think it likely that most readers, be they antenna specialists, supporting mathematicians, physicists, or otherwise, will have some connection with electronic system design, and this of course includes many postgraduates who no longer dwell on one facet of antenna design, but encounter electromagnetic radiation as a subset of modern system design.

Our own personal experiences in both industry and academia encompass many developments in antennas. One of the Editors (K. Fujimoto) has been engaged in the research and development of vehicular communication since 1953, and has increasingly felt that a consolidated volume on mobile antenna systems would be much appreciated by designers. In fact, it is felt that the full significance of antenna design has, on occasion, been undervalued in the past, not only by some communications engineers, but by some antenna engineers as well. The other Editor (J.R. James) has researched widely in electromagnetic defence systems, especially, recently, printed antennas. Having written the monograph *Small Antennas* in 1987 (with coauthors Ann Henderson and Kazuhiro Hirasawa), both of us were keenly aware of the influence of system requirements on antenna design and the fact that this evolutionary process was being driven by the demands of mobile communication. The need for a more systematic design approach is now of paramount importance, and the promotion of this concept has been one of our Editorial priorities. We first discussed the outline of a new book on mobile antenna systems at the International Symposium on Antennas and Propagation in Tokyo in 1989, and in 1990 we invited respected and experienced experts from Europe, the United States, and Japan to join the authoring team.

We aimed to cover the important current areas of interest in land, maritime, satellite, and aeronautical mobile systems, including navigation, but some previous, conventional maritime and aeronautical systems have been omitted, having been replaced by satellite versions. When setting out the organization of chapters, we were mindful of the many ways of classifying antennas by type, frequency, application, and so on. For a systems exposition, classification by application was a natural choice, and the main chapters separately address land, maritime, satellite, and aeronautical mobile systems, while the final chapter, a glossary, gives a wealth of classifications and other details that have a strong coordination role in the book. Within each chapter particular attention is given to design factors relating to propagation problems, operational requirements, and environmental conditions. The most important propagation problems, fading and the delay time effects caused by multipath propagation, are discussed in considerable detail, and a space diversity system, designed for small portable terminals to cope with multipath fading, is introduced. Also discussed is the relevance to antenna design of various system parameters, such as communication zone, modulation system, frequency spectrum, interference, system signal-to-noise ratio (S/N), and bit error rate (BER). Personalization of mobile systems requires the reduction in size of mobile equipment, which in turn may impose the requirement that the antenna be reduced in size; the antenna may sometimes need to be electrically small. The environmental conditions that inevitably affect mobile system performance directly, are also involved. Proximity effects caused by the interaction between the antenna and the body of equipment, front-end circuits, or the human operator are also treated as an important factor which must be included in antenna design.

The book is organized into eight chapters. Chapter 1 presents an overview of antenna systems, including a historical perspective on mobile communications and related antenna technologies, trends, and antenna design concepts for modern mobile systems. Chapter 2 discusses the essential techniques of mobile antenna systems. Problems related to propagation, radio transmission, choice of frequencies, communication zones, interference, and so forth, are treated, followed by the requirements for antenna systems, involving problems of proximity effects and diversity schemes and the evaluation of antenna performance in mobile environments. Chapter 3 deals with the fundamental problems of land mobile antenna systems, such as propagation, the design and applications of antennas for both base and mobile stations, and diversity systems.

Chapter 4 describes antenna systems for pagers and portable phones. The first part presents the fundamentals and performance of the types of antennas for paging receivers. Antenna design for various types of pager receivers, with shapes such as the conventional rectangular box, pencil, credit card, and so forth, are introduced. In the second part, antennas for portable phones are described. Finally, the third part discusses safety aspects for portable and mobile antenna systems.

Chapter 5 describes antenna systems for various land mobile systems concerned with broadcast reception in a car, and communications in train and city bus systems. The design of a diversity antenna system for car broadcast reception needs technology similar to that in other mobile antenna systems. In addition, there has been recent development whereby car radios are adapted to receive traffic information broadcast by an FM multiplex system. The challenge of receiving TV programs in a moving car is also addressed. In this chapter, another unusual mobile system, operated in low-frequency bands using ferrite antennas, is introduced. Ferrite antennas are usually used for receiving only; however, in the bus operating systems described here, ferrite antennas are used for both transmission and reception.

Chapter 6 presents antenna systems used for mobile satellite systems, covering land, maritime, and aeronautical applications. In the first half of the chapter, antenna systems deployed in satellites such as ETS-V, INMARSAT, and MSAT are discussed by introducing their structure, performance, and characteristics. The second half of the chapter treats antenna systems used on trains and in cars for navigation and reception of broadcasting via satellite.

Chapter 7 describes a wide variety of airborne antenna systems for communications, navigation, and other applications. Finally, the Appendix is a glossary of antenna systems cataloged in reference tables by listing them according to types of antenna and mobile terminals associated with land, maritime, satellite, and aeronautical systems, together with a reference table relating frequency bands, typical antenna types, and their applications.

As to the future, there is evidence that the potential of mobile communication is so great that it may change the infrastructure and scale of communications worldwide. Global individual communications may be realized by personalizing mobile equipment so that people will be able to communicate with each other regardless of time, place, or distance. Small and highly sophisticated antenna systems would certainly be required for such an advanced concept, and by that time antennas may be made "smart" by embodying integrated intelligence faculties. It is hoped that this book will contribute to the development of a comprehensive understanding of the design of mobile antenna systems, and will promote the further development of advanced antenna systems by providing material and data in an easy-to-access form.

Our sincere thanks are due to the authors for taking time from their busy schedules to contribute to the book. We are most grateful to the reviewers for their patient and skillful advice, and to the authors for their attentive responses. In particular, we thank Dr. Julie Lancashire, commissioning editor for the publishers, for her unfailing enthusiasm and energy throughout in keeping continuity, without which the book would not have been kept on schedule. Finally, we acknowledge the many colleagues who have provided technical material throughout, most of which is disclosed for the first time, and who have contributed to this work in various ways.

K. Fujimoto
J. R. James
April 1994

Chapter 1
General View of Antennas in Mobile Systems

J. R. James and K. Fujimoto

1.1 INTRODUCTION

Some 40 years ago telecommunications was a somewhat uninteresting utility industry showing only gradual and conservative change. The industry was heavily entrenched in the machinery of national governments worldwide, and the relevance to the promotion of business activity had yet to be recognized. Equipment design and operation was costly and remained static for long periods. The revolution that has occurred is common knowledge, and the thrust of both technology and economics has transformed the industry into one of the most, if not the most, exciting fields. Communications has become the key to momentous changes in the organization of businesses and industries worldwide as they themselves adjust to the shift toward an information economy: information is indeed the lifeblood of modern economies.

Data links now have vast capacities; for instance, a fiber-optic transmission line can absorb the information flow from an entire city. It is, however, the processing of information flow between people that continues to be cumbersome and wasteful in both time and energy; much research effort continues to be expended on ways of inputting speech commands and visual data directly into computer systems without the usual keyboard operations. Office automation using complex yet easy-to-use communication networking is now seen as a way toward making people's daily office interaction less onerous, while for people on the move, mobile communications is rapidly becoming the fastest developing aspect of communications, and there seems to be no limit to what can be achieved in the future. With further sophistication of visual displays, it may be thought that it is unnecessary to travel so much in one's working life to hold meetings, negotiate, and so forth, but there is no indication of this at present; in fact, if anything, there is mounting evidence that people will always travel and want to communicate on the move.

Hence, given the opportunity for mobile communications, it will be exploited to the full.

Mobile communication, it would seem, has a very big future, and for equipment designers it represents one of the greatest opportunities of the era. The rapid increase in mobile communication is expected as a worldwide trend, as predicted in Figure 1.1, which illustrates the anticipated subscriber growth of cellular mobile systems in the U.S., Europe, and Japan [1–3]. The number of mobile terminals, including personal mobile systems such as cordless phones and data terminals may reach many times the number of cellular system subscribers. In Europe, the Groupe Special Mobile (GSM) cellular standard, initiated in 1982, will allow users to operate across frontiers using their own unique dialing number. This system aims to create greater capacity and compatibility, with the analog equipment being superceded by digital. Personal communication networks (PCN) at 1.8 GHz using hand or pocket sets rather than car telephones will further boost usage. Whether these adventurous systems will give the customer the degree of integration of mobile services demanded by the year 2000 remains to be seen, but the prospects are good.

Like many emerging areas in electronics, these new equipment concepts have been made possible by the revolutionary semiconductor chip products which are

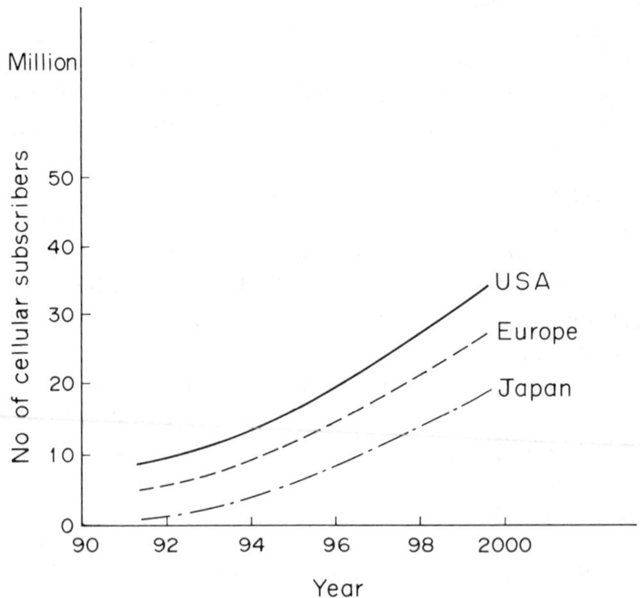

Figure 1.1 Predicted number of cellular system subscribers in Europe, the U.S., and Japan.

now freely available to exploit, and the electronics packed into a portable telephone headset or pager are indeed an impressive sight. For the antenna designer, there is the unprecedented demand to create compact or even electrically small antennas that are compatible with modern technology, will operate on a small handheld ground plane, and satisfy the performance specification, particularly with respect to bandwidth and efficiency. This of course is a familiar theme [4,5], whereby electronic equipment is now so reduced in size that the use of a conventional antenna would not be acceptable to the user and would in any case make equipment miniaturization rather pointless.

However, the challenge for the mobile antenna designer goes further because there is now an awareness that with clever design the antenna can give added value by embodying additional system functions such as diversity reception capability, reduction of multipath fading, or selectivity of polarization characteristics. Mobile antenna design is therefore no longer confined to the design of small, lightweight, low-profile or flush-mounted, omnidirectional antennas on a well-defined flat ground plane, but is rather the creation of a sophisticated electromagnetic configuration that plays a significant role in signal processing while operating in a generally ill-defined time-varying environment. The antenna is now an integral part of the system design, and this is described pictorially in Figure 1.2, encompassing propagation behavior, the local environmental conditions, system constitution and performance, signal-to-noise and bandwidth aspects, physical disposition of the antenna itself, its compatibility with the manufacturing technology, and the user's ease of operation. The nature of the mobile system itself greatly influences the antenna design, and several distinctions can be made between land, maritime, aeronautical, and satellite mobile systems, for instance, and the type of mobile platforms such as vehicles, ships, aircraft, and portable equipment. Frequency reuse capabilities, the type of information, modulation, and personalization of mobile terminals are some of the many factors that are the concern of antenna designers. In zoned systems, radiation patterns have to match the zone patterns to avoid interference, and performance is also subject to variations in the field strength according to the movement of mobile terminals and environmental conditions in the propagation path. In urban area communications, diversity systems are adopted to cope with multipath fading problems. Personalization of mobile terminals demands downsizing mobile systems as well as antennas. In small portable units, an antenna and the radio frequency (RF) front circuits of the transmitter/receiver are usually unified into a system that performs as a radiator. Also, the equipment onto which an antenna element is mounted may itself act as the radiator, so that the antenna element and the body of the equipment must be treated together as an antenna system. Proximity effects caused by obstacles near an antenna element affect the antenna performance and must be allowed for in the design. The operator who holds a portable mobile terminal can significantly perturb antenna performance, and the human hazard problem must also be kept in mind. This wider role

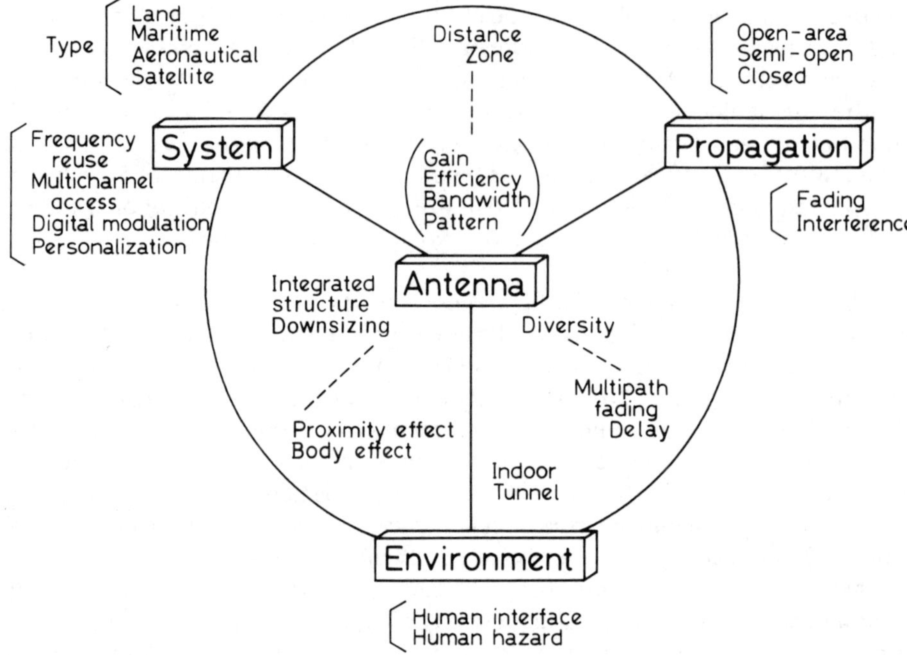

Figure 1.2 The antenna as an integral part of a mobile system.

for antennas, as briefly summarized in the list below, is very significant, and the system aspect is incapsulated in the title of this book. An antenna for mobile systems cannot be designed in isolation from the system, but is indeed an integral part.

Mobile Antenna System Requirements

Antenna as a System. Not as an isolated receive/transmit terminal.

Designed to Accommodate Propagation Effects. Some degree of polarization or pattern diversity control embodied.

Compatible With Environmental Conditions. Pattern characteristics to match zone requirements and allow for nearby obstacles.

Integration of Antenna With Vehicle or Platform. To include hand and body effects and possible hazard considerations.

Latest Manufacturing Technology. Exploitation of new composite materials and integrated electronic technology.

User-Friendly and Reliable Performance. Minimum of moving parts and switches and high reliability of mechanical design.

Very-large-scale integration (VLSI) techniques and microwave integrated circuit (MIC) technologies have been instrumental in producing very small equipment such as pendant-size pagers and portable telephones of 125 cc in volume. High-frequency materials with low-loss characteristics are required for achieving good antenna performance in these designs. The human interface is one of the most important issues in operating mobile terminals, where ease and safety in handling are essential. Before concentrating on modern antenna design concepts, it is important to note the various changes and trends associated with mobile antenna systems and the implications for antennas.

1.2 TRENDS

1.2.1 Communications

One of the biggest changes in electronics has been the onset of digital techniques, which were made possible by the progress in the large-scale integration of semiconductors. Analog components are now in the minority, and this change, in turn, has enabled an order increase in processing power to be packaged in a much reduced volume of equipment space. Mobile equipment now includes portable handsets and inconspicuous pager units worn on the body. Designers have been successful in compacting the electronics into wristlet watches, pendants, and so forth, and this was a quantum jump in antenna design. A distinct bonus that we now take for granted is the increased reliability of integrated semiconductor chip technology. Another striking feature of communications today is the operational extremes, with satellites providing global coverage for speech, video data transmission, and navigational systems, while cellular and microcell arrangements enable mobile telephone services to be highly optimized in dense urban communities. Many variations on these themes are continually being demonstrated to suit some consumer requirement, and the comment that soon anyone can communicate with anyone else, whatever the distance and location, is seemingly becoming realizable. One possible constraint that might limit the growth of mobile systems concerns the ambient level of man-generated noise, but recent electromagnetic compatibility specifications [6] are likely to hold this constraint at bay at least into the near future. More serious constraints to mobile systems are likely to be the finite limitation of the frequency spectrum, unification of systems into those with common standards, and security problems.

Table 1.1 itemizes the trends in mobile antenna systems during this century and is subdivided into five generations of design. In the early days of mobile communications, the services were provided by single-channel systems covering unspecified wide areas on the earth and ocean. The antennas themselves have made

Table 1.1
Trends in Mobile Antenna System Concepts

Year	*1900~*	*1950~*	*1970~*	*1990~*	*2000~*
	Over earth/sea		Urban, highway, rail	In buildings/tunnels, satellites	
Propagation	Reflection/diffraction		Multipath	Delay spread	
Radiation	Omnidirectional		Shaped	Circular polarization	
Antenna	Single element		Composite/integrated	Phased array	Signal processing
Size	Electrically small		Low profile	Built-in	Downsizing
Function			Diversity	Adaptive control, intelligence	
Service	Wide coverage		Zone	Microzone	
Channel	Single		Multizoned, multichannel access, digital modulation		
Problems		Intermodulation, multipath fading, EM noise, delay, proximity effects			

progress from single-element antennas to various composite and integrated configurations and arrays. In the future, mobile antennas will have sophisticated adaptive control and signal processing facilities for enhanced performance. In urban areas, multipath fading has become a serious problem, while zone arrangements and multichannel access networks improve the efficiency of frequency spectrum management. The personal use of mobile terminals has shown a rapid growth, and various small antennas for these systems have been developed. Digital systems and satellite mobile systems demand new types of antennas, such as the short-backfire antennas for ships and microstrip antenna phased arrays for vehicles.

1.2.2 Nature of Information

The dramatic advances in equipment design have led to the use of higher frequency bands and, in turn, the availability of larger allocations of channel bandwidth. Mobile communications provide for people in motion, and at present speech is the major requirement, with specific types of data handling the next priority. However, there will be an ultimate demand to make mobile channels compatible with the type of information handled by conventional static communication systems, and there will be an increasing need to accommodate limited forms of video images to support telephone conversations, facsimile transmissions, computer data, and so forth. Mobile services are expected to be integrated within integrated services digital networks (ISDN) in the future, and in Europe the integration will embody GSM standards. All this points to the use of even higher frequencies to obtain the bandwidth and greater use of mobile satellite links to serve wider areas without the attendant problems of propagation in urban environments. The design of antennas for higher frequency bands introduces both additional problems and increased design freedom, so the techniques for incorporating system functions such as diversity will be somewhat different to those at lower frequencies.

1.2.3 Propagation Environment

The upsurge in mobile communications has revitalized many topics in propagation, and much effort is now devoted to refining propagation mathematical models for urban and other scenarios, together with substantiation by field trials. A summary of some of the models is given in Table 1.2.

The field structure becomes, in most cases, complicated because it is often composed of reflected and diffracted waves produced by multipath propagation environments. Propagation in open areas free from obstacles is the simplest to treat, but, in general, propagation over the earth and the sea invokes at least one reflected wave. Ray theory was used for deriving these fields in the early days, and later the full-wave theory was developed. Sommerfeld [7], van der Pol, and

Table 1.2
Typical Propagation Models

Environment	*Incident Waves*	*Models*	*Mobiles*
A. *Open area*			
Free space	Direct	Free space; Ray van der Pol	Vehicles, portables
Earth	Direct + reflected + diffracted*	Sommerfeld; Burrows	Vehicles, portables
Rural	Direct + reflected†	Ray theory; Okumura; Nakagami-Rice	Vehicles, portables
Suburban	Direct & reflected†	Ray theory; Okumura; Nakagami-Rice	Vehicles, portables
Urban	Direct + reflected + diffracted†	Okumura; Rayleigh; log-normal	Vehicles, portables
Ocean	Direct + reflected + diffracted†	Free space; reflection; Nakagami-Rice	Ships
Air	Direct + reflected + diffracted†	Free space; reflection; Nakagami-Rice	Aircraft
B. *Semi-Open Area*			
Mountain	Diffracted, reflected	Diffraction	Vehicles
Highway	Direct + reflected	Free space LOS; Nakagami-Rice	Cars
Rail	Direct + reflected	Free space LOS; Nakagami-Rice	Train
Underground street	Direct + reflected + diffracted†	Free space LOS; Ray model; Rayleigh	Portables
Indoor to outdoor	Direct + reflected + diffracted†	Free space LOS; Ray model; Rayleigh	Portables
C. *Closed area*			
Indoor	Direct + reflected, diffracted	Transmission line waveguide; Rayleigh (short distance)	Portables
Underground passage	Direct + reflected, diffracted	Transmission line waveguide; Rayleigh (short distance)	Portables
Tunnel	Standing wave, guided wave	Transmission line waveguides; Rayleigh (short distance)	Vehicles

Note: LOS = line of sight.
*Single.
†Multiple.

others [8] contributed immensely to the understanding of propagation over the earth and seas. For land mobile propagation, charts developed by Bullington [9], based on the Burrows theory [10], were used for estimating the VHF/UHF field strength. Various models for land mobile, maritime, aeronautical, and mobile satellite systems have been developed [11–14] where theory is well supported by experimental data. In urban areas, the field is commonly composed of a multiplicity of waves, and multipath propagation problems occur. Analytical solution of the urban area propagation problem is almost impossible, and statistical treatment features strongly. Okumura provided useful charts, known as *Okumura curves*, for estimating propagation loss in both urban and suburban areas [15–22]. Propagation in maritime and aeronautical mobile systems is associated with similar problematic phenomena. In closed areas such as indoors, tunnels, and underground passages, no established models have been developed yet, since the field has a complicated structure [23–25]. However, when the field structure is random, the Rayleigh model used for urban area propagation may be applied. When the propagation path is on the line of sight, as in tunnel and underground passages, the field may be treated by lossy transmission line or waveguide theory. Direct-wave models may be used for propagation in a corridor.

Multipath propagation can occur in satellite communication systems if the receiver is near a reflecting surface (e.g., an aircraft flying over the ocean), but in urban mobile communications it is the dominant effect. Both tall and small buildings can totally block the view of a satellite, while the use of lower frequencies below 2 GHz and typically below 1 GHz leads to high multipath content in urban areas, and the fading characteristics are highly dependent on the nature of the local environment, time of day, and other factors. Not surprisingly, there is a demand for adaptive systems to cope with ambient conditions as they occur, and here lies an opportunity for the antenna designer to create new radiating configurations that offer diversity functions based on ingenious algorithms, discrimination in polarization, control of patterns, and so forth. Closer to the antenna for the lower frequency bands, material effects and obstacles will cause near-field coupling to the antenna; for electrically small antenna platforms, the proximity of the body and hand effects will also create additional variations in antenna efficiency and pattern characteristics. It is tempting to try to couple the operator more precisely into the radiation characteristics in some way to stabilize the electromagnetic behavior, but although unsubstantiated, the possibility of creating an electromagnetic hazard for the operator while transmitting has to be considered.

1.2.4 Maritime Systems

Traditional high-frequency telegraphy and voice has been supplemented by global satellite INMARSAT-type systems [26] offering superior voice and data transmission. An international paging service is also planned. Global positioning systems (GPS) provide precise navigation data, and much work is being carried out on reliable distress and rescue systems. The Global Maritime Distress and Safety System (GMDSS) implemented in 1992 is one such example. GPS is also being applied to land vehicle navigation and fleet vehicle management in Japan, Europe, and the U.S. Rugged antenna installations with added-value system functions compatible with shipborne operation and compact antennas for small-boat operation are some of the fascinating requirements for the antenna designer.

1.2.5 Aeronautical Systems

The aerodynamic constraints are significant, and antennas for both satellite and radio systems must conform to minimum drag and reliability requirements. International services are currently available which carry data and voice communication, together with navigation information on a global basis, but higher data rates to accommodate image and computer data are being developed. ETS/V (Engineering Test Satellite-V) experiments [27] for the transmission of telephone, image, and low- and high-speed digital data between satellite and land vehicles, aircraft, and ships were successful. A test of low-speed data transmission by using attache-case-type transportable equipment, onto which two small printed antennas were mounted, was among the experiments.

1.3 MODERN ANTENNA DESIGN CONCEPTS

The progress in mobile antenna design is listed in Table 1.3, which shows the related technical issues over the five generations in the time periods cited in Table 1.1. In the beginning, mobile communications was conceived for the transmission of telegraphy from trains, ships, and, eventually, aircraft. Since the frequencies used were mainly in low-, medium-, and high-frequency bands, the antennas were monopoles or modified versions, made of either flexible wires or rigid bars. Whip antennas were used on various mobile terminals. The introduction of VHF and FM technologies have brought significant progress in vehicular communications for both public and private use, such as taxi communications, business applications, and navigation systems for ships and aircraft. Automobile radios for receiving AM and FM broadcasts and later TV broadcasts have significantly progressed. Most antennas used on cars were monopoles regardless of frequencies, and those at base stations in VHF were dipole types. For portable equipment in the high-frequency and VHF/UHF bands, monopoles or whips were mainly used, but some of them were equipped with built-in ferrite antennas and normal-mode helical antennas (NMHA), depending on the operating frequency and type of mobile terminals.

Table 1.3
Progress in Antennas and Mobile Systems

	~1900	*~1950*	*~1970*	*~1990*
Frequency	Low to high	<400 MHz	<1,500 MHz	<3 GHz
System	Telegraph/ telephony for train, ship, aircraft, police cars, portable receive and transmit	Voice system for business, navigation, taxis, tone pagers		Satcom, voice and data channels for aircraft, personal phones, microzones, facsimile, TV-type images, wireless local area network
Antenna	Monopole/dipole, whip, top-loaded monopole, inverted-L, loop	Blades, coil loaded, ferrite, helical	Corner reflector, leaky coaxial cable, diversity configuration, body integrated planar inverted-F, bifilar helix, microstrip arrays, parallel plate pager, base station, printed wire on glass	Adaptive signal processing

As the system capacity reached its limit with the increase in the mobile system users, the frequencies allocated for mobile communications were gradually raised from 30 MHz to 50, 150, 250, and then 450 MHz. Currently, frequencies from 800 MHz to 1.5 GHz have been assigned for mobile telephones, and an allocation of even higher frequency bands is being considered.

Various new kinds of antennas, such as inverted-F (IFA) and very small rectangular loops, were developed and applied to pagers and portable telephones. From the early 1980s further advances in antenna design were observed as a consequence of the introduction of cordless phones, multichannel access systems, and navigation systems. In urban mobile communications, diversity antennas with space or polarization schemes have evolved to reduce multipath fading, while the personalization of mobile terminals has demanded electrically small antennas. The concept of integrated antenna systems [28] has been applied to antenna systems for portable equipment. The development of even smaller antennas will be required as more advanced personal communication systems, particularly like the Personal Handy Phone (PHP) in Japan and the Future Public Land Mobile Telecommunication Systems (FPLMTS), are introduced. Adaptive control and intelligent control by means of signal processing will be featured strongly in the future.

It is of course understood that in any particular design only some of the objectives will be achievable, and each case must be treated as a separate entity. Some aspects are always likely to be a consideration, such as ease of operator control and making the best use of the new materials that are available. Both these aspects are intimately related to the styling and manufacture of a product and to some extent its sales appeal, given of course the specified communication performance. An illustration of the size and weight reduction achieved for pagers and portable telephones is in Figure 1.3. The dimensions of portable telephones and pagers has been downsized year by year, and the smallest portable telephone in Japan is about 125 cc in volume, while the smallest pager is the size of a pendant, about $7 \times 7 \times 9.9$ mm^3.

Antenna design is increasingly dependent on computer-aided design (CAD) based on well-known mathematical methods, and one of the latest techniques is the finite difference time domain (FDTD) method, which shows much promise for radiating structures having arbitrary shape and composed of layers of heterogenous material, and this can include components within a dielectric outer case. Mobile antennas, which are essentially bent wire sections, can, however, be modeled with wire-grid modeling provided the equipment case is conducting. Some examples of computational methods applied to analyze and design mobile antennas and others are (1) moment method: a monopole antenna or an IFA on portable equipment

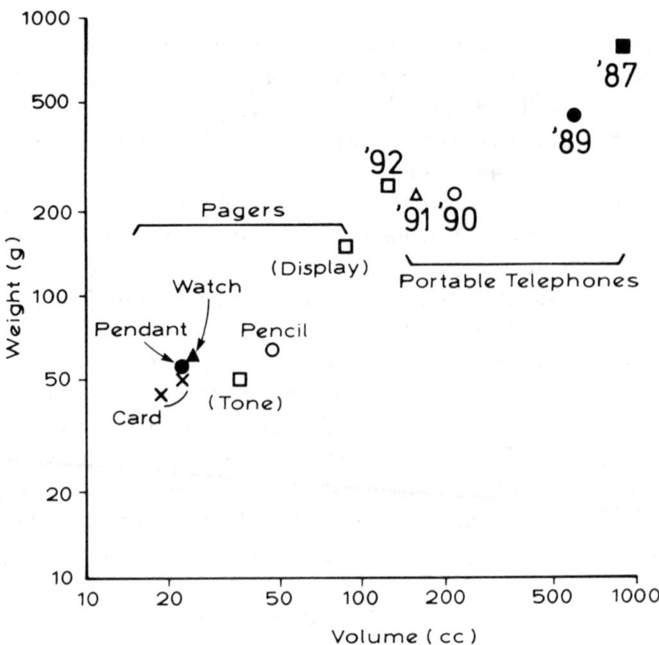

Figure 1.3 Volume and weight of pagers and portable telephones.

[29–31] and a monopole on an automobile at VHF [32]; (2) geometrical theory of diffraction (GTD): a monopole on an automobile at UHF [33]; (3) hybrid methods [34]; (4) FDTD: an IFA on portable equipment [35] and simple antennas [36]; and (5) spatial network method: impedance characteristics of an IFA on an infinite ground plane [37].

1.4 OBJECTIVES OF THIS BOOK

There are many new types of antennas that have been specifically designed for mobile systems, and detailed technical coverage of these antennas is clearly timely. We have, however, embarked on the wider task of examining antenna systems for numerous mobile applications, and our scope necessarily encompasses all relevant aspects as listed in Table 1.1 to 1.3. Since we are dealing with a system, it was felt inappropriate to break the text down into headings as in these tables, but rather address the main application areas of land, maritime, satellite, and aeronautical mobile systems. These main areas are in themselves further subdivided into numerous ways to reflect current developments. Most antenna systems described arise from communication requirements, but there are exceptions, where the acquisition of data relates directly to navigation, identification, and so on. For completeness, Chapter 2 collates background material on technology, propagation, and antennas to support the subsequent chapters. For readers requiring additional information on antennas, a glossary of antenna properties and relevant data, with the emphasis on mobile applications, is included at the end of the book in the Appendix. Land mobile systems are divided into three parts, and the first part, dealing mainly with fundamental techniques, is dealt with in Chapter 3. Chapter 4 covers the second land mobile part on pagers and portable phone systems, while the third and final part, on a variety of landmobile systems for cars, trains, and buses, is addressed in Chapter 5. Chapter 6 is devoted to mobile satellite systems embracing vehicle, shipborne, and broadcast applications. Finally, Chapter 7 concentrates on the variety of airborne systems for communication, navigation, and other related aspects of avionics.

REFERENCES

[1] "Carrier News," *Cellular Business*, Feb. 1992, pp. 134–136.

[2] Sibers, R. and G. Kirby, Increase in Cellular Expected With European/U.S. Transition to Digital Standard, No. 579, Japan: Nikkei Electronics, Nikkei Publications, Inc., April 1993, pp. 121–138 (in Japanese).

[3] "*New Mobile Communication Era*", No. 555, Japan: Nikkei Electronics, Nikkei Publications, Inc., May 1992, pp. 163–192.

[4] Fujimoto, K., A. Henderson, K. Hirasawa, and J. R. James, *Small Antennas*, Research Studies Press, 1987, distributed by Wiley & Sons.

[5] James, J. R., "What's New in Antennas," *IEEE Antennas and Propagation Society Magazine*, Feb. 1990, pp. 6–18.
[6] European Community's Directive on Electromagnetic Compatibility starting 1 Jan. 1992.
[7] Sommerfeld, A., "Uber die Ausbreitung der Wellen in der Drahtlosen Telegraphie," *Ann. Phys.*, Vol. 28. 1909, pp. 665.
[8] van der Pol, B. and H. Bremmer, "The Diffraction of Electromagnetic Waves From an Electrical Point Source Round a Finitely Conducting Sphere With Application to Radiotelegraphy and the Theory of Rainbow," *Phil. Mag.*, Vol. 24, 1937, pp. 141, 825.
[9] Bullington K., "Radio Propagation at Frequencies Above 30 MC," *IRE*, Vol. 35, 1947, pp. 1122–1136.
[10] Burrows, C. R. and S. S. Attwood, *Radio Wave Propagation*, New York, 1949.
[11] Akeyama, T., S. Sakagami, and K. Yoshizawa, "Propagation Characteristics of Air-Ground Paths, 900 MHz Band," *Trans. IECIE*, Vol. J73-B-II, No. 8, Aug. 1990, pp. 383–389.
[12] CCIR Rept. No. 567-3 "Methods and Statistics for Estimating Field-Strength Values in the Land Mobile Services Using the Frequency Range 30 MHz to 1 GHz," *Rec. and Reports of the CCIR*, Vol. 5, ITU, Geneva, 1986.
[13] Karasawa, Y., "Complex Frequency Correlation Characteristics of L-Band Multipath Fading Due to Sea Surface Scattering," *Trans. of IECIE*, Vol. J72-B-II, No. 12, Dec. 1989, pp. 633–639.
[14] CCIR Rep. AL/5, "Propagation Data for Aeronautical Mobile-Satellite Systems for Frequency Above 100 MHz," *Rec. and Reports of the CCIR*, Vol. 5, ITU, Geneva, 1991.
[15] Suzuki, H., "A Statistical Model for Urban Radio Propagation," *IEEE Trans. Comm.*, Vol. 25, 1977, pp. 673–679.
[16] Turin, G. L., et al., "A Statistical Model for Urban Multipath Propagation," *IEEE Trans.*, Vol. VT-21, 1972, pp. 1–8.
[17] Hata, M., "Empirical Formula for Propagation Loss in Land Mobile Radio Service," *IEEE Trans.*, Vol. VT-29, No. 3, 1980, pp. 317–325.
[18] Okumura, Y., et al., "Field Strength and Its Variability in VHF and UHF Land-Mobile Radio Service," *Rev. Elec. Comm. Lab.*, Vol. 16, Sept./Oct. 1968, pp. 825–873.
[19] Tsuruhara, T., et al., "Mobile Radio Propagation Characteristics in Urban Areas in UHF Bands," *Trans. IECE*, Vol. E66, 1983, pp. 724–725.
[20] Okumura, Y., and T. Akeyama, "Propagation in Mobile Communications," Ch. 2 in *Fundamentals of Mobile Communications*, Y. Okumura and M. Shinji, eds., IEICE, 1986.
[21] Akeyama, T., "Propagation in Mobile Communication Networks," Ch. 12 in *Radiowave Propagation*, M. Shinji, ed., 1992, pp. 203–248.
[22] Lee, W.C.Y., *Mobile Cellular Communications Systems*, McGraw-Hill International edition, 1990.
[23] Lafortune, J. F., and M. Lecours, "Measurement and Modeling of Propagation Losses in a Building at 900 MHz," *IEEE Trans.*, Vol. VT-39, No. 2, May 1990, pp. 101–108.
[24] Turkmani, A.M.D., and J. D. Parsons, "Measurement of Building Penetration Loss on Radio Signals at 442,900 and 1400 MHz," *J. IERE*, Vol. 58, No. 6, Sept./Dec. 1988, pp. S169–S174.
[25] Kozono, S., "Experimental Test Results of 800 MHz Band Mobile Radio Propagation in Highway Tunnels," *ISAP '78*, Aug. 1978, pp. c-3–c-5.
[26] Miya, K., ed., *Satellite Communication Technology*, KDD Engineering and Consulting, Inc. (KEC), 1982, pp. 5, 37.
[27] Hamamoto, N., et al., "Results on CRL's Mobile Communication Experiments using ETS-V Satellite," *Space Communications*, No. 7, 1990, pp. 483–493.
[28] Fujimoto, K., "A Treatment of Integrated Antenna Systems," *IEEE AP-S Int. Symp.*, 1970, pp. 120–123.
[29] Nakano, H. and S. R. Kemer, "The Moment Method Solution for Printed Wire Antennas of Arbitrary Configuration," *IEEE Trans.*, Vol. AP-36, 1988, p. 1667.

[30] Taga, T. and K. Tsunekawa, "Performance Analysis of a Built-in Planar Inverted-F Antenna for 800 MHz Band Portable Radio Units," *IEEE Trans. Selected Areas in Communication*, Vol. SAC-5, No. 5, June 1987, pp. 921–929.

[31] Sato, K., et al., "Characteristics of a Planar Inverted-F Antenna on a Rectangular Conducting Body," *Electronics and Communications in Japan*, Vol. 71-B, Scripta Publishing, Aug. 1989, pp. 43–51.

[32] Nishikawa, K., "Effect of Automobile Body and Earth on Radiation Patterns of Antennas for FM Radio," *Trans. IECE Japan*, Vol. E67, Oct. 1984, pp. 555–562.

[33] Nishikawa, K. and Y. Asano, "Vertical Radiation Patterns of Mobile Antenna in UHF Band," *IEEE Trans. Vehicular Technology*, Vol. VT-35, May 1986, pp. 57–62.

[34] Thiele, G. A. and T. H. Newhouse, "A Hybrid Technique for Combining Moment Methods With the Geometrical Theory of Diffraction," *IEEE Trans.*, Vol. AP-17, 1969, pp. 62–69.

[35] Kagoshima, K., A. Ando, and K. Tsunekawa, "FD-TD Analysis of a Planar Inverted-F Antenna Mounted on a Conducting Box," *Proc. Int. Symp. Ant. Propag.* (Sapporo, Japan), Sept. 1992, pp. 713–716.

[36] Maloney, J. G., et al., "Accurate Computation of the Radiation From Simple Antennas Using the Finite-Difference Time-Domain Method," *IEEE Trans.*, Vol. AP-38, 1990, p. 1059.

[37] Taga, T., "Analysis of Planar Inverted-F Antennas and Antenna Design for Portable Radio Equipment," Ch. 5 in *Analysis, Design, and Measurement of Small and Low-Profile Antennas*, K. Hirasawa and M. Haneishi, eds., Norwood, MA: Artech House, 1991.

Chapter 2

Essential Techniques in Mobile Antenna Systems Design

K. Kagoshima, W.C.Y. Lee, K. Fujimoto, K. Hirasawa, and T. Taga

2.1 MOBILE COMMUNICATION SYSTEMS

2.1.1 Technologies in Mobile Communications

Throughout the history of mobile communication systems, various kinds of systems have been developed and commercialized. The depth and breadth of the technologies supporting the systems reflect system size and complexity. For example, public mobile phone systems, such as the automobile telephone system, are one of the largest and most complex systems; the outline of the system configuration is shown in Figure 2.1. A complete understanding of wave propagation, radio transmission, channel control, and hardware devices is essential to construct a mobile phone system [1].

To create antenna hardware, we must first consider propagation and radio transmission characteristics. For example, the antenna pattern used at the base station affects the propagation characteristics, and the arrangement of the antenna pair in a reception diversity system virtually determines the transmission characteristics of the radio channels. Antenna engineers must consider all related factors in order to realize the full potential of any antenna hardware.

The main features and recent trends of the related factors are summarized in the following sections. Readers interested in more detail may refer to individual textbooks [2,3] or the papers appearing in the special issue of [4].

Propagation

In all radio communication systems, the study of radio wave propagation is inevitable, especially in mobile communication systems, where the propagation path

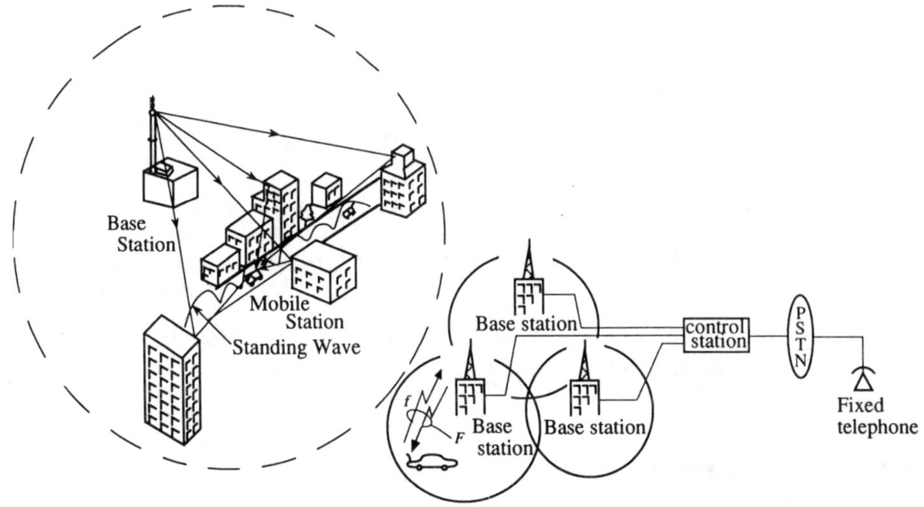

Figure 2.1 System configuration of mobile communications (cellular radio telephone systems).

is seldom line-of-sight within the propagation region. This means that transmitted waves are much affected by buildings, towers, and objects within the spatial environment before they reach the receiver. Their propagation characteristics differ greatly from those observed in free space. Although much research concerning mobile propagation has been published, Okumura was first to present the entire range of mobile propagation characteristics needed in mobile communication system design [5]. In 1970, he clarified propagation-path loss characteristics in the propagation environments of urban, suburban, and open areas across wide-frequency ranges from the 200-MHz band to the 2.0-GHz band. In his monumental paper, he also demonstrated the influence of antenna height on the path loss characteristics. He summarized his results in graphic form as *Okumura's curve*, which was adopted by CCIR Recommendation 370 of SG5 [6]. This has become the basis of mobile communication system design throughout the world. Following his seminal research, propagation studies were continuously conducted to increase prediction accuracy, and some improvements in propagation characteristics have been achieved. The predicting formulas for path loss, which took into account building density in the horizontal plane, were obtained [7]. Hata derived the predicting formulas for propagation-path loss, which are now used by system designers [8]. Recently, theoretical predictions for determining field strength were presented based on the data of buildings and geographical features using geometrical optics and/or geometrical theory of diffraction [9,10]. In the future, it will be possible to estimate propagation characteristics by using computer simulations.

One of the significant achievements in the field of radio wave propagation concerns reception diversity. The correlation coefficient between two antennas was measured in various propagation environments, and the received levels achievable through diversity reception were clarified. Digital mobile systems are being developed throughout the world to provide enhanced services, but the problems facing wide-frequency-band transmission within a severe fading environment are formidable. For example, measurements of wide-band mobile propagation were carried out for the development of the GSM, and the important results obtained through these measurements and evaluations are summarized by Lorenz [11]. Diversity reception or equalization to reduce multipath delay is being actively pursued throughout the world. In the study of propagation delay, the characteristics of propagation and transmission cannot be considered in isolation. Antenna characteristics such as the radiation pattern and correlation coefficient are related to propagation delay. Therefore, it should be pointed out that a total study, taking into account propagation, transmission (system), and the antenna, is necessary [12].

Radio Transmission

Frequency modulation/demodulation (FM) is the most used modulation scheme in mobile communication systems, because it is robust against thermal noise and/or interference and is easy to realize as hardware. From the viewpoint of frequency-effective use, an important point in developing an appropriate FM transmission technique is determining how to narrow the bandwidth. In the Japanese 400-MHz band mobile communication system, channel separation was successfully narrowed from 50 to 25 KHz and then from 25 Hz to 12.5 KHz. Moreover, an 800-MHz automobile telephone system has been created around the interleave channel allocation method and offers the narrow channel separation of 6.25 KHz. The achievement of this narrow channel separation is due to the development of a highly stabilized oscillator and high-performance filter, as well as a syllabic compandor [13].

For analog transmission, single-sideband (SSB) is an attractive modulation scheme, which can realize frequency-effective use. SSB equipment has been substantially improved, but this has not prevented the emergence of digital systems.

Digital mobile systems have been actively pursued in order to realize enhanced mobile communication services. Commercial services started in Europe and Japan in 1991 and 1993, respectively [14]. For these systems, the technologies of voice codecs and narrow-band digital modulation were greatly advanced. Diversity reception, as well as equalization, continues to be studied to improve the bit error rate;

error correction also plays an important role in digital transmission systems. Moreover, the effect of all the above techniques on transmission system performance was investigated.

Control

In mobile communication systems, numerous control technologies such as channel connection, register of position, and zone switching are required to establish a link between the transmitter and receiver. Special switching technology is also necessary to link the mobile radio network to public telephone networks. These control technologies differ widely in many points from those of the ordinary fixed radio communication system. Although the quality of the control technology significantly affects the performance of mobile communication, it is peripheral to the subject of this book and readers interested in control technologies should refer to [15].

Hardware

The hardware technologies needed for establishing the radio transmission network cover a wide area, including antennas, active circuits, and batteries. Nippon Telegraph and Telephone Corporation's (NTT) automobile telephone service was put into commercial use in 1979, and its portable telephone service began in 1987. Since that time, the volume and weight of the portable equipment unit have been decreased as shown in Figure 2.2. Cutting size and weight by 75% has been achieved in under five years. These results are due mainly to the use of large-scale integration (LSI) parts, especially in the intermediate frequency (IF) circuits, baseband circuits,

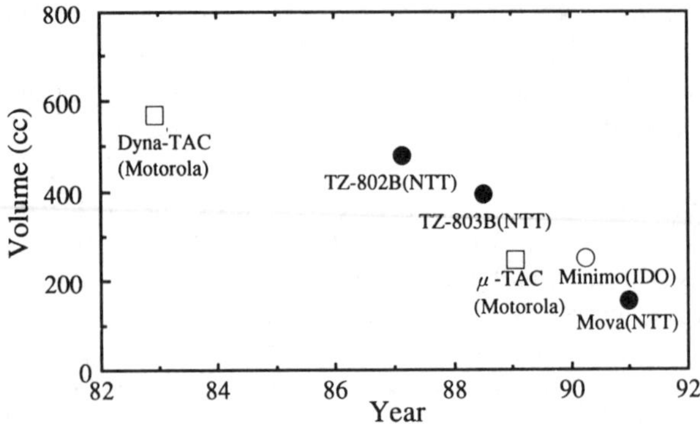

Figure 2.2 Change in volume of portable radio phone equipment.

and control circuits. Reducing the power consumption of RF circuits is also an important breakthrough, which has lengthened the operation time of radio units, as have highly efficient batteries.

As for the passive circuits, the high ϵ dielectric filter with relative permittivity of more than 90 and surface acoustic wave (SAW) filters contribute to the downsizing of radio units [16]. As the radio units become smaller, the antennas of the mobile units should be equally small or even integrated within the units' bodies. Research and development are being carried out to obtain a suitably small antenna with high performance [17].

Major research items concerning the four technologies mentioned above are summarized in Table 2.1. The development of these technologies are aimed to achieve three goals: (1) effective use of radio frequency resources, (2) a wider variety of enhanced services, and (3) cost effectiveness. Even when we pursue individual technical developments, we should keep these goals clearly in mind to ensure that the developments will contribute to an effective and successful system.

By downsizing and increasing the gain of the portable radio antenna, very effective radio units can be created that will have long operation times and will be easy to use. Increasing the gain and lowering the sidelobe levels of the base station antenna [17] will enhance system economy, because the transmitting power of the base station can be reduced and frequency reuse can be pursued more aggressively. The creation of a more sophisticated system that can handle dramatically increased levels of service requires the development of a base station antenna that offers higher performance and expanded functions. The relationship between antenna technology and its impact on mobile communication systems is shown in Figure 2.3.

The above discussion focuses mainly on land mobile telephone systems. The aeronautic and maritime mobile telephone systems need almost the same technologies, since only the service areas are different from that of the land mobile

Table 2.1
Major Study Items for Mobile Communications

	Study Items
Propagation	Propagation-path loss prediction, delay profile measurement, diversity
Transmission	Modulation/demodulation, signal processing, code/decode, error correction, diversity reception
Control	Power control, channel allocation, channel switching, location registration, file access
Hardware	Antennas, filter, oscillator, synthesizer, amplifier, codec, LSI, battery

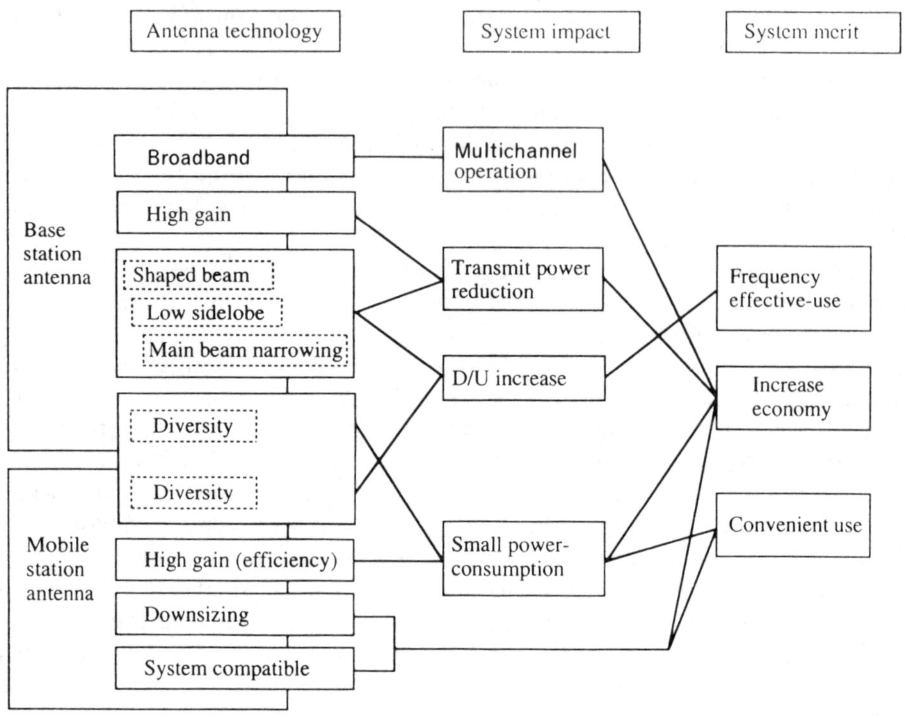

Figure 2.3 Relation between antenna technology and its system impact.

systems. However, the antenna patterns required for each system depend on the expanse and shape of the service area. Thus, different types of antennas are needed, as discussed elsewhere in this book. For example, in aeronautic mobile telephone systems, the service area exists from ground level up, and the base station antenna, which is located on the ground, should radiate upwards. In order to compensate for differences in field intensity, cosecant squared shaped beams are used.

Private mobile radio systems like the multichannel access (MCA) system in Japan [18] (which were common in the early period of mobile communication systems) have much simpler control station functions than cellular systems because they (the private systems) are not connected to the public service telephone network (PSTN). However, they require virtually identical technologies for propagation, radio transmission, and hardware.

As will be seen in Chapter 6, mobile satellite communication systems have been commercialized through INMARSAT. In the near future, advanced mobile satellite systems will be realized in some countries using domestic satellites. In these systems, the antennas for mobile stations usually need higher gain and satellite tracking. Moreover, they should be economical. Therefore, the design of those antennas becomes more complicated, and notable progress has been made in many countries.

2.1.2 Frequencies Used in Mobile Systems

Technical Aspects

Frequencies used in mobile communication systems have undergone a slow shift from the lower frequency bands to the higher frequency bands. The optimum frequency band is basically determined by the following technological demands: (1) Frequencies at which small and light mobile terminals can operate, (2) Frequencies yielding the largest propagation distances, and (3) Frequencies offering adequate bandwidths to meet for system requirements.

The frequencies are not permanently fixed, but change with system enhancement or the appearance of new systems. Figure 2.4 shows the changes in frequencies used in major Japanese mobile communications systems [19], and Figure 2.5 illustrates the change process. If a system is introduced for commercial service and grows rapidly, it soon exhausts its frequency resources. The typical response is to develop technologies that maximize the utilization of frequency resources. If the frequency shortage cannot be overcome, new frequency bands are employed. This generally requires the establishment of a new system based on the largest technologies. Frequencies used by the mobile communication systems have expanded to almost completely occupy the VHF and UHF bands, because these frequencies

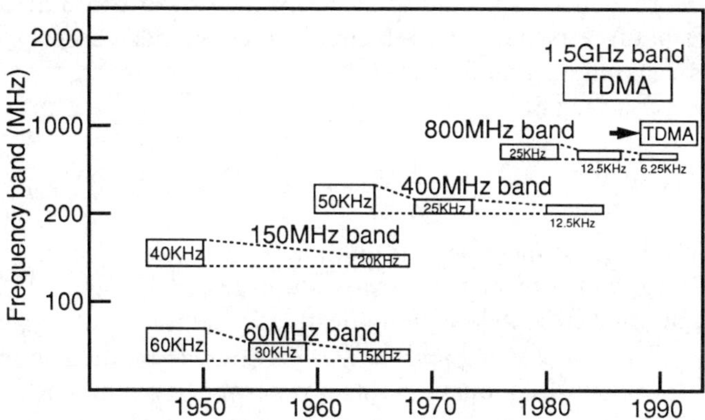

Figure 2.4 Frequency used and channel separation for major mobile systems in Japan.

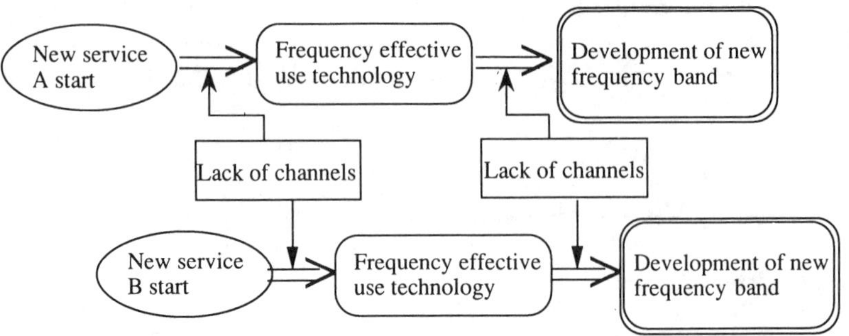

Figure 2.5 Channel increase process due to frequency-effective use technology and higher frequency band development.

best satisfy the above three demands, as shown in Figure 2.6 [19]. This continued development mirrors the development experienced by fixed radio relay systems, which use frequencies below 10 GHz, the so-called *microwave frequency band*.

The first mobile communication systems were private networks, and service demands were not so high. The recent introduction of public systems and convenient equipment such as automobile telephones, portable telephones, cordless telephone, and pager systems has led to a dramatic increase in demand such that frequency resources are rapidly exhausted. The response is to utilize new bands; for example, because the 800-MHz band is becoming congested, the 1.5-GHz band or 2-GHz band should be developed for mobile communication system use. As the frequency increases, the radio waves attenuate more with distance. It is necessary therefore to increase antenna gain or to narrow the cell size to maximize system economy. If the higher frequency bands are to be used, the power consumption and size of the mobile station, especially the RF circuit, must be decreased to achieve a convenient and economical mobile station.

Regulation Aspects

The frequency band adopted depends on the frequencies permitted by Radio Regulation (RR), as well as technical considerations. The VHF and UHF bands, which are commonly used in mobile communication systems, are also allocated for FM broadcasting, TV broadcasting and various private fixed radio communication systems. The principle of frequency sharing is determined by the RR of each country. Under RR, an administrative body in each country determines allocation of the radio frequency within the country.

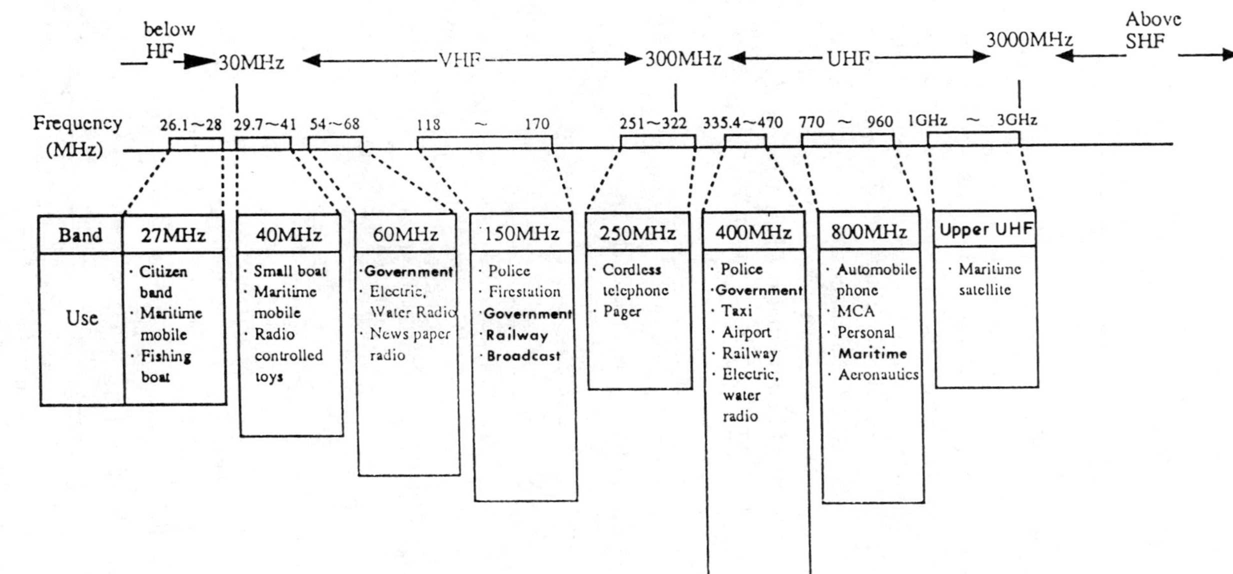

Figure 2.6 Use of frequency band for mobile communications in Japan. (After [19].)

The most recent World Administrative Radio Committee (WARC) session was held in 1992 at Torremolinos, Spain, and new frequency bands were allocated for mobile communication systems, as shown in Table 2.2 [20]. This will result in the rapid increase of mobile services all over the world.

2.1.3 System Design and Antennas

The word *system* is commonly used in engineering today and throughout this book. The word implies that many factors are considered in the design, and it is instructive here to briefly discuss system design in relation to antennas. Several decades ago, equipment was often designed by assembling a collection of component parts that had been optimized in isolation from one another. For instance, communication apparatus would be fitted with an antenna of standard optimized design and subsequent adjustments made in situ. As equipment complexity increased, this bottom-up design process gave way to today's top-down system approach, where component parts are optimized in relation to all factors that influence the equipment function. An obvious yet seldom appreciated fact is that a component that is highly optimized

Table 2.2
Results of WARC'92 Related to Mobile Service

	Allocation	*Frequency (MHz)*
Mobile satellite:		
Low earth orbit	New	137–138 (down)*
		148–149.9 (up)
		149.9–150.05 (up): land mobile satellite
		312–315 (up)†
		312–315 (down)†
		400.15–401 (down)
1.6/2.4-GHz band	New	1,610–1,625.5 (up)
		2483.5–2560 (down)
L band	Extended	1525–1530 (down)
		Region 1: maritime and land mobile satellite
		Region 2, 3: mobile satellite
S band	Extended (region)	2500–2520 (down)
		2670–2690 (up)
FPLMTS	Terrestrial: new	1885–2010
		2110–2185
	Terrestrial/Satellite: new	2010–2025
		2185–2200

Note: Down = downlink from satellite to earth; up = uplink from earth to satellite; FPLMTS = future public land mobile telecommunications system.
*Partly including secondary service
†Including secondary service.

in performance may not be the best choice indicated by the system approach. A good example is the printed patch antenna, which can have a lower efficiency than a conventional wire monopole antenna, yet it is the printed patch with its low profile and compatibility with printed technology that has made many new types of systems possible, particularly in mobile communications, radar, and navigation equipment. Antennas today cannot be designed in isolation from their host equipment, and systems design is an essential technological approach in the realization of high-performance radio equipment operating to a critical specification. System design is amply demonstrated by the various antenna systems described in this book, and it serves no useful purpose to generalize further. However, a few comments on how system design is carried out are in order and illustrated by the following example of an urban mobile communication system for vehicles and portable units.

The factors that a system designer would list include the following.

- Zone configurations—defining signal coverage and antenna patterns;
- Base station antennas—height, physical constraints, and requirements for beam down tilting;
- Noise levels—thermal and environmental;
- Interference—its level and nature and cochannel and adjacent channel effects;
- Signal requirements—optimal frequency of operation, bandwidth, intermodulation effects, and effective utilization of frequency spectrum;
- Cost of development and subsequent manufacture;
- Reliability—servicing required and ease of access, and costs;
- Vulnerability to damage—exposure to weather, corrosion, and wear and tear;
- Network operation requirements;
- Customer appeal.

There are, no doubt, many other considerations, but the list is sufficient to illustrate the various factors that will influence the antenna design to some greater or lesser degree. The point we are leading to here is the actual process of design that enables the designer to embody these factors and translate them into physical constraints on the hardware. It is tempting to think that the constraints can be translated into equations and the requirements computed, but this is most unlikely. It is true that in academic circles there are topic areas called *system analysis*, where every facet is described by equations and numerical optimization methods, but this assumes that all unknowns have given functional forms and, likewise, noise and interference effects. In reality, the mobile antenna system designer will have incomplete data, much in subjective form. Furthermore, the system requirements will be time-varying and likely to change significantly with the fashion of customer demand. The process of practical system design amounts to an iterative procedure geared to a commercial business plan with limited time scales. The procedure will embrace computation, measurement, and field trials and will involve a team of individuals responsible for different aspects of the system. One might enquire how such a procedure could guarantee that an optimum design is achieved, and the answer lies in the existence of competition from other manufacturers and the influence of the customer, who has the ultimate experience with the system and the choice. Most of the antenna systems in this book have been through this pragmatic design process, and the work of a variety of manufacturing companies worldwide is presented.

2.2 FUNDAMENTALS AND PREDICTIVE MODELS IN LAND MOBILE PROPAGATION

As already mentioned, the propagation mechanism of radio waves in a mobile environment is a vital aspect of design. A book such as this that concentrates on antennas does not allow the space to give a comprehensive coverage of the subject of propagation, but some outline is necessary to introduce the terminology, physical behavior, mathematical models, and predictive processes encountered by system designers. Both introductory and advanced texts on propagation are readily available in the literature, so we have chosen here to present a compact outline of established techniques for land mobile systems, due mainly to W. C. Y. Lee, to illustrate the role of propagation in antenna system design. Propagation methodologies pertaining specifically to maritime, aerospace, and satellite systems are available in the literature, but are generally not as complex and convoluted as those in land mobile systems.

2.2.1 Propagation Problems in Land Mobile Communications

In the mobile radio environment, there are five unique factors [21].

- Natural terrain configurations such as flat ground, hills, water, mountain, valley, and desert;
- Manmade structures such as open areas, suburban and urban areas, and metropolitan areas;
- Manmade noise such as automotive ignition noise and machine noise;
- Moving medium brought about by the mobility of the mobile and portable units;
- Dispersive medium causing frequency-selective fading and time-delay spread.

The above five factors are made significant because the mobile antenna is very close to the ground. When the mobile antenna is around 1.5m to 3m above the ground, the signal received by the mobile unit comprises a direct path signal and a strong reflected wave due to the closeness with the ground. These two waves, when combined, result in an excessive path loss at the mobile reception. Also, because the mobile antenna is close to the manmade structures and manmade noise sources, the path losses, multipath fading, and interference will have profound effects.

2.2.2 Propagation Models and Field Strength

There are essentially two kinds of progagation models. One is the propagation prediction model and the other is the multipath fading model. The propagation prediction model is used to predict the average field strengths (also called the *local means*) and as a tool to design a mobile radio communication system in different geographical areas. The antenna height, gain, and directivity at the base station play a big role in the model. The multipath fading model is derived from the natural randomness of multipath wave arrival. The model can predict the behavior of instantaneous field strengths in the field.

2.2.2.1 Propagation Prediction Models

Value of the Prediction Model

Why is the prediction model so important? The value of the prediction model is to save manpower, cost, and time. Before planning a cellular (or mobile radio) system in an area, selecting the cell site (or base station) locations for signal coverage that are mutually interference-free is a big task. Without prediction tools, the only way is to use cut-and-try methods by actual testing. This involves the measurement of the coverage of the cell sites following each proposed plan and selection of the best one, which is very costly. With a fairly accurate prediction tool and computer manipulation, we can easily pick up the optimal cell site locations of a plan after comparing and evaluating the performances of all plans from the computer outputs. As we all know, anyone can write a prediction model. However, a reasonably accurate prediction model that has been verified by measured data in a mobile radio environment over the years is the one that should be chosen.

Requirements for Selecting the Right Model for Mobile Radio

A mobile radio prediction model has to be able to distinguish various natural terrain contours, such as flat areas, hilly areas, and valleys, from the signal prediction results. Also, the model has to be able to distinguish various manmade environments, such as open, suburban, urban, and metropolitan areas, from the signal prediction results. There are many variables involved, and they all have their appropriate roles to play. Therefore, a good mobile radio prediction model is very hard to form. The manmade structures are different in different cities; hence, to predict the signal strength received in a city of interest one cannot use the measured data collected and averaged from one or many different cities elsewhere. The radio signal propagating in a mobile radio environment, in general, follows wave propagation theory; but because many environment-generated variables are involved, a statistical methodology is necessary to complete the prediction tool.

A good prediction model should be simple to use. The model should be specified very clearly and should not provide the user with any subjective judgment or interpretation, which could draw different predicted values in the same area. It should also allow a clear explanation of why the theory of wave propagation and the properties of communication statistics are representive of the physical action.

Free-Space Path Loss Formula

$$\text{FSPL} = -20 \log_{10}\left(\frac{4\pi R}{\lambda}\right) \text{dB},$$

where λ = wavelength.

The free-space path loss (FSPL) is the minimum loss that can be found at the same distance R. To achieve free-space loss is to place both the transmitting and receiving antennas high above the ground or away from nearby scatterers. Thus, the reflected waves from the ground or scatterers become weak and can be neglected.

Reciprocity Theory in Mobile Radio Environment

If the receiving antenna is made to transmit and the transmitting antenna receive, then the received signal strength will be the same as before, given the same transmitted power and matching conditions. This is known as the *reciprocity theory*. However, the received carrier-to-interference ratio (CIR) is not reciprocal because the interference level I received at the vehicle and the base station is different.

2.2.3 Formula for the Two-Wave Theory on a Flat Ground

The received power at the mobile antenna is obtained by summing up two waves: a direct wave and a reflected wave from a direct path and a reflected path, respectively, as shown in Figure 2.7.

$$P_r = P_0\left(\frac{1}{4\pi d/\lambda}\right)^2 |1 + a_v \exp(j\Delta\Phi)|^2 \tag{2.1}$$

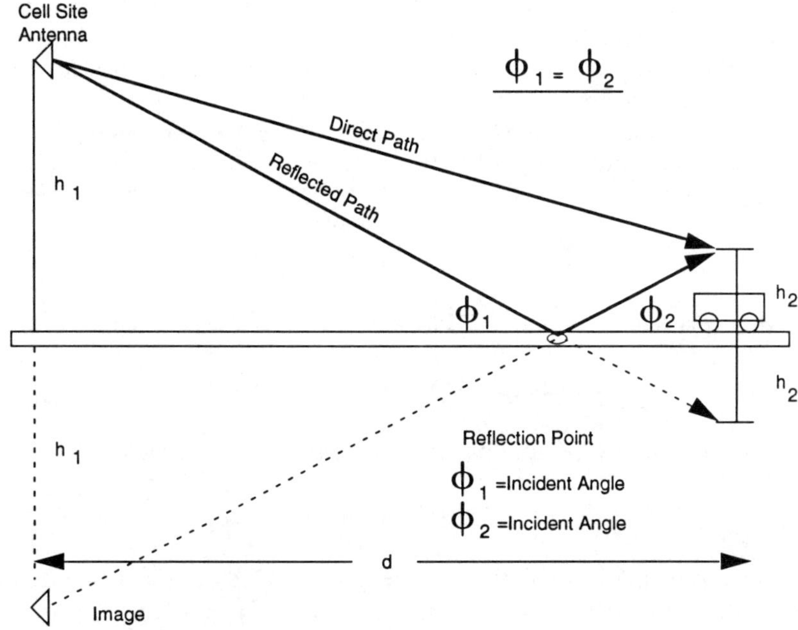

Figure 2.7 A two-wave model.

where $P_0 = P_t G_t G_m \alpha_e$; P_t = transmitted power; G_t = gain of the cell site antenna; G_m = gain of the mobile antenna; α_e = additional loss due to propagation in different manmade environments; a_v = reflection coefficient = -1 when the incident angle of the reflected wave is very small;

$$\Delta\Phi = \beta\Delta d \tag{2.2}$$

equals the phase difference between a direct path and a reflected path; β = wave number = $2\pi/\lambda$; Δd = the difference in path lengths between the direct path and the reflected wave path;

$$\Delta d = \sqrt{(h_1 + h_2)^2 + d^2} - \sqrt{h_1 - h_2)^2 + d^2} \tag{2.3}$$

h_1, h_2 = antenna height of the cell site and the mobile unit, respectively; d = the distance between the cell site and the mobile unit; and $[1/(4\pi d/\lambda)]$ = the free-space path loss formula.

Equation (2.1) is an equation of the exact solution of the two-wave model, which can be calculated by a computer. This simple model does not represent the close-in mobile radio environment very well, but when the distance from the cell site becomes 1 km or greater, the two-wave model curve fairly matches the measured data in a statistical sense. For this reason, we are interested in a distance greater than 1 km. Under this condition, $h_1 + h_2 \ll d$, and (2.3) can be approximately expressed by

$$\Delta d \approx (2/d) h_1 h_2 \tag{2.4}$$

Also, $\Delta\Phi$ is very small, so that $\sin \Delta\Phi \approx \Delta\Phi$ and $\cos \Delta\Phi \approx 1$. Hence,

$$\exp(\mathrm{j}\Delta\Phi) = 1 + \mathrm{j}\Delta\Phi = 1 + \mathrm{j}\beta(\Delta d) \tag{2.5}$$

Substituting (2.4) into (2.5) and putting the result into (2.1) yields

$$P_r = P_0(h_1^2 h_2^2/d^4) \tag{2.6}$$

which is an approximation with the wave length canceled out by the approximation process, and α_e is an unknown shown in P_0 of (2.1). Without the wavelength information, (2.6) cannot be used for calculating the absolute path loss as the free-space loss formula can. However, (2.6) has a simpler expression than (2.1) and agrees with the measurement of approximately 40 dB/dec (dec = decade) for the path loss and 6 dB/oct (oct = octave) for the cell site antenna height gain for the values of d and h_1 and h_2 of interest. Unfortunately, a value of 3 dB/oct, not 6 dB/oct, for the mobile unit antenna height gain was found from the measurements taken when the mobile antenna height is close to 3m. This finding does not agree with (2.6). Therefore, (2.6) has to be modified as follows.

$$P_r = P_0(h_1^2 h_2^c/d^4) \tag{2.7}$$

Where c is in the range of $1 \leq c \leq 2$. When $h_2 \geq 30$m, $c = 2$; when $h_2 \leq 10$m, $c = 1$. When $10\text{m} \leq h_2 \leq 30\text{m}$, the value of c varies because of the different effects caused by the manmade environments. Based on measured data, a linear formula may be used to estimate the value of c, thus: $c = h_2/20 + 1/2$.

Since (2.7) is fairly good at predicting the path loss and the base station antenna height gain, the two-wave model (Figure 2.7) used to derive (2.7) is thus regarded as the correct model. This is because the mobile antenna height is close to the ground and the incident angle of the reflected wave arriving at the reflection point is very small, so the reflection coefficient approaches −1. This means that the direct and reflected waves have approximately the same signal strength, but due to the properties of electromagnetic waves, the reflected wave always has a

180-deg phase shift after reflecting from the ground. Therefore, the two waves are actually subtracting rather than adding. With a slight phase difference due to the different path lengths of the two waves, a loss of 40 dB/dec is obtained, as compared to a free-space loss of 20 dB/dec due to a single-wave path.

2.2.4 Some Established Models

2.2.4.1 Path Loss Prediction Model Versus Local Mean Prediction Model

Most models predict the path loss over the radio path, which is the distance measured from the mobile site to the base station [21–23]. The standard deviations of the prediction errors against the measured data for these models are about 6 to 18 dB. This indicates that the path loss curves used for the prediction of signal strength are too loose to use to design a cellular system. Another prediction model [24,25] is to predict the local means along the mobile path (i.e., along individual streets and roads).

The Path Loss Model

We present the Okumura Model [5], which was formed from averaging measured data in Japan and expressed as a statistical average as follows:

$$L = 69.55 + 26.16 \log_{10} F - 13.82 \log_{10} h_b + (44.9 - 65.5 \log_{10} h_b) \log_{10} R - A_{h_m} \tag{2.8}$$

where

L = path loss from the base station to the mobile unit;
F = carrier frequency in megahertz;
h_b = base station antenna height in meters;
R = distance between the base station and mobile unit in kilometers;
A_{h_m} = $(1.1 \log_{10} F - 0.7) \cdot h_m - (1.56 \log_{10} F - 0.8)$;
h_m = mobile antenna height in meters.

Local-Mean Prediction Model—Lee's Model

Lee's Model is a local-mean prediction model. The predicted local mean is used as a comparison with the measured local mean at any given location. It should be noted that the local-mean prediction is different from the path loss prediction. The philosophy of Lee's model is to try to separate the effects of the received signal on the natural terrain configuration and on the manmade structures. The model is described in more detail in the following section.

2.2.5 Outline of Lee's Method and Various Forms

2.2.5.1 Effect of the Manmade Structures

Since the terrain configuration of each city is different, and the manmade structure of each city is also unique, we have to find a way to separate these two. The way to factor out the effect due to the terrain configuration from the manmade structures is to work out a way to obtain the path loss curve for the area as if the area were flat, even if it is not. The path loss curve obtained on virtually flat ground indicates the effects of the signal loss due to solely manmade structures. This means that the different path loss curves obtained in each city show the different manmade structure in that city. To do this, we may have to measure signal strengths at those high spots and also at the low spots surrounding the cell sites, as shown in Figure 2.8(a). Then the average path loss slope (Figure 2.8(b)), which is a combination of measurements from high spots and low spots along different radio paths in a general area, represents the signal received as if it is from a flat area affected only by a different local manmade structured environment. We are using 1-mi intercepts (or, alternatively, 1-km intercepts) as a starting point for obtaining the path loss curves. The reasons are:

- There are fewer streets within a mile of the cell site. Statistically, the signal-strength data collected within 1 mi of the cell sites are not enough.
- The streets oriented in line with the radio path always have stronger signals than the ones oriented perpendicular to the radio path in the area closer to the cell site, as shown in Figure 2.9. Sometimes the received signals can be different by 20 dB within a 1-mi radius. As soon as the distance is larger than 1 mi, this phenomenon will begin to disappear. Therefore, the near-in data can be biased up or down and should not be used.
- Certain noticeable objects close to the cell site could affect the signal strengths at the near-in areas, and should not be considered from a statistical sense.

Now, a path loss should be obtained from measurements in each city following the rule previously described. The general representation of a flat-area path loss slope is as follows.

$$P_r = P_{r0}(d/d_0)^{-\gamma}\alpha_0 \qquad \text{mW} \tag{2.9a}$$

$$= P_{r0} - 10\,\gamma\log(d/d_0) + \alpha_0 \qquad \text{dBm} \tag{2.9b}$$

(a)

(b)

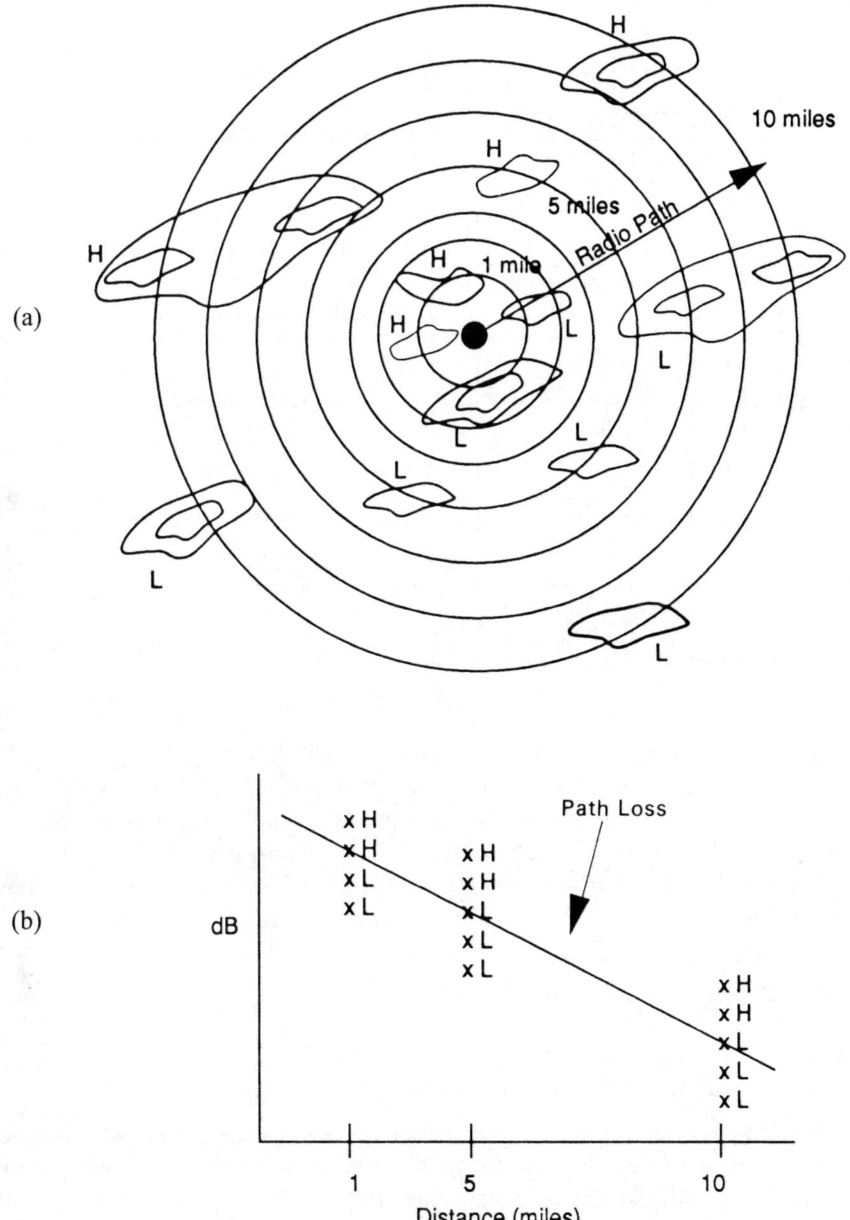

Figure 2.8 Terrain contours: (a) for selecting measurement areas; (b) path loss phenomenon.

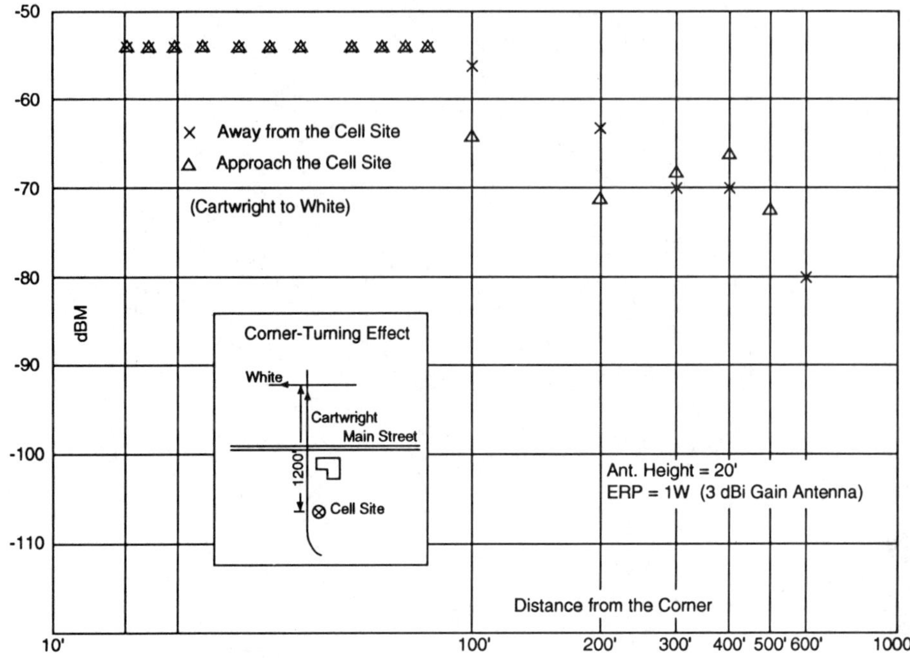

Figure 2.9 Corner-turning effect tested in Irvine, CA [37].

where P_{r0} is the received power in milliwatts at the distance of d_0, which can be 1 mi or 1 km, and γ is the slope of the path loss in mobile radio environment. γ shown in (2.9a) is closer to 4 rather than 2, which is proper for free-space loss. γ, shown in (2.9b), represents the decibel value per decade. α_0 is the correction factor when the actual condition is different from the reference conditions, which may be specified as:

- Antenna height at the cell site = 100 ft (or 30m);
- Antenna gain at the cell site = 6 dBd;
- Transmitter power = 10W;
- Antenna height at the mobile unit = 10 ft (or 3m);
- Antenna gain at the mobile unit = 0 dBd.

All of the curves shown in Figure 2.10 [26] are normalized to this set of reference conditions and illustrated in each geographical area as if they were obtained from flat ground. Their differences are solely due to the manmade structures in their corresponding cities. We call these path loss slopes the *flat-area path loss slopes*. These slopes have to be measured according to the special rule mentioned previ-

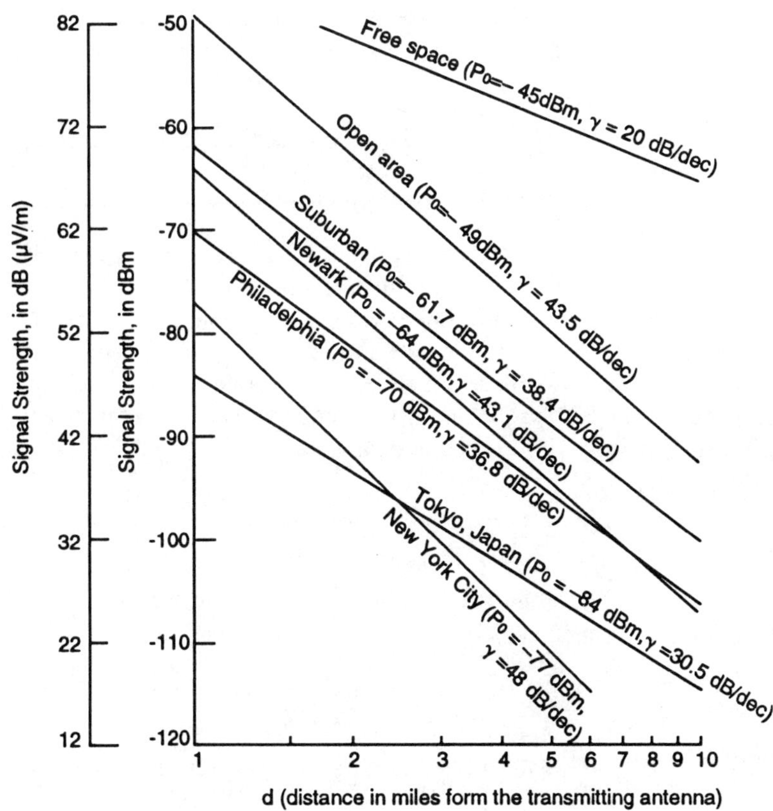

Figure 2.10 Propagation path loss in different areas [26].

ously, so the path loss slopes obtained from the other published sources may not be appropriate. Each of the slopes has to be measured in each corresponding city, since no manmade structures are alike.

Prediction of Land-to-Mobile

Sometimes, no actual measured data in a city is available. Then a standard form of path loss curve which fits most of the measured data in U.S. cities can be applied. The reason is that the path loss curves reveal a standard deviation of 6 to 8 dB, which is very loose and easy to fit in a standard form. The standard form is derived from the path loss curve of the suburban areas shown in Figure 2.8 and expressed in a general condition different from the reference condition as follows [27].

$$P_r = (P - 46) - 61.7 - 38.4\log d + 20\log\left(\frac{h_1}{100'}\right) + 10\log\frac{h_2}{10'} + G_m \quad (2.10a)$$

Where in (2.10a), P is the effective radiated power (ERP) at the transmit site, P_r is the received power, G_m is the mobile antenna gain, h_1 and h_2 are the antenna heights at the base and at the mobile, respectively, and d is the path distance. P and P_r are in dBm, d in miles, h_1 and h_2 in feet, and G_m in dBi. In reality, a mobile van is normally used to collect the data. The mobile antenna height of a van is generally set about 10 ft from the ground and the mobile antenna gain G_m is set to be 0 dBi. Equation (2.10a) can be simplifed as

$$P_r = P \cdot 10^{-14.77} \cdot d^{-3.84} \cdot h_1^2 \quad (2.10b)$$

Where P_r and P are in milliwatts, d is in miles, and h_1 is in feet. The cell boundary can be obtained from (2.10b) by

$$d = \left(\frac{P}{P_r} \cdot 10^{-14.77} \cdot h_1^2\right)^{1/3.84} \quad (2.11)$$

For example, if P_r should be +32 dBu [28] (see Figure 2.10), then apply the following equation.

$$\text{dBu} = 132 + \text{dBm} \quad (2.12)$$

Equation (2.12) is valid [29] providing that the signal is received at 850 MHz, with a 50Ω terminal and a 1/2-wave dipole. Then 32 dBu is equivalent to −100 dBm. Inserting $P_r = -100$ dBm into (2.11) yields

$$d = 0.346(h_1^{0.5208} \cdot P^{0.2604}) \quad (2.13)$$

where d is in miles. In (2.13), h_1 is in feet and P is in watts. h_1 is the effective antenna height shown in (2.10a). In (2.13), we may treat h_1 as the antenna height above the average terrain.

Prediction of Land-to-Boat

The situation of land-to-boat over water is equivalent to propagation over an open area because there are no manmade structures on the water. In this case, the experimental curve of an open area shown in Figure 2.10 should be used [30].

$$P_r = (P - 46) - 49 - 43.5 \log d + 20 \log(h_1/100) \tag{2.14}$$

Where P is the ERP in dBm at the transmit site, P_r is the power in dBm, d is the path distance in miles, and h_1 is the effective antenna height, which can be treated as the antenna height above average terrain in feet at the base station. We may follow the same derivation steps as for land-to-mobile cases, to predict land-to-boat, using a noise level of −120 dBm [31], which is from measured data in a noise-limited area. Based on the required C/I = 18 dB, the received signal should be −102 dBm, which is equivalent to the 30-dBu contour. Substituting $P_r = -102$ dBm in (2.14) yields

$$d = 0.853(h_1^{0.86} \cdot P^{0.23}) \tag{2.15}$$

where d is in miles, h_1 is in feet, and P is in watts.

2.2.5.2 *Effects of the Natural Terrain Contour*

After we determine the path loss curves in the cities or any suburban areas solely due to manmade structures (assuming the areas are flat), we next need to add the effect of the natural terrain contour to complete the prediction model. There are three conditions to consider: a nonobstructive condition, an obstructive condition, and an over-water condition.

Nonobstructive Condition

In a nonobstructive condition [24], the signal is not obstructed by hills or mountains. Although the signals are blocked by buildings or homes, they can still reach the mobile units or portable units from multiple reflected waves. We call this wave path a *direct-wave path*. The direct-wave path is distinguished from the line-of-sight path, along which the cell site antenna can be seen from the mobile unit.

A Modified Two-Wave Theory. When the ground is not flat, it creates not only a direct-wave path but more than one reflected-wave path, as shown in Figure 2.11(a). A two-reflection ground situation can be illustrated by either the cell site antenna

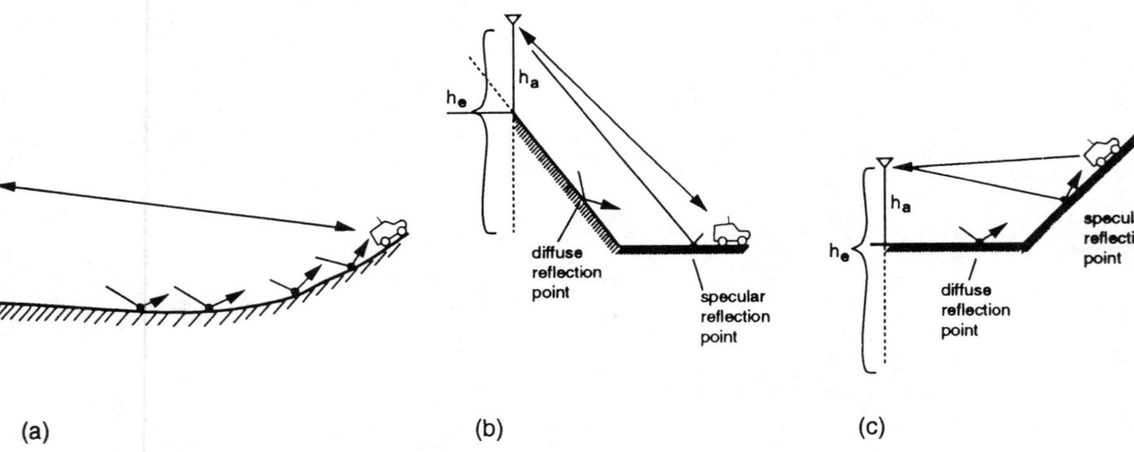

Figure 2.11 Identification of specular reflection points: (a) multiple; (b) case A; (c) case B.

located at the foothill, as in case A shown in Figure 2.11(b), or on the hill, as in case B shown in Figure 2.11(c). Based on the wave theory, there are two reflection points corresponding to two reflected waves on two reflected grounds, as shown in Figure 2.11(b,c). The one closest to the mobile unit is called the specular reflection point, and the one away from the mobile unit is called the diffused reflection point. The reflected wave that can deliver most of the energy to the mobile unit is from the one reflecting from the specular reflection point. Therefore, in reality, there is always one reflected wave dominating the reflected energy and the others can be neglected.

Reflection Points on a Nonflat Ground. To illustrate the reflected waves on a drawing (see Figure 2.7), the vertical scale and the horizontal scale usually have a difference of two orders of magnitude; that is, the vertical scale is usually in meters or feet and the horizontal scale is usually in kilometers or miles. This is just for the purpose of having a clear illustration. Nevertheless, the readers' perception could be misled. The slopes of the hills as shown on the drawings are always large. Actually, the slopes of the hills are small. The actual incident angle of a reflected wave on the hillside is also small. Therefore, we can use the same method for finding the reflection point as that used for flat areas. The antenna masts shown in any drawings should be perpendicular to the x-axis, not perpendicular to the slope of the ground in the drawing. Therefore, drawing the antennas perpendicular to the x-axis in a drawing of the mobile radio environment with different scales in two dimensions is the best approximation, because of the actual small slopes. Since the wave propagates on nonflat ground, the reflection points obtained from the base antenna image and from the mobile antenna image may not coincide. In this case, pick the points that are closer to the mobile unit. This is based on the specular reflected energy concept.

Effective Antenna Height. The specular reflection point occurring on a nonflat ground would be used to measure the effective antenna height. The effective antenna height is measured at the cell site antenna location from an extended ground plane where the specular reflection point is. The effective antenna height h_e is different from the actual antenna height h_a. This is illustrated in Figure 2.11. Sometimes the effective antenna height is less than the actual antenna height, and sometimes it is greater. The effective antenna height is used in conjunction with the flat-ground path loss curves shown in Figure 2.10. The antenna height gain can be calculated by

$$G_e = 20 \log(h_e/h_a) \tag{2.16}$$

The antenna height gain G_e will be changed along the mobile path as the mobile unit travels from place to place because the effective antenna heights h_{eA}, h_{eB}, and h_{eC} are changing, as shown in Figure 2.12.

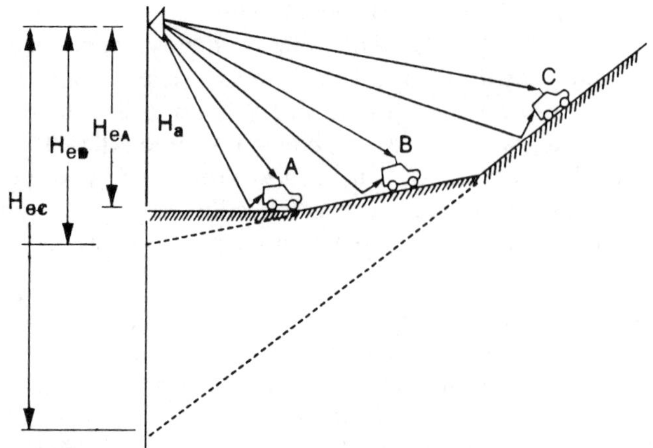

Figure 2.12 Change of effective antenna height with vehicle position.

Radio Path and Mobile Path. Predictions are made by applying the correction given by (2.16) to values read from an appropriate path loss curve like those of Figure 2.10. Predicted values and measurements made at the mobile unit may then be plotted as shown in Figure 2.13. Many measured data points can be collected from different sites in the vicinity of a given location, and they are averaged and taken as the signal strength at the location. Standard deviations of 6 to 8 dB due to the undulation of nonflat ground and the variation of building distribution are typical.

The mobile path is the path on which the mobile unit is traveling (i.e., streets and roads). The signal strength along the mobile path, obtained by averaging over a length of 20 to 40 wavelengths, is called the *local-mean signal* [32]. This local-mean signal is the one that this model is predicting, not the path loss. To predict a local-mean signal, the natural terrain configuration between the cell site and the mobile unit, while the mobile unit is traveling along a street, should be known. From the terrain elevation, the effective antenna height can be found. Then the antenna height gain can be applied to the flat-ground path loss to predict the local-mean signal along the mobile path. An illustration of the measured data and the predicted values on the same mobile path is shown in Figure 2.13. The difference between the measured and predicted values, on average, as shown in [25] is about 2 dB due to errors in loran-C used for site location. There is also a time delay due to processing.

Obstructive Condition

When hills or mountains are standing between the cell site and the mobile unit, no direct wave path exists, because the wave transmitting from the cell site cannot

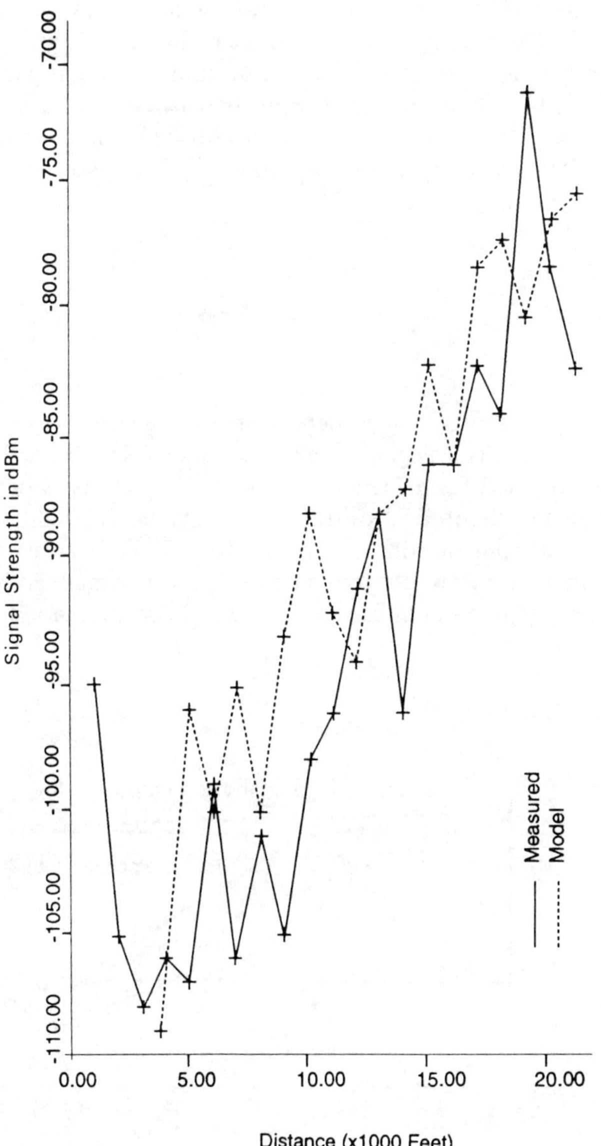

Figure 2.13 Measured and predicted local-mean signal strength in Orange County, CA.

bounce sideways to reach the mobile unit due to the extended dimension of the hills. The wave can only be diffracted at the top of the hill and reach the mobile unit, as shown in Figure 2.14. Since the dimension of the hill is large compared with the wavelength, the knife-edge diffraction formula [33] can be adapted fairly accurately to reality. The height of the knife-edge is shown in Figure 2.14. As soon as the obstructive condition occurs, the effect of antenna height gain disappears. The diffraction loss, sometimes called *shadow loss*, is based on four parameters: the distance from the cell site to the knife edge d_1, the distance from the mobile unit to the knife edge d_2, the height of the knife edge h_p, and the wavelength λ, embraced in to form a new parameter V:

$$V = h_p \sqrt{\frac{2}{\lambda}\left(\frac{1}{d_1} + \frac{1}{d_2}\right)} \tag{2.17}$$

h_p can be a positive or negative value, depending on the shadow region, or in the nearly shadow region (h_p is a negative value), as shown in Figure 2.14. The diffraction loss L is obtained from Figure 2.15, with a given value of V calculated from (2.17). When the theoretical diffraction loss value is small, the predicted signal levels deduced from the diffraction loss values agree fairly well with the measured data. The theoretical diffraction loss value sometimes can be very large; thus, the predicted signal strength can be very weak and may not agree with the

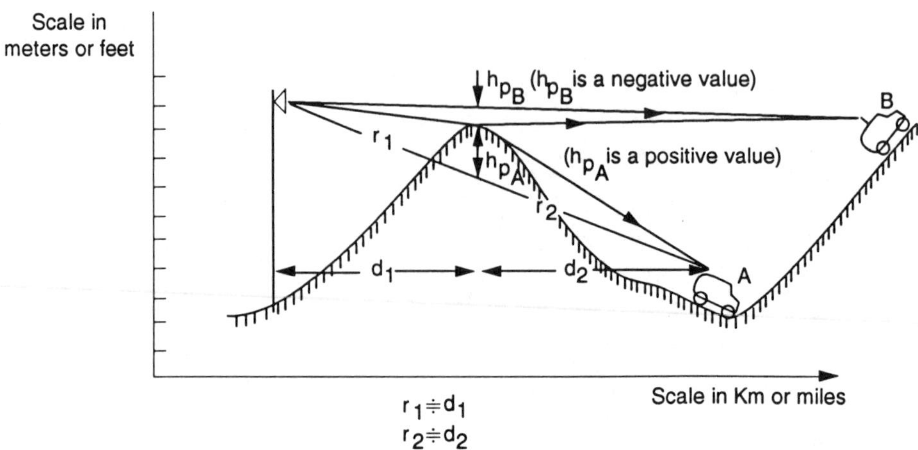

Figure 2.14 Knife-edge height calculation.

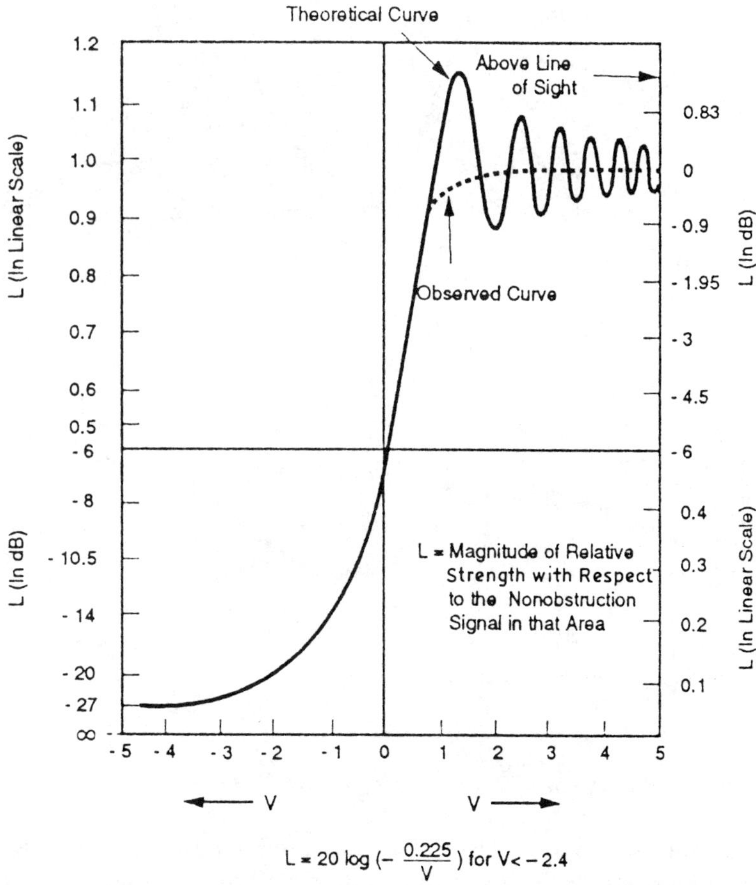

Figure 2.15 Shadow-loss prediction [26].

measured data. Fortunately, as soon as the received signal is below the accepted level, the disagreement becomes unimportant.

Double Knife-Edge Condition. In reality, the condition of having double knife edges does exist. However, a reasonable theoretical calculation of a double knife-edge diffraction loss is very difficult. Therefore, a methodology without theoretical backup is applied based on the two paths delineated by the triangles shown in Figure 2.16. The two diffraction losses L_1 and L_2, obtained from two triangles ΔACB and ΔCDB, are summed up in decibels to represent the total loss. Although this methodology does not have a theoretical base, it agrees fairly well with the measured data for a diffraction loss less than 20 dB in most cases.

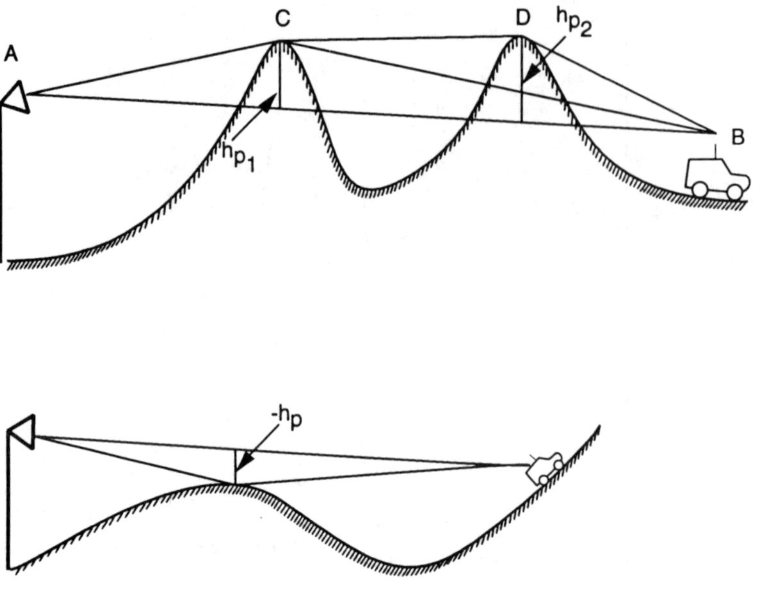

Figure 2.16 Multiple knife-edge calculation.

Land-to-Mobile Propagation Over the Water [24]

When a transmitted wave propagates over the water and is received by the mobile unit on the other side of the water, many generated reflected waves can occur as shown in Figure 2.17. Only one specular reflected wave on the ground surface is chosen normally. But in this case, the wave reflected from the open water surface (no manmade structures around) has no diffuse effect and also delivers the energy to the mobile receiver. Therefore, the signal received at the mobile unit will be three waves, the direct wave, the specular wave, and the wave reflected from the water, as follows.

$$P = P_0\left[\frac{1}{4\pi d/\lambda}\right] \left|1 - \exp(j\Delta\Phi_1) - \exp(j\Delta\Phi_2)\right|^2 \tag{2.18}$$

Because the phases $\Delta\Phi_1$ and $\Delta\Phi_2$ are very small, we may apply the same techniques as those used in deriving (2.6). This result becomes

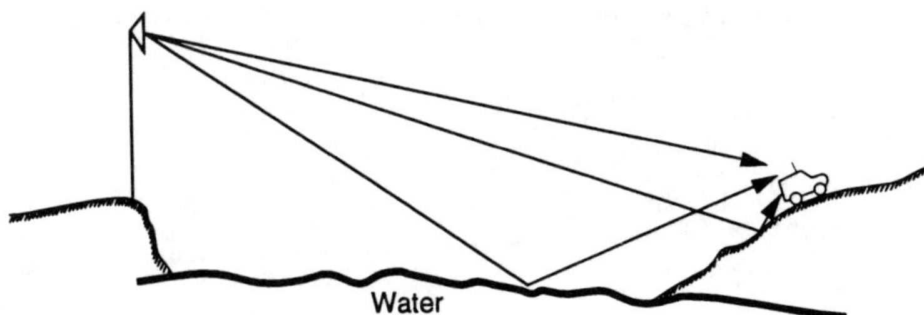

Figure 2.17 Land-to-mobile over water.

$$P_r = P_0 \frac{1}{(4\pi d/\lambda)^2} \tag{2.19}$$

Equation (2.19) shows that the received signal over the water can be treated as reception in a free-space condition. This is why the signal is strong crossing over the water. However, the area in which the received signal is strong has to have two strong reflected waves occurring as in Figure 2.17.

2.2.5.3 *Lee's Model for Macrocell [24]*

The macrocell model is used for a cell greater than 1 km in diameter and can be formed by the following. For a nonobstructive path,

$$P_r = L_F + G_e = P_{r0} - 10(\gamma) \log \frac{d}{d_0} + 20 \log \frac{h_e}{h_a} + \alpha_0 \quad \text{dBm} \tag{2.20a}$$

where L_F is the flat-ground path loss (in a specified city). For an obstructive path,

$$P_r = L_F + L + \alpha_0 = P_{r0} - 10(\gamma) \log \frac{d}{d_0} + L + \alpha_0 \quad \text{dBm} \tag{2.20b}$$

where L is the defraction loss. For land-to-mobile over water, use a free-space reception formula:

$$P_r = P_0 \left(\frac{1}{4\pi d/\lambda} \right)^2 \tag{2.20c}$$

For land-to-boat over water, use an open area formula:

$$P_r = -49 - 43.5 \log d + \alpha_0 \quad \text{dBm} \tag{2.20d}$$

Remarks

- The received signal strength P_r cannot be higher than that from the signal received from the free space. When G_e is very large, be sure to check that the predicted P_r is capped by the received signal strength from the free-space loss.
- α_0 is the corrected factor (the gain or loss) that is obtained from the reference condition.
- The foliage loss should be added in a heavy foliage area. The loss is affected by the size of the leaves, the density of the branches, the height of the trees, the season of the year, and so forth. One thing can be certain when predicting the coverage in a heavy-foliage area: the antenna height at the cell site should be higher than the top of the trees.
- When we feed in data on the natural terrain contour of an area of interest for calculating (2.20), we have to be aware of the resolution or the accuracy of the terrain contour map, especially when we are designing small cells. In general, the quarter-million-to-one contour map is not a proper one to use for (2.20), but the twenty-four-thousand-to-one (7.5-minute map) contour map is the one to use. That is because we are trying to predict the local means along the streets. The poor-resolution map can produce an incorrect answer. For this reason, if we are using the quarter-million-to-one contour map, we may have to predict the local means by dropping the last term of (2.20a), which is 20 log (h_e/h_a). Other terms in (2.20) remain unchanged. The point is that if the terrain data is incorrect, imposing an accurate adjustment according to terrain configuration will create more errors.
- The two-wave model expressed in (2.1) cannot be used to predict a received signal within 1 km in a manmade environment. Therefore, a microcell model is filled in for applying this range.
- The difference between the path loss prediction over the radio path and the local-mean prediction over the mobile path is shown in Figure 2.18. In Figure 2.18(b), the difference at location F can be 15 dB. That is why the path loss prediction has a large standard deviation of ± 8 dB. The local-mean prediction follows the solid curve. Its standard deviation along the local-mean prediction can be reduced to ± 2 dB, as shown in Figure 2.19.

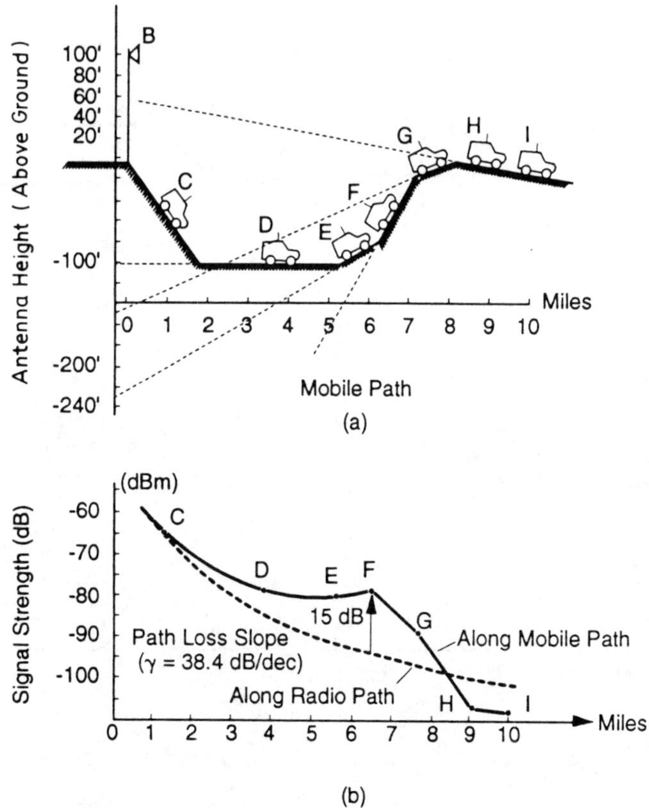

Figure 2.18 Terrain effect on the effective antenna gain at each position: (a) hilly terrain contour; (b) point-to-point prediction [26].

2.2.5.4 Lee's Model for Microcell [24]

When the size of the cells is small, less than 1 km, the street orientation and individual blocks of buildings make a difference in signal reception, as mentioned previously. Those street orientations and individual blocks of buildings do not make any noticeable difference in reception when this signal is well attenuated at a distance over 1 km. Over a large distance, the relatively great mobile radio propagation loss of 40 dB/dec is due to the situation that two waves, direct and reflected, are more or less equal in strength. The local scatterers (i.e., buildings surrounding the mobile unit) reflect this signal, causing only the multipath fading and not the

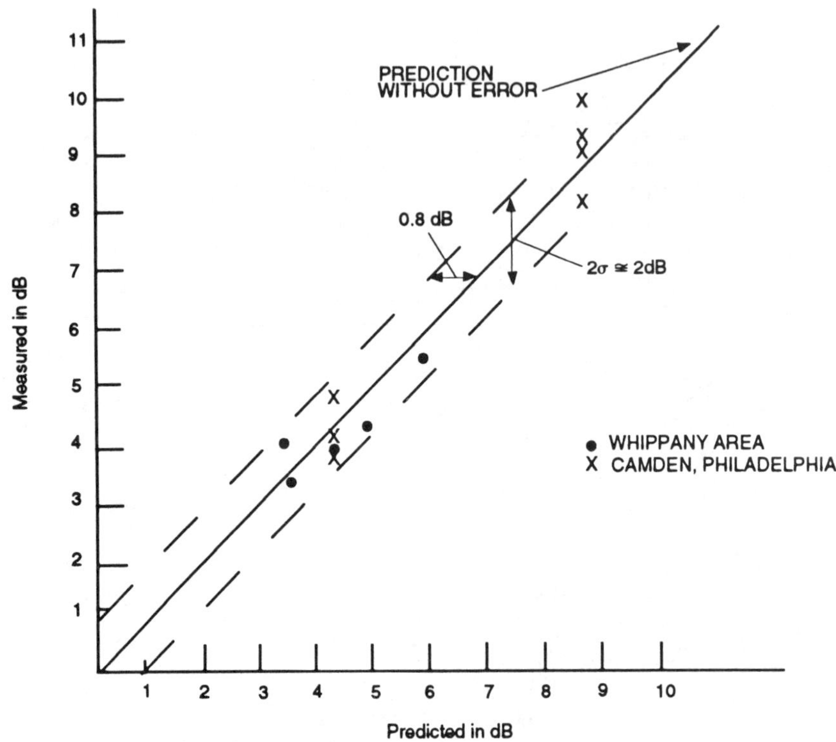

Figure 2.19 Errors in point-to-point predictions under nonobstructed conditions.

path loss at the mobile unit. When the cells are small, the signal arriving at the mobile unit is blocked by the individual buildings, which weaken the signal strength and are considered as part of the path loss. Therefore, we have to take another approach in our prediction, as will be described in this section. Although the waves received by the mobile unit are multipath waves reflected from the building walls and are not the waves penetrating through the buildings, we find a strong correlation between the signal strength from the reflected waves and the thickness of the building blocks along the direct path. In small cells, we are calculating the loss based on the ground dimensions of the building blocks. Since the ground incident angles of the waves are, in general, small due to the low antenna heights used in small cells, the exact height of buildings in the middle of the propagation paths is not important.

Therefore, only a two-dimensional plan map is used. We can use aerial photographs to calculate the proportional length of the direct-wave path being attenuated by the building blocks. When the wave is not being blocked by the building, it is a line-of-sight condition. From measurement data along streets under an open line-of-sight condition, the line-of-sight signal reception P_{los} is formulated. Also, from the measured signal P_{os} along streets in out-of-sight conditions within the cells, we form an additional signal attenuation α_B formula due to the portion of building blocks over the direct path by subtracting the received signal from P_{los}. The steps for forming an additional signal attenuation formula α_B are as follows.

- Calculate the total blockage length B by adding the individual building blocks. For example, $B = a + b + c$ at point A shown in Figure 2.20.
- Calculate the signal strength P_{los} for line-of-sight conditions.
- Measure the signal strength P_{os} for out-of-sight conditions.
- The local mean at point A is P_{os}. The distance from the base to the mobile unit is d_A. The blockage length B at point A is $B = a + b + c$. Then the value of α_B for a blockage of B can be expressed as α_B $(B = a + b + c) =$

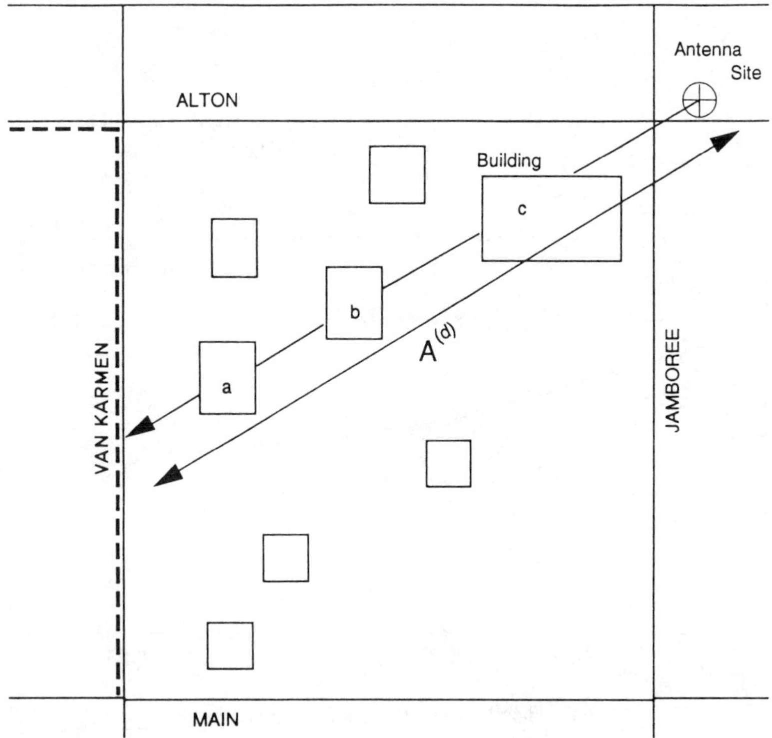

Figure 2.20 Building block occupancy at location A [37].

$P_{\mathrm{los}}\ (d = d_{\mathrm{A}}) - P_{\mathrm{os}}$ (at d_{A}). Therefore, the microcell (small cell) model can be formed as

$$P_r = P_{\mathrm{los}} - \alpha_B \qquad \mathrm{dB} \tag{2.21}$$

where P_{los} is a statistically formed line-of-sight path loss curve obtained from the measured data, and α_B is the additional loss due to the length of the total building blocks B along the radio paths. Up to 500 ft, P_{los} would be based on the free-space formula regardless of the height of the base station antenna. When the distance exceeds 1,500 ft, which is of the order $4h_1h_2/\lambda$, the 40-dB/dec rule should be used. An example for a particular base station height is shown in Figure 2.21(b).

The additional signal attenuation curve based on the building blockage found experimentally is shown in Figure 2.21(a). The α_B curve was obtained at Irvine, California. The curve shows that rapid attenuation occurred while B is less than 500 ft. When B is greater than 1,000 ft, a nearly constant value of 20-dB attenuation is observed. It can be explained by the street corner phenomenon as shown in

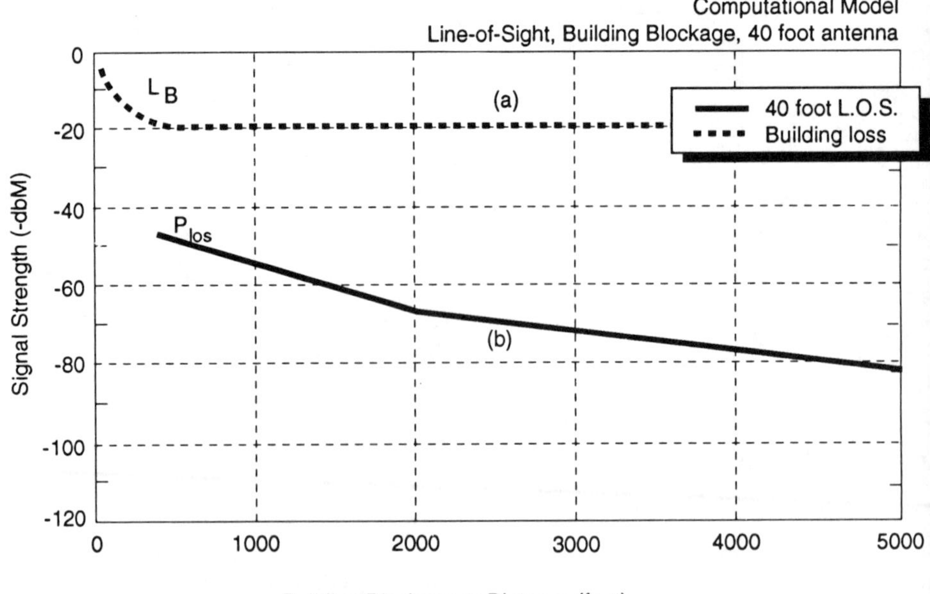

Figure 2.21 Microcell prediction parameters: (a) L_B due to building blockage; (b) line-of-sight P_{los} (ERP = 1W) [37].

Figure 2.9. The rapid attenuation was seen on the mobile signal during the turning from one street to another as B starts from 0 ft and increases. After B reaches 500 ft, the received signal strength P_{os} will remain 18 dB below the P_{los} as the distance d increases. The path losses due to a line-of-sight condition for a series of antenna heights have been measured along many streets. The 9-dB/oct (or 30-dB/dec) antenna height gain over an antenna height change is usually observed in a small cell, as shown in Figure 2.22. It is due to the fact that the incident angle in the small cell is usually larger than 10 deg. The 6-dB/oct rule only applies to a large cell with a small incident angle. In the small-cell prediction model, we use the two curves P_{los} and α_B to predict the received signal strength.

The expression to be evaluated is

$$P_{los} = P_0 + \gamma_0 \log \frac{d_1}{d_0} \tag{2.22}$$

from Figure 2.21 and

$$\alpha_B = \gamma_1 \log(B/B_0) \tag{2.23}$$

from Figure 2.21, where P_0 is the intercept point at a distance d_0 and d_1 is the total distance. Sometimes d_1 is smaller than d_0 in the small-cell prediction. γ_0 is the line-of-sight path loss slope, B is the length of blocking, and γ_1 is the slope of additional loss due to blocking length. The dimension of γ_0 and γ_1 is dB/dec.

$$P_r = P_0 + \gamma_0 \log(d_1/d_0) - \gamma_1 \log(B/B_0) \tag{2.24}$$

This microcell model has been verified in the areas of Irvine and San Diego, California, with good results, as shown in Figures 2.23 and 2.24. In a hilly area, (2.21) can be modified by adding the term antenna height gain obtained from Figure 2.22 as

$$P_r = P_{los} - \alpha_B + 30 \log(h_e/h_a) \tag{2.25}$$

The prediction from the microcell model is not as accurate as that from the normal-cell model. This is due to the fact that we are using a statistical prediction tool to predict the signal in more or less deterministic conditions where the propagation distance is short.

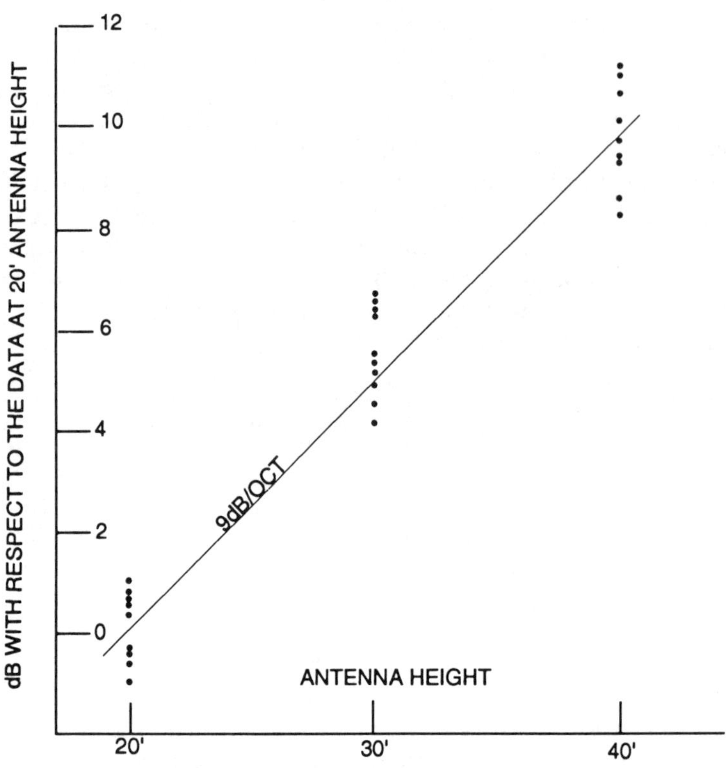

Figure 2.22 Antenna height effects in microcell systems [37].

2.2.6 Full Prediction Model

A full prediction of Lee's model [24] will be superimposed on the normal-cell model and the microcell model to predict the received signal strength in a range up to 15 mi. Beyond this distance, two conditions may have to be addressed. One is the signal reception over the radio horizon and the other is the nonobstructive case over the water.

2.2.6.1 Prediction of Signals Beyond the Radio Horizon

For calculating the interference in a distance greater than 15 mi, we still recommend the use of (2.20) for prediction, taking into account that the interference would be strong in a worse case. This is a conservative approach for designing a reliable

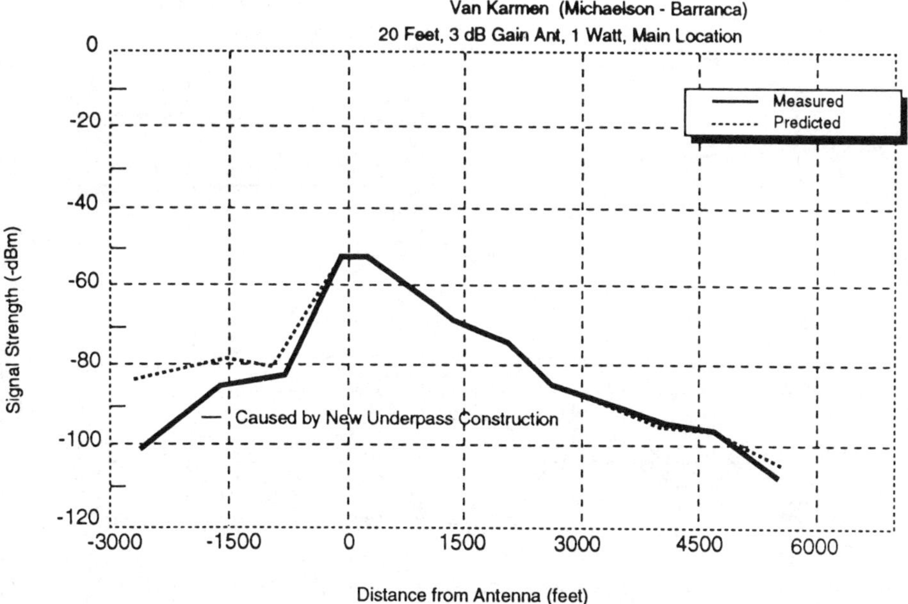

Figure 2.23 Comparison of measured and predicted data at Van Karmen, Irvine, CA [37].

system. When an additional loss is considered due to the radio horizon which would further reduce the radio interference, the distance beyond the radio horizon should be

$$d = \sqrt{2h_1} + \sqrt{2h_2} \tag{2.26}$$

Where h_1 and h_2 are the antenna height and mobile antenna height, respectively, in feet and d in miles is the distance spanning the radio horizon.

The parameters shown in (2.26) are indicated in Figure 2.25. The additional loss L_a should be added to P_r in (2.20). In general, there is a 20-dB loss in the transition region where

$$\sqrt{2h_1} \leq d \leq \sqrt{2h_1} + \sqrt{2h_2} \tag{2.27}$$

Therefore, the additional losses of the two cases are expressed as follows. For case 1,

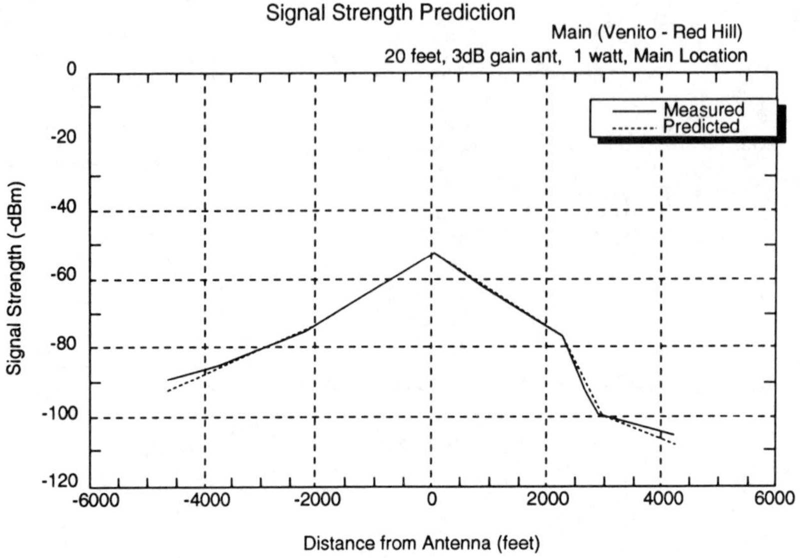

Figure 2.24 Comparison of measured and predicted data at Main Street, Irvine, CA [37].

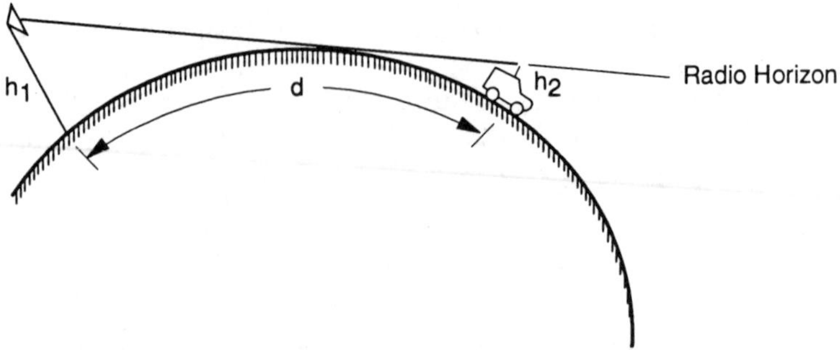

Figure 2.25 Finding the radio horizon.

$$L_a = 0 \tag{2.28a}$$

when $d \leq \sqrt{2h_1}$, and, for case 2,

$$L_a = 20 \cdot \log\left(\frac{d_2}{2h_1 + 2h_2}\right) \tag{2.28b}$$

when $\sqrt{h_1} \leq d \leq \sqrt{2h_1} + \sqrt{2h_2}$. When the distance d from the interference is beyond the radio horizon, the interference is so weak that it can be neglected in the mobile radio system. These equations are not critical because the interference obtained from (2.28) is always weaker than that predicted by (2.20).

2.2.6.2 Illustrations of the Nonobstructive Case Over the Water

Below are four conditions that can be illustrated while the signal propagates crossing over the water (see Figure 2.26). The A wave is reflected from the water, while the B wave is a specular reflection from the coast line or ground.

- Condition I: land-to-mobile over the water condition. Both reflected waves A and B are not blocked. Use (2.20c).
- Condition II: land-to-mobile over the land condition. A is blocked, but B is not. Use (2.20a).
- Condition III: land-to-mobile condition. Both A and B are blocked. Using (2.20c) for the nearly shadow (negative knife-edge) case.
- Condition IV: land-to-boat condition. A single reflected wave is on the water. Using (2.20d) with the open-area slope case shown in Figure 2.10.

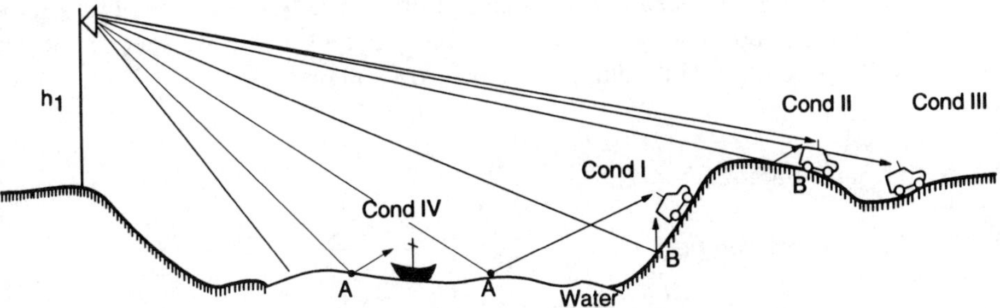

Figure 2.26 Wave propagation for various nonobstructive conditions.

2.2.6.3 Discussion of Lee's Model

This prediction model can be used not only for the narrow-band signal propagation but also for wide-band. The general expression of the received power would be based on [34]

$$P_r = \frac{K}{(4\pi r^2)^2} \cdot \frac{1}{f_0^3[1 - (B/2f_0)^2]^2} \tag{2.29}$$

where K is a constant, r is the distance, f_0 is the carrier frequency, and B is the channel bandwidth. When $B = 0.5f_0$, the difference in receiving a wide-band signal as compared with receiving a narrow-band signal $B \ll f_0$ is 0.56 dB, which is a small difference and can be neglected.

This prediction model can be modified to predict the signal strengths in different carrier frequencies. Another term is added to (2.20) [35] for this purpose:

$$P_{r_f} = P_r - 30 \log(f/f_0) \tag{2.30}$$

where f is the new carrier frequency and $f_0 = 850$ MHz.

In designing satellite or microwave communication, the concept of using a signal margin is applied because the variation due to the atmosphere and various causes will affect the received signal between the two fixed points. However, in designing a cellular system, we do not use a signal margin because the propagation distance of mobile paths is short and the carrier frequency is in UHF. The atmospherical variation effect can be neglected. In addition, when a mobile unit is moving, the received signal at the mobile unit may vary over 70 dB. Therefore, the signal margin concept cannot be applied. We design a good mobile radio system using the prediction model and verify the system performance with the measured data, which remain unchanged over the time period. A signal margin would only be applied for areas that have heavy foliage loss.

The prediction model can be applied not only to cellular systems, but also to personal communication service (PCS) systems, paging systems, special mobile radio (SMR) systems, mobile data, or vehicle tracking systems.

2.2.7 Multipath Fading Model

2.2.7.1 Long-Term Fading

The raw data shown in Figure 2.27 has been artificially broken down into two parts, the long-term fading and the short-term fading [36]. The long-term fading

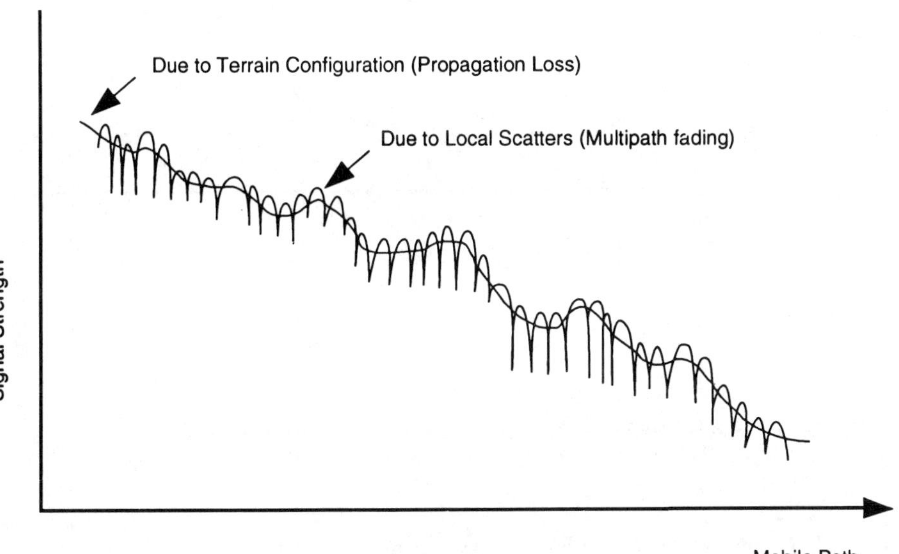

Figure 2.27 The nature of fading.

is affected by the terrain configuration. It is the local-mean (L) variation and follows the log-normal distribution [37]:

$$p(L) = \frac{1}{2\pi\sigma} \exp\left(-\frac{b^2}{2\sigma^2}\right) \tag{2.31}$$

where L is in dBm and σ is the standard deviation in decibels. σ is different depending on the terrain variation: in a hilly area, σ is large, and in a flat area, σ is small. The value of σ from a log-normal plot shown in Figure 2.28 can be found by the following method. Find the level $\overline{L}$ at a percentile of 90%:

$$P(L \leq \overline{L}) = 90\% \qquad \overline{L}/\sigma = 1.29 \tag{2.32}$$

Then

$$\sigma = \frac{\overline{L}}{1.29} \quad \text{dB} \tag{2.33}$$

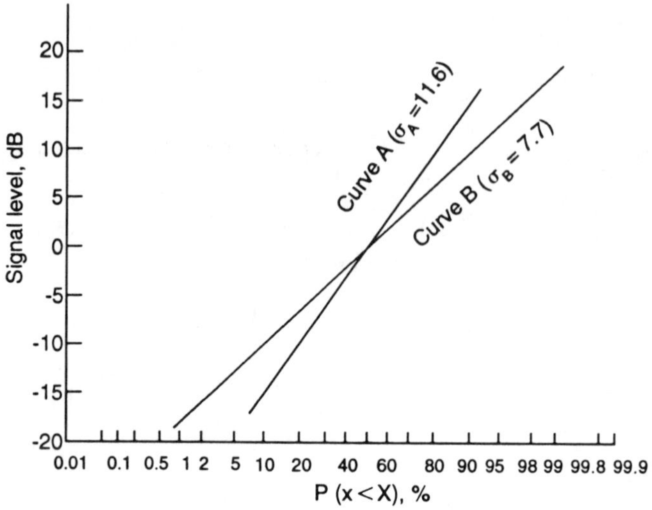

Figure 2.28 Cumulative probability distribution curves for lognormal fading in two different areas.

2.2.7.2 Short-Term Fading

Short-term fading is fast fading. It is caused by the structures surrounding the mobile receiver. The signal is transmitted from the base station and arrives at the mobile unit through the multipath reflections. If a strong line-of-sight path also exists, then the signal received by the mobile unit will perform a Rician fading [38]. If there is no line-of-sight path, then the signal received will perform a Rayleigh fading [36]. The Rayleigh fading is formed from many reflected waves. The components of the Rayleigh fading can be expressed as:

$$X_k = \sum_{j=1}^{9} a_j \cos\left[2\pi \frac{k}{100} \cos \theta_j\right] + \sum_{j=1}^{9} b_j \sin\left[2\pi \frac{k}{100} \cos \theta_j\right] \tag{2.34}$$

$$Y_k = -\sum_{j=1}^{9} a_j \sin\left[2\pi \frac{x_i}{100} \cos \theta_j\right] + \sum_{j=1}^{9} b_j \cos\left[2\pi \frac{k}{100} \cos \theta_j\right] \tag{2.35}$$

and

$$r_k = [X_k^2 + Y_k^2]^{1/2} \tag{2.36}$$

where a_j and b_j are Gaussian variables with mean 0 and variance 1. θ_j has nine values ranging from 0, 40, 80, . . . , 320 deg. The Rayleigh fading can be generated by (2.36) and is shown in Figure 2.29. The distribution of Rayleigh fading is

$$p(r) = (1/\bar{r}) \exp[-(r^2/\bar{r}^2)] \tag{2.37}$$

and

$$P(r \leq R) = 1 - \exp[-(R^2/\bar{r}^2)] \tag{2.38}$$

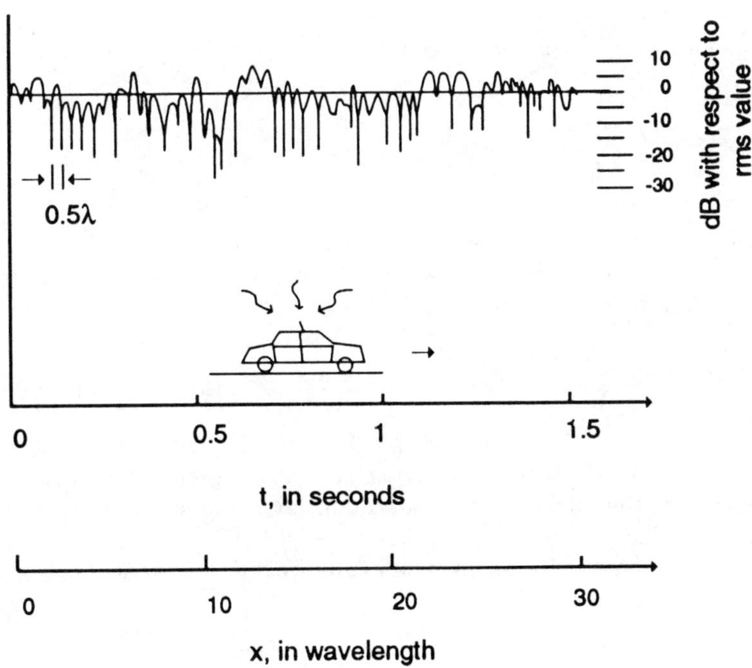

Figure 2.29 Typical signal fading while vehicle is in motion.

Fading Depending on Frequency

In (2.37) we are assuming that all the waves are arriving at the mobile unit at the same time. This is called *flat fading*. In reality, the wave arrivals are at different times and cause time-delay spread. However, the flat-fading model is usually applied to analog-modulated signals on voice systems or to a slow transmit bit rate on digital systems. Under these circumstances, waves arriving at different times do not cause noticeable degradation in voice quality or bit error rate in data transmission. When the transmit bit rate becomes high, the waves arrive at different times, causing selective frequency fading. Selective frequency fading will affect the digital modulation or signaling transmission over an analog system. We have to indicate the cause of the selective frequency fading as follows.

2.2.7.3 Selective-Fading Model

We can modify the flat model shown in (2.34) and (2.35):

$$X_k = \sum_{j=1}^{9} a_j \cos\left[2\pi \frac{k\Delta_j}{100} \cos\theta_j\right] + \sum_{j=1}^{9} b_j \sin\left[2\pi \frac{k\Delta_j}{100} \cos\theta_j\right] \tag{2.39}$$

$$Y_k = -\sum_{j=1}^{9} a_j \sin\left[2\pi \frac{k\Delta_j}{100} \cos\theta_j\right] + \sum_{j=1}^{9} b_j \cos\left[2\pi \frac{k\Delta_j}{100} \cos\theta_j\right] \tag{2.40}$$

where Δ_j is the time delay of the jth wave. Δ_j will be different due to different environments. The relationship between the interval of Δ_j and the difference in frequency Δf can be expressed as

$$\Delta_j = \frac{1}{2\pi(\Delta f)} \tag{2.41}$$

which indicates that when the frequency changes, the time delay of wave arrival also changes. The time delay of wave arrival forms a frequency-selective fading as shown in Figure 2.30.

Then, in a fading environment, we can use two different frequencies to achieve a frequency diversity if

$$\Delta f \geq \frac{1}{2\pi\Delta} \tag{2.42}$$

where Δf is the separation between two carrier frequencies. In the mobile radio environment:

- Suburban: $\Delta = 0.5\ \mu s$, and $\Delta f = 300$ kHz;
- Urban: $\Delta = 3\ \mu s$, and $\Delta f = 50$ kHz.

Applying frequency diversity in suburban areas requires two frequencies separated by 300 kHz, and in urban areas separated by 50 kHz. If the frequency diversity has to apply in both areas, then take the 300-kHz separation. Selective fading is harmful to system reception.

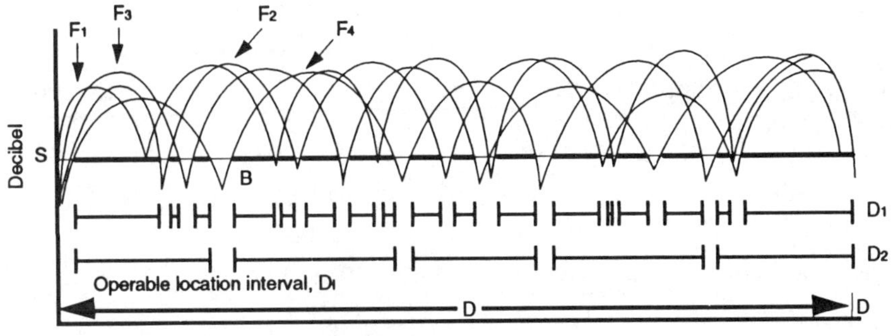

F1 — Forward link setup channel
F2 — Reverse link setup channel
F3 — Forward link voice channel
F4 — Reverse link voice channel
Dl Distance interval where the signal is above level S; operable location interval
D Total distance of interest
D1 Total distance that all four frequencies are above level S
D2 Total distance that only F4 is above level S

Figure 2.30 Portable cellular four-frequency selective fading system. (Lee, W. C. Y, "In Cellular Telephone, Complexity Works," *IEEE Trans. on Circuits and Devices*, Vol. 7, No. 1, January 1991, pp. 26–32.)

Selective-Fading Problems

For example, in cellular systems, we need four frequencies to complete a call. Two frequencies (because of a duplex system) are used for setup calls and two for connecting to voice. When the mobile unit is moving, the average signal strengths of all four frequencies are the same. If one frequency signal is strong, all four frequency signals will be strong. The calls do not depend on the condition of any particular frequency. But if the mobile or a portable unit is standing still, then the antenna is located at one spot. Due to frequency-selective fading, all four frequency signals are not always above a certain threshold; then the system only needs the signal strength of one frequency below the threshold to cause dropped calls. In order to calculate the percentage that the signals of all four frequencies are above a certain threshold, the following formula is used.

$$P_4(r > R) = [P_1(r > R)]^4 \tag{2.43}$$

where $P_1(r > R)$ is the probability that one frequency is above the threshold R. In a Rayleigh fading environment,

$$P_1(r > R) = \exp(-R/R_0) \tag{2.44}$$

where R_0 is a reference level at which the threshold R can be referred to. Then (2.43) becomes

$$P_4(r > R) = [\exp(-R/R_0)]^4 \tag{2.45}$$

The four-frequency phenomenon can be illustrated as shown in Figure 2.30.

2.2.8 Maritime and Aeronautical Mobile Systems

Maritime and aeronautical mobile systems basically operate in different environments. As mentioned in Section 2.2.1, ground mobile systems are very much affected by natural terrain configuration and manmade structures.

However, the transmitted media of maritime and aeronautical mobile systems are affected little by the natural terrain configuration and manmade structures. Usually, frequencies below 30 MHz (referred to as HF) are used for maritime and aeronautical mobile systems. For aeronautical mobile satellite systems, the allocated frequency is about 1.5 to 7.3 GHz.

HF is used for communication distances greater than 150 km. The transmission depends on the sky wave reflected from the ionosphere. Depending on the distance, the E-layer, which is at a height of about 110 km, can communicate over a distance of less than 150 km during the day, and less at night. The F_1-layer is at a height of 175 to 250 km, only exists during the day, and does not actively reflect the energy from the incident wave. The F_2-layer is at a height of about 200 to 400 km and can have a greater propagation distance of up to 3,000 km.

Because of constant changes in ionospheric conditions, maritime mobile and aeronautical mobile communication systems should use space and frequency diversities to compensate for fading caused by these ionospheric changes.

Now, maritime satellite and aeronautical mobile satellite systems use satellites to replace the ionospheric layers. There are two kinds of satellites, which are differentiated by their altitudes. One is a stationary satellite (2,200 mi above the earth) and the other is the low-elevation orbit (LEO) satellite (200 to 500 mi above the earth). These satellite systems are also developed for coverage of rural areas on land. The propagation mechanisms for satellite communication are generally less problematical than those for land mobile systems, particularly when the latter is in an urban environment.

2.3 ANTENNA DESIGN

2.3.1 Requirements for Mobile Antennas

In general, antennas for mobile terminals have been required to be small, lightweight, and low-profile, and to have an omnidirectional radiation pattern in the horizontal plane. In addition, antennas must exploit ambient propagation characteristics and be robust against mechanical and environmental hazards encountered while moving. In the early days of mobile communications, wire antennas such as whip antennas, monopole antennas, and inverted-L antennas were commonly used for vehicles and mobile equipment. These antennas can be simply mounted onto the body of the vehicle or mobile unit and still meet the requirements for mobile terminal use.

With the evolution of mobile communication systems, antenna technology has also progressed, and the design concept has changed as well, although the fundamentals essentially remained the same. A typical example is the development of small antennas. The rapid growth in civil applications of mobile communications, particularly the increased use of personal mobile terminals, has generated a need for the development of small mobile terminals and small-sized radiating systems. It is well known that the smaller the antenna size, the lower the antenna efficiency and the narrower the bandwidth; but, in addition, the design concept for portable equipment must include the body of the equipment as a part of the radiator [39], since currents flowing on the body contribute to radiation [40,41]. The radiation characteristics of such an antenna system differ much from those of the antenna element solely in free space. The radiation pattern varies depending on the size and shape of the portable unit and the location of the antenna element [16,41,42].

Another example is antennas used in zone systems, which were adopted in order to increase the utilization of the frequency spectrum. The radiation pattern of base station antennas in cellular land mobile systems is not always omnidirectional, but designed to conform to the pattern of specified zones in the horizontal plane and to be tilted downward in the vertical plane to minimize the cochannel interference. Base station antennas in aeronautical mobile systems must have either a conical beam to cover a specified zone or a satellite tracking facility.

Today, antennas used in mobile communication systems have been recognized as critical elements that can either enhance or constrain system performance. Accordingly, antenna performance and characteristics have come to be studied extensively, specifically taking mobile environmental and propagation conditions into account.

Requirements for antennas depend on the types of mobile systems, as summarized in Table 2.3 [43–45], where the types are categorized into land, maritime, aeronautical, and satellite mobile systems. There are various types of antennas, from simple ones, such as whip antennas for a single-channel operation, to com-

Table 2.3
Requirements for Mobile Antenna Systems

	Base Station Antenna		Mobile Station Atenna	
System (Band)	*Requirement*	*Example*	*Requirement*	*Example*
Maritime telephone system (250-MHz)	Cover wide area (50 to 100 km)	Corner reflector Plane reflector	Omnidirectional in horizontal plane High gain	Dipole (sleeve)
Aeronautical telephone system (800-MHz)	Cover wider area (400 km) Suppress ground reflection wave	Collinear array Broadside array (shaped beam)	Omnidirectional Extremely low aerodynamic resistance	Blade antenna (flush mount)
Train telephone system (400-MHz)	Cover belt-shape zone Cover inside tunnel	Grid paraboloid reflector Coaxial leaky cable	Low aerodynamic resistance	Monopole
Automobile/portable (800-MHz)	Efficient illumination for cell configuration Multichannel operation Reduction of fading	Collinear array Broadside array with beam tilt and/or shaped beam	Vehicle: Omnidirectional, high gain, diversity	Sleeve Monopole (on glass)
		Yagi	Portable: high efficiency, small volume, diversity	Monopole Planar IFA
Pager (250-MHz)	Cover wide area Efficient illumination inside the area	Corner reflector Brown antenna	High efficiency Inhouse antenna	Small loop

plicated ones, such as beam-shaped array antennas used in multichannel systems. Table 2.4 [44] shows some typical antennas used in both base and mobile stations [45] in various mobile communication systems. Antennas used in base stations have a different role from those in mobile stations and are designed under different requirements. The electrical performance and mechanical configuration of a base station antenna mainly depends on the size and the shape of the service area.

For instance, in maritime and aeronautical mobile systems, the base station antenna should cover a wide area, and hence high directivity (7 to 11 dBd) is required. In train communication systems, service areas are restricted within long, narrow areas along the tracks. In order to cover such areas, a parabolic reflector antenna has been used. To solve problems in mountainous areas and in tunnels, where the radio wave is obstructed, the LCX communication system has been developed and used. This system has been applied to the train telephone systems in Japan (Chapter 5). Since antennas installed on the body of vehicles such as trains, automobiles, and particularly aircraft must be lightweight and tolerate severe aerodynamic conditions, thin, short monopoles or low-profile antennas (such as inverted-L) are usually used. Types of flush-mounted or conformal antennas such as blade and microstrip antennas are preferred for aircraft. In automobile telephone systems, either a sleeve antenna or a monopole antenna is commonly used and is mounted either on the trunk lid or on the roof of a car. They are designed so that the antenna element will not disturb the driver's view and the car body will not have very much of an effect on the radiation pattern. Various other types of antennas, such as wire antennas printed on the window glass for receiving broadcasts (Chapter 5) and monopole antennas fed by electromagnetic coupling through the window glass (called an *on-glass antenna*) for the automobile telephone system, have been developed and used. Antennas for portable radio phones, cordless telephones, and pagers are required for personal convenience and comfort, and consequently antenna dimensions should be small enough to be mounted either on or built into the equipment. In order to meet these requirements, antennas such as short whip, normal mode helix, and small loop antennas are mainly used. In Japanese portable telephone systems, two elements, a monopole element and a planar IFA, which can be housed inside the equipment, have been used, thus achieving space diversity performance [46].

Since pagers are usually carried in the bearer's vest pocket or attached to his or her belt, small, built-in rectangular loop antennas have been employed. It is usually necessary to eliminate proximity effects, which generally degrade antenna performance but for pager loops, antenna performance is actually enhanced by using both the proximity and body effects as follows: (1) conductors existing near an antenna element are used to generate dipole mode currents, so that nearly omnidirectional receiving patterns can be obtained, and (2) the body effect is used to enhance the receiver sensitivity as a result of superposition of the field produced by the image of the loop [47]. Magnetic-type antennas such as loop, microstrip,

Table 2.4
Typical Antennas Used in Practical Mobile Communication Systems

System	Mobile Station		Base Station	
	Antenna Type	*Requirements*	*Antenna Type*	*Requirements*
Pager: 150 MHz, 280 MHz, 450 MHz, 900 MHz	Small square loop Multiturn loop Ferrite coil antenna Micro strip antenna Parallel plate (magnetic current loop)	Limited space to mount antenna element Built-in, lightweight Nondirectional sensitivity by combining H- and E-components produced in antenna system Sensitivity enhanced by using image loop Low cost	Brown antenna Corner reflector Plane reflector Array (broadside, low sidelobe)	Wide-area coverage or specified area coverage with shaped beam Homogeneous field distribution in service area Suppression of field near base station
Mobile telephone: Vehicles, 800 MHz	$\lambda/4$ monopole $\lambda/2$ sleeve dipole Printed dipole 2 monopoles: horizontal, vertical	Omnidirectional pattern in horizontal plane Low-elevation angle in vertical plane Space diversity	Collinear array MSA Cylindrical parabola Broadside array Corner reflector 2 dipoles	Same as above Pattern tilt downward Low sidelobe in vertical plane Space or polarization diversity
Portable, 800 MHz	$\lambda/4$ monopole $\lambda/4$ whip normal mode helix Planar IFA	Limited space to mount antenna elements Body or portable unit included in antenna system Space diversity	Same as above Cross dipole	Same as above
Cordless telephone: 280/400 MHz	Short dipole Small loop $\lambda/4$ monopole	Built-in or mounted on body of telephone	Short dipole Short monopole	Inhouse use (20m–100m)
Maritime telephone: 250 MHz	Dipole Brown antenna	Omnidirectional horizontal plane pattern Antenna mounting space not severely limited	Corner reflector Plane reflector	Long-distance transmission (50–100 km)
Train telephone: 400 MHz	$\lambda/4$ monopole ILA	Antennas usually mounted on roof Low profile	Parabola with grid reflector	Transmission along railway

		Rigid structure Resistant to weather and aerodynamic conditions		
	Monopole buried on train body	Flush-mounted onto lower side of train	LCX	Along railway
Airplane telephone: 800 MHz	Blade antenna MSA array	Low-profile, flush-mount Conformal structure Lightweight Dragless structure Rigid Resistant to environmental conditions	Collinear array Broadside array (shaped beam)	Long distance transmission (~400 km) Conical zone coverage
Aircraft communication	Wire Tail cap Sleeve Monopole Notch Blade MSA	Almost same as above LF/HF/MF bands VHF/UHF bands		
ETS/V				
Portable	2-circular MSA	1.5-GHz band Placed on lid of T-unit		
Ship	Improved SBF	Low-speed data transmission, compact, high efficiency, wide bandwidth		
Aircraft	16-MSA phased array	Sequential array		
INMARSAT:				
Ship Land	Drooping-dipole Quadrifiler helix	1.5-GHz band		
PROSAT	5-turn helix 16-crossed dipole array SBF	1.5 GHz band Narrow bandwidth		

Note: ETS/V = Engineering Test Satellite V; IFA = inverted-F antenna; ILA = inverted-L antenna; INMARSAT = International Maritime Satellite Organization; LCX = leaky coaxial cable; MSA = microstrip antenna; PROSAT is a European Space Agency program for the promotion of small terminal techniques for satellite systems; SBF = short backfire antenna.

and planar IFA are compatible with the above requirements [2]. Radiation patterns of such a radiating system must be taken in three dimensions. Polarizations in this case may become composite; for instance, it may be not only vertical, but also horizontal, even when a vertical monopole element is used, as a result of radiation from the body of the unit. Antenna gain must also be evaluated in three dimensions.

The gain of a portable unit should be as high as possible so that the transmitter power can be made lower and the use of a smaller battery becomes possible. This will result in making the size of the mobile unit smaller and lighter. In turn, it may also allow a longer time between recharging the battery, thus extending the transmitter operation time. When a transmitter is operated very close to the human body, the effect of electromagnetic energy on the human organs such as brain and eyes must be carefully taken into account in the antenna design (Chapter 4).

Environmental conditions must also be seriously considered in the design and practical use of portable terminals. In mobile communication environments, especially in urban areas, the diversity reception technique has been applied to both base station and mobile station antennas in order to overcome multipath fading problems. Diversity antenna systems are required to have branches with small correlation factors and must be economically constructed to be of compact size. In Japanese mobile telephone systems, diversity reception has been adopted in mobile stations as well as in base stations [17]. Very compact diversity antennas, which are composed of two elements but have an appearance of a single element, have been developed for mobile stations [48]. One of them is the vertically spaced sleeve antenna for the vehicle mount and another is the antenna system composed of a whip and an inhouse planar IFA for a portable radio unit.

In digital modulation systems, problems due to delay spread in incident waves may be mitigated by angle or directivity diversity schemes. Further study and the development of new antenna systems having a capability to cope with problems caused by multipath propagation, interference, and so forth are urgently required.

Various antennas have been developed for mobile satellite systems, which require either a conical beam or a satellite tracking facility in order to link continuously with a satellite when mounted on a moving vehicle. Mechanical, electronic, and combination tracking systems have been developed [49]. MSA phased arrays, for instance, have been introduced for receiving satellite broadcasts on trains, ships, and buses and for data reception on cars.

2.3.2 Antenna Near Conducting Body

Antennas are usually installed on a body with good conductivity such as a portable radio or automobile, and sometimes other antennas or conducting bodies are nearby, as on a tower [50]. Thus, it is important to know the effects of other antennas or nearby conducting bodies on the antenna performance [51–54]. Nowadays, antenna performance prediction by computers gives satisfactory results quickly, and many books and papers on this subject have been published [55,56]. When a conducting body is less than about 5λ (λ = operating free-space wavelength), the method of moments (MOM) can successfully be used to calculate the characteristics of an antenna and the conducting body. At higher frequencies the geometrical theory of diffraction becomes more suitable.

2.3.2.1 Uncertain Interference Between Close Antennas

Consider two closely located antennas operating at different frequencies, as shown in Figure 2.31(a). It is assumed that antennas #1 and #2 operate at frequencies $f = f_1$ and f_2, respectively. At $f = f_1$, antenna #2 can be considered as a loaded scatterer, as shown in Figure 2.31(b), and the loading impedance Z_L ($= R_L + jX_L$) at the feed point of antenna #2 is the input impedance looking toward the transmitter or the receiver. Although this loading impedance is usually close to the characteristic impedance Z_0 of the transmission line connected to the feed point of antenna #2 at $f = f_2$, it is really a function of frequency and may be quite different from Z_0 at $f = f_1$. For the extreme case when the transmitter or the receiver is not connected to the transmission line, Z_L may become almost reactive. Thus, the value of Z_L is uncertain, and the antenna #1 characteristics may change greatly with the change of Z_L. Therefore, it is very important to find bounds on

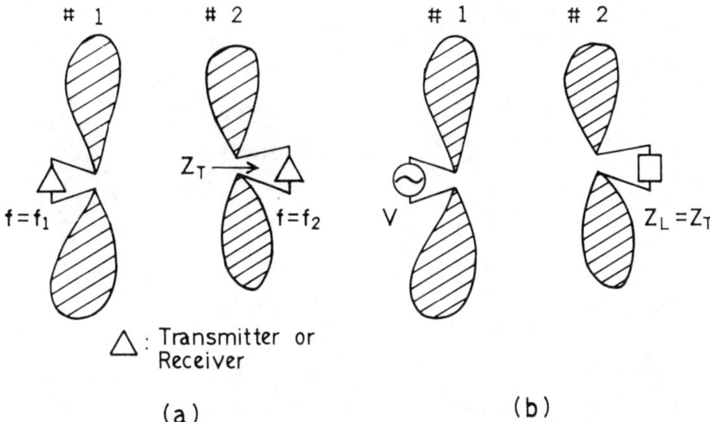

Figure 2.31 Antenna configuration: (a) actual situation (operating frequency $f_1 \neq f_2$, and Z_T: input impedance looking toward transmitter or receiver); (b) operating at $f = f_1$. (From [57], © 1984 IEEE.)

antenna characteristics at f_1 or f_2 with respect to Z_L. The bounds are shown in [57,58] for transmitting antennas and in [59,60] for receiving antennas.

An example is shown in Figure 2.32, where a dipole antenna operating at 250 MHz and a loop antenna operating at 400 MHz are separated by d [57]. In Figure 2.33, the calculated bounds of the radiation field magnitude by the MOM are shown with respect to Z_L in all horizontal plane directions when d is fixed to a half-wavelength at the operating frequency. The blocking effect by the other antenna is clearly shown.

2.3.2.2 An Antenna on a Portable Radio

An antenna element and a conducting body are treated in one unified system; that is, the conducting body is not taken as a ground plane but as a part of the antenna system itself. The portable radio equipment is modeled as an idealized rectangular metal box with an antenna operating in frequency ranges of 900 MHz, which is widely used for mobile communications at present. In this frequency range, the antenna and the conducting body are not necessarily small compared with λ. Characteristics of a wire antenna attached on a conducting rectangular box are shown in [61].

An example is an IFA attached to a conducting rectangular box shown in Figure 2.34, where $L_x = \lambda/8$, $L_y = \lambda/4$, $L_z = 3\lambda/8$, $l = 0.14\lambda$, $w = 0.092\lambda$, and $h = 0.037\lambda$ at the operating frequency of 920 MHz [62]. The length l_m is adjusted to get the input impedance equal to 50Ω. A conducting body is modeled by wire grids (wire radius: 0.0015λ), as shown in Figure 2.35, and the MOM is used to

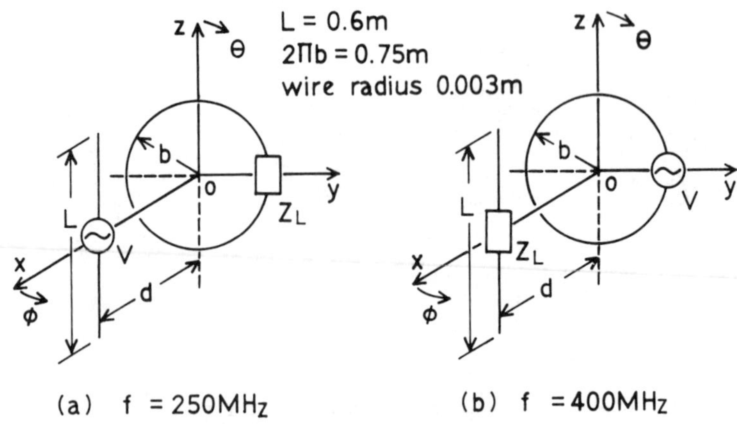

Figure 2.32 A dipole (a) and a loop (b) antenna. (From [57], © 1984 IEEE.)

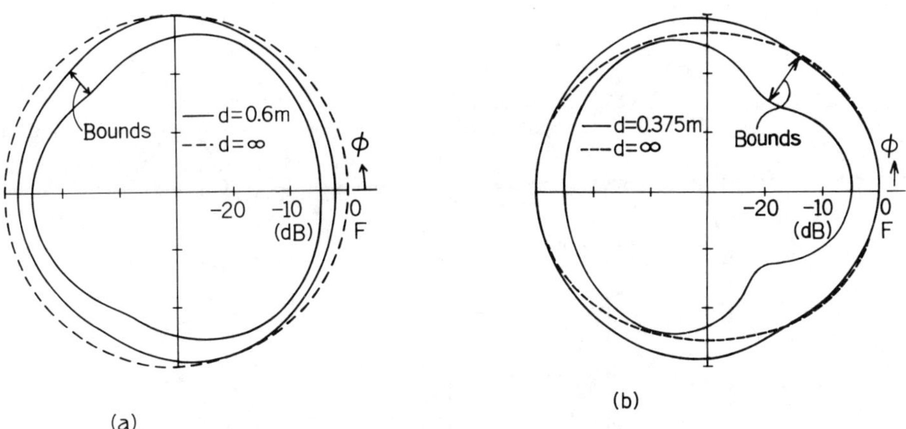

Figure 2.33 Horizontal $|E_\theta|$ pattern: (a) f = 250 MHz; (b) f = 400 MHz. (From [57], © 1984 IEEE.)

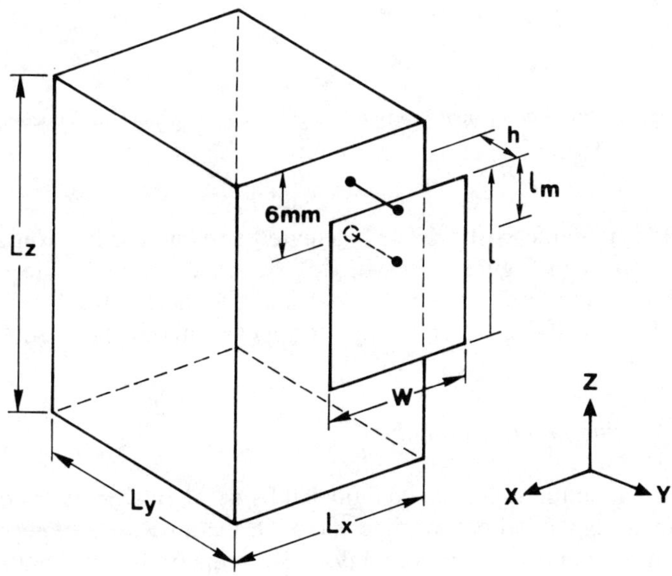

Figure 2.34 A planar inverted-F antenna on a rectangular conducting body. (From [62], © 1988 IECEJ.)

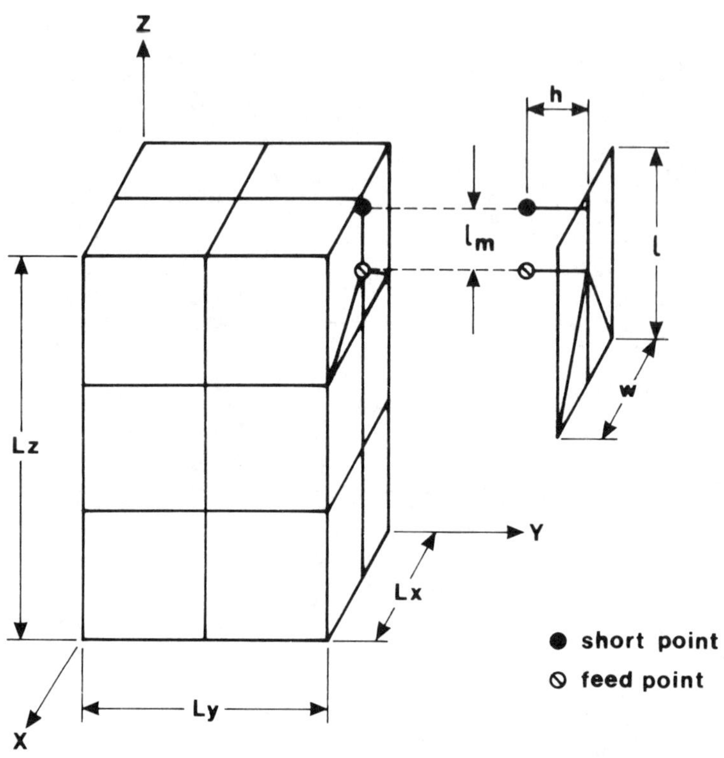

Figure 2.35 Wire grid model of a planar inverted-F antenna on a rectangular conducting body. (From [62], © 1988 IECEJ.)

calculate antenna characteristics. The calculated vertical and horizontal radiation patterns are shown in Figure 2.36 with the measured results. The effects of a conducting body can most clearly be seen in the y-z plane pattern, where there is a null around $\theta = 230$ deg which could not appear without the conducting body.

2.3.2.3 An Antenna on an Automobile

In VHF bands, an automobile can be modeled by wire grids similar to the portable radio equipment explained before, and the MOM can be used to get the characteristics of an antenna on an automobile [63]. A quarter-wavelength monopole antenna operating in UHF bands mounted on the roof of an automobile is considered to show the effects of a conducting body on radiation patterns [64,65].

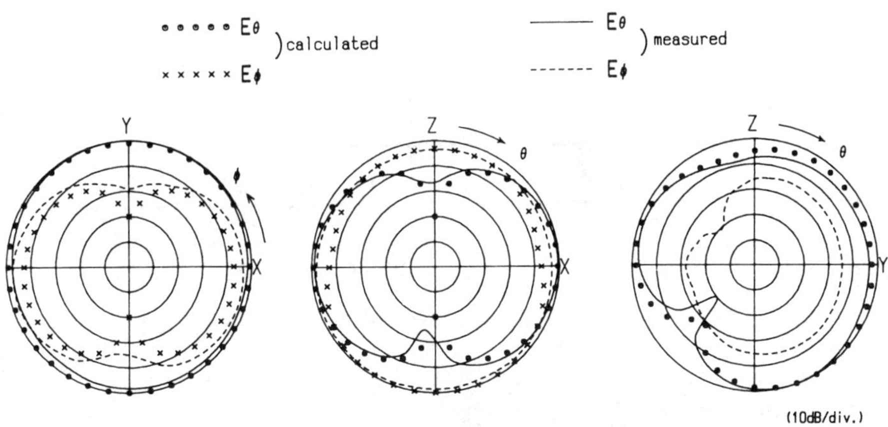

Figure 2.36 Radiation pattern of a planar inverted-F antenna on a rectangular conducting body. (From [62], © 1988 IECEJ.)

Since an automobile body is larger than 5λ, the GTD is an appropriate method to calculate radiation patterns. A model of an automobile is shown in Figure 2.37. The GTD is applied to a three-plate model as shown in Figure 2.38. The diffraction points are A, B, C, D, E, and F for an x-z plane pattern, and G and H for a y-z plane pattern. The earth is considered as an imperfect conductor with a relative dielectric constant ϵ_g. The calculated and measured radiation patterns are shown in Figure 2.39. The effects of the automobile body are clearly shown.

2.3.2.4 A Conducting Body Near a VOR Antenna

The VHF omnidirectional range (VOR) is an air navigation antenna system that gives azimuthal information to an aircraft in the 108 to 118-MHz carrier frequency range. Bearing errors of the airborne VOR receiver become satisfactorily small without any obstacles in the vicinity of the VOR [66]. The bearing errors due to wire scatterers such as wire fences, power and telephone lines [67,68], and water tanks [69] are significant in certain directions compared with those introduced by the equipment itself.

VOR antennas as shown in Figure 2.40 consist of four square-loop antennas located b_1 above ground. These loops are above a circular counterpoise (radius R), which is b_2 above ground. The reference phase signal, constant in phase and independent of the aircraft azimuth, is radiated by driving all four loops simultaneously in phase with the RF carrier frequency f_c. The RF carrier is amplitude-modulated by a subcarrier of 9.96 kHz, which is frequency-modulated at 30 Hz.

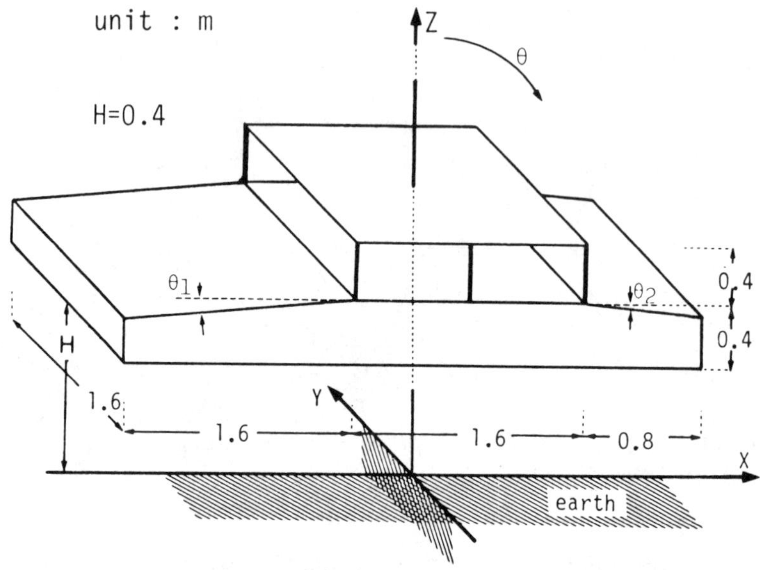

Figure 2.37 Automobile model. (From [65], © 1986 IEEE.)

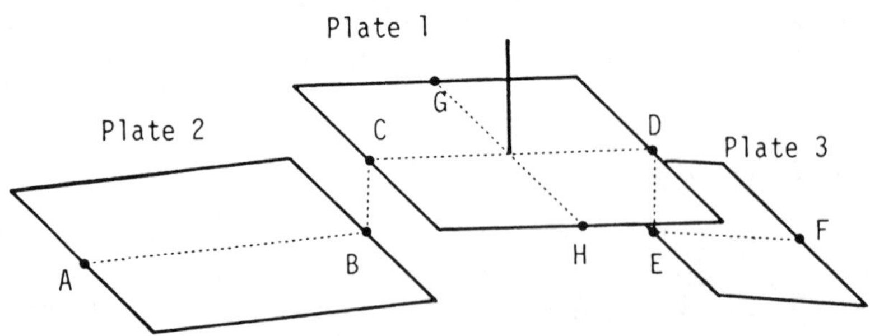

Figure 2.38 Theoretical model of an automobile for the GTD. (From [65], © 1986 IEEE.)

The variable phase signal, which varies in phase with the bearing of the aircraft, is radiated by the north-south and east-west pairs of the loops. Each pair is fed by the RF carrier, frequency-modulated by 30 Hz. Each pair makes a horizontal figure-eight pattern, rotating at 30 Hz.

The calculations are done using the following parameters: d = 0.406m, b_1 = 4.88m, b_2 = 3.66m, R = 7.92m, f_c = 109 MHz, airborne receiver bandwidth

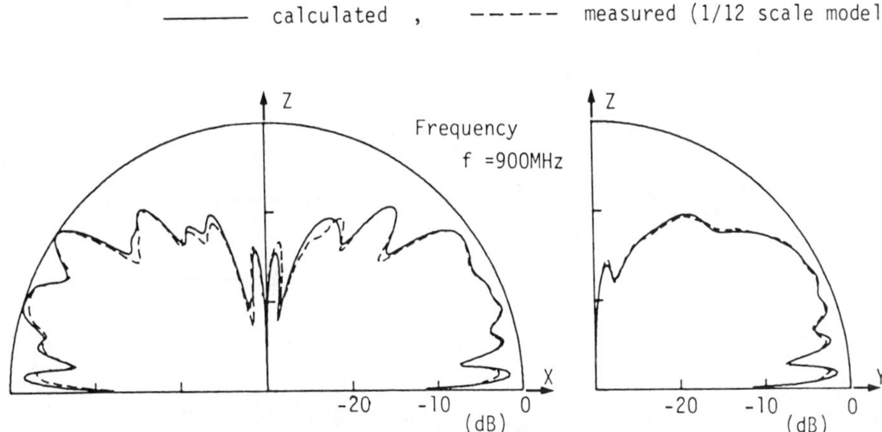

Figure 2.39 Vertical radiation pattern ($\theta_1 = \theta_2 = 0$ deg, $\epsilon_g = 1.3 - j0.1$). (From [65], © 1986 IEEE.)

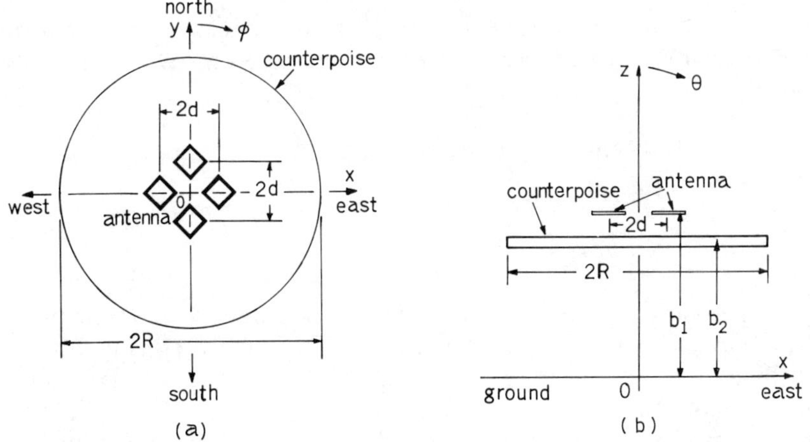

Figure 2.40 VOR antenna configuration: (a) upper view; (b) front view. (From [68], © 1982 IEEE.)

$f_0 = 0.35$ Hz, aircraft range $r = 46.3$ km and $\theta = 88.87$ deg, aircraft speed is 351.8 km/hour, and the radius and height of a wire scatterer are 0.00206m and 7.86m, respectively [68]. It is assumed that the VOR antenna and the wire scatterer are located above a perfect ground plane. The MOM is used to obtain the scattered fields from which the bearing error $\Delta\phi$ is calculated. Figure 2.41 shows the envelope

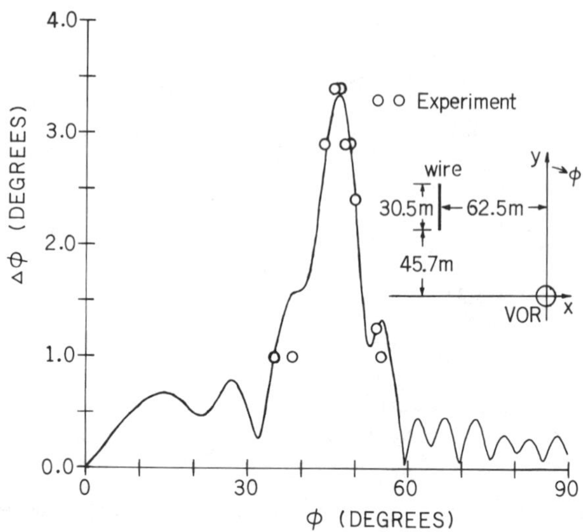

Figure 2.41 Dynamic bearing error due to a horizontal straight wire. (From [68], © 1982 IEEE.)

of the calculated dynamic bearing error due to a horizontal straight wire scatterer. When compared with the measured dynamic bearing error [69], both errors correspond very well.

2.3.3 Diversity Techniques

2.3.3.1 Diversity Performance

Since signal fading in the mobile radio environment causes severe reception problems, diversity techniques are used to reduce fading effects. Usually the diversity is applied at the receiving site. It is therefore a passive device and does not cause any interference, and it can take the form of space diversity, field component diversity, polarization diversity, frequency diversity, and directivity diversity, each of which will be described later in this section. The diversity performance is dependent on the number of diversity branches and on the correlation coefficient between the received branches. Each diversity scheme can reach the same performance if the branch correlation coefficients are the same. Two diversity signals are received from two antennas if space, polarization, and directivity schemes are used, or from one antenna if frequency and field diversity are used. We also have to consider how to combine two signals which are received from the diversity scheme, and an appropriate combining technique can yield better performance. There are four general combining techniques. Maximal ratio combining (MRC) maximizes the signal-to-noise ratio after the combining. Equal-gain combining (EGC) brings the received signals to a common phase and combines them in voltages. Selective combining (SEC) selects the strongest of the two received signals. Switch combining (SWC) is based on a threshold level below which the signal switches to the other antenna. The MRC, EGC, and SEC require more complex equipment, perform better, and cost more than SWC, which needs only one front end. Among the MRC, EGC, and SEC, the performance of MRC is the best, but the equipment is complicated. The performance of SEC is lower than that of MRC and EGC, but the equipment is simple to build. Although combining techniques are not within the scope of this book, the implications for antenna design are relevant.

This section illustrates the performance of a two-branch SEC diversity signal and its variants. Assume that the average CIR Γ of two-branch signals are the same. Then the probability that the combined signal CIR is γ and is less than the threshold R is [70]:

$$P(\gamma \leq R) = [1 - \exp(R/\Gamma)]^2 \tag{2.46}$$

Equation (2.46) is valid for two uncorrelated-branch signals. Sometimes the two-branch signals received are correlated because the required separation of two base stations cannot be physically wide enough to achieve the uncorrelated condition between two signals [71]. Then the combined signal γ of two correlated signals can be expressed as [70]

$$P(\gamma \leq R) = 1 - \exp(-R/\Gamma)[1 - Q(a, b) + Q(b, a)] \tag{2.47}$$

where

$$\begin{aligned} Q(a, b) &= \int_b^{\infty} \exp[-(1/2)(a^2 + x^2)] I_0(ax) \cdot x \, dx \\ a &= \sqrt{\frac{2R}{\Gamma(1 + |\rho|^2)}} \\ b &= \sqrt{\frac{2R}{\Gamma(1 - |\rho|^2)}} \end{aligned} \tag{2.48}$$

when the correlation ρ approaches zero, $Q(a, b) = Q(b, a)$, and then (2.47) reduces to (2.46). Equation (2.47) is plotted in Figure 2.42 with different values of correlation coefficient ρ. At a level of 10 dB below the mean CIR level, 10% of the signal is below the level for $\rho = 1$ (no diversity), 2% of the signal is below the

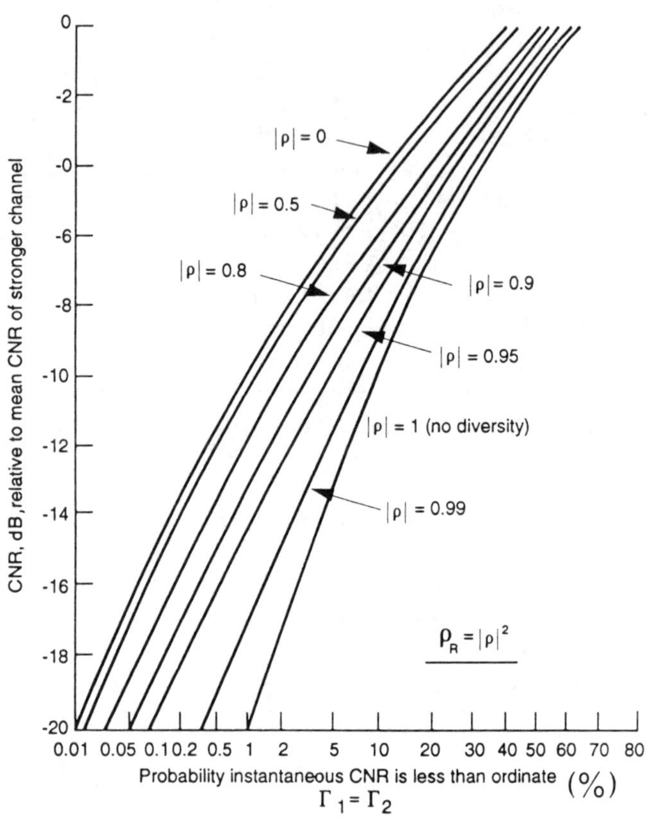

Figure 2.42 Selective combining of two correlated signals [70].

level for $\rho = 0.7$, and 1% of the signal is below the level for $\rho = 0$. The best performance of a diversity receiver is trying to make the correlation coefficient ρ of two signals approach zero.

2.3.3.2 Space Diversity

Many cellular design engineers have asked: Why does horizontal antenna separation lead to better diversity performance than vertical separation at a cell site? There is an important reason why. The greater the antenna horizontal separation, the less likely the fades of the two received signals will occur simultaneously. Thus, the diversity gain for reducing the effect of the fades increases as the separation increases and relies on the concept that the signal strength of two signals should be nearly equal. If the two received signal strengths are not equal, as is generally the case for vertical separation, then the diversity gain cannot be achieved, regardless of the requirements of antenna separation.

Designing a diversity antenna scheme is based on the parameter η, which depends [71] on the real antenna height (h) and the antenna separation (D):

$$\eta = h/D \tag{2.49}$$

Horizontal Separation

It has been determined experimentally that the optimum value of η is 11 for horizontal antenna separation [71]. For example, if the antenna h is 100 ft, the optimum D is 9 ft. Therefore, the higher the antenna, the more separation will be needed for optimum diversity gain. Incidentally, the h is the antenna height at the cell site when the system is designed.

In a real-world environment, vehicles and portable units are at various ground elevations in all directions from the cell site. Because of this, the effective antenna height (h_e) measured at the cell site varies based on the real-time locations of vehicles and portables. The same actual antenna height may have two different effective antenna heights according to two different vehicle or portable units. For a specific base-to-mobile/portable transmission, when h_e is less than the h, the signal received by the vehicle will be weaker [72]. This is expressed as a gain or loss ΔG thus:

$$\Delta G = 20 \log_{10}(h_e/h) \qquad \text{dB} \tag{2.50}$$

During system operation, the value of η_e, obtained from $\eta_e = h_e/D$, will vary depending on the present location of the vehicle or portable units. When h_e is less (greater) than the h, the value of η_e becomes less (greater) than 11 and the diversity gain typically varies as in Figure 2.43.

Vertical Separation

Usually the vertical antenna separation is greater than the horizontal antenna separation for achieving a given diversity gain [71]. Let the antenna height of the lower antenna be h_1 and the antenna height of the higher antenna be h_2. The vertical separation is $D_v, = h_2 - h_1$. The difference in reception gain Δg between the two effective antenna heights of the two antennas can be found by using the following equation [72].

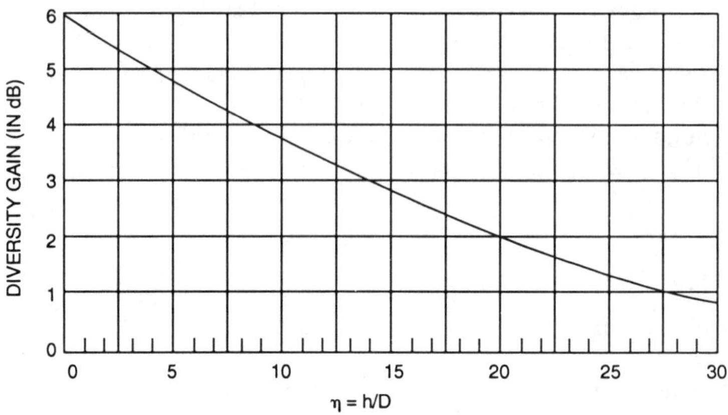

Figure 2.43 Diversity gain versus η [73].

$$\Delta g = 20 \log_{10}(h_e'/h_e) = 20 \log_{10}[1 + (D_v/h_e)] \tag{2.51}$$

where h_e and h_e' = effective antenna heights of two vertically separated antennas (Figure 2.44). For example, when the difference in Δg is 4 dB, $1 + (D_v/h_e) = 10^{4/20}$ or $D_v = 0.58h_e$, in which case the values of D_v and h_e could be 15 and 26 ft, respectively, for a given vehicle location. Although $\eta_e = h_e/D$ is 1.73, there will be no diversity gain, since Δg is too large [73].

Although the horizontal-separation antennas can provide higher diversity gain, there can sometimes be limitations when vehicles are in certain areas with respect to the base station, when no diversity gains are observed [71] because of the relative antenna heights.

2.3.3.3 Polarization Diversity Schemes

Two polarizations, vertical and horizontal, from two antennas can carry two signals on one radio frequency over satellite-ground links or microwave links. Because there is no coupling effect introduced by the medium, no mutual interference would occur. However, in the mobile radio environment, strong mutual coupling effect occurs. This means that after the signal propagates through the mobile radio medium, the signal energy in the vertical polarization wave can be cross-coupled (leaking) into the horizontal polarization wave and vice versa. Let us define the symbols. Γ_{11} = transmit vertical, receive vertical. Γ_{12} = coupling vertical into horizontal (from base to mobile). Γ_{21} = coupling horizontal into vertical (from base to mobile). Γ_{22} = transmit horizontal, receive horizontal. The two differently polarized waves (vertical E_v and horizontal E_h) received by the two polarization antennas at the mobile unit can be expressed as

$$E_v = \Gamma_{11} + \Gamma_{21} \qquad E_h = \Gamma_{22} + \Gamma_{12} \tag{2.52}$$

See Figure 2.45. Since the principle of reciprocity is applied to the polarization components, the two differently polarized waves, E_v' and E_h', arriving at the base station can be expressed as

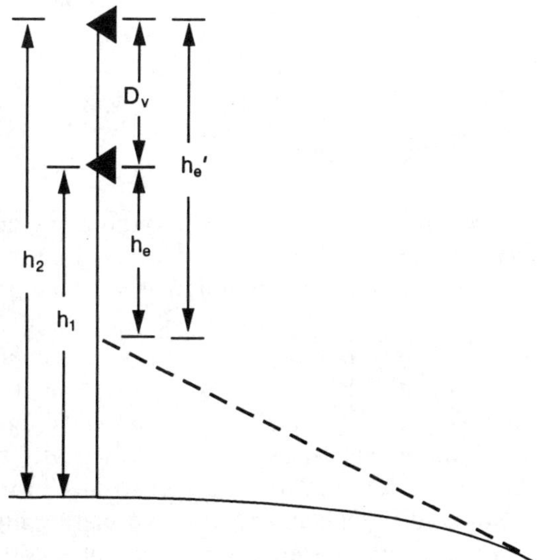

Figure 2.44 Vertical antenna separation [73].

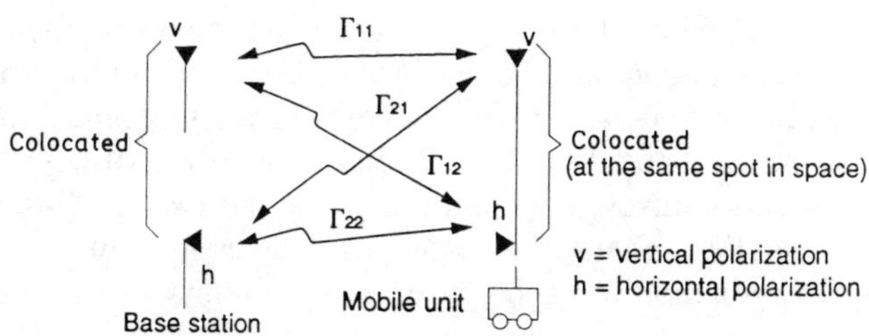

Figure 2.45 Reciprocity of horizontally and vertically polarized waves. (Lee, W. C. Y., "Polarization Diversity for Mobile Radio," *IEEE Trans. on Comm. Tech.*, No. 5, October 1972, p. 921.)

$$E'_v = \Gamma_{11} + \Gamma_{12} \qquad E'_h = \Gamma_{22} + \Gamma_{21} \tag{2.53}$$

If in some cases $\Gamma_{12} \neq \Gamma_{21}$, then $E_v \neq E'_v$ and $E_h \neq E'_h$. However, the leaking energy is very small compared to the main stream; that is, $\Gamma_{11} \gg \Gamma_{12}$ and $\Gamma_{22} \gg \Gamma_{21}$. Therefore, from the measurement we found that $E_v \approx E_h$. Also, it can be found that $E_v \approx E'_v$ and $E_h \approx E'_h$.

For antennas on an infinite ground plane, the array pattern in the elevation angles could be [74]

$$P_{\text{vertical polarization}}(\theta) = \left[1 + \cos\left(\frac{4\pi h}{\lambda} \sin\theta\right)\right]^2 \tag{2.54}$$

$$P_{\text{horizontal polarization}}(\theta) = \left[1 - \cos\left(\frac{4\pi h}{\lambda} \sin\theta\right)\right]^2 \tag{2.55}$$

where h is the height of the antenna from the ground plane and the angle θ is the elevation angle from the horizon. Notice that for $h = 1.5\lambda$, the vertical pattern is maximum at $\theta = 0$ deg, but goes into a null at $\theta = 8$ deg and picks up again at 19 deg.

The horizontal pattern is nulling at $\theta = 0$ deg, peaks up to $\theta = 8$ deg, and is nulled again at $\theta = 19$ deg. When they are in a line-of-sight condition, the waves come toward the mobile at one particular elevation angle and the powers of two polarization waves are not equal. In an actual mobile environment, the multipath waves arrive at the mobile terminal from a spread in elevation angles, and the ground plane also works to the advantage of decorrelating the received signals in the sense that the two antennas are aiming at different elevation angles.

Typical estimated mean signal strengths as a function of antenna height are shown in Figure 2.46 for the two polarizations. It was observed that for $h \geq \lambda$, the two branches of power were about the same. However, as the antenna height decreases, the power of the horizontal polarization wave decreases. The rapid Rayleigh fadings received by the two polarization antennas are uncorrelated. The local means (the average signal strength seen as the envelope of the Rayleigh fading signal, sometimes called *long-term fading*) received from two polarization antennas have the same strength. The advantage of polarization diversity in mobile radio is that the two polarization diversity antennas, unlike the two space-diversity antennas, can be placed as close as we wish to achieve full diversity gain. The disadvantage is the 3-dB power reduction at each transmitting antenna due to the split of the transmitter power into the two polarization antennas.

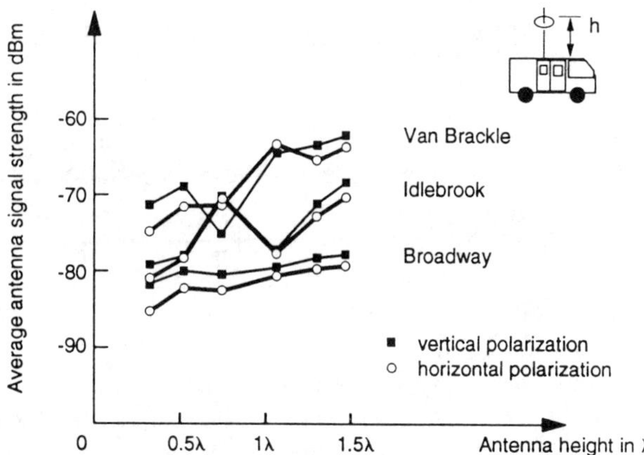

Figure 2.46 Average received signal strength as function of mobile antenna height. (Lee, W. C. Y., "Polarization Diversity for Mobile Radio," *IEEE Trans. on Comm. Tech.*, No. 5, October 1972, p. 921.)

Field Component Diversity [75]

Multiple scattering and fading effects can create cross-polarization such that it is desirable for the antenna to receive both E- and H-field components. This is called an *energy diversity antenna*; an example consisting of two cross semiloops is the basis of many pager antennas. An advantage of this technique is that only one location is needed and the transmit antenna need only have E-field.

Directional Diversity System [76]

Fading signals arriving from different angles at the receiving antenna come via different paths. In directivity diversity, the arriving angles of two fading signals should be separated far enough to avoid correlation among them. Also, it is preferred that the two antennas be in-line with the motion of the vehicle rather than perpendicular to its motion [74]. Then the two antennas can be placed at 0 deg (heading of the vehicle motion) and 180 deg. The effect of antenna directivity has been tested in the mobile radio environment. In an out-of-sight condition, the directivity of a received antenna does not increase the average power, but signal fading is reduced for a narrower beam. This technique is feasible at microwaves, where the directional aperture antennas have a small physical size [75].

2.4 ANTENNA PERFORMANCE EVALUATIONS IN MOBILE ENVIRONMENTS

The performance of a mobile antenna is strongly dependent on its effective gain in a multipath propagation environment, since the gain influences the size of the radio zone, transmitting power of mobile radio equipment, and, in particular, battery capacity in the case of portable radio equipment. Furthermore, in an antenna diversity system, the diversity gain is determined not only by correlation characteristics between signal envelopes received in the antenna elements, but also by the effective gain characteristics of antenna elements. Therefore, for designing a mobile antenna system, it is essential to maximize the effective antenna gain for the anticipated mobile radio environment. However, it is common knowledge that the effective gain of mobile antennas in land mobile propagation environments cannot be evaluated with sufficient accuracy from just antenna directive gain, since the received signals undergo Rayleigh-like fading, and random multipaths exist due to reflection, diffraction, and scattering. Hence, it is common to use the experimental method [77] that measures the mean signal level received over a certain route for evaluating the effective gain. With this method, the mean power levels of the unknown antenna and a reference antenna are obtained by averaging the signal levels received while each antenna moves along the same selected route. The mean effective gain (MEG) of the unknown antenna can, therefore, be related to the reference antenna by comparing the mean power level of the unknown antenna with that of the reference antenna. Half-wavelength dipole antennas have been used as the reference antenna. This method is useful for measuring the MEG of mobile antennas in practical environments, and it has been used for evaluating the MEG of several mobile antennas [16,77,78]. However, measured MEG reflects the interaction between the antenna power gain pattern and the propagation characteristics along the route; it depends on the measurement route. Even though the reference antenna also experiences the same interaction effects, the conventional experimental method is not accurate enough to evaluate the MEG of mobile antennas in general.

On the other hand, antenna diversity reception is an effective technique for mitigating multipath fading in a mobile antenna system. Diversity reception attains effective fading mitigation by the synthesis, selection, and switching of several received signals that are correlated to the smallest possible extent. In the case of antenna diversity, reduction of the correlation coefficient between the receiving antenna branches leads to an enhanced diversity effect. Correlation coefficients have been analyzed mainly by considering only copolarization [79,80]. However, when a diversity antenna is mounted on a portable transceiver, the antenna polarization is varied randomly. As a result, there are situations in which the antenna directivity for the cross-polarization component predominates or the main beam direction of the pattern is oriented in an arbitrary three-dimensional direction.

Hence, it is necessary when designing an antenna diversity system to consider the effect of the cross-polarization component of the arriving wave and the effect of the three-dimensional dispersion of the incident wave direction on the correlation characteristics.

This section introduces a theoretical method for analyzing the MEG of mobile antennas and the correlation characteristics of the antenna diversity branches. The method can treat, in general, the contribution of both vertically polarized (VP) and horizontally polarized (HP) radio waves, the dispersion of incident waves in elevation, and the variations of antenna polarization. Furthermore, the MEG characteristics of half-wavelength dipole antennas are discussed theoretically. The theoretical results are extremely interesting, since dipole antennas are usually used as the reference antennas in MEG measurement. A novel experimental method is described that can accurately evaluate the MEG of any mobile station antenna. We shall also discuss the correlation characteristics of orthogonally crossed dipoles, one of the polarization diversity antenna schemes.

2.4.1 Theoretical Expression of Antenna Performance in the Mobile Environment

2.4.1.1 Mean Effective Gain of Mobile Antennas

First it is necessary to establish a theoretical MEG expression that takes into account the VP and HP incident radio waves in multipath environments. Figure 2.47 illustrates the notation where the signal transmitted from a base station antenna passes through a multipath propagation environment and arrives at a mobile antenna. P_V and P_H are, respectively, the mean incident powers of the VP and HP incident radio waves received while the antenna moves over a random route in the environment. Thus, the total mean incident power arriving at the antenna, averaged over the same route, is $P_V + P_H$. The ratio between the mean received power of antenna over the random route P_{rec} and the total mean incident power $P_V + P_H$ can be considered the MEG of the mobile antenna in the environment. It is assumed that the average over a random route in an environment equals the average over the environment. This ratio is defined as the MEG of the mobile antenna in the environment and is denoted by the symbol G_e:

$$G_e = \frac{P_{\text{rec}}}{P_V + P_H} \tag{2.56}$$

The mean incident power ratio P_V/P_H represents the cross-polarization power ratio (XPR):

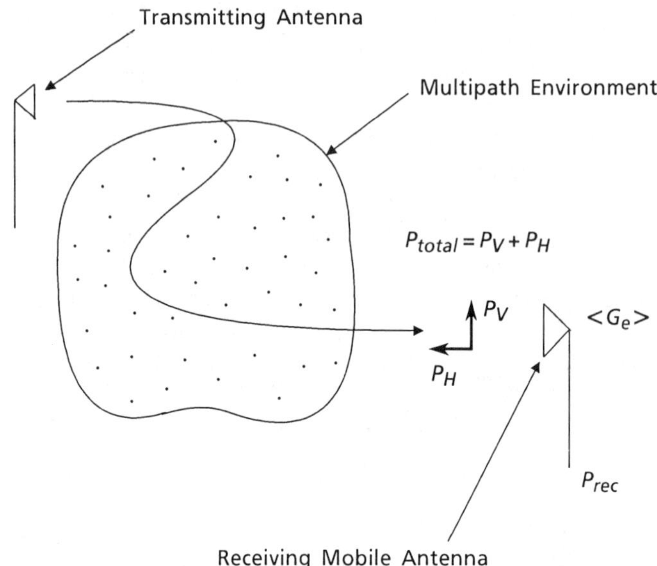

Figure 2.47 Average power arriving at receiving mobile antennas in multipath environments (From [81], © 1990 IEEE.)

$$\text{XPR} = P_V/P_H \tag{2.57}$$

XPR corresponds to the *cross-polarization coupling* [82] when the polarization of the transmitted radio waves is horizontal, and to the reciprocal of the cross-polarization coupling when the polarization of the transmitted waves is vertical. In a spherical coordinate system, as shown in Figure 2.48, the mean antenna received power P_{rec} is expressed by the following [83, pp. 133–140].

$$P_{\text{rec}} = \int_0^{2\pi} \int_0^{\pi} \{P_1 G_\theta(\theta,\phi) P_\theta(\theta,\phi) + P_2 G_\phi(\theta,\phi) P_\phi(\theta,\phi)\} \sin\theta \, d\theta d\phi \tag{2.58}$$

where $G_\theta(\theta,\phi)$ and $G_\phi(\theta,\phi)$ are the θ and ϕ components of the antenna power gain pattern, respectively, and $P_\theta(\theta,\phi)$ and $P_\phi(\theta,\phi)$ are the θ and ϕ components of the angular density functions of incoming plane waves, respectively. These functions satisfy the following conditions.

$$\int_0^{2\pi} \int_0^{\pi} \{G_\theta(\theta,\phi) + G_\phi(\theta,\phi)\} \sin\theta \, d\theta d\phi = 4\pi \tag{2.59}$$

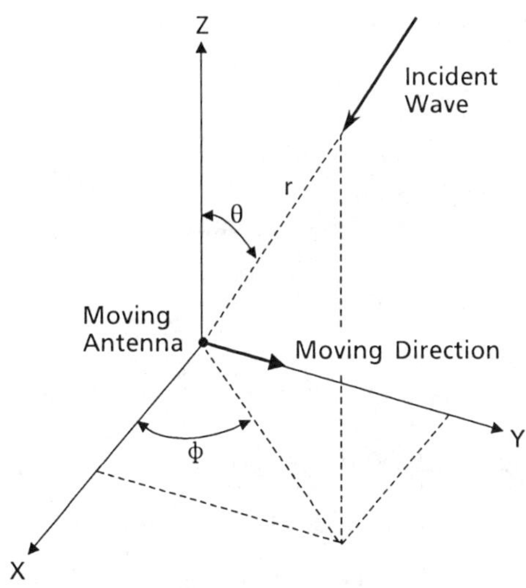

Figure 2.48 Spherical coordinates in mobile radio environments. (From [81], © 1990 IEEE.)

$$\int_0^{2\pi}\int_0^{\pi} P_\theta(\theta,\phi)\sin\theta\, d\theta d\phi = 1 \tag{2.60}$$

$$\int_0^{2\pi}\int_0^{\pi} P_\phi(\theta,\phi)\sin\theta\, d\theta d\phi = 1 \tag{2.61}$$

where P_1 is the mean power that would be received by an $\mathbf{i}_\theta$-polarized isotropic antenna in the mobile radio environment. Similarly, P_2 is the mean power that would be received by an $\mathbf{i}_\phi$-polarized isotropic antenna. $\mathbf{i}_\theta$ and $\mathbf{i}_\phi$ are unit vectors associated with θ and ϕ, respectively.

In Figure 2.48, since the mobile antenna moves in the X-Y plane, the θ and ϕ components correspond to the VP and HP components. Thus, the terms P_1 and P_2 are, respectively, the mean received power of VP isotropic antennas and that of HP isotropic antennas, and XPR is equal to the ratio P_1/P_2. By using (2.58) and the XPR notation, the expression for MEG can be rearranged to yield the following equation.

$$G_e = \int_0^{2\pi}\int_0^{\pi}\left[\frac{\text{XPR}}{1+\text{XPR}} G_\theta(\theta,\phi)P_\theta(\theta,\phi) + \frac{1}{1+\text{XPR}} G_\phi(\theta,\phi)P_\phi(\theta,\phi)\right]\cdot\sin\theta\, d\theta d\phi \tag{2.62}$$

When only a VP wave (XPR = ∞) is incoming from a single (θ_s,ϕ_s) direction, which corresponds to line-of-sight propagation with VP wave transmission, the angular density functions in (2.62) are represented as

$$P_\theta(\theta,\phi) = \frac{\delta(\theta - \theta_s)\cdot\delta(\phi - \phi_s)}{\sin\theta_s} \tag{2.63}$$

and

$$P_\phi(\theta,\phi) = 0 \tag{2.64}$$

where $\delta(x)$ is the delta function. It then follows from (2.62) to (2.64) that MEG becomes

$$G_e = G_\theta(\theta_s,\phi_s) \tag{2.65}$$

This means that MEG corresponds to the antenna directive gain in the (θ_s,ϕ_s) direction when incoming signals are centered on the (θ_s,ϕ_s) direction.

If the characteristics of incoming signals in various environments can be represented as statistical distribution functions P_θ, P_ϕ, the MEG given in (2.62) is the mean power gain of the antenna in each environment.

2.4.1.2 Correlation Coefficient of Antenna Diversity Branches

In the spherical coordinate system in Figure 2.48, the incident wave is described by

$$\mathbf{F}(\theta,\phi) = F_\theta(\theta,\phi)\mathbf{i}_\theta + F_\phi(\theta,\phi)\mathbf{i}_\phi \tag{2.66}$$

where F_θ and F_ϕ indicate the random amplitude and phase of the incident electric field in $\mathbf{i}_\theta$ or $\mathbf{i}_\phi$ direction, respectively. Also, the electric field pattern of the antenna n (n = 1,2) can be given as follows.

$$\mathbf{E}_n(\theta,\phi) = E_{\theta n}(\theta,\phi)\mathbf{i}_\theta + E_{\phi n}(\theta,\phi)\mathbf{i}_\phi \tag{2.67}$$

where $E_{\theta n}$ and $E_{\phi n}$ are the complex expressions of the θ and ϕ components of the electric field pattern. Thus, the received voltages of the two antennas shown in Figure 2.49 are as follows.

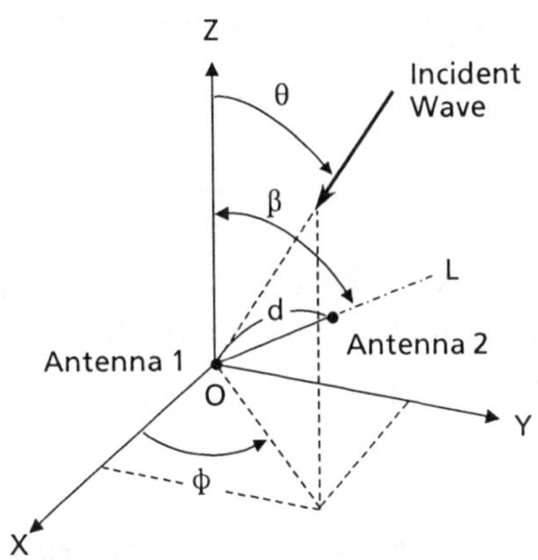

Figure 2.49 Antenna diversity and its coordinate system. (From [84], © 1990 IEICE, Japan.)

$$V_1(t) = C_1 \int_0^{2\pi} \int_0^{\pi} \mathbf{E}_1(\theta,\phi) \cdot \mathbf{F}(\theta,\phi) e^{-jk\mathbf{u}\cdot\mathbf{r}t} \sin\theta \, d\theta d\phi \tag{2.68}$$

$$V_2(t) = C_2 \int_0^{2\pi} \int_0^{\pi} \mathbf{E}_2(\theta,\phi) \cdot \mathbf{F}(\theta,\phi) e^{-jk\mathbf{u}\cdot\mathbf{r}t} e^{jkx} \sin\theta \, d\theta d\phi \tag{2.69}$$

where C_n $(n = 1,2)$ is the proportionality constant, $e^{-jk\mathbf{u}\cdot\mathbf{r}t}$ is the Doppler shift caused by the velocity $\mathbf{u}$ of the antenna, k is the wave number, $\mathbf{r}$ is the unit vector in the radiating direction, and x is the phase difference of the incident waves as seen at the two antennas. As shown in Figure 2.49, $x = d(\sin\theta \sin\phi \sin\beta + \cos\theta \cos\beta)$ if the diversity antenna system lies on the Y-Z plane and the line that connects the antenna branches is inclined by the angle of β from the vertical direction. If it is assumed that there is no correlation between incident waves, then the following equations hold, since the phases of F_θ and F_ϕ are independent for incident waves from different directions $\Omega = (\theta,\phi)$ and $\Omega' = (\theta',\phi')$.

$$\langle F_\theta(\Omega) F_\theta^*(\Omega') \rangle = \langle F_\theta(\Omega) F_\theta^*(\Omega) \rangle \, \delta(\Omega - \Omega') \tag{2.70}$$

$$\langle F_\phi(\Omega) F_\phi^*(\Omega') \rangle = \langle F_\phi(\Omega) F_\phi^*(\Omega) \rangle \, \delta(\Omega - \Omega') \tag{2.71}$$

Furthermore, since the phases of F_θ and F_ϕ are independent and uniformly distributed between 0 and 2π, the next relationship is valid.

$$\langle F_\theta(\Omega) F_\phi^*(\Omega') \rangle = 0 \tag{2.72}$$

where the brackets indicate the ensemble average, the asterisk indicates the complex conjugate, and δ is the delta function. $V(t)$ is approximated by the complex Gaussian process with average of zero and satisfies

$$\langle V(t)\rangle = 0 \tag{2.73}$$

Hence, the cross-covariance of the two received voltages can be obtained as follows if (2.66) and (2.67) are substituted into (2.68) and (2.69) and the conditions of (2.70), (2.71), and (2.72) are applied.

$$\begin{aligned} R_{12} &= \langle V_1(t)V_2^*(t)\rangle \\ &= 2KP_H\int_0^{2\pi}\int_0^{\pi}[\mathrm{XPR}\cdot E_{\theta 1}(\theta,\phi)E_{\theta 2}^*(\theta,\phi)P_\theta(\theta,\phi) \\ &\quad + E_{\phi 1}(\theta,\phi)E_{\phi 2}^*(\theta,\phi)P_\phi(\theta,\phi)]e^{-jkx}\sin\theta\, d\theta d\phi \end{aligned} \tag{2.74}$$

where K is a proportionality constant. Similarly, the standard deviations σ_1 and σ_2 of the complex envelopes of the first and second antennas are

$$\begin{aligned} \sigma_1^2 &= \langle V_1(t)V_1^*(t)\rangle \\ &= 2KP_H\int_0^{2\pi}\int_0^{\pi}[\mathrm{XPR}\cdot E_{\theta 1}(\theta,\phi)E_{\theta 1}^*(\theta,\phi)P_\theta(\theta,\phi) \\ &\quad + E_{\phi 1}(\theta,\phi)E_{\phi 1}^*(\theta,\phi)P_\phi(\theta,\phi)]\sin\theta\, d\theta d\phi \end{aligned} \tag{2.75}$$

$$\begin{aligned} \sigma_2^2 &= \langle V_2(t)V_2^*(t)\rangle \\ &= 2KP_H\int_0^{2\pi}\int_0^{\pi}[\mathrm{XPR}\cdot E_{\theta 2}(\theta,\phi)E_{\theta 2}^*(\theta,\phi)P_\theta(\theta,\phi) \\ &\quad + E_{\phi 2}(\theta,\phi)E_{\phi 2}^*(\theta,\phi)P_\phi(\theta,\phi)]\sin\theta\, d\theta d\phi \end{aligned} \tag{2.76}$$

In general, if the complex correlation coefficient is ρ and the correlation coefficient for the observed envelope is ρ_e, then ρ_e is approximately equal to $|\rho|^2$ [79]. Since

$$\rho_e \approx |\rho|^2 = \frac{|R_{12}|^2}{\sigma_1^2\sigma_2^2} \tag{2.77}$$

from (2.73), a theoretical expression of the correlation coefficient of antenna diversity can be obtained if (2.74), (2.75), and (2.76) are substituted into (2.77). This expression is a general theoretical equation which takes into account the effect of XPR and the incident wave distribution of the θ and ϕ components. The expression indicates that the correlation coefficient ρ_e becomes zero regardless of XPR or the angular density function of the incident wave if the complex radiation patterns of the θ and ϕ components of the two antennas do not overlap.

2.4.2 Statistical Distribution Model of Incident Waves

2.4.2.1 Theoretical Model

In general, multipath propagation in a land mobile communication environment is caused by the reflection, diffraction, and scattering from topography and buildings. Let us define the secondary wave source generically as the points at which the wave is reflected, diffracted, or scattered immediately before reaching the antenna. The secondary wave sources are widely distributed on the side surfaces and edges of buildings, the ground plane, and objects on the ground (such as trees and vehicles) around the antenna. Their number and location vary widely depending on the style of the city. Hence, if the antenna moves randomly in a metropolitan area with buildings of various heights, shapes, dimensions, and materials, the secondary wave sources observed from the antenna can be assumed to be statistically independent and to be distributed uniformly in the azimuth direction. This assumption corresponds to the scattered ring model [85]. When the antenna moves randomly within a metropolitan area, the building heights and distances from the antenna are considered to be independent variables that are distributed around certain average values. Hence, the secondary wave sources can also be assumed to be distributed around the average elevation direction. If it is assumed that an extremely large number of secondary wave sources are experienced during a random move, this elevation angle distribution can be assumed to be a Gaussian distribution according to the central limit theorem [86].

The preceding assumptions make it quite reasonable to adopt a statistical model in which angular density functions P_θ and P_ϕ are assumed to be Gaussian in elevation and uniform in the azimuth direction, as shown in Figure 2.50. Note that the elevation can take on negative values, since mobile antennas are usually operated above the ground. The distribution functions of incident plane waves are expressed as follows.

$$P_\theta(\theta,\phi) = \mathrm{A}_\theta \exp\left\{-\frac{\left[\theta - \left(\frac{\pi}{2} - m_V\right)\right]^2}{2\sigma_V^2}\right\} \qquad (0 \leqq \theta \leqq \pi) \qquad (2.78)$$

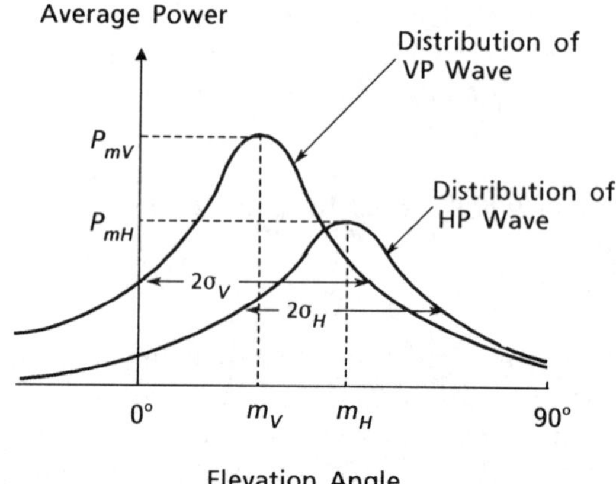

Figure 2.50 Gaussian distribution model of incident waves. (From [81], © 1990 IEEE.)

$$P_\phi(\theta,\phi) = A_\phi \exp\left\{-\frac{\left[\theta - \left(\frac{\pi}{2} - m_H\right)\right]^2}{2\sigma_H^2}\right\} \qquad (0 \leqq \theta \leqq \pi) \tag{2.79}$$

where m_V and m_H are, respectively, the mean elevation angle of each VP and HP wave distribution observed from the horizontal direction, and σ_V and σ_H are, respectively, the standard deviation of each VP and HP wave distribution. A_θ and A_ϕ are constants determined by (2.60) and (2.61). If the mean power strengths of VP and HP waves in the directions of $\theta = \pi/2 - m_V$ and $\theta = \pi/2 - m_H$ are, respectively, P_{m_V} and P_{m_H}, then

$$P_{m_V} = P_1 \cdot A_\theta \tag{2.80}$$

$$P_{m_H} = P_2 \cdot A_\phi \tag{2.81}$$

Therefore, XPR is also determined by P_{m_V} and P_{m_H}, as shown in the following equation.

$$\text{XPR} = \frac{P_V}{P_H} = \frac{P_1}{P_2} = \frac{P_{m_V}}{P_{m_H}} \cdot \frac{A_\phi}{A_\theta} \tag{2.82}$$

2.4.2.2 *Validity of the Statistical Model*

In order to confirm the validity of the statistical model shown in Figure 2.50, the incident wave distribution had been measured in a Tokyo urban area using a vertically polarized radio signal in the 900-MHz band. The transmitting antenna was a vertically polarized omnidirectional antenna set 87m above the ground. Two measurement routes were selected for the Ningyo-cho and Kabuto-cho areas. Both routes were some 1.2 km from the transmitting antenna, and all receiving points on the routes were out of sight of the transmitting antenna. The receiving antenna was mounted 3.1m above the ground on the roof of a van.

The received power pattern of incident radio waves was measured with a 0.9m-diameter parabolic reflector antenna with a dipole element for the primary radiator. The half-power beamwidth and the first sidelobe level of the antenna were 22 deg and less than −9 dB, respectively. Measurements were taken every 5m for a total of 34 points along the Ningyo-cho route, and every 7m for a total of 30 points along the Kabuto-cho route. At each receiving point, the received power patterns for both VP and HP incident waves were measured by rotating the reflector antenna 360 deg in azimuth, at elevation angles of −10, 0, 20, and 45 deg.

It can be found through the measurements that the number of principal waves is up to five or six at each point, but the incident power and arrival direction vary considerably from point to point even though the measuring points are separated by only 5m. As a result, it can be considered that an extremely large number of incident waves are observed while the antenna moves over a random route, and that the assumption of the model that incident waves arrive from numerous and random azimuthal directions has been confirmed.

Figure 2.51 shows the average power distributions of the incident waves in elevation obtained from the measurement. In this figure, the solid lines express the best approximation of a Gaussian distribution function corresponding to the mean power levels. Figure 2.51 shows, therefore, that the statistical distribution of incident waves can be accurately estimated by the model proposed in this section.

In the distributions of the VP waves, the mean elevation angle m_V is about 20 deg for both routes. This value is consistent with the former experimental results for elevation angles somewhat larger than zero but less than 39 deg [83, p. 149], [2, p. 158]. However, it is found that the dispersion in elevation of the HP waves is larger than that of the VP waves; in fact, the distributions of the HP waves happen to have nearly uniform elevation in this case. This is probably because most buildings are considerably higher than the width of the road along the measurement routes, so that the diffraction over and the multiple reflection from the buildings seem to produce HP waves at relatively high elevation angles. Although further investigation is needed to confirm the statistical distributions for various mobile environments, it is expected that the HP wave distribution will be concentrated

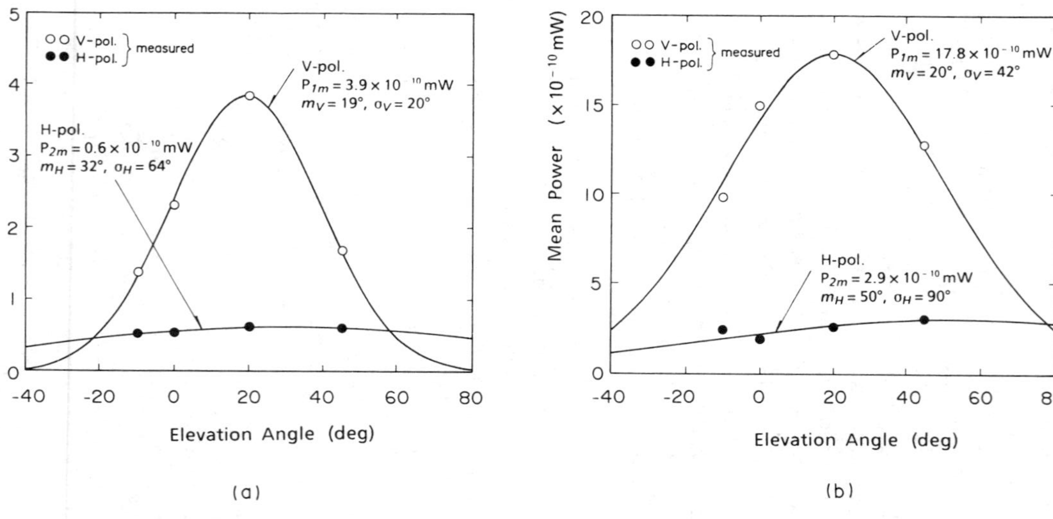

Figure 2.51 Mean power distribution of incident waves versus elevation angles in (a) Ningyo-cho route and (b) Kabuto-cho route. (From [81], © 1990 IEEE.)

around low elevation angles in suburban areas where there are few high buildings. The distribution parameters for the proposed model, m_V, σ_V, m_H, σ_H, can still be obtained empirically, even in the closely uniform distribution case, by approximating the measured values with Gaussian distributions, as shown in Figure 2.51. The distribution parameters for the two routes are shown in Table 2.5. From these parameters, the constants in (2.78) and (2.79), A_θ, A_ϕ, can be determined by (2.60) and (2.61). Thus, XPR can be evaluated by substituting these constants and the maximum mean power levels for both VP and HP waves into (2.82), as shown in Table 2.5. The XPR is evaluated as 5.1 dB in the Ningyo-cho route and as 6.8 dB in the Kabuto-cho route. These XPR values are quite reasonable considering the measured results that show the cross-polarization coupling in urban areas to lie between -9 and -4 dB [74]. This means that XPR in urban areas is larger than about 4 dB but less than 9 dB, because, the VP wave transmission, XPR corresponds to the reciprocal of the cross-polarization coupling.

2.4.3 MEG Characteristics of Dipole Antennas

2.4.3.1 Power Gain Pattern

The half-wavelength dipole antenna and the spherical coordinate system considered here are shown in Figure 2.52. The feeding point of the dipole antenna is situated at the origin of the coordinate system, and the antenna elements are on the L-axis inclined at angle α from the Z-axis in the vertical Z-X plane. A thin dipole is assumed and the element radius ignored. The mismatching and ohmic losses in the antenna are also ignored.

The three-dimensional power gain patterns of the dipole antenna with respect to the inclination angle α, from 0 to 90 deg, are shown in Figure 2.53. It is found that when $\alpha = 0$ deg, the antenna is only VP wave sensitive; however, at other inclination angles, it is both VP and HP wave sensitive. In particular, it should be noted that the horizontally oriented half-wavelength dipole antenna is not just HP wave sensitive.

2.4.3.2 MEG Characteristics of Vertical Dipole Antennas

When the inclination angle α is equal to 0 deg the power gain pattern for HP waves G_ϕ is nonexistent, as shown in Figure 2.53. MEG is, therefore, obtained by integrating only (2.62). If, moreover, the incident waves are only VP waves (i.e., XPR is infinitely large), the coefficient XPR/(1 + XPR) of the first term in the integrand in (2.62) becomes 1, and thus MEG is independent of the XPR value. However, when the incident waves have both VP and HP components, MEG is further reduced by the factor XPR/(1 + XPR). This gain degradation due to XPR is

Table 2.5
Experimental Results of Statistical Distribution Parameters

Measurement Route	m_v *(deg)*	σ_V *(deg)*	m_H *(deg)*	σ_H *(deg)*	*XPR (dB)*
Ningyo-cho	19	20	32	64	5.1
Kabuto-cho	20	42	50	90	6.8

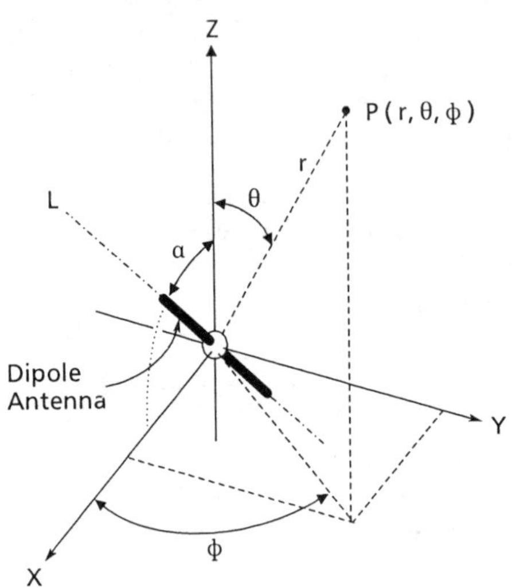

Figure 2.52 Half-wavelength dipole antenna and its coordinate system. (From [81], © 1990 IEEE.)

represented by the solid line in Figure 2.54. Thus, to clarify the MEG characteristics of vertical dipole antennas, it is sufficient to consider just the MEG characteristics for XPR $= \infty$.

The dependence of MEG (of the vertical dipole) on the standard deviation σ_V of VP incident waves is shown in Figure 2.55. When $\sigma_V = 0$ deg, MEG equals the directive gain with respect to the incoming direction of incident waves. Furthermore, when $m_V = 0$ deg, and $\sigma_V = 0$ deg (i.e., all incident waves lie completely in the horizontal plane), MEG is equal to the directivity of the half-wavelength dipole antenna at 2.15 dBi, because the incoming direction of incident waves corresponds to the direction with the maximum gain. The more the standard deviation increases,

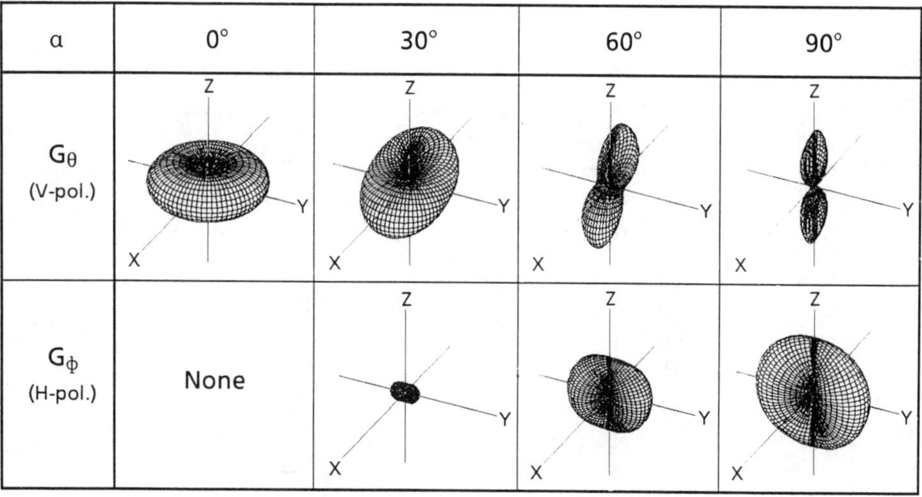

Figure 2.53 Power gain patterns of half-wavelength dipole antennas with inclination angle of 0, 30, 60, and 90 deg. (From [81], © 1990 IEEE.)

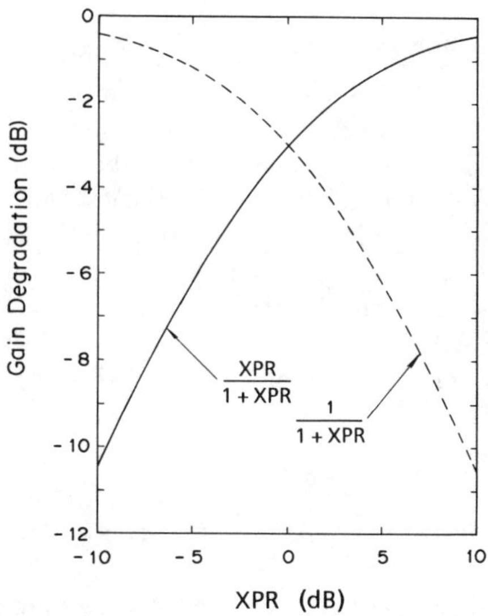

Figure 2.54 Mean effective gain degradation due to the cross-polarization power ratio. (From [81], © 1990 IEEE.)

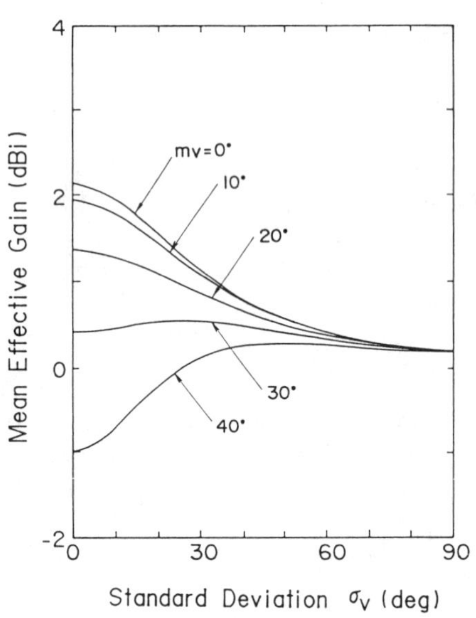

Figure 2.55 Mean effective gain of vertically oriented half-wavelength dipole antennas for vertically polarized incident waves: XPR = ∞. (From [81], © 1990 IEEE.)

the closer MEG is to the isotropic antenna gain (0 dBi). At $\sigma_V = \infty$ (i.e., when the statistical distribution of the incident waves is completely uniform), MEG is equal to the isotropic antenna gain. As m_V increases from $m_V = 0$ deg, with $\sigma_V = 0$ deg (i.e., incident waves lie in the azimuth plane at the m_V elevation), MEG decreases in proportion to the power gain at the mean elevation angle m_V. However, the more the standard deviation increases, the closer MEG again approaches the isotropic antenna gain (0 dBi). In the actual propagation environments seen in mobile communications, the distribution of the incident waves seems to be spread in elevation. Thus, the effective gain of the vertical half-wavelength dipole antenna becomes lower than the directivity of 2.15 dBi. MEG is reduced further by a lower XPR, as shown in Figure 2.54.

2.4.3.3 MEG Characteristics of Inclined Dipole Antennas

Several interesting results are obtained by investigating the MEG characteristics of inclined dipole antennas. Figure 2.56 shows the typical MEG characteristics calculated for several XPR values.

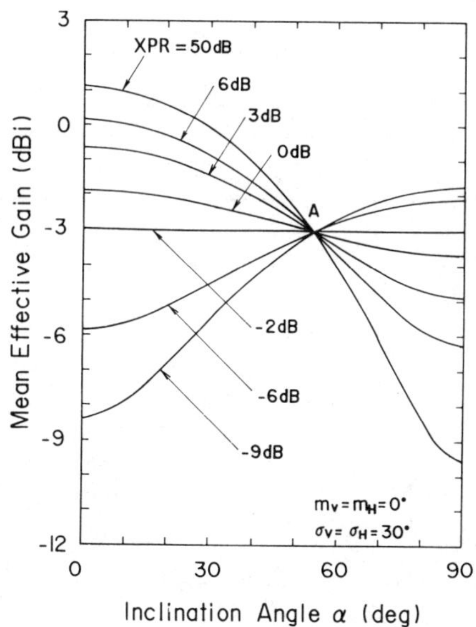

Figure 2.56 Mean effective gain of inclined half-wavelength dipole antennas: $m_V = m_H = 0$ deg, $\sigma_V = \sigma_H = 30$ deg. (From [81], © 1990 IEEE.)

First, it is found that there is a particular inclination angle at which MEG is -3 dBi regardless of the XPR value. This inclination angle is 55 deg and is shown as point A in Figure 2.56. At this angle, the VP radiation power of the antennas is equal to the HP radiation power, as shown in Figure 2.57. MEG variation at point A with the given incident wave distribution parameters (XPR, m_V, m_H, σ_V, σ_H) is less than 0.2 dB. It is proposed that the sum of the incident power of the VP and HP waves, $P_V + P_H$, can be measured by using this property. The power $P_V + P_H$ is twice the average received power measured by a dipole antenna inclined at 55 deg. Therefore, this power level can be used as the reference signal level in the MEG measurement, instead of the mean power level of the vertically oriented dipole antenna, because the latter is strongly affected by the propagation conditions.

Second, in Figure 2.56 the MEG characteristic for XPR $= -2$ dB shows that there are incident wave parameters that make the MEG constant (-3 dBi) regardless of the antenna inclination angle. According to this theoretical analysis, there are many propagation parameters that yield constant MEG (-3 dBi) characteristics regardless of the antenna orientation, in addition to the unusual environment of uniform illumination from all directions. In other words, there is a possibility of developing an artificial propagation environment in which the average received signal level of antennas can be made constant regardless of the variation of antenna

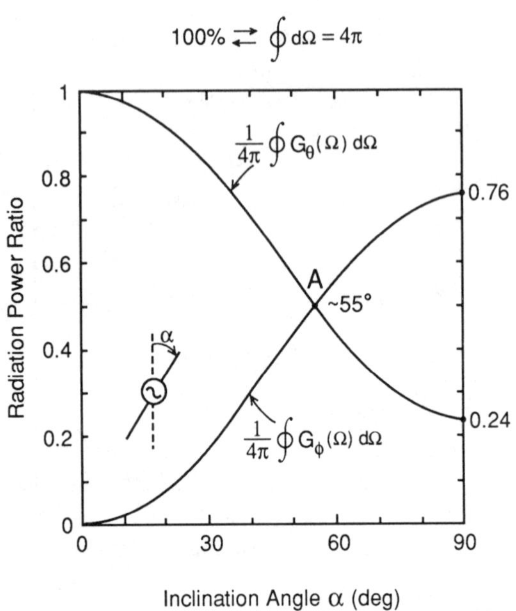

Figure 2.57 Radiation power ratio of vertical or horizontal polarized component of inclined half-wavelength dipole antenna.

pattern and polarization. It is expected that these environments could be developed by controlling the polarity of the transmitting antenna for XPR, the height of the transmitting antenna for mean elevation angle, and the beamwidth of the transmitting antenna for standard deviations, but further experimental investigation is required to confirm this.

2.4.3.4 Experimental Results

According to the theoretical consideration described in Section 2.4.3, the reference signal level in the MEG measurement cannot be evaluated definitely by measuring the mean received power level of a vertically oriented half-wavelength dipole antenna; however, it can be evaluated by measuring the mean received power level of a half-wavelength dipole antenna with an inclination angle of 55 deg from the vertical.

To confirm the validity of the theoretical considerations, a 900-MHz band experiment had been performed in the urban area of Tokyo. In this experiment, measurements of the received signal level of a test antenna were carried out over the same routes used in the statistical distribution measurements described in Section 2.4.2. A half-wavelength dipole antenna mounted 3.1m above the ground on the roof of a van was used as the test antenna. To evaluate the variation of received power with the direction of the antenna radiation pattern, the dipole antenna was inclined in the vertical planes with azimuthal angles of 0, 90, +45, and −45 deg from the forward direction of the moving van. The received signal level was digitized by an analog-to-digital (A/D) converter at distances of about 1 cm, and the average received signal level was calculated by averaging all data sampled over the route.

Table 2.6 shows the measurement results for average received signal levels of the dipole antenna with 55-deg inclination and the reference signal levels for the MEG measurement. The average received signal levels were obtained by averaging all measured values digitized over each measurement route. The variation of these values is caused by the mutual influence of the antenna pattern and the lack of the uniformity in azimuth of wave distribution for measurement routes. To evaluate the mean received signal level without such influences, the average of the measured signal levels for all antenna azimuthal orientations must be used as the mean received signal level of the 55-deg inclined dipole. Thus, in this case, the average of measured signal levels for four different antenna orientations was adopted as the mean level. The mean was 35.0 dBμV for the Ningyo-cho route and 33.5 dBμV for the Kabuto-cho route. The mean received signal level corresponds to one-half the reference signal level, because the MEG of the 55-deg inclined half-wavelength dipole antenna is −3 dBi. Therefore, the reference signal levels were obtained by adding 3 dB to the mean received signal levels and were evaluated as 38.0 dBμV for the Ningyo-cho route and 36.5 dBμV for the Kabuto-cho route. The MEG value of the test antenna is calculated by normalizing the antenna's average received signal level with the corresponding reference signal

Table 2.6
Measured Signal Level of a Dipole Antenna with 55-deg Inclination and Reference Signal Level for Isotropic Mean Effective Gain

	Orientation of Antenna Inclination	*Signal Level (dBμV)*		
Route	*Plane (deg)*	*Average Received*	*Mean Received*	*Reference for $P_V + P_H$*
Ningyo-cho	0	35.0	35.0	38.0
	90	34.6		
	+45	35.3		
	−45	35.3		
Kabuto-cho	0	33.5	33.5	36.5
	90	35.2		
	+45	32.4		
	−45	32.2		

Source: [81], © 1990 IEEE.

level. Figure 2.58 shows the normalized measurement results and the theoretical curve of MEG for half-wavelength dipole antennas. The solid line shows the theoretical curve calculated using the empirical distribution parameters shown in Table 2.5. The open circles, crosses, open triangles, and open squares show the measurement results for the antenna orientations of 0, 90, +45, and −45 deg, respectively. The closed circles show the average values of the measured results for each antenna inclination.

The most significant result is that the theoretical curves show excellent agreement, within about 1 dB, with the average values of the measured results. This indicates that the MEG analysis, described in Sections 2.4.1 to 2.4.3, evaluates not only the MEG of mobile antennas (if their movements are sufficiently random so that the statistical distribution of incident waves can be considered completely uniform in the azimuth direction), but also the average of the MEG variation of directional antennas operating in environments whose statistical distribution of incident waves shows a lack of uniformity in the azimuth direction. Furthermore, the experimental results confirm the validity of the presented model and the derived theoretical expression for the MEG.

2.4.4 Correlation Characteristics of Polarization Diversity

It is well known that a cross-dipole antenna made of two orthogonally placed half-wavelength dipole antennas constitutes a polarization diversity branch [74,87,88]. This section describes the analytical derivation of correlation characteristics of this antenna diversity using the theoretical expression described in Section 2.4.1 and the statistical model described in Section 2.4.2.

2.4.4.1 Theoretical Analysis

Figure 2.59 shows the coordinate system for the polarization diversity branch using a cross-dipole antenna. The distance d between the two dipole antennas is zero so that the phase difference x in (2.74) becomes zero. Figure 2.59 exhibits the situation in which the cross-dipole antenna is inclined at an angle α from the vertical (Z) direction in the X-Z plane. The power gain patterns of each dipole antenna for various inclination angles α are shown in Figure 2.60(a) to (d).

The characteristic properties of this polarization diversity antenna are shown in Figures 2.61 through 2.64. The first characteristic is that the correlation coefficient becomes zero (uncorrelated) if one of the dipole antennas is placed vertically and the other horizontally so that $\alpha = 0$ deg. This characteristic can be explained as follows. In the case of $\alpha = 0$ deg, it can be seen from Figure 2.60(a) that the horizontal dipole antenna has radiation patterns of both θ and ϕ components whereas the vertical dipole has the pattern of only the θ component. Hence, accord-

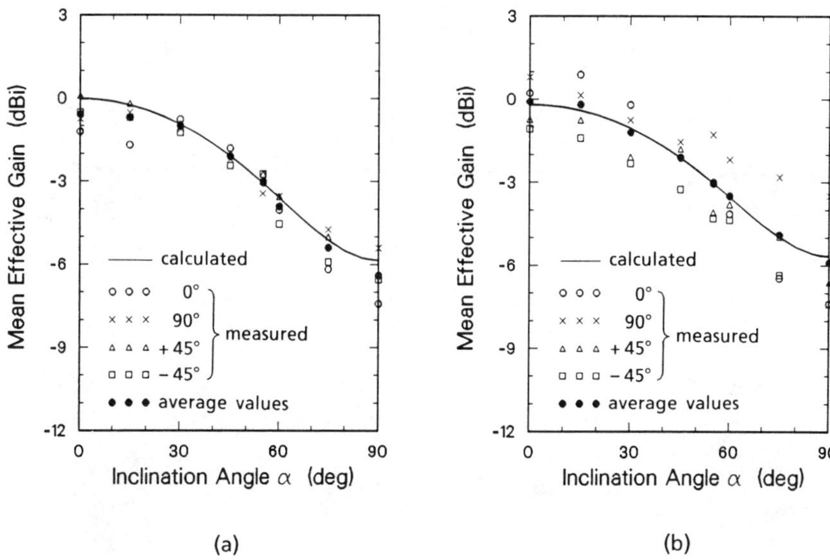

Figure 2.58 Comparison of calculated and measured mean effective gain of an inclined half-wavelength dipole antenna in (a) Ningyo-cho route and (b) Kabuto-cho route. (From [81], © 1990 IEEE.)

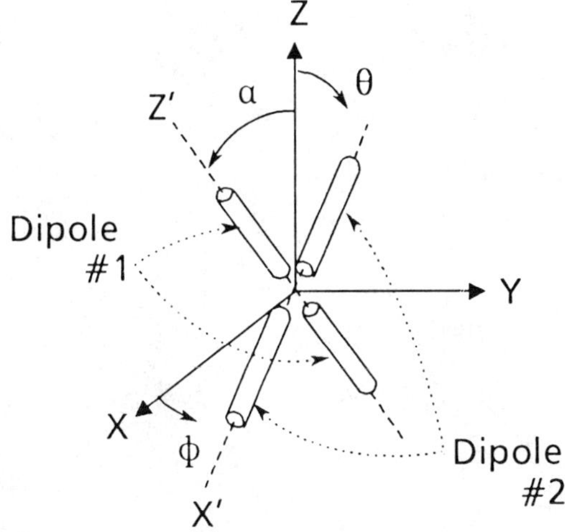

Figure 2.59 Polarization diversity using cross-dipole antennas and their coordinate system. (From [84], © 1990 IEICE, Japan.)

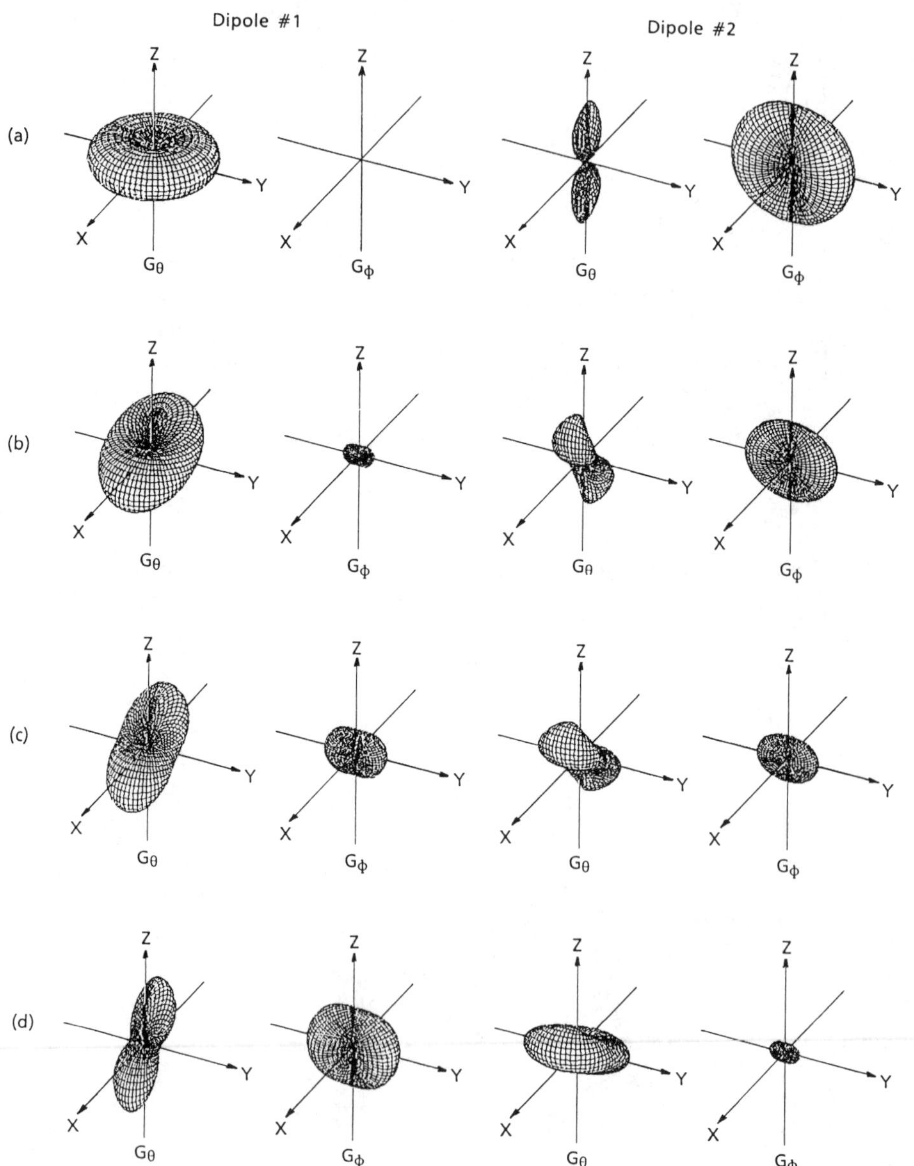

Figure 2.60 Power gain patterns of cross-dipole antennas: (a) $\alpha = 0°$; (b) $\alpha = 30°$; (c) $\alpha = 45°$; (d) $\alpha = 60°$. (From [84], © 1990 IEICE, Japan.)

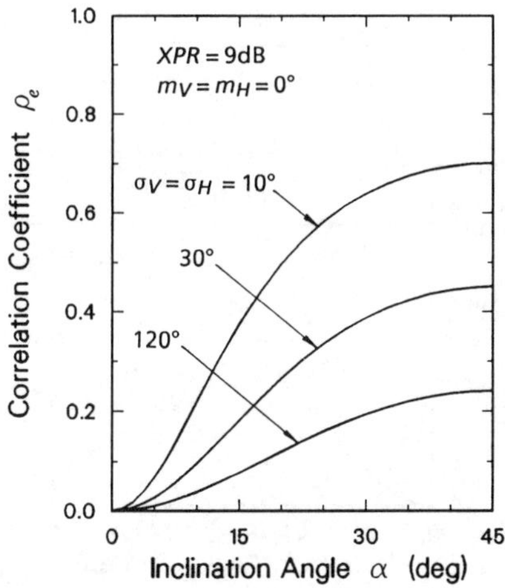

Figure 2.61 Correlation coefficient of polarization diversity using cross-dipole antennas. (From [84], © 1990 IEICE, Japan.)

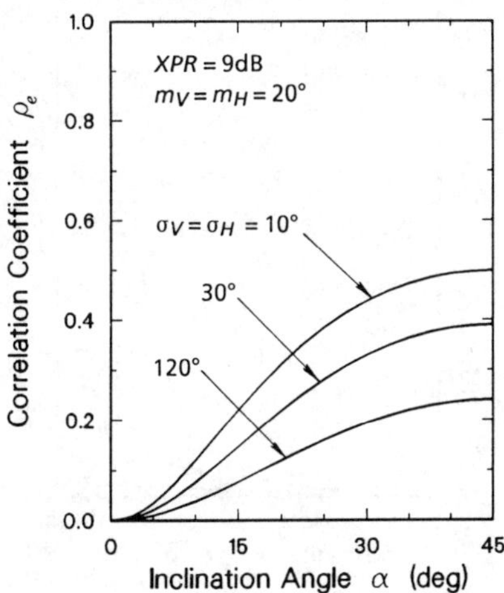

Figure 2.62 Correlation coefficient of polarization diversity using cross-dipole antennas. (From [84], © 1990 IEICE, Japan.)

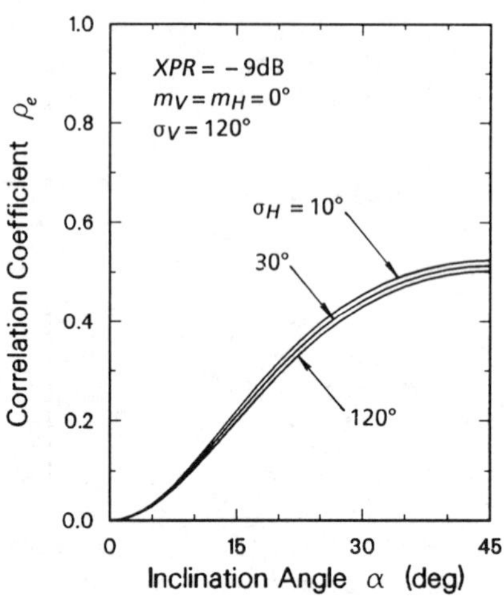

Figure 2.63 Correlation coefficient of polarization diversity using cross-dipole antennas. (From [84], © 1990 IEICE, Japan.)

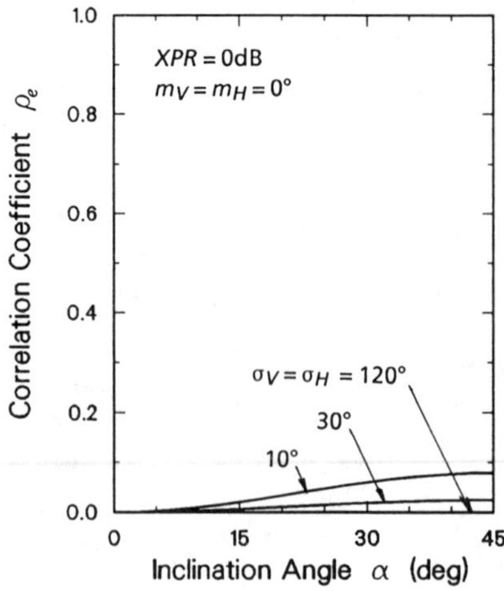

Figure 2.64 Correlation coefficient of polarization diversity using cross-dipole antennas. (From [84], © 1990 IEICE, Japan.)

ing to (2.74), only the radiation patterns of the θ components are related to the correlation coefficient. The radiation patterns of the θ components are, accordingly, spatially orthogonal. Therefore, the integral of the numerator term in (2.74) is equal to zero so that the diversity branches become theoretically uncorrelated. This lack of correlation does not depend on XPR or the variation in incident wave distribution. According to the experimental results reported to date [74], the correlation characteristics are almost zero for $\alpha = 0$ deg and hence the validity of this analysis appears to be confirmed.

When the antenna system is inclined, the perfect orthogonality of the antenna radiation patterns described above cannot be maintained, and hence the correlation coefficient increases with the inclination angle α. Such a situation is common in an antenna system installed on a portable transceiver. The second characteristic is the fact that the correlation coefficient becomes maximum for the inclination angle of 45 deg. This is because the radiation patterns of the θ components of the two dipoles in the horizontal (X-Y) plane become identical omnidirectional patterns, while the radiation patterns of the ϕ components are almost identical three-dimensionally. (See Figure 2.60(c).) As a result, if the environment yields an XPR much larger than 0 dB, the correlation increases as the incident wave distribution concentrates more into the horizontal plane, and decreases as it is spread in the elevation angle direction, as shown in Figure 2.61. As shown in Figure 2.62, as the mean elevation angle m_V of the VP waves moves away from the horizontal direction, the correlation is reduced. In an environment where XPR is much smaller than 0 dB, the effect of the radiation pattern of the ϕ component becomes significant. However, since the variation of the ϕ component pattern on the elevation angle is small, as shown in Figure 2.60, the variation of the correlation coefficient for the divergence of the incident wave distribution in the elevation direction is negligible. (See Figure 2.63.) Moreover, since the degree of ϕ component pattern overlap of both antennas is significant, the correlation increases as XPR is reduced.

According to the characteristics clarified above, the integral terms for each polarization component in (2.74) can be reduced, and the correlation becomes extremely small in environments whose XPR values are nearly equal to 0 dB. This low correlation is almost independent of the parameter values of the incident wave distribution and inclination of the diversity antenna. Figure 2.64 shows the correlation characteristics in such cases.

It is clear that the correlation characteristics of a polarization diversity antenna depend on the radiation pattern of the antenna branch, XPR, and incident wave distribution. This dependence is, in fact, also true for all antenna diversity schemes. Hence, to realize optimum antenna diversity, we need to take into account antenna radiation patterns and the propagation environment characteristics. If the dependence of the correlation characteristics on the propagation environment parameters can be used positively, realization of a polarization diversity branch with an extremely low correlation can be expected. According to a theoretical analysis, it is found that this low-correlation diversity branch is realized at XPR $= -1.5$ dB in the diversity branches of a cross-dipole antenna. Furthermore, its realization has been confirmed through an indoor experiment [89].

2.4.4.2 Experimental Results

To confirm the validity of these theoretical considerations, a 900-MHz-band experiment was performed in the urban area of Tokyo. In this experiment, measurements of the correlation characteristics of a cross-dipole antenna were carried out on the Ningyo-cho route passing through all the measurement points described in Section 2.4.2. A cross-dipole antenna was installed on the roof (height: 3.1m above the ground) of a mobile van. Since the radiation pattern of an inclined dipole antenna has an azimuthal deviation as shown in Figure 2.60, it is considered that the measured values might be affected by the azimuthal deviation of the incident wave distribution in the experimental environment. Hence, the direction of the receiving antenna (equal to the direction of the X-axis in Figure 2.59) took four directions 0, 90, +45, and −45 deg, with respect to the moving direction of the van. For each direction, the inclination angle α of the receiving antenna was varied with respect to the vertical direction and the receiving signal strength between the two branches was measured simultaneously as the van moved at about 20 to 30 km/hr along the route. The measured data were digitized by an A/D converter at intervals of about 1 cm. The correlation coefficient of the envelope of the received signal for one circuit of the measurement route was derived by numerical processing on a computer.

Figure 2.65 shows the correlation coefficient of the cross-dipole antenna calculated from the incident wave distribution parameters in Table 2.5 and the corresponding experimental data. The symbols of open circle, cross, open triangle, and open square are measured values for the antenna azimuth directions of 0, 90, +45, and −45 deg, respectively. The closed circles indicate the average of the values in the four directions, whereas the solid line is the theoretical curve. If the azimuth distribution of the incident wave is uniform, the measured values should not depend on the azimuth direction. Hence, the dependence of the measured values on the antenna azimuth direction indicates that the incident wave distribution on the measurement route is not uniform in the azimuth direction. Also, since the incident wave distribution parameters in Table 2.5 were obtained from the mean level of the measured patterns in the azimuth direction, the theoretical curve in Figure 2.65 represents the average value of the correlation coefficient over various antenna azimuth directions. Hence, to discuss the validity of the analytical results, comparison with the average value of the correlation coefficient measured for many antenna azimuth directions is important. The average values indicated by closed circles are those over only four directions; however, the values agree closely with the theoretical results, and hence the effectiveness of the analysis results is confirmed.

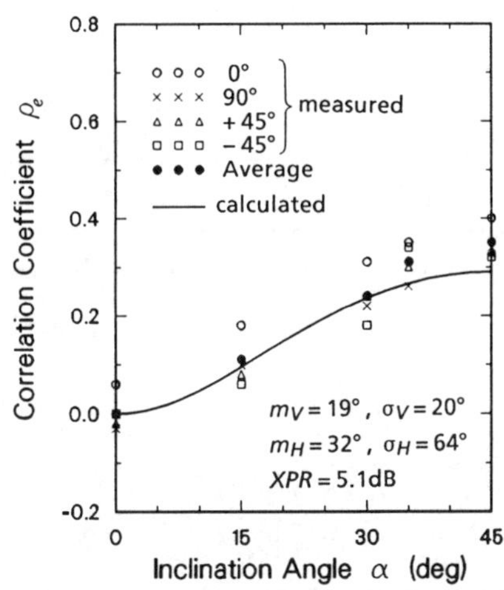

Figure 2.65 Comparison of theoretical correlation curve with measurement value. (From [84], © 1990 IEICE, Japan.)

REFERENCES

[1] Kuramoto, M., and H. Tohyama, "General View of Mobile Communication," Ch. 1 of *Basic Technology of Mobile Communications*, Y. Okumura and M. Shinji, IEICE Press (Japanese), 1986, pp. 1–23.

[2] Lee, W.C.Y., *Mobile Communications Engineering*, McGraw-Hill, 1982.

[3] Jakes, W. C., Jr., *Microwave Mobile Communications*, New York: John Wiley, 1974.

[4] Special issue on "Mobile Communications," *IEICE Trans.*, Vol. E74, No. 6, 1991.

[5] Okumura, Y., E. Ohmori, T. Kohno, and K. Fukuda, "Field Strength and Its Variability in VHF and UHF Land-Mobile Radio Service," *Rev. Elec. Comm. Lab.*, Vol. 16, No. 9–10, 1968, pp. 825–873.

[6] "VHF and UHF Propagation Curves for the Frequency Range From 30 MHz and 1000 MHz," CCIR SG-5, Recommendation 370.

[7] Kozono, S., and K. Watanabe, "Influence of Environmental Buildings on UHF Band Mobile Radio Propagation," *IEEE Trans. Com.*, Vol. COM-25, No. 10, 1977, pp. 1133–1143.

[8] Hata, M., "Empirical Formula for Propagation Loss in Land Mobile Radio Service," *IEEE Trans.*, Vol. VT-29, No. 3, 1980, pp. 317–325.

[9] Ikegami, F., S. Yoshida, T. Takeuchi, and M. Umehira, "Propagation Factors Controlling Mean Field Strength on Urban Streets," *IEEE Trans.*, Vol. AP-32, No. 8, 1984, pp. 822–829.

[10] Sakagami, S., "Mobile Propagation Loss Prediction for Arbitrary Urban Environments," *IEICE Trans.*, Vol. J74-B-II, 1991, pp. 17–25.

[11] Fujimoto, K., "Overview of Antenna Systems for Mobile Communications and Prospects for the Future Technology," *IEICE Trans.*, Vol. E74, No. 10, 1991, pp. 3191–3201.

[12] Special issue on "Mobile Radio Communications," *J. IECE Japan*, Vol. 68, No. 11, 1985.

[13] Special issue on "Digital Mobile Telephone System," *J. IEICE*, Vol. 73, No. 8, 1990, pp. 799–844.

[14] Sekiguchi, H., "Switching of Mobile Communications," Ch. 8 of *Mobile Communications*, M. Shinji, eds., Maruzen, 1989.
[15] Nishikawa, T., "RF Front End Circuit Components Miniaturized Using Dielectric Resonators for Cellular Portable Telephone, *IEICE Trans.*, Vol. E74, No. 6, 1991, pp. 1556–1562.
[16] Taga, T., and K. Tsunekawa, "Performance Analysis of a Built-in Planar Inverted F Antenna for 800 MHz Band Portable Units," *IEEE JSAC*, Vol. SAC-5, No. 5, 1987, pp. 921–929.
[17] Yamada, Y., Y. Ebine, and K. Tsunekawa, "Base and Mobile Station Antennas for Land Mobile Radio Systems," *IEICE Trans.*, Vol. E74, No. 6, 1991, pp. 1547–1555.
[18] Ishihara, H., "Mobile Radio Communications for Business Use and Multi-Channel Access System," *J. IECE Japan*, Vol. 68, No. 11, 1985, pp. 1183–1187.
[19] Kanzaki, K., "Current Status and Future Trends on Efficient Spectrum Utilization for Mobile Radio Communications," *J. IECE Japan*, Vol. 68, No. 11, 1985, pp. 1251–1253.
[20] Reinhart, E. E., R. M. Taylor, and A. O. Heyward, "WARC's Last Act?" *IEEE Spectrum*, Vol. 29, No. 2, 1992, pp. 20–33.
[21] Bullington, K., "Radio Propagation for Vehicular Communications," *IEEE Trans. Vehicular Technology*, Vol. VT-26, No. 4, 1977, pp. 295–308.
[22] Shepherd, N. H., "Radio Wave Loss Deviation and Shadow Loss at 900 MHz," *IEEE Trans. Vehicular Technology*, Vol. VT-26, Nov. 1977, pp. 309–313.
[23] Young, R. W., "Comparison of Mobile Radio Transmission at 150, 450, 900 and 3700 MC/s," *BSTJ*, Nov. 1952, pp. 1068–1085.
[24] Lee, W.C.Y., "Lee's Model," Propagation Ad Hoc Committee of IEEE Vehicular Technology Society appeared in a special issue of IEEE Transactions on Vehicular Technology, Feb. 1988, pp. 68–70. Also *IEEE VTS '92 Proc.*, pp. 343–348.
[25] Lee, W.C.Y., "Studies of Base-Station Antenna Height Effects on Mobile Radio," *IEEE Trans. Vehicular Technology*, Vol. VT-29, No. 2, May 1980, pp. 252–260.
[26] Lee, W.C.Y., *Mobile Cellular Telecommunications Systems*, McGraw-Hill, 1989, p. 102, 126–131.
[27] Ibid., p. 109.
[28] Ibid., p. 442.
[29] Ibid., p. 150.
[30] Ibid., p. 102.
[31] Ibid., p. 23.
[32] Lee, W.C.Y., "Estimate of Local Average Power of a Mobile Radio Signal," *IEEE Trans. Vehicular Technology*, Vol. VT-34, No. 1, Feb. 1985, pp. 22–27.
[33] Anderson, L. J., and L. G. Trolese, "Simplified Method for Comparing Knife-Edge Diffraction as the Shadow Region," *IEEE Trans. Ant. Propag.*, July 1958, pp. 281–286.
[34] Lee, W.C.Y., "Overview of CDMA," *IEEE Trans. Vehicular Technology*, Vol. 40, No.2, May 1991, pp. 291–302.
[35] Lee, W.C.Y., *Mobile Communications Design Fundamentals*, H. W. Sams, ed., 1986, p. 277.
[36] Lee, W.C.Y., *Mobile Communications Engineering*, McGraw-Hill, 1982, p. 229, p. 46.
[37] Lee, W.C.Y., *Mobile Communications Design Fundamentals*, 2nd edition, John Wiley, 1993, p. 30, pp. 89–93.
[38] Ibid., p. 30.
[39] Fujimoto, K., "A Loaded Antenna System Applied to VHF Portable Communication Equipment," *IEEE Trans.*, Vol. VT-17, 1968, pp. 6–13.
[40] Hirasawa, K., and K. Fujimoto, "Characteristics of Wire Antenna on a Rectangular Conducting Body," *Trans. IECE Japan*, Vol. J65-B, 1982, pp. 1133–1139.
[41] Sato, K., K. Matsumoto, K. Fujimoto, and K. Hirasawa, "Characteristics of Planar Inverted-F Antenna on a Rectangular Conducting Body," *Trans. IECIE*, Vol. J71-B, 1988, pp. 1237–1243.
[42] Fujimoto, K., A. Henderson, K. Hirasawa, and J. R. James, *Small Antennas*, Research Studies Press, 1986, pp. 131–151.

[43] Fujimoto, K., "Overview of Antenna Systems for Mobile Communications and Prospects for the Future Technology," *IEICE Trans. Comm.*, Vol. E74, No. 10, 1991, p. 3196.
[44] Ibid., p. 3193.
[45] Kagoshima, K., and Y. Yamada, "Mobile Communication Antennas," Ch. 9 in *Mobile Communications*, M. Shinji, ed., Maruzen, 1989.
[46] Tsunekawa, K., "Diversity Antennas for Portable Telephone," *IEEE Proc. VTC '89*, 1989, pp. 50–56.
[47] Fujimoto et al. [42], op. cit., pp. 89–112.
[48] Ebine, Y., and Y. Yamada, "A Vehicular-Mounted Vertical Space Diversity Antenna for a Land Mobile Radio," *IEEE Trans.*, Vol. VT-40, No. 2, 1991, pp. 420–425.
[49] Nishikawa, K., K. Sato, and K. Fujimoto, "Phased Array Antenna for Land Mobile Satellite Communications," *IEICE Trans.*, Vol. J72 B-II, No. 7, 1989, pp. 323–329.
[50] Hirasawa, K., "Small Antennas for VHF and UHF Mobile Communications," National Science Foundation (NSF) Workshop on Future Directions in Electromagnetics Research, July 1989, pp. 199–201.
[51] Perini, J., and K. Hirasawa, "Antenna Pattern Distortion and Mutual Coupling in Antenna Farms," *IEEE Int. Symp. Electromagnetic Compatibility*, June 1973, pp. 201–208.
[52] Katagi, T., "Mutual Coupling Effects Between Antenna Elements on Antenna Performance," Ch. 8 in *Analysis, Design and Measurement of Small and Low-Profile Antennas*, K. Hirasawa and M. Haneishi, eds., Norwood, MA: Artech House, 1991.
[53] Zhang, et al., "Opened Parasitic Elements Nearby a Driven Dipole," *IEEE Trans. Ant. Propag.*, Vol. AP-34, May 1986, pp. 711–713.
[54] Hirasawa, K., "Effects of Mutual Coupling on the Performance of Closely Spaced Antennas," *Nordic Antenna Symp.*, May 1985, pp. 1–15.
[55] Harrington, R. F., "Field Computation by Moment Methods," New York: Macmillan, 1968; republication by R. E. Krieger, 1982.
[56] Kouyoumjian, R. G., and P. H. Pathak, "A Uniform Geometrical Theory of Diffraction for an Edge in a Perfect Conducting Surface," *Proc. IEEE*, Vol. 62, Nov. 1974, pp. 1448–1461.
[57] Hirasawa, K., "Bounds of Uncertain Interference Between Closely Located Antennas," *IEEE Trans. Electromagnetic Compatibility*, Vol. EMC-26, Aug. 1984, pp. 129–133.
[58] Hirasawa, K., "Computer Programs for Calculating Bounds of Interference Between Arbitrarily Shaped Wire Antennas," *6th Symp. & Technical Exhibition on Electromagnetic Compatibility*, March 1985, pp. 321–326.
[59] Hirasawa, K., "Bounds of Uncertain Interference Between Closely Located Antennas," *IEEE Int. Symp. Electromagnetic Compatibility*, Oct. 1984, pp. 16–18.
[60] Hirasawa, K., "Computer Programs for Calculating Bounds of Interference Between Receiving Wire Antennas," *7th Symp. & Technical Exhibition on Electromagnetic Compatibility*, March 1987, pp. 35–38.
[61] Fujimoto, K., et al., Ch. 4 in *Small Antennas*, London: Research Studies Press, 1987.
[62] Sato, K., et al., "Characteristics of a Planar Inverted-F Antenna on a Rectangular Conducting Body," *Electronics and Communications in Japan*, Part 1, Vol. 72, Scripta Publishing, Oct. 1989, pp. 43–51.
[63] Nishikawa, K., "Effect of Automobile Body and Earth on Radiation Patterns of Antennas for FM Radio," *IECEJ Trans.*, Vol. E67, Oct. 1984, pp. 555–562.
[64] Nishikawa, K., and Y. Asano, "Vertical Patterns of Mobile Antenna at UHF Frequencies," *IEEE Vehicular Technology Conference*, May 1985, pp. 44–49.
[65] Nishikawa, K., and Y. Asano, "Vertical Radiation Patterns of Mobile Antenna in UHF band," *IEEE Trans. Vehicular Technology*, Vol. VT-35, May 1986, pp. 57–62.
[66] Anderson, S. R., "VHF Omnirange Accuracy Improvements," *IEEE Trans. Aerospace and Navigational Electronics*, Vol. ANE-12, Jan. 1965, pp. 26–35.

[67] Gruenberg, H., and K. Hirasawa, "Effects of Scattering by Obstacles in the Field of VOR/DVOR," Federal Aviation Administration, Report 6700.11, Aug. 1976.
[68] Hirasawa, K., "Effects of Finite Wire Scatters in the Field of VOR," *IEEE Trans. Aerospace and Electronic Systems*, Vol. AES-18, Sept. 1982, pp. 668–674.
[69] Hirasawa, K., "VOR Bearing Errors Due to Scattering From Conducting Bodies," *IEEE Int. Symp. Electromagnetic Compatibility*, Oct. 1984, pp. 688–692.
[70] Lee, W.C.Y., "Mobile Communication Engineering," McGraw-Hill, 1982, p. 300.
[71] Lee, W.C.Y., "Mobile Radio Signal Correlation vs. Antenna Height and Spacing," *IEEE Trans. Vehicular Technology*, Aug. 1977, pp. 290–292.
[72] Lee, W.C.Y., "Mobile Communications Design Fundamentals," Howard W. Sams and Co., 1986, p. 204.
[73] Lee, W.C.Y., "Vertical vs. Horizontal Separations for Diversity Antennas," *Cellular Business*, Dec. 1992, pp. 56–60.
[74] Lee, W.C.Y., and Y. S. Yeh, "Polarization Diversity System for Mobile Radio," *IEEE Trans. Communications*, Vol. COM-20, No. 5, Oct. 1972, pp. 912–923.
[75] Lee, W.C.Y., "Statistical Analysis of the Level Crossings and Duration of Fades of the Signal from an Energy Density Mobile Radio Antenna," *Bell System Technical J.*, Vol. 46, Feb. 1967, pp. 416–440.
[76] Lee, W.C.Y., "Preliminary Investigation of Mobile Radio Signal Fading Using Directional Antennas on the Mobile Unit," *IEEE Trans. Veh. Comm.*, Vol. 15, Oct. 1966, pp. 8–15.
[77] Bach Andersen, J., and F. Hansen, "Antennas for VHF/UHF Personal Radio: A Theoretical and Experimental Study of Characteristics and Performance," *IEEE Trans. Vehicular Technology*, Vol. VT-26, No. 4, Nov. 1977, pp. 349–357.
[78] Davidson, A. L., and W. J. Turney, "Mobile Antenna Gain in Multipath Environment at 900 MHz," *IEEE Trans. Vehicular Technology*, Vol. VT-26, No. 4, Nov. 1977, pp. 345–348.
[79] Clarke, R. H., "A Statistical Theory of Mobile-Radio Reception," *Bell System Technical J.*, Vol. 47, No. 6, July-Aug. 1968, pp. 957–1000.
[80] Awadalla, K. H., "Direction Diversity in Mobile Communications," *IEEE Trans. Vehicular Technology*, Vol. VT-30, No. 3, Aug. 1987, pp. 121–123.
[81] Taga, T., "Analysis for Mean Effective Gain of Mobile Antennas in Land Mobile Radio Environments," *IEEE Trans. Vehicular Technology*, Vol. VT-39, No. 2, May 1990, pp. 117–131.
[82] Cox, D. C., R. R. Murray, H. W. Arnold, A. W. Norris, and M. F. Wazowicz, "Cross-Polarization Coupling Measured for 800 MHz Radio Transmission in and Around Houses and Large Buildings," *IEEE Trans. Ant. Propag.*, Vol. AP-34, No. 1, Jan. 1986, pp. 83–87.
[83] Jakes, W. C., *Microwave Mobile Communications*. New York: Wiley & Sons, 1974.
[84] Taga, T., "Analysis for Correlation Characteristics of Antenna Diversity in Land Mobile Radio Environments," *IEICE Trans. (B-II) of Japan*, Vol. J73-B-II, No. 12, Dec. 1990, pp. 883–895, (in Japanese). (This paper was translated by Scripta Technica, Inc., to *Electronics and Communications in Japan*, Part 1, Vol. 74, No. 8, 1991, pp. 101–115.)
[85] Gans, M. J., "A Power Spectral Theory of Propagation in the Mobile Radio Environment," *IEEE Trans. Vehicular Technology*, Vol. VT-21, No. 1, Feb. 1972, pp. 27–38.
[86] Ikegami, F., and S. Yoshida, "Analysis of Multipath Propagation Structure in Urban Mobile Radio Environments," *IEEE Trans. Ant. Propag.*, Vol. AP-28, No. 4, July 1977, pp. 531–537.
[87] Kozono, S., H. Tsuruhara, and M. Sakamoto, "Base Station Polarization Diversity Reception for Mobile Radio," *IEEE Trans. Vehicular Technology*, Vol. VT-33, No. 4, Nov. 1984, pp. 301–306.
[88] Cox, D. C., "Antenna Diversity Performance in Mitigating the Effects of Portable Radiotelephone Orientation and Multipath Propagation," *IEEE Trans. Communications*, Vol. COM-31, No. 5, May 1983, pp. 620–628.
[89] Taga, T., K. Tsunoda, and H. Imahori, "Correlation Properties of Antenna Diversity in Indoor Mobile Communication Environments," *Proc. 39th IEEE Vehicular Technology Conf.*, San Francisco, 1–3 May 1989, pp. 446–451.

Chapter 3

Land Mobile Antenna Systems I: Basic Techniques and Applications

K. Kagoshima and T. Taga

3.1 ANTENNAS

Land mobile communication systems employ base stations and mobile stations. These stations employ different types of antennas, and the design criteria also differ. Figure 3.1 shows important items to be considered in designing a base station antenna. Although antenna design in the narrow sense means electrical design, in reality it includes a wider area, and it is important to derive the antenna hardware specifications from system requirements. In order to determine the hardware specifications, it is necessary to perform an evaluation that compares electrical and mechanical characteristics and the tradeoff between performance and cost, as shown in Figure 3.2. Performance and cost considerations are sometimes the first step, while determination of the electrical and mechanical design are the second step [1].

In designing the practical antenna, it is important to assess how the antenna hardware will be installed after manufacturing. For base station and mobile station antennas (e.g., vehicular antennas), installation fees may be larger than the cost of the antennas themselves. For this reason, it is important to consider not only the reduction of construction costs, but also the creation of an antenna design which makes installation easy. In this sense, the on-glass antenna shown in Figure 3.3, which enables easy mounting, is superior.

Figure 3.4 shows some items to be considered in designing a mobile station antenna. Mobile station antennas are classified into two categories: antennas for mobile mounting, such as on vehicles, and antennas for mounting on portable radio equipment. Mobile station antennas are more independent of system parameters than are base station antennas, and should be designed for easy handling and customer convenience.

Because mobile communication antennas are not used in free space, but within a multipath environment, antenna gain or radiation patterns specified for the system

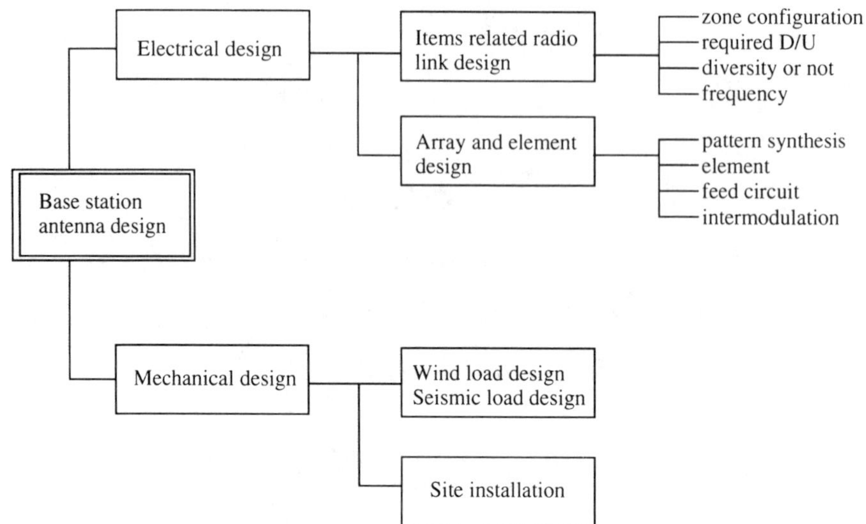

Figure 3.1 Key items in designing base station antennas.

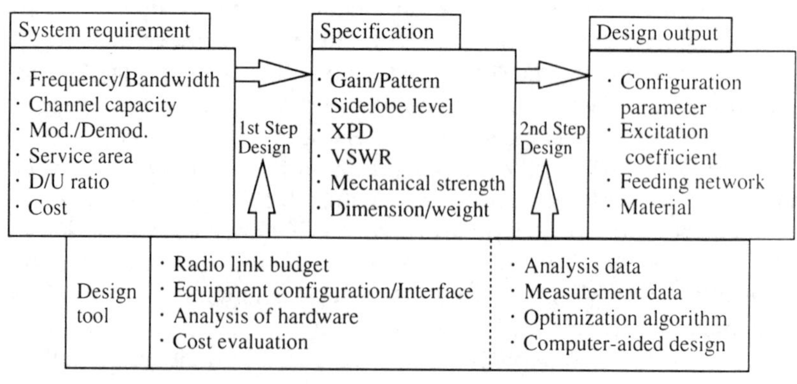

Figure 3.2 Steps of antenna design.

design should be designed for their environment. Therefore, it is not necessarily meaningful for an antenna to have superior performance in free space if its performance seriously degrades in a multipath environment. Although it has been

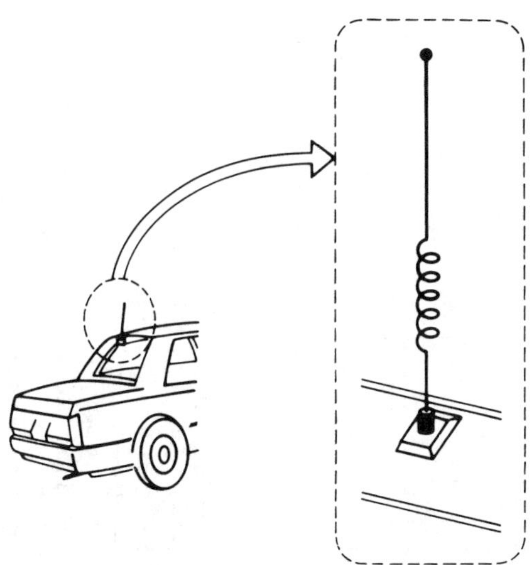

Figure 3.3 Installation example of an on-glass antenna. (Courtesy of Harada Industry Co., Ltd.)

pointed out that the prediction of antenna performance should be carried out under the multipath environment, a method to predict antenna gain and radiation patterns has only recently been proposed. Taga and Ebine [2,3] developed the prediction method by assuming the amplitude distribution of the incoming wave is a Gaussian distribution, while the plane wave is incident in free space. Since this was discussed in more detail in Chapter 2, further description is abbreviated here.

Analysis and measurement technologies for land mobile communication antennas are also very important, as is true for other kinds of antennas. Since mobile antennas are usually made of wires or metal plates, the method of moments is highly effective for analyzing these antennas. An antenna, consisting of metal plates or box(es), can be analyzed by simulating the box(es) with a multiple-wire grid [4], or by dividing the surface of the box(es) into small triangular patches [5]. The method of moments has been modified and extended to analyze antennas made of dielectric materials, and computer simulations can be effectively carried out in lieu of direct experiments. Moreover, the progress of computer technology has allowed remarkable advances in time-domain analysis, such as transmission line matrix method (TLM) and finite difference time domain (FDTD). This has allowed the development of more complex antennas and models, which more closely emulate real antennas. Thus, many valuable results have been reported [6–8]. Several tutorial books on antenna analysis have been published, so interested readers can refer to these for more details [9,10]. As for measurement, mobile communication antennas should be measured in a real propagation environment. Therefore, eval-

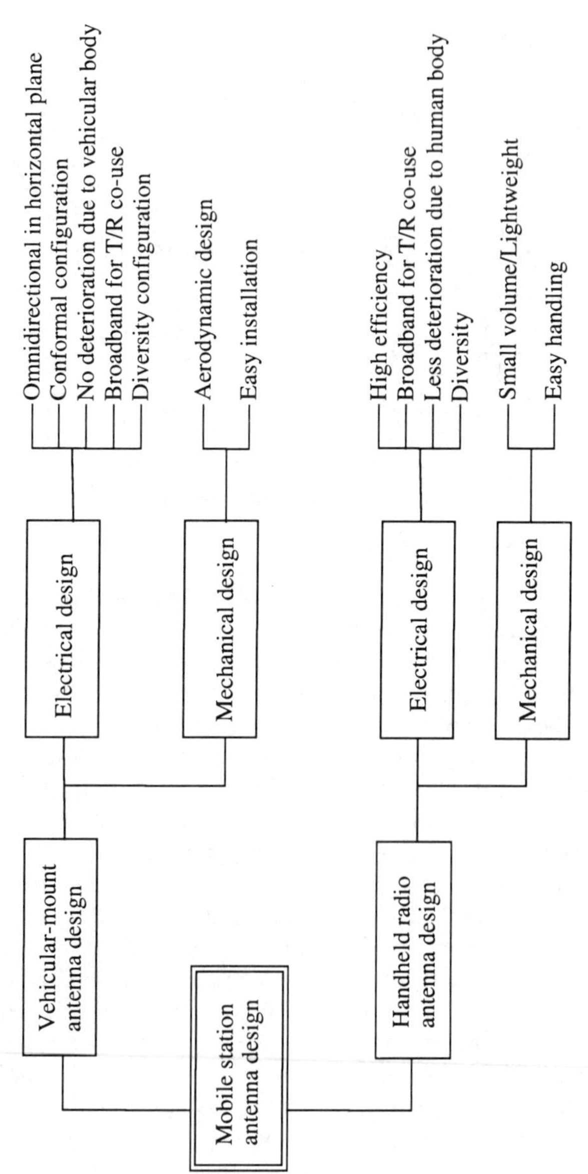

Figure 3.4 Key items in designing mobile station antennas.

uation of antenna performance under such circumstances is more difficult and laborious than is usual for communication antennas used in line of sight.

3.2 PROPAGATION PROBLEMS

3.2.1 Antenna Height and Propagation Environment

Propagation in mobile communication occurs within a diffraction region and is not free-space propagation. Propagation path loss in free space is simple and is proportional to the square of the distance, but in mobile communications it depends on various factors, such as the propagation environment, antenna height, and frequency. It is important to define such terms clearly in order to discuss propagation problems.

The height of a base station antenna, the h_{te}, is defined as $h_{te} = h_{ts} - h_{ga}$, as shown in Figure 3.5, where h_{ts} is the sea level height of the base station antenna, and h_{ga} is the averaged sea level height of the geometrical profile within 3 to 15 km around the base station [11]. The height of a mobile station antenna is defined as the height measured from the ground level of the mobile station.

Mobile propagation environments are very complex, but are roughly categorized into three kinds of environments [11].

- Open area: There are few obstacles, such as high trees or buildings, in the propagation path. Roughly speaking, free spaces of about 300m to 400m in length lie between the base and mobile stations.
- Suburban area: There are some obstacles around the mobile stations, but they are not dense. Roughly speaking, it is an area of trees and low houses.
- Urban area: There are many buildings or other high structures. Roughly speaking, it is an area with high, close buildings, or a densely mixed area of buildings and high trees.

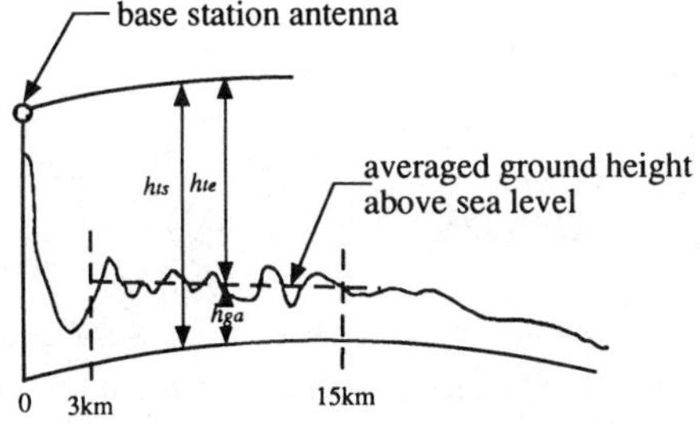

Figure 3.5 Definition of effective height of a base station antenna. (After [11].)

Propagation-Path Loss Characteristics

Although there are many aspects of mobile propagation, path loss characteristics are most closely related to the antenna design. Path loss in a mobile propagation environment is larger than that in free space due to the existence of various obstacles. Figure 3.6 shows measured field strength versus distance from the antenna with the parameter of base station antenna height. These measurements were recorded at a frequency of 453 MHz with vertical polarization and a mobile station height of 3m [11].

Frequency dependence of path loss in an urban area was measured by Okumura and is shown in Figure 3.7, where the base station antenna height and mobile station antenna height are 200m and 3m, respectively. The ordinate of Figure 3.7 shows the path loss increase from the free-space value [12]. Suburban areas have smaller path loss than urban areas. The suburban path loss is usually expressed by the decrease in attenuation as compared to the urban area value and is shown in Figure 3.8. In Figure 3.8, the path loss difference between the values in urban and

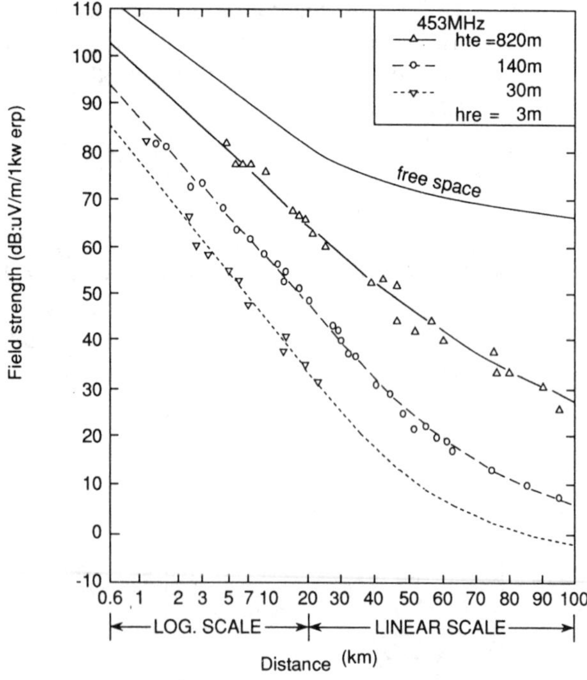

Figure 3.6 Measured mean value of field strength versus distance from a base station (f = 453 MHz urban area). (After [11].)

open areas is also shown. As previously mentioned, the path loss greatly depends on the antenna height. Figure 3.9 gives the path loss difference from the value in free space when the antenna height is varied with the parameter of transmission distance. These values are for urban areas. The suburban open-area values can be obtained using Figure 3.8.

The discussion in this section is limited to path loss characteristics because these directly relate to the design of a base station antenna. If readers are interested in further details concerning mobile propagation, they should refer to [13], as well as to Chapter 2.

3.2.2 Recent Trends in Propagation Research

Existing land mobile communications systems are rapidly advancing in both capacity and quality while moving toward digitalization and increased portability. Mobile propagation studies have contributed their support to the development of these features, which will result in the enhancement of present systems, leading to new and more diverse services, and entirely new mobile communication systems [14].

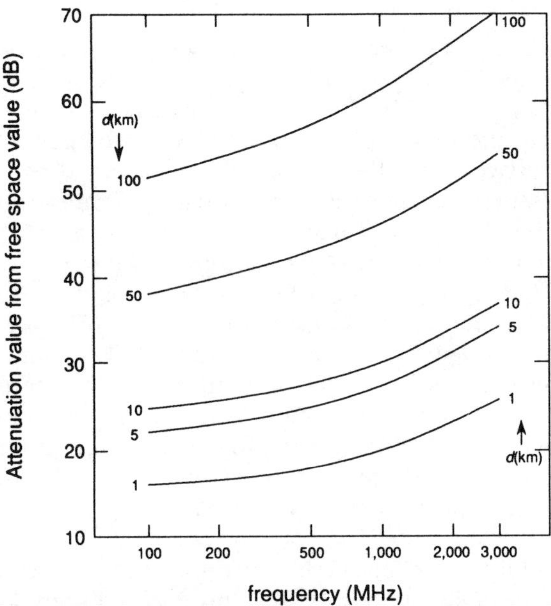

Figure 3.7 Frequency characteristic of propagation path loss at urban area (base station height = 200m, mobile station height = 3m). (After [12].)

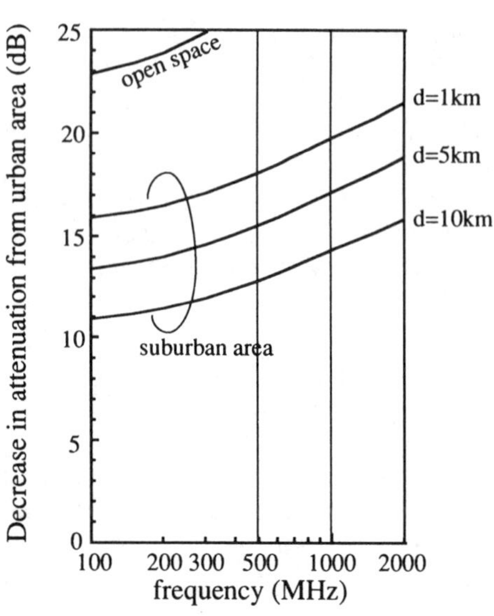

Figure 3.8 Modification of path loss of suburban and open area from urban area. (After [11].)

In order to increase the number of subscribers, methods have been developed for increasing the number of radio channels and realizing more effective frequency spectrum utilization. To increase the number of radio channels, some new radio frequency bands in the region of 1.5 to 3.0 GHz have been assigned to mobile communication services, and the propagation losses at these frequencies have already been measured under various conditions. The first system using a new frequency band was the 1.5-GHz trunk system put into service in 1990 in Japan. For more effective frequency spectrum utilization, radio waves need to be confined in small areas. This can be accomplished by means of the base station antenna beam tilting and the utilization of sector zones. The effects of these techniques were clarified and applied to the present automobile telephone system in urban areas. This system employs 2- to 5-deg beam tilting, 1.5-km radius radio cells, and 120-deg sector zones. Diversity reception, which is effective in reducing thermal noise and cochannel interference, was investigated, and the correlation coefficients between antenna branches for space and polarization diversity were obtained both theoretically and experimentally. The system uses diversity in both base and mobile stations.

In digital mobile communications, a great deal of research is currently being done in the areas of multipath power delay profiles and prediction and reduction methods of delay spread [15]. Power delay profiles have been measured in urban,

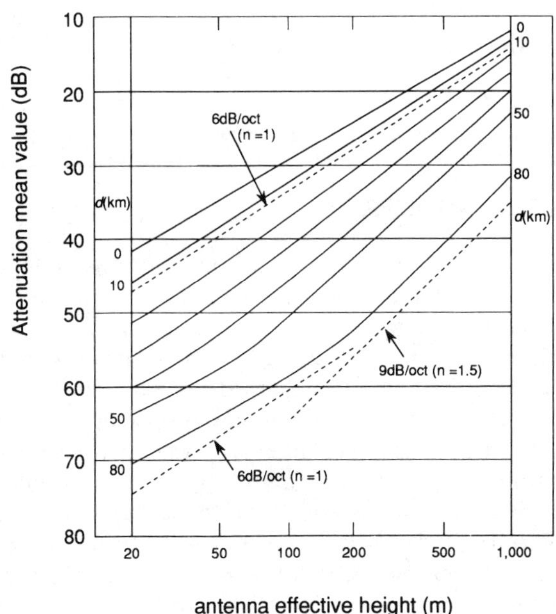

Figure 3.9 Antenna height characteristics at urban area (f = 453 MHz). (From [11], © 1986 IEICE.)

suburban, and mountainous areas at 400, 800, 1,500, and 2,300 MHz, and the delay spread was calculated using these profiles. These investigations clarified the delay spread characteristics such as distribution, dependence on distance, and various environmental conditions. Furthermore, a method of predicting delay profiles using a building-height and position database was developed. By comparison with measured delay profiles, it was shown that accurate predictions are possible. The accuracy of measurement equipment has been steadily advanced, and a simple delay-time measurement method using multiple signal classification to analyze a spectrum with high resolution and a delay-spread prediction method to measure coherent bandwidth were also proposed. Since the received-signal envelope variation with high-rate digital signal transmission is entirely different from that in analog systems because of the wide-band reception, the signal variation has been studied both theoretically and experimentally, and it was proved that the level distribution is not a Rayleigh distribution.

Concerning portable telephone systems, much research has been done on propagation loss, and delay profile measurements have been made both indoors and outdoors at frequencies over 1 GHz. The dependency of loss and delay spread values on distance over small areas, the effect of base station antennas sited next to roads, and the effect of reception diversity have been clarified.

3.3 BASE STATION ANTENNA TECHNIQUES

3.3.1 Antenna System Requirements

The role of antennas in mobile communication systems is to establish a radio transmission line between radio stations, at least one of which is moving. There are two types of mobile communication systems: one where a transmitter and receiver communicate directly, and the other where they communicate through a base station. It is the latter type that has advanced around the world in recent years. Examples include automobile telephone systems, portable telephone systems and MCA, which is a multichannel access system for private use. Automobile telephone and portable telephone systems adopt a cellular structure, and the relation between system requirements and the necessary antenna technology is illustrated in Figure 3.10. In order for the base station to communicate with the mobile stations located in the service area, radio wave energy must be radiated uniformly inside the area. Moreover, antenna gain should be as high as possible. Since the width of the service area is already specified, antenna gain cannot be increased by narrowing the beam in the horizontal plane. Therefore, it is necessary to narrow

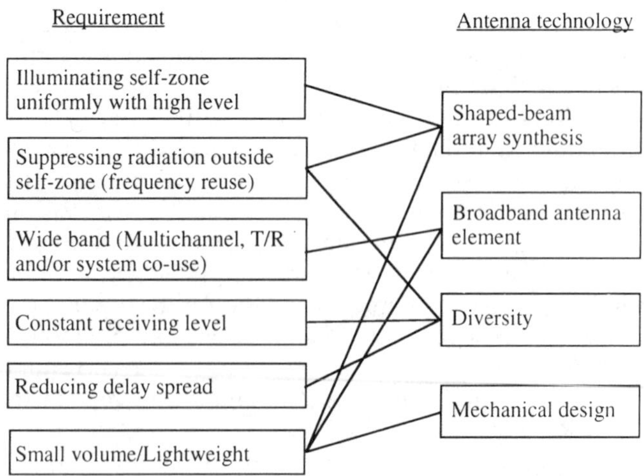

Figure 3.10 System requirement and antenna technology.

the antenna beam in the vertical plane to increase gain; a vertically arrayed linear array antenna is effective to achieve this. Normal cellular systems use antennas with a gain from 7 to 15 dBd as the base station antennas.

In order for the base station antenna to communicate with many mobile stations simultaneously, multiple channels must be handled. This requires wide-frequency characteristics and a function for branching and/or combining the channels. For example, the base stations of a Japanese cellular system using the 800-MHz band use one antenna for both transmitting and receiving. The required bandwidth of the antenna is more than 7% where the specified voltage standing wave ratio (VSWR) is less than 1.5. Moreover, if the antenna is shared by several systems (e.g., an analog land mobile telephone and a digital land mobile telephone), a wider antenna frequency bandwidth is required. To comply with Radio Regulation allocations, the frequency bandwidth for land mobile around 800 MHz runs from 810 to 960 MHz. Therefore, in order to cover this bandwidth with one antenna, a 17% bandwidth is necessary. When the antenna both transmits and receives, passive intermodulation arises and this increases interference. This problem is treated in detail in Section 3.3.5.

Due to the rapid growth of demand, the lack of communication channels has become a serious problem for metropolitan areas in the United States, Europe, and Japan, and technologies for effective frequency reuse are strongly needed. Although the cellular system has an advantage in terms of reusing frequency, its efficiency significantly depends on the radiation pattern of the base station antenna. Technologies for main-beam tilting and beam shaping have been developed and effectively contribute to frequency reuse. These technologies are described in detail in Section 3.3.3.

One of the most common features of mobile communication is that the base station and the mobile station do not fall within line of sight of each other. Moreover, the mobile station moves within a complex propagation environment. As a result, fading occurs constantly at the base and mobile stations, and the receiving levels may fluctuate by 10 dB or more. If system design takes the minimum receiving level into account, the load on the devices is excessive and system cost becomes too high.

One technology for overcoming fading is diversity reception, which has been studied since the 1960s. Its effectiveness has been confirmed both experimentally and theoretically. Reception diversity was first used commercially in the Advanced Mobile Phone System (AMPS) of the United States in 1982. It is also used in the Large-Capacity System of Japan. This technology is also described in detail in Section 3.3.4.

Various base station antenna technologies have been developed to enhance the system through the achievement of high performance and new functions. Figure 3.11 shows the historical trends of base station antenna technology.

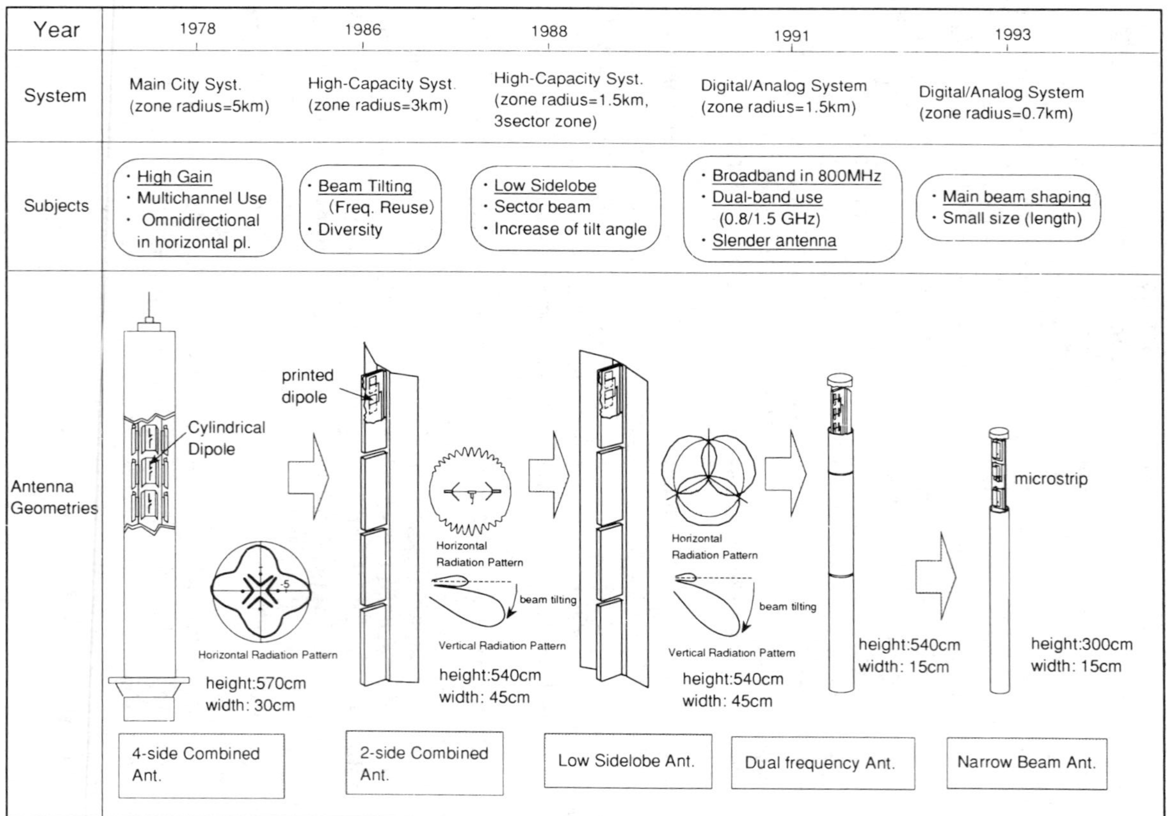

Figure 3.11 Historical trends in base station antenna technology.

3.3.2 Types of Antennas

Base station antenna configurations depend on the size and shape of the service area and the number of cells and channels. In a private mobile communication system whose service area is small, the base station antenna is as small as the vehicular-mount antenna used in automobile telephone systems. If the service area is limited within a restricted angle in the horizontal plane, a corner reflector antenna is often used. When the service area is wide, as in a pager system, maritime telephone system, or aeronautical telephone system, a linear array antenna, which has large directivity in the vertical plane, is used. It is generally uniformly excited, and examples of this type are shown in Figure 3.12. The linear array antenna is often used in cellular systems, too. Since the base station antenna of a cellular system must simultaneously handle 30 to 60 channels, it is important to realize a feeding circuit with low loss. NTT used an omnidirectional pattern antenna in its early commercial system. It consisted of four sector beam antennas combined with 3-dB hybrid circuits [16]. A multibeam antenna in which a Butler matrix is used for the feeding circuit was also introduced, but was not used commercially.

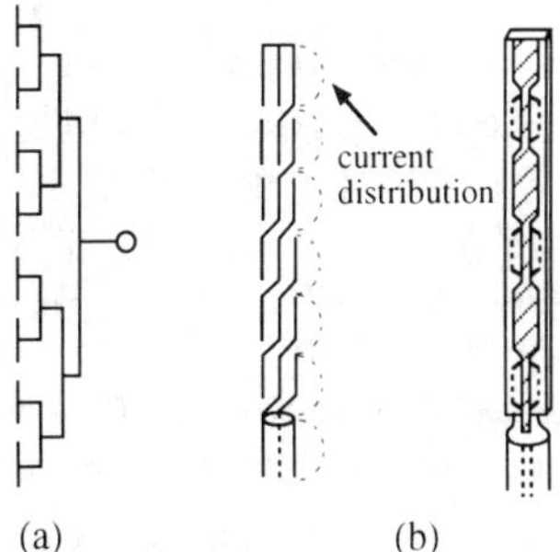

Figure 3.12 Uniformly excited array antenna: (a) parallel feed; (b) series feed.

In the early stage of cellular system development, the length of the base station antenna was determined by the required gain. To achieve higher gain, an array antenna was usually excited uniformly. However, in order to reuse frequency more effectively, cells should be subdivided. Given this situation, it is more important for the base station antenna to have a large ratio of desired-to-undesired signal strength (D/U ratio) than to have high antenna gain. Therefore, main-beam tilting, either mechanically or electrically, has been adopted throughout the world. Experiments have determined that cochannel interference can be reduced by about 10 dB, as shown in Figure 3.13, and it is recognized that beam tilting is essential for

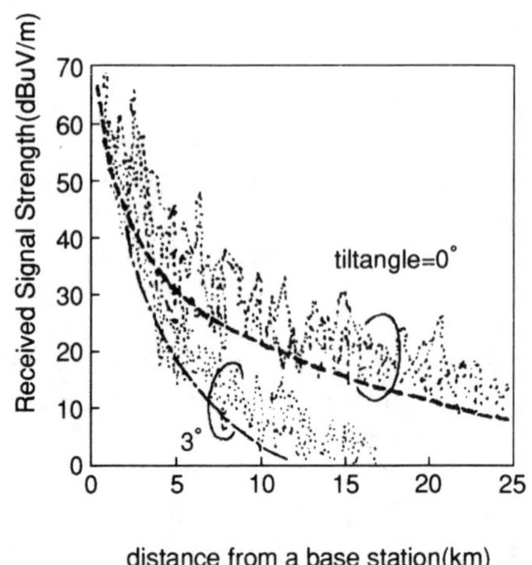

Figure 3.13 Effect of beam tilting for frequency reuse: — - — and -------- calculated value under the assumption of free-space pattern. (After [17].)

enhancing frequency reuse. Moreover, as shown in Figure 3.14, sidelobe suppression adjacent to the main beam, achieved by synthesizing appropriate array antenna patterns, is also effective to decrease the frequency reuse distance.

As for diversity antennas, space diversity, in which two antennas are separated by 5 to 10 wavelengths, is commonly used. A special diversity antenna, such as the pattern diversity antenna [17] or the polarization diversity antenna [17], have also been developed and are being used in commercial systems as base station antennas. Figure 3.15 categorizes the types of base station antennas from the viewpoints of functions and antenna characteristics.

3.3.3 Design of Shaped-Beam Antennas

The shaped-beam technique, which enhances spatial frequency reuse, is described in this section. A base station antenna in a cellular system is required to radiate energy at as low a level as possible toward the cell where the same frequency is used; conversely, it is required to illuminate the service area at as high a level as possible. There are two types of shaped-beam antennas. One shapes its radiation pattern in the horizontal plane so that a sector beam is needed, and the other shapes the pattern in the vertical plane so that a cosecant beam is desired.

Strictly speaking, main-beam tilt is not really a shaped-beam technique, but since the purpose is the same, the simplified configuration of a beam-tilting antenna and its measured performance are presented. First, the relation between frequency

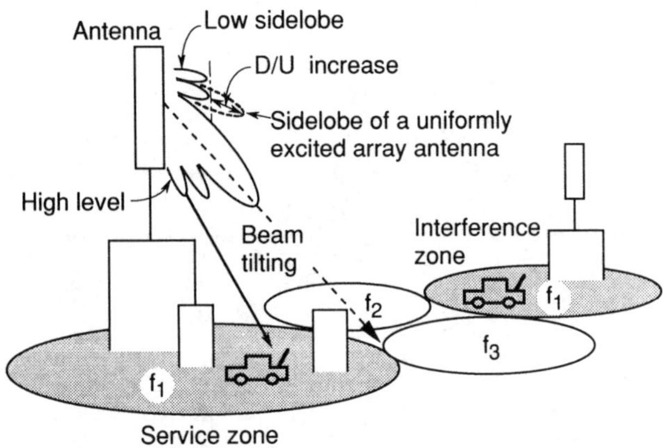

Figure 3.14 Effect of sidelobe reduction for frequency reuse.

reuse distance and antenna radiation pattern is explained. Next, the design of a sector beam antenna is described, and finally, the shaped-beam antenna design in the vertical plane for a linear array antenna is described. This section limits discussion on shaped-beam design to the specific application of cellular mobile systems. Readers interested in array antenna pattern synthesis or the numerical technique to obtain the excitation coefficient of the array, should refer to the publications listed in the reference section.

Frequency Reuse Distance

From the cellular system shown in Figure 3.16 [18], the following formula can be derived.

$$N = 1/3(D/R)^2 \tag{3.1}$$

where N is the number of cells, R is the radius of a cell, and D is the distance between the centers of adjacent cells [19]. The worst value of the CIR, expressed in decibels, appears at the edge of the cell and is given by:

$$\text{CIR} = -\alpha \cdot 10 \log\{R/(D - R)\} \tag{3.2}$$

where α is the attenuation constant of the path loss characteristic curve. In (3.2), the antenna pattern difference between the desired wave direction (θ_d, ψ_d) and the interference wave direction (θ_i, ψ_i), $(C/I)_{\text{ANT}}$, expressed in decibels, is not included.

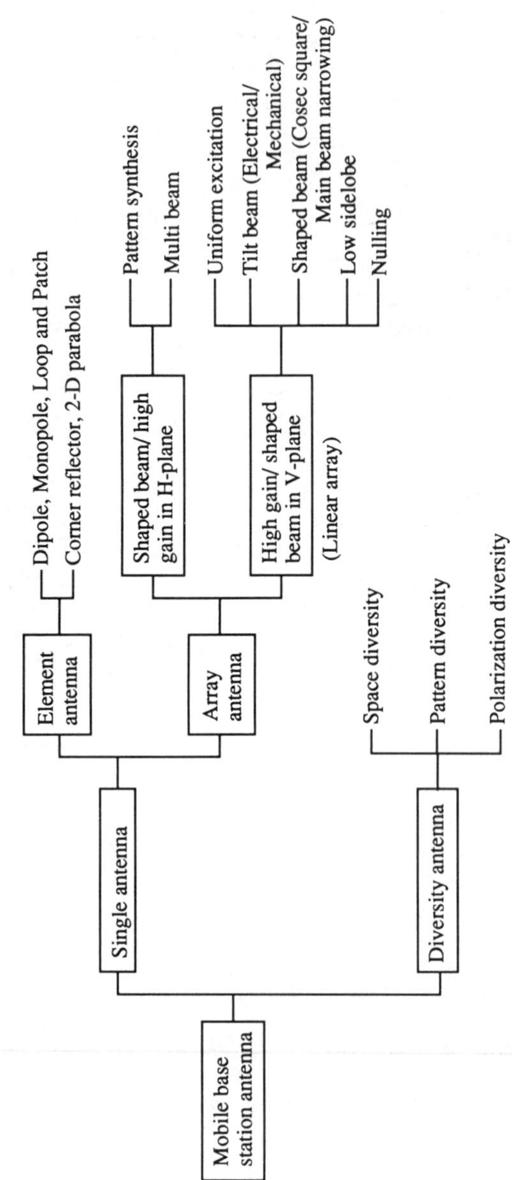

Figure 3.15 Classification of base station antennas.

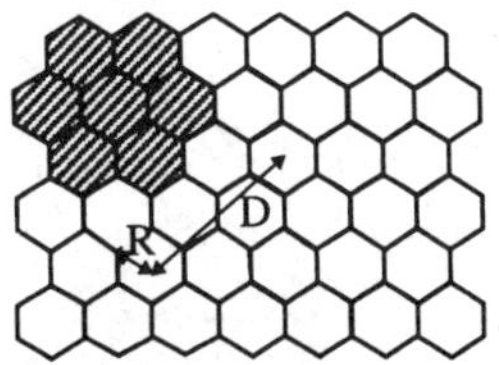

Figure 3.16 Zone configuration in cellular system (cluster number = 7). (From [18], © 1991 IEICE.)

Therefore, the total CIR of antenna pattern $P(\theta, \psi)$, expressed in decibels, is obtained by:

$$\text{CIR} = -\alpha \cdot 10 \log \{R/(D - R)\} + (C/I)_{\text{ANT}} \tag{3.3}$$

$$(C/I)_{\text{ANT}} = P(\theta_d, \psi_d)/P(\theta_i, \psi_i) \tag{3.4}$$

Figure 3.17 shows the relation between D/R and CIR when $(C/I)_{\text{ANT}}$ is varied and $\alpha = 3$ [18]. From this figure, the significance of $(C/I)_{\text{ANT}}$ can be understood.

Figure 3.18 shows a typical radiation pattern of a base station antenna in the vertical plane and radiation levels at the edge of the cell and toward the interference direction. If the two directions are separated enough compared to 50% of the main beamwidth of the antenna, interference points exist in the sidelobe region of the radiation pattern, and the sidelobe level must be decreased to increase $(C/I)_{\text{ANT}}$.

If the interference direction approaches the desired direction (N becomes small), interference lines may exist in the main beam as well as the desired direction. To increase $(C/I)_{\text{ANT}}$ in this case, it is necessary to increase the length of the antenna or to narrow the main beamwidth with beam-shaping techniques without increasing the length of the antenna.

When sector beams with the angle of θ_s are used in place of omnidirectional (circular) beams, the interference distance $(D - R)$ in (3.2) lengthens to $(D^2 + R^2 - 2DR \cos \theta_s)^{1/2}$, as shown in Figure 3.19. These are more advantageous in frequency reuse than omnidirectional beams.

Sector Beams

The horizontal pattern of a base station antenna is usually omnidirectional. However, a sector beam may effectively cover the service area if the service area is not a circle, but rather a hemicircle or a sector. Moreover, in a cellular system, the frequency reuse distance with a sector zone arrangement is shorter than that with a circular zone arrangement, as described in the previous section. Consequently,

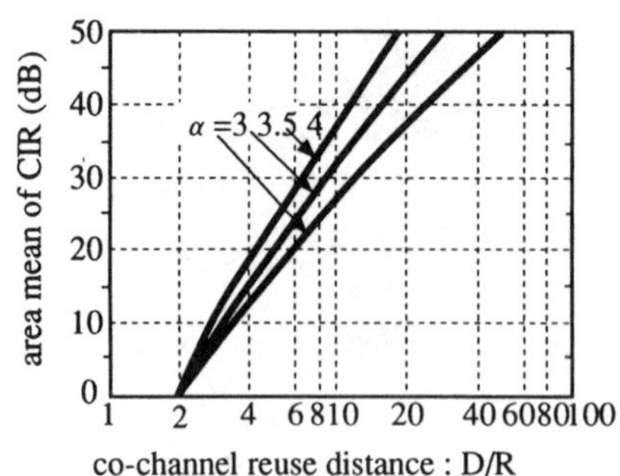

Figure 3.17 Relation between D/R and CIR. (From [18], © 1991 IEICE.)

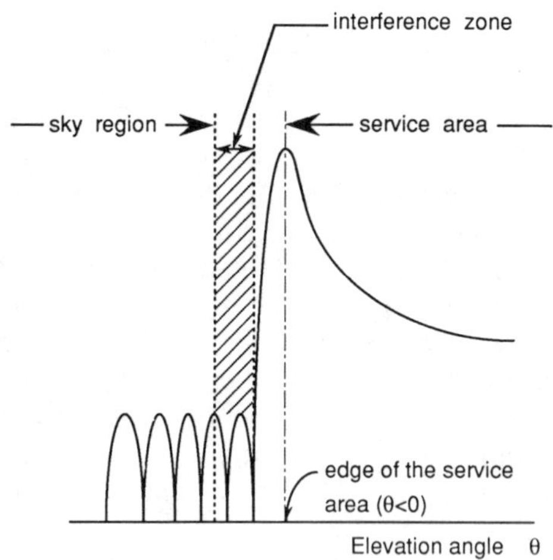

Figure 3.18 Typical radiation pattern of a base station antenna that is effective in reducing interference.

the sector zone arrangement is used in several recent automobile telephone systems. A typical sector-beam antenna is of the corner reflector type, while a two-dimensional parabolic reflector antenna fed by two primary radiators has also been developed [20].

A corner reflector antenna has the advantage that the beamwidth can be adjusted by controlling the aperture angle of the reflector. Figure 3.20 shows the fundamental configuration of the corner reflector antenna. In fact, a base station

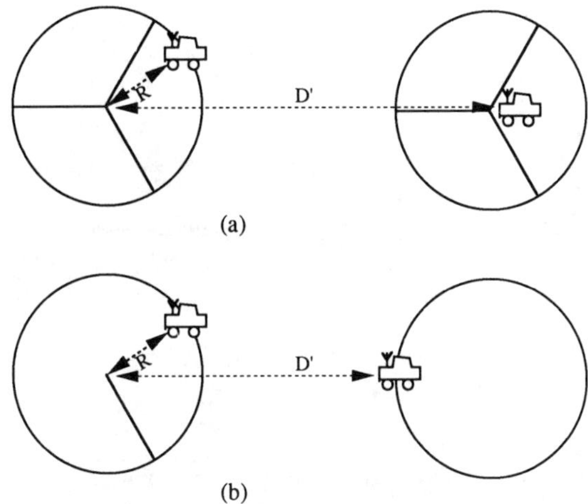

Figure 3.19 Interference distance of (a) sector zones and (b) omni zones.

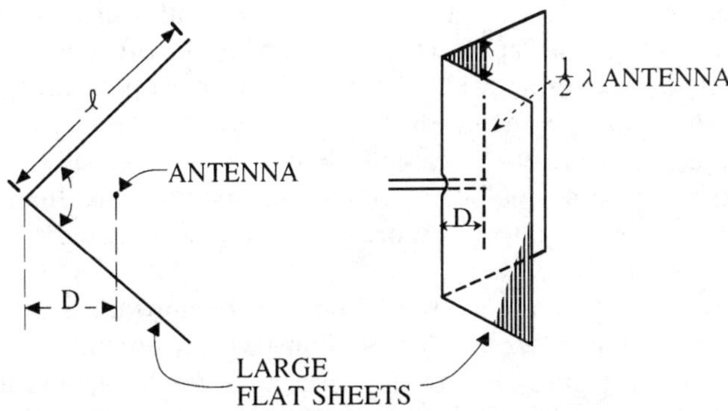

Figure 3.20 Fundamental geometry of a corner reflector antenna.

antenna is a vertically arrayed antenna consisting of the elements shown in Figure 3.20. It achieves high directivity by narrowing the main beam in the vertical plane and can radiate a shaped radiation beam by controlling the excitation coefficient, as described in the next section. In this section, discussion is confined to a corner reflector antenna with one element for easier understanding.

Figure 3.21 shows the relationship between the aperture angle of the corner reflector and a half-power beamwidth in the horizontal plane, as well as the relationship between the aperture angle and the directivity when the primary radiator

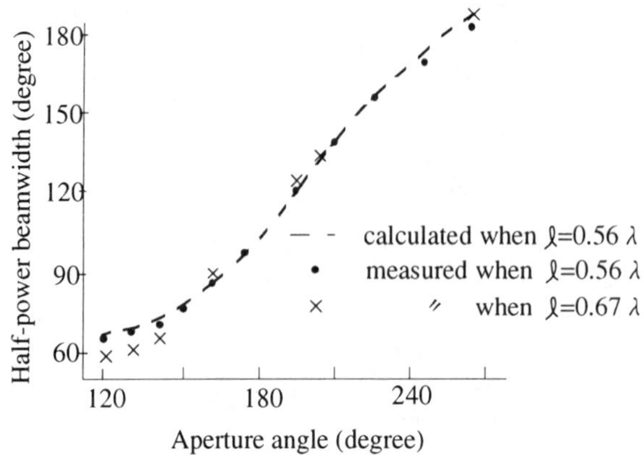

Figure 3.21 Relation between aperture angle and half-power beamwidth of a corner reflector antenna (f = 900 MHz, D = 0.28λ). (After [21].)

is a half-wavelength dipole. Sector beams with beamwidths from 60 to 180 deg can be obtained by setting the aperture angle from 60 to 270 deg [21]. The results shown in Figure 3.21 are somewhat different from those presented by Kraus [22], who assumes an infinite corner width. When the aperture angle is 180 deg, corresponding to a flat plate, the beamwidth becomes approximately 120 deg. If one would like to obtain a sector beam with a beamwidth more than 120, a corner reflector with an aperture angle of more than 180 deg (i.e., the so-called *superior angle corner reflector*) is needed.

If two sector beam antennas of 180 deg beamwidth are combined with moderate spacing (approximately larger than 6λ), an omnidirectional pattern is realized, as shown in Figure 3.22. In this case, large ripples appear in the direction of ±90 deg due to the interference of the two patterns. However, measurements confirm that these ripples disappear in a multipath environment, as shown in Figure 3.22.

Concerning a corner reflector sector beam antenna, a dual-frequency design method has been reported [23]. To realize a dual-frequency corner reflector antenna, it is necessary to prepare a dual-frequency primary radiator. This is possible with a dipole with a closely spaced parasitic element as shown in Figure 3.23. Such an antenna has the advantage of being very compact. When the primary radiator shown in Figure 3.23 is used, sector beams of equal beamwidth with frequencies f_1 and f_2 can be obtained by determining the aperture angle α, distance d between reflector and primary radiator, and the width of the corner l. Typical parameter curves are shown in Figure 3.24 for frequencies in the 900- and 1,500-MHz band. An effective parameter for controlling beamwidth is not aperture angle

α, but corner width l, and a dual-frequency sector beam with angles from 60 to 150 deg can be realized by adjusting l.

Shaped Beams in the Vertical Plane

When a limited horizontal area is to be illuminated with equal received signal level from an antenna fixed at a certain height, as shown in Figure 3.25, it is known that this can be achieved by a cosecant squared shaped-beam power pattern in the vertical plane. If the path loss is larger than that in free space ($\propto r^{-2}$), as in the case of mobile communication systems, the pth power order of a cosecant shaped-beam power pattern is necessary to achieve equal received signal level at all points in the area. However, the significance of a shaped beam in mobile communication

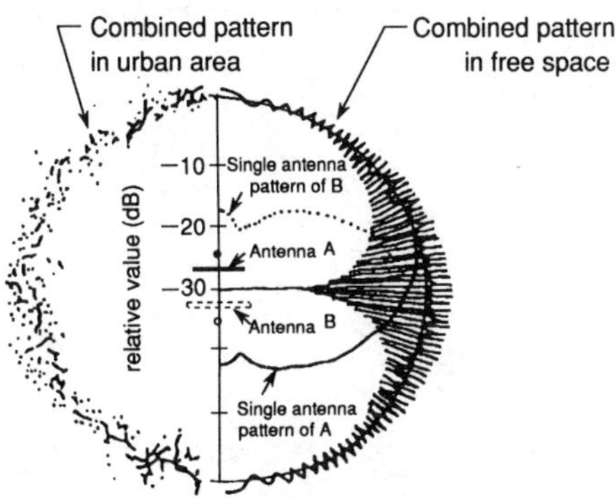

Figure 3.22 Omnidirectional pattern by combining two 180-deg sector beam antennas. (After [17].)

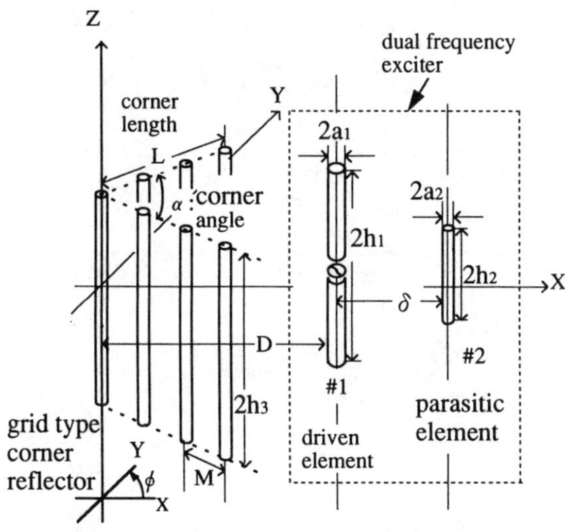

Figure 3.23 Geometry of a dual-frequency corner reflector antenna. (After [23].)

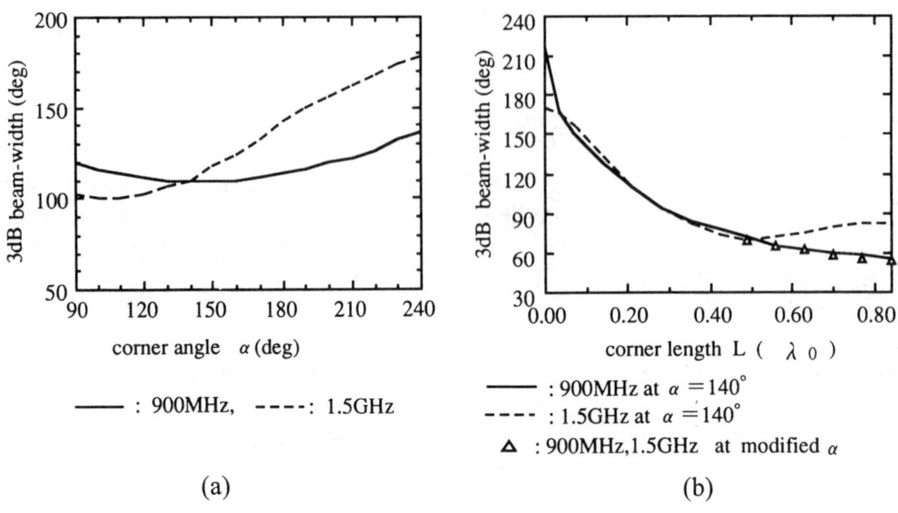

Figure 3.24 Beamwidth characteristics of a dual-frequency corner reflector antenna: (a) beamwidth versus corner angle; (b) beamwidth versus corner length. (After [23].)

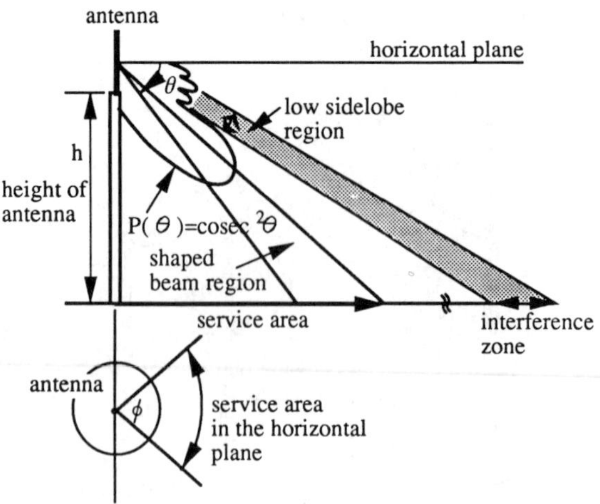

Figure 3.25 Shaped beam for illuminating the service area uniformly with low interference.

systems, especially cellular systems, may be more in suppressing the radiation toward the cell where the same frequency is reused than for illuminating the self-zone uniformly, as illustrated in Figure 3.26. If frequencies are reused, and the cells are closely packed, a part of the main beam illuminates the reuse cell. Therefore, it may be effective to tilt the main beam down to suppress the interference, even if the received signal level within the self-zone weakens. By reducing the main beamwidth while maintaining the length of the antenna, it is possible to increase $(C/I)_{ANT}$.

The remainder of this section describes major techniques and results obtained from shaped beams, with particular emphasis on practical use.

Beam Tilt. The principal idea of the beam tilt-down technique is to tilt the main beam in order to suppress the direction level toward the reuse cell and to increase $(C/I)_{ANT}$. In this case, the carrier level also decreases in the zone edge. However, the interference level decreases more than the carrier level, so the total $(C/I)_{ANT}$ increases. This is an advantage from the viewpoint of system design, and this technique is used in most cellular systems in the world. Figure 3.27 shows the comparison between antennas with and without beam tilt, which verify the effectiveness of the beam tilt technique [22]. It can be easily understood from this figure that the distance from the base station inside which the interference level exceeds

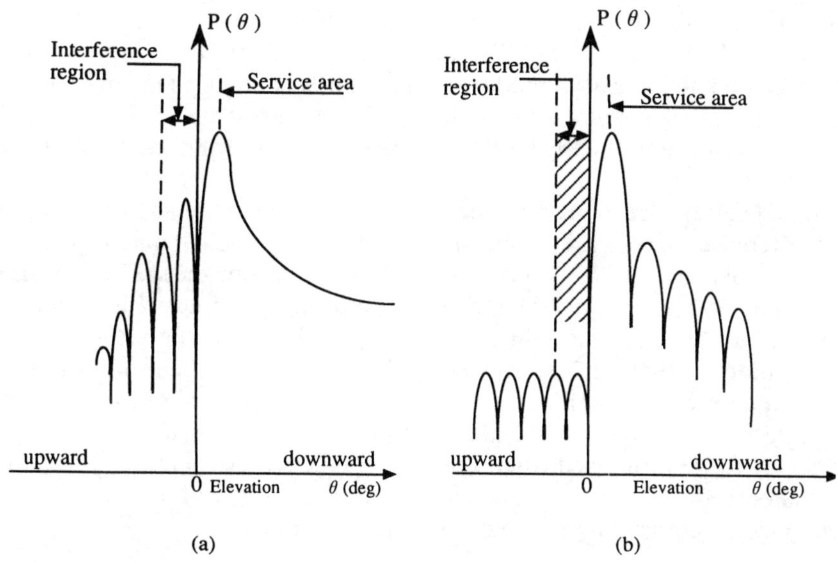

Figure 3.26 Comparison of synthesized pattern: (a) smooth cosecant beam with high sidelobe level; (b) ripple cosecant beam with low sidelobe level.

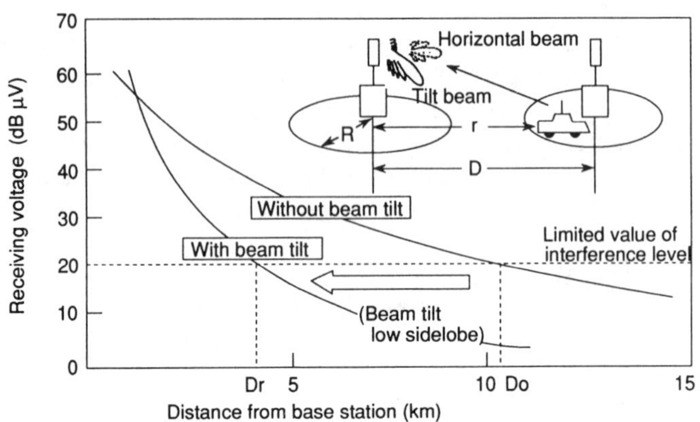

Figure 3.27 Beam tilt effect to reduce the frequency reuse distance.

the threshold level of the system can be significantly reduced. In Japan, beam tilt is achieved electrically by adjusting the excitation coefficient of the array, while in Europe it is mainly achieved mechanically.

Rippled Radiation Pattern With Low Sidelobe Level. As shown in Figure 3.22, the sharp null in the radiation pattern is nearly filled in the urban area due to multipath wave incidence. Noticing this point, Kijima derived a synthesized pattern expression with finite null depth by using a modified Schelknoff's unit circle [25] and obtained the excitation coefficients of the linear array [26]. Pattern synthesis to achieve the desired pattern shown in Figure 3.28(a) was carried out, and the results are shown in Figure 3.28(b). From this figure, it is understood that the excitation coefficients of the rippled pattern are almost symmetrical, both in amplitude and in phase, and that the difference between the maximum value and minimum value of the excitation amplitude is decreased. This is advantageous in realizing the feeding network.

If a smoothed pth power order cosecant pattern is required to illuminate the self-zone uniformly, the precise difference between the maximum value and minimum excitation coefficients becomes large, and it may be difficult to realize the desired excitation coefficients. As a result, the realized pattern is worse than the correct pattern. On the other hand, when a rippled radiation pattern, whose envelope is $\text{cosec}^p\ \theta$, is synthesized, the excitation coefficient distribution is smoother than the smoothed radiation pattern. The above result is a typical example taking propagation characteristic into account for antenna design.

Shaped Beam With a Locally Suppressed Sidelobe Level. The necessary directions for suppressing sidelobe level to reduce the interference do not exist in all sidelobe regions, but only in limited angle regions. A pattern synthesized with locally suppressed sidelobe levels was first studied in the field of radar [27]. However, since this is located closely to the main beam direction, and sidelobe level is closely related to the main beamwidth, sidelobe suppression must be carried out carefully. As shown in Figure 3.29, Kijima et al. obtained a main beam 30% narrower than that of a uniformly excited array by suppressing only several sidelobes near the main beam and setting the other sidelobes at a comparatively high level [28]. An antenna with this radiation pattern can increase the level at the zone edge by approximately 1.5 dB if the interference level is constant, a great advantage in system design.

Dual-Frequency Shaped Beams in a Vertical Plane. The previous section described a dual-frequency shaped beam in the horizontal plane, and the feasibility of creating

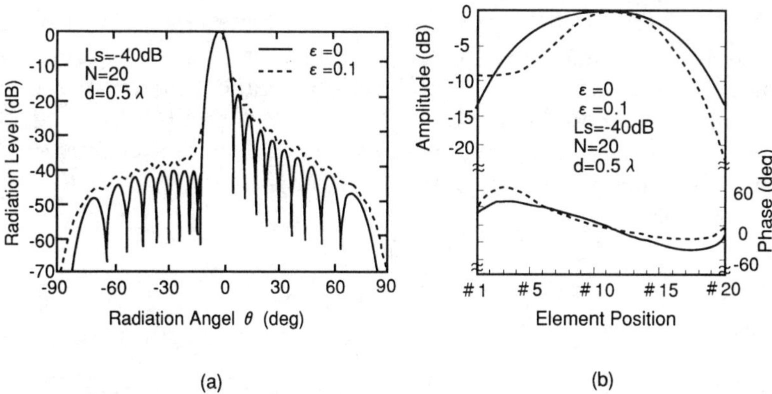

Figure 3.28 Synthesized excitation coefficient depending on the pattern shape with or without nulls: (a) synthesized pattern; (b) excitation coefficients. (After [26].)

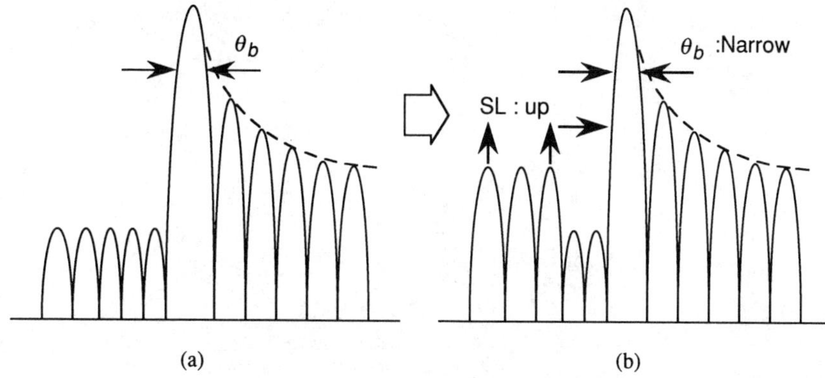

Figure 3.29 Main-beam narrowing technique by controlling sidelobe levels: (a) uniformly suppressed sidelobe pattern; (b) restricted angle low-sidelobe pattern.

a dual-frequency shaped beam in the vertical plane with a linear array antenna was presented by Kijima [29]. The difficulty of designing a shaped beam with dual-frequency operation results in a compromise of the performance at each frequency, and Kijima found that the optimum frequency for dual-frequency design tends to occur at lower frequencies.

3.3.4 Diversity Antenna Systems

Effect of Reception Diversity

The effect of reception diversity of the base station was first reported in 1965 by [30]. They showed that fading reduction could be achieved by placing two antennas approximately ten wavelengths apart in the horizontal plane. Figure 3.30 shows the cumulative probability of received level using either one isolated antenna or two antennas with a correlation coefficient of 0, 0.5, and 0.8, respectively [31]. From this it can be understood that the received level at a probability of 1% with the diversity antenna is larger by 8 dB than that with a single antenna.

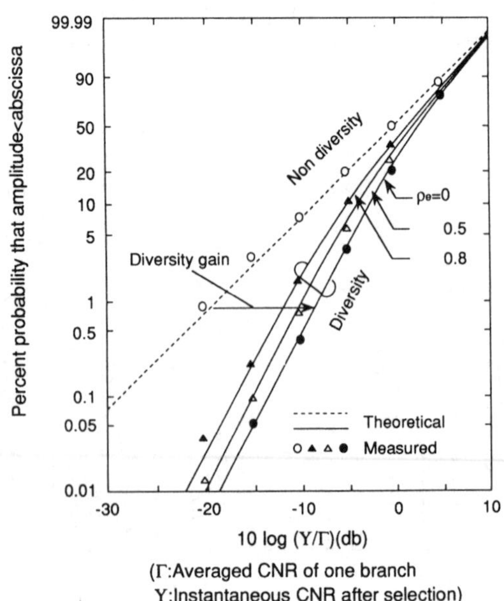

Figure 3.30 Diversity effect versus correlation coefficent; ρ_e is the correlation coefficient assumed in the theoretical prediction. (From [31], © 1991 IEICE.)

Although two or more ports are necessary to carry out reception diversity, it significantly reduces fading. As a result, the transmitting power of the mobile station is reduced, and the quality of the transmission is enhanced. This is a great advantage from the total-system point of view. Reception diversity in the base station has been in commercial use in AMPS since 1982 in the United States and in NTT's Large-Capacity System since 1985 in Japan. Since the theory of diversity reception is discussed in detail in Chapter 2 of this book as well as in other publications listed in the reference section, only those items pertaining to the design of base station antennas are described in this section.

Configuration of Base Station Diversity Antennas

Figure 3.31 shows antennas that are in actual use in commercial systems in the United States, England, and Japan [18]. In each base station antenna, sector beam antennas with a 3-dB beamwidth of 120 deg are used. Diversity antennas in the United States and Japan are arranged at angular increments of 120 deg, antennas in England are composed of six sector beam antennas with a 3-dB beamwidth of 60 deg, and at the border line of the sector zone, the port with the higher receiving level is chosen.

There are three types of diversity antenna configurations: space diversity, pattern diversity, and polarization diversity. In these three configurations, space diversity is most commonly used, and the configurations in Figure 3.31 are all space diversity designs. Since the design method and the characteristics of the space

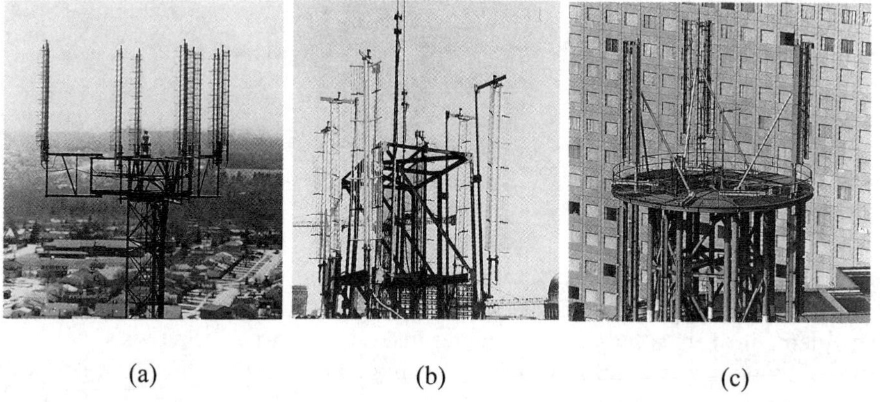

(a) (b) (c)

Figure 3.31 Practical diversity base station antennas in the world: (a) United States; (b) United Kingdom; (c) Japan. (From [18], © 1991 IEICE.)

diversity antenna are described in detail in the next section, examples of a pattern diversity antenna and a polarization diversity antenna are briefly introduced in the following.

Figure 3.32 depicts the pattern diversity antenna for an omnizone. This configuration places the two antennas shown in Figure 3.22 at 90 deg to each other [17]. Whereas 180-deg sector beam antennas for synthesizing an omnidirectional pattern are placed apart in space, the centers of the two omnidirectional antennas coincide with the center of the platform, and the spacing between the two antennas is regarded as zero. Therefore, the difference of the received power for each antenna is considered to be the cause of the difference in the radiation pattern. When the 180-deg sector beam antenna spacing is six wavelengths, as in Figure 3.32, measurements found that the correlation coefficient between two antennas is less than 0.2 in urban areas.

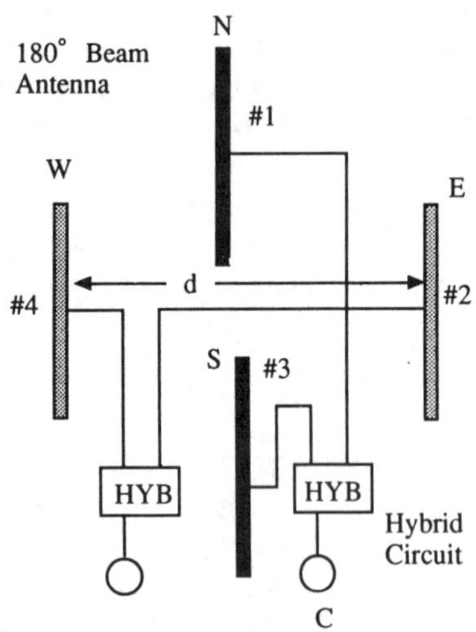

Figure 3.32 Configuration of pattern diversity antenna with omnidirectional pattern ($d = 6\lambda$ typical).

Figure 3.33 presents an example of a polarization diversity antenna developed by NTT [32]. It appears as a single antenna and has the advantage of small volume. The element of this antenna is a circular disk microstrip antenna with two feeding ports for diversity reception which are orthogonal to each other. The correlation coefficient is sufficiently low at 0.2 in urban areas. However, because the incident wave is mostly vertically polarized and the average received power in each port differs considerably, improvement in the received power is not as large as that

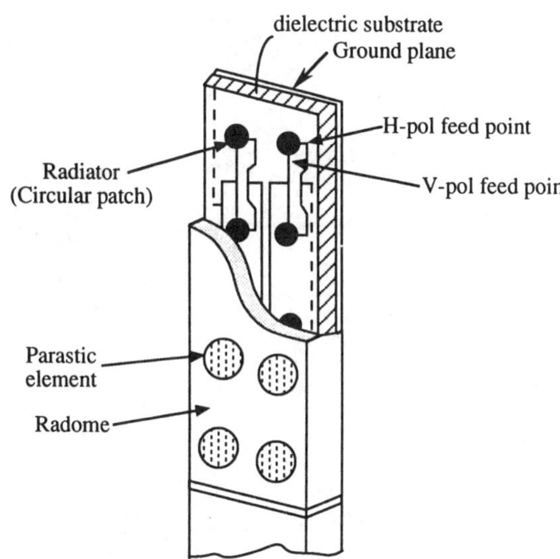

Figure 3.33 Configuration of polarization diversity antenna (60-deg sector beam). (After [32].)

achieved by other diversity schemes. Therefore, this type of diversity antenna is used only where the installation space is limited.

Design of the Base Station Diversity Antenna

When the received power level at the receiving terminals of the antennas take a Rayleigh distribution, the relationship between the correlation coefficient of the diversity terminals and carrier-to-noise ratio (CNR) level at the cumulative probability of 1% is as shown in Figure 3.34. From this figure, it can be understood that the ideal improvement of CNR is 9.5 dB (when the correlation coefficient is 0), and that the improvement of CNR remains at 8 dB even if the correlation coefficient rises to 0.6. It is usual, therefore, to design the diversity antenna using such antenna spacings or antenna radiation patterns in order to achieve a correlation coefficient of less than 0.6. In order to provide design data for the diversity antenna, the relationship between the configuration of the diversity antenna and its correlation coefficient is described below.

A Horizontally Spaced Diversity Antenna. Figures 3.35 (a,b) shows the relationship between antenna spacing and the correlation coefficient in urban and suburban areas. Antenna heights are 120m (□) and 45m (●) in (a) and 65m in (b). It can be understood from this figure that the antenna spacing should be larger than five

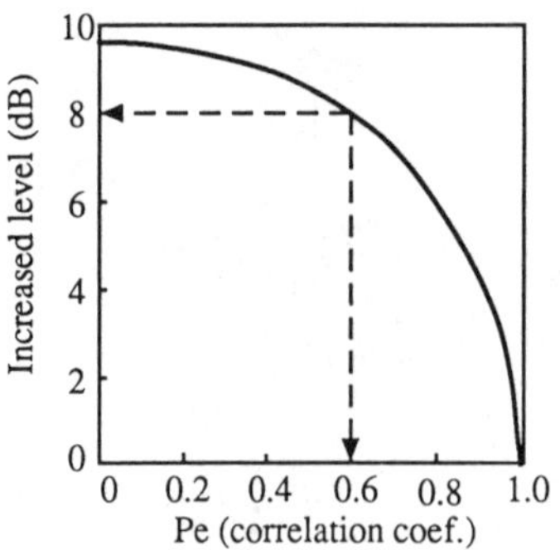

Figure 3.34 Increased level due to diversity reception with selection scheme.

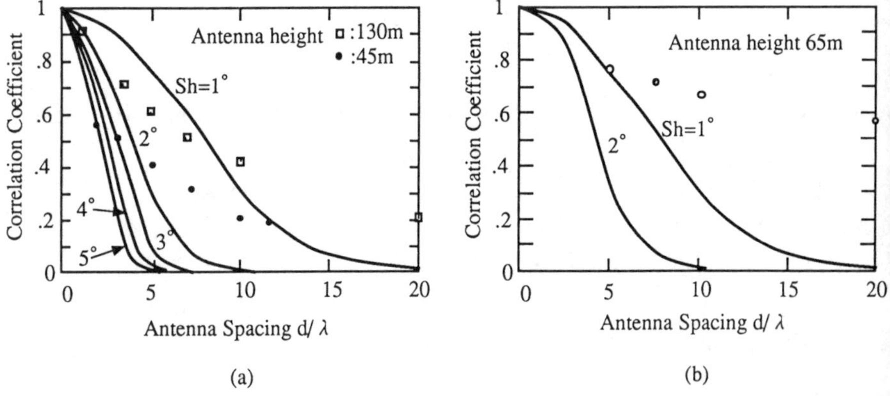

Figure 3.35 Relation between correlation coefficient and antenna spacing: (a) urban area; (b) suburban area. *Sh* is the standard deviation of the incoming move profile in the horizontal plane.

wavelengths in urban areas to achieve a correlation coefficient of less than 0.6 while more than 20 wavelengths are required in suburban areas. It is also understood that the correlation coefficient increases with the antenna height. In this figure the calculated value of the correlation coefficient, derived from the equation given in Chapter 2 is also shown, where S is the standard deviation of the amplitude distribution of the incoming wave. The value of S in the horizontal plane has not yet been obtained, while its value in the vertical plane was determined from a comparison of measured antenna gain dependent on antenna height, and calculated antenna gain using an incoming wave of Gaussian distribution with standard deviation S, as shown in the next section.

A Vertically Spaced Diversity Antenna. No vertically spaced diversity antenna has been commercially used as a base station antenna yet, although this type is used for mobile station antennas. However, a sufficiently low correlation coefficient can be obtained when the antenna height is just above the average height of surrounding buildings. This may be practical in an urban area. Figure 3.36 shows the correlation coefficient between antennas composed of 24 element arrays as a function of their center spacing [33]. In this figure, the solid line shows the calculated correlation coefficient under the assumption that the incident wave is a Gaussian distribution with a standard deviation *S*. The measured value and calculated value agree well with each other when an adequate value of *S* is assumed. Ebine found from this and the antenna gain characteristic of the antenna height obtained by the Okumura Curve that standard deviation *S* is determined by antenna height, as shown in Figure 3.37 [33]. The value of *S* gained from this curve can be used to estimate the correlation coefficient of a vertically spaced diversity antenna.

3.3.5 Intermodulation Problems in Antennas

Passive intermodulation (PIM) must be considered when an antenna is used for both transmitting and receiving. Multiple transmitting channels are intermodulated due to the nonlinear effect of the metal heterojunctions that exists between the

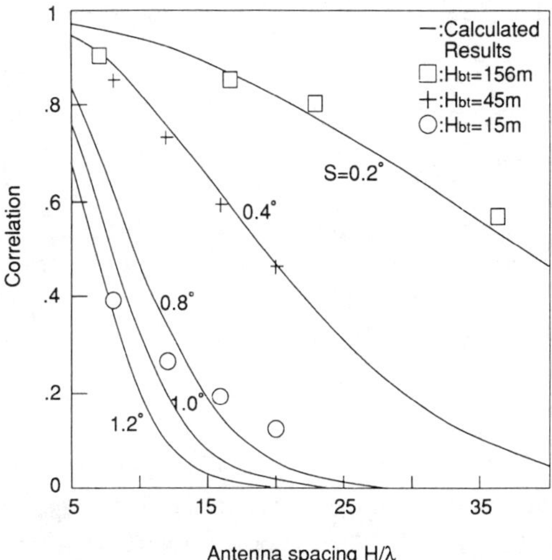

Figure 3.36 Relation between antenna spacing and correlation coefficient in vertically spaced diversity antenna. (After [33].)

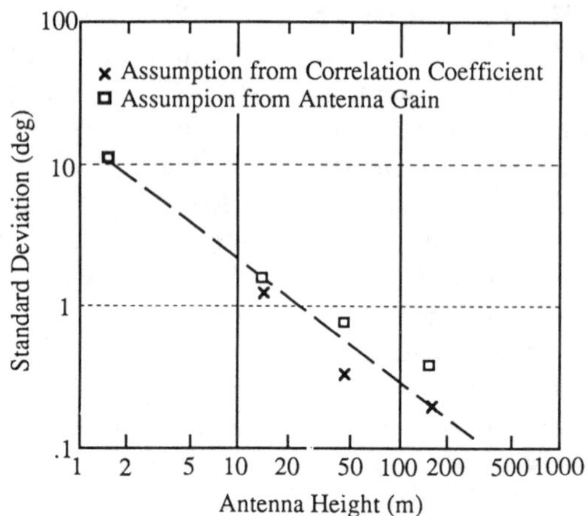

Figure 3.37 Prediction curve of standard deviation of Gaussian distribution incident wave. S is the standard deviation of incoming move profile in the vertical plane. (After [33].)

antenna radiating elements and the feed line; an interference wave with the same frequency as that of receiving wave may occur in the receiving circuit. Therefore, in order to serve simultaneously for transmitting and receiving, the antenna should be designed and manufactured in such a way that the intermodulation power is less than the specified value (usually -10 dBμW in the case of an automobile telephone system) [18].

Principle of Passive Intermodulation

Since the theory of the occurrence mechanism for PIM has already been published [34], only minimum material need be presented in this section. Assuming that the frequencies of the two transmitting waves are f_i and f_j, frequency f_{IM} of the intermodulated wave of the $(m + n)$th order is given by the following equation.

$$f_{IM} = mf_i \pm nf_j \tag{3.5}$$

where m, n are positive odd integers. The possibility of f_{IM} being an interference wave on the receiving frequency band depends on the frequency distance between transmitting and receiving and the value of $(m + n)$. For example, in frequencies used in the Japanese analog automobile telephone system, the lowest order intermodulation that can be considered as interference is the seventh order, as shown in Figure 3.38, and the interference power is limited to -10 dBμW, which is below

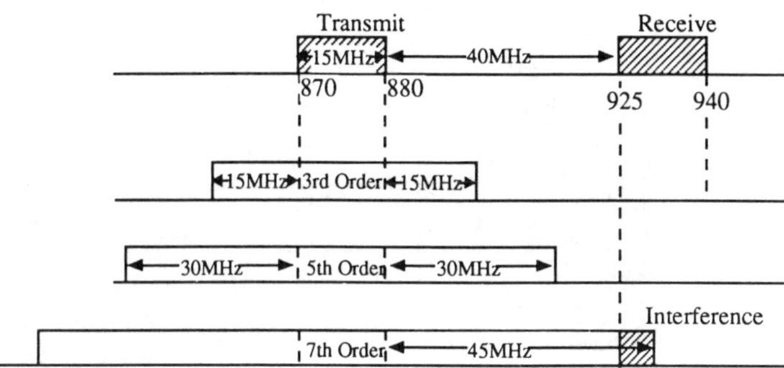

Figure 3.38 Higher order intermodulation and interference.

the threshold level for transmitting and receiving antenna co-use, even if the total input power to the antenna is 200W.

Studies of the relationship between the order of PIM and the generated power have found that power is approximately given by 10 dB/$(m + n)$. Therefore, if the frequency separation between the transmitting wave and the receiving wave is narrow, fifth or third order PIM may become interference with levels 20 or 40 dB higher than that of the seventh order, and may exceed the threshold level permitted for antenna co-use.

Occurrence Points and the Suppressing Technique of PIM

Generated power of PIM depends on the kinds of metal in contact and their configuration. Figure 3.39 shows a typical configuration of a co-use antenna for transmitting and receiving in a frequency band using a branching filter. Major points at which PIM is generated are also shown. Table 3.1 shows techniques used to suppress PIM. Although the fundamental theory of PIM has been established, it cannot be said that the relationship between contact point structure and PIM generation has been investigated quantitatively. On the other hand, due to the rapid growth of demand in mobile communications, the number of base stations will continuously increase. Co-use antennas offer great advantage to the system, especially from the standpoint of system economy, and it is expected that co-use antennas will be used more often. Thus, it is becoming more urgent to investigate PIM, especially the suppression techniques.

3.4 MOBILE STATION ANTENNA TECHNIQUES

Mobile antennas should be designed to reduce the required transmitting power of mobile radio equipment while ensuring that the required service qualities (i.e.,

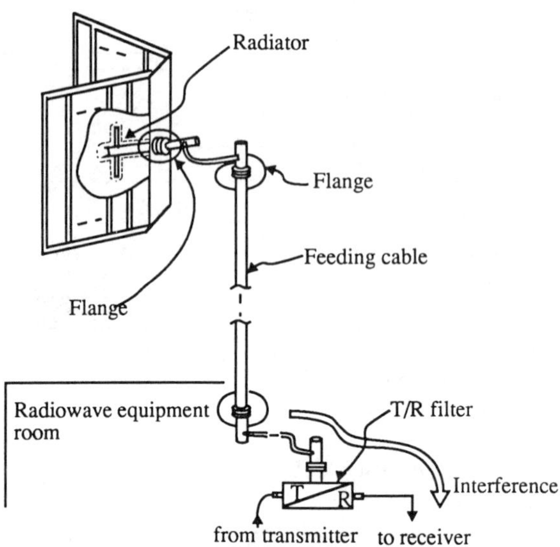

Figure 3.39 Generating points of passive intermodulation (PIM).

Table 3.1
Suppression Methods of Passive Intermodulation

Items	*Methods*
Antenna:	
Radiator	Printed instead of wire element
Connection between radiator and flange	Increase contact area between flange and printed element; tightly fixed with silver material
Welding	Put paste to surrounding contact-point area
Cable:	
Heterojunction of metal contact	Prohibit using contact of aluminum and nickel
Rust	Use grooves in case of manufacturing and integration to prevent oxide generation

speech quality, outage probability within a whole cell (service coverage), and so on) are satisfied. The optimum design realizes small but economical mobile telephone units. This section describes the vehicular antenna characteristics of cellular mobile systems presently in commercial service and the antenna technologies used in their design. The requirements of and technologies for portable telephone equipment are also described.

3.4.1 System Requirements and Antenna Performance

Vehicular Antennas

The first generation of cellular mobile radio systems were analog mobile radio systems that used the 400-MHz band or 800- to 900-MHz band. They are still in commercial use in the United States, the United Kingdom, the Nordic nations, Japan, and other countries, and the number of subscribers has been rapidly increasing. Table 3.2 shows the system specifications of several cellular mobile radio systems using the 800- to 900-MHz frequency band [35]. The characteristics of vehicular antennas for three of these systems are shown in Table 3.3 [36–38]. The requirements for vehicular antennas include the operating frequency, bandwidth, directivity, pattern characteristics, polarization, and the support of diversity recep-

Table 3.2
General Land Mobile Communication Systems' Specifications

Feature	*North American*	*Japanese MCS-L1*	*Japanese MCS-L2*	*NMT-900*	*United Kingdom (TACS)*
Transmit frequency bands (MHz):					
Base station	870–890	870–885	870–885	935–960	935–950
Mobile station	825–845	925–940	925–940	890–915	890–905
Duplex separation (MHz)	45	55	55	45	45
Channel spacing (kHz)	30	25	12.5	12.5	25
Number of duplex channels	666*	600	1,200	1,999	666*
Maximum base station ERP (W)	100†	50	20	100	100
Nominal mobile station transmitter power (W)	3	5	1	Mobile: 6 Hand-held: 1‡	Class 1 mobile: 7
Typical cell radius (km)	2–20	3◇, 10§	3◇, 10§	0.5–20	2–20

Source: [35].
*333 channels in each of two subbands, including 21 signaling channels.
†Exceptions may be allowed, depending on the circumstances.
‡With autonomous power control.
◇Urban area.
§Suburban area.

Table 3.3
Vehicular Antenna Characteristics of Cellular Mobile Systems

	Japanese (NTT)	*United States (AMPS)*	*United Kingdom (TACS)*
Antenna type	Sleeve	Wire	Wire
Frequency range (MHz)	865–945	825–890	890–950
Bandwidth for 2:1 VSWR (MHz)	80*	65†	60*
Horizontal pattern (deg)	360	360	360
Gain (dBd)	0	3	3
Diversity reception	Possible	Not possible	Not possible
Antenna size			
Diameter (mm)	13	10	10
Height (mm)	560	330–610	310–570

*Transmit and receive.
†Transmit and receive; maximum VSWR is 1.9:1.

tion. The requirements of operating frequency, bandwidth, and directivity of antennas are different for the system of each country. In the Japanese system (NTT system), the required directivity is 0 dB relative to a reference half-wavelength dipole antenna. The corresponding value is 3 dB in the U.S. system (AMPS system) and the U.K. system (TACS system).

Since the typical mobile station moves randomly in a radio zone, an omnidirectional azimuth pattern is required for mobile antennas. Particularly in suburban areas, the base station and the mobile station are in line of sight, and, hence, if the antenna pattern were not omnidirectional, the received signal level would vary considerably. Therefore, omnidirectional antennas are usually required in a mobile radio system. Experimental results from suburban and urban environments [39–42] indicate that angular distributions of arriving waves extending from 0 to 50 deg in elevation are quite common. Since the mean elevation angle of their angular distribution depends on the environment, a mobile antenna whose elevation angle of maximum radiation could be aligned to match the mean elevation angle of the incident wave distribution seems to be effective for ensuring the maximum received power. At present, however, practical concerns require the radiation pattern to be maximum in the horizontal direction. Vertical polarization is usually used in most mobile systems because it makes broadband omnidirectional antennas very easy to develop (e.g., whip and dipole antennas).

It is commonly known that diversity reception is very effective for mitigating multipath fading [43,44]. However, in most cellular mobile systems, diversity reception is an optional technology in the system specifications. NTT's system adopts diversity reception for both the base and mobile stations, and postdetection selection diversity is used. To develop the diversity branches, space diversity has been adopted, and the correlation between the antenna branches has been held to less than 0.6.

Antennas for Portable Telephones

Most analog portable telephone services are provided by using the radio channels of the vehicular telephone system; hence, the requirements for operating frequency and bandwidth are the same as those of vehicular antennas. However, because of limited battery capacity, the transmitting power of portable telephone units must be less than that of vehicular telephone units. Furthermore, antenna gain is generally less than that possible with vehicular antennas, because only small antennas can be used and there is gain degradation due to the proximity of the human body. Under these conditions, the antenna's requirements for portable telephone units are to develop the highest possible gain over the required bandwidth. The effective gain means the mean effective gain in a multipath mobile radio environment, and is the same as the mean effective gain described in Chapter 2. Improving the antenna effective gain is very effective in reducing the size and weight of portable telephone units.

There are some distinctive features of antennas to be mounted on portable telephone units. The first feature is that the polarization direction and radiation pattern of the antennas are not fixed, since the portable telephone unit is randomly directed when used. The second feature is that their radiation pattern and radiation efficiency vary considerably when close to the human body. This means that it is impractical to require the antennas of portable units to have vertical polarization and omnidirectional radiation pattern. In the design of portable telephone antennas, it is important to try to optimize the effective gain. Theoretical analyses must consider variation of the radiation pattern due to the effect of the unit's housing, the degradation of the antenna's radiation efficiency due to the proximity effect of the human body, and the variation of pattern and polarization due to human operation. It is necessary to determine the following characteristics when designing an antenna: (1) the effective gain in the multipath propagation environment; (2) the effective gain when the portable unit is operated in the speaking, carrying, and dialing positions.

It should be noted that an evaluation based on the free-space radiation pattern is not accurate enough to assess the practical performance of an antenna because the effect of the human body is not considered. Finally, it also should be noted that the influence of the RF power radiated from the antenna on the human body is very important for designing antennas for portable telephone units.

3.4.2 Types of Antennas

Dipole Antenna

The basic dipole antenna structure is a center-fed linear cylindrical antenna as shown in Figure 3.40, the same basic structure used by most linear antennas. The current distribution on the cylindrical antennas has been calculated by Hallen et al., and the input impedance characteristics have also been calculated [45]. Impedance spirals based on Hallen's data are represented in Figure 3.41 for a center-fed cylindrical antenna whose ratio of total length to diameter (l/a) is 60 and 2,000. When the length l approximately equals half a wavelength ($\lambda/2$), the impedance becomes a pure real value, and impedance matching can be done using a transmission line with a characteristic impedance of 73.13Ω. Figure 3.41 also indicates that the variation in impedance with frequency of the thicker antenna is much less than that of the thinner antenna, and that the thicker antenna supports more broadband operation.

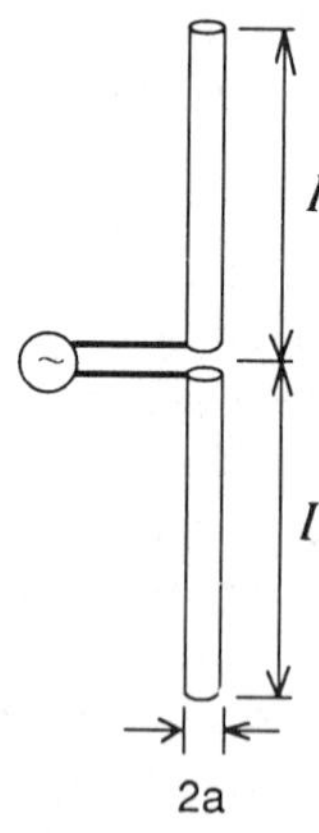

Figure 3.40 Symmetrical center-fed cylindrical antenna.

The radiation pattern of a center-fed cylindrical antenna generally depends on its length and thickness. However, for thin antennas, whose diameter is less than, say, $\lambda/100$, the sinusoidal current distribution is a good approximation of the actual pattern. Thus, if the current distribution is assumed as follows.

$$I(z) = I_0 \sin k(I - |z|) \qquad |z| \leqq I \tag{3.6}$$

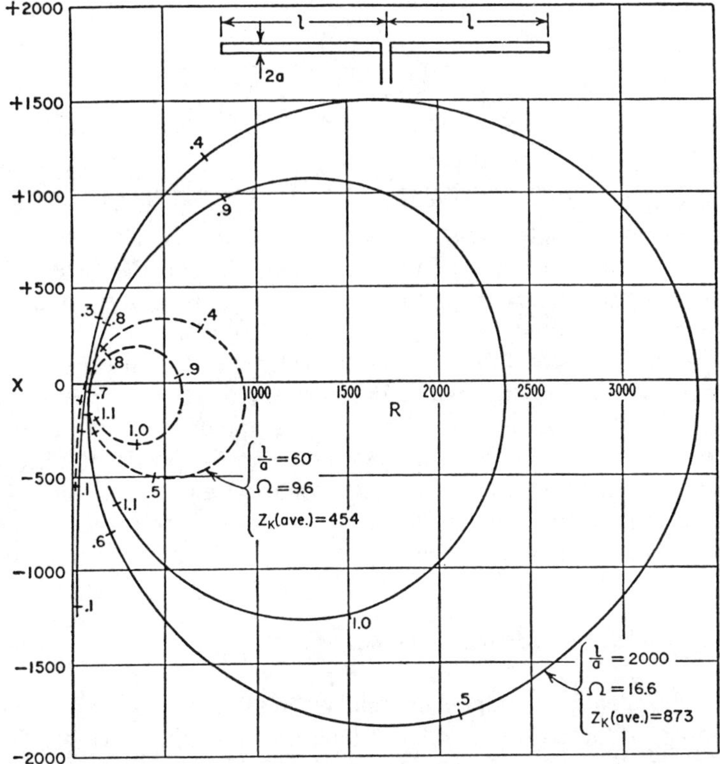

Figure 3.41 Calculated input impedance ($R + jX$) in ohms for cylindrical center-fed antennas with ratios of total length to diameter ($2l/2a$) of 60 and 2,000. (After Hallen. From [45], © 1950 McGraw-Hill.)

where I_0 is the peak value in time of the current, k is a free-space wave number, and l is the half length of the dipole element. The radiation field, expressed in a spherical coordinate system, is given by equations (3.7) and (3.8) [46].

$$E_\theta = j60I_0 \cdot \frac{\exp(-jkR)}{R} \cdot \frac{\cos(kl \cos \Theta) - \cos(kl)}{\sin \Theta} \tag{3.7}$$

$$H_\phi = \frac{E_\theta}{120\pi} \tag{3.8}$$

Particularly, for a half-wavelength antenna (i.e., $l = \lambda/4$), it is given by

$$E_\theta = \text{j}60I_0 \cdot \frac{\exp(-\text{j}kR)}{R} \cdot \frac{\cos[(\pi/2)\cos\Theta)]}{\sin\Theta} \tag{3.9}$$

Figure 3.42 shows several radiation patterns of center-fed thin dipoles of various lengths when a sinusoidal current distribution is assumed. The directivity with respect to an isotropic antenna is also shown. The directivity of a half-wavelength dipole antenna is 2.15 dBi. The dBi abbreviation indicates the decibel referred to the isotropic antenna gain. In the mobile communications field, the half-wavelength dipole antenna is commonly used as a reference antenna when the effective gain of an antenna in a multipath field is to be evaluated. The effective gain is often given by the abbreviation dBd, which indicates the decibel referred to the gain of the half-wavelength dipole antenna. Figure 3.42 also indicates that a dipole antenna with length l of 5/8 wavelength, is quite directive in the horizontal direction.

However, it should be noted that the principal dipole antenna is practically never used in vertical polarization applications, due to the problem of mounting and feeding while maintaining symmetry.

The Quarter-Wavelength Monopole (Whip) Antenna

The quarter-wavelength monopole antenna is the fundamental mobile antenna and has the simplest structure. Figure 3.43 shows this antenna structure. This type of

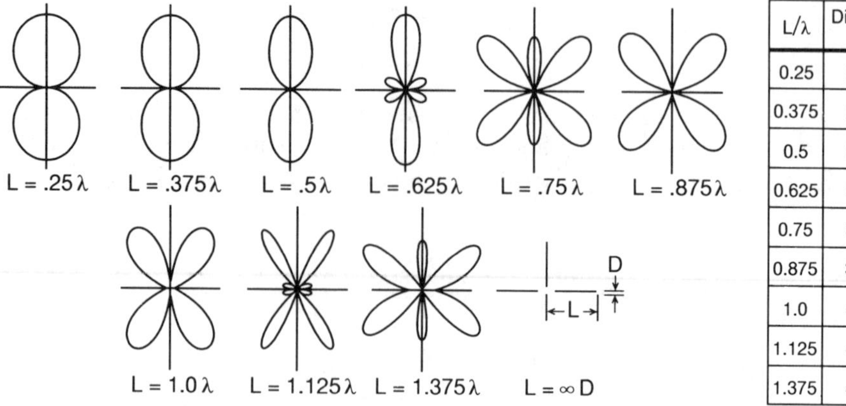

L/λ	Directivity (dBi)
0.25	2.151
0.375	2.746
0.5	3.822
0.625	5.162
0.75	3.474
0.875	3.748
1.0	4.026
1.125	4.867
1.375	4.914

Figure 3.42 Radiation patterns of center-driven dipoles assuming sinusoidal current distribution. (From [51], © 1961 McGraw-Hill.)

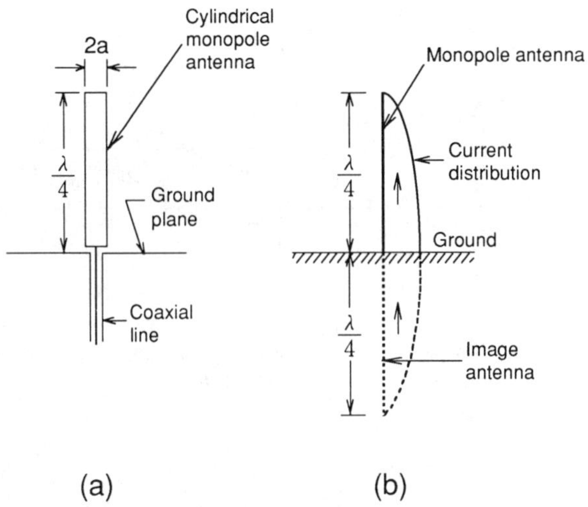

Figure 3.43 1/4-wavelength monopole antenna: (a) typical structure; (b) antenna image and current distribution.

antenna normally employs a flexible antenna element, and thus it is also called the *quarter-wavelength whip antenna*. The radiation element is mounted on a ground plane, which is, in practice, the roof of a car. If the ground plane were infinitely large and were a perfect conductor, the radiation pattern and bandwidth characteristics of the antenna would be the same as a half-wavelength dipole antenna, due to the effect of an image of the element formed by the ground plane. However, the input impedance is only half that of a half-wavelength dipole antenna (i.e., about 36Ω). Theoretically, the directivity is 3 dB larger than that of a half-wavelength dipole antenna, because the radiation power is radiated only to the upper half space of the ground plane. However, because of the size and the conductive loss of real-world ground planes, the practical directivity cannot be improved by this amount. Figure 3.44 shows the radiation patterns of a quarter-wavelength monopole antenna mounted on a circular ground plane of radius r [47]. In actual usage, the size of the ground plane is finite, and the direction of maximum radiation tilts somewhat upwards from the horizontal plane. Therefore, the effective gain of this antenna is usually lower than that of a half-wavelength dipole antenna.

Sleeve Antenna

Figure 3.45 shows the structure of a sleeve antenna. The center conductor of a coaxial cable is connected to an element whose length is one-quarter wavelength,

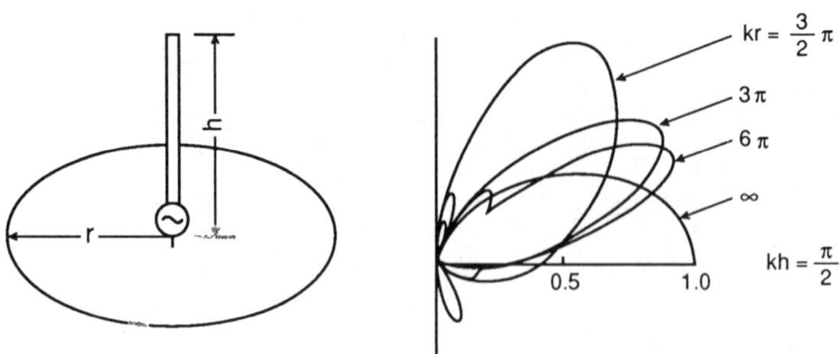

Figure 3.44 Normalized radiation patterns of 1/4-wavelength monopole antenna on circular ground plane of 0.75λ, 1.5λ, 3λ, ∞ (λ: wavelength). (From [47], © 1973 IEEE.)

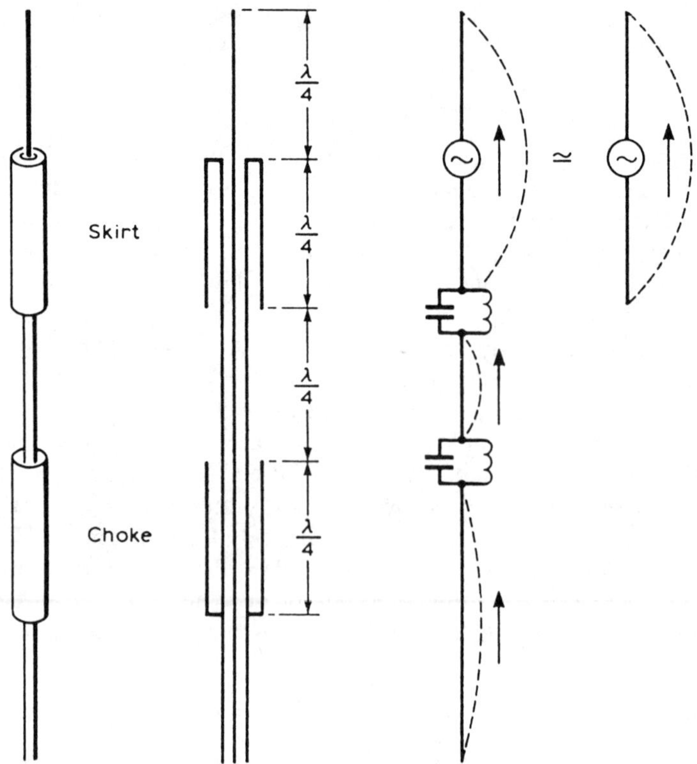

Figure 3.45 Coaxial center-fed half-wave dipole with choke. (From [51], © 1961 McGraw-Hill.)

and the outer conductor of the coaxial cable is connected to a cylindrical skirt whose length is one-quarter wavelength. This coaxial cylindrical skirt behaves as a quarter-wavelength choke, and thus most of the antenna current does not leak into the outer surface of the coaxial cable. Adding chokes on the lower part of the coaxial cable is usually done to improve the radiation pattern by further suppressing current leakage from the skirt. As a result, this antenna has almost the same characteristics as a half-wavelength dipole antenna. This antenna does not require a ground plane, so that the gain degradation due to the mounting location is less than that experienced with quarter-wavelength monopole antennas. Furthermore, the feeding structure is more suitable for developing practical antenna configurations for vehicle mounting than the center-fed half-wavelength dipole antenna; hence, this antenna, the trunk-lid-type antenna, is often used for 800-MHz band mobile radio systems in Japan. This type of antenna has also been selected for other portable telephone systems (i.e., Dyna-TAC, Motorola).

The 5/8-Wavelength Monopole (Whip) Antenna

As shown in Figure 3.42, 5/8-wavelength dipole antennas have comparatively large directivity in the horizontal direction. Thus, the 5/8-wavelength monopole antenna is often used as a *high-gain antenna* in combination with a ground plane. The antenna element is usually a flexible material, and it is thus also called the *5/8-wavelength whip antenna*. The reason why its gain is higher than that of a quarter-wavelength monopole antenna is that its antenna aperture is larger, and the rise of radiation beam due to the finite ground plane is less than that of the quarter-wavelength monopole antenna. The typical antenna structure is shown in Figure 3.46. The series coil placed between the 5/8-wavelength radiator element and the ground plane has an effective length of 1/8 wavelength, and the antenna's input impedance is roughly 50Ω.

Collinear Antenna

A fundamental way of realizing a high-gain antenna is to form a collinear array in which several collinear radiator elements are fed in phase. Figure 3.47 shows the typical structures of a collinear antenna. In Figure 3.47(a), the length of radiator elements is usually a half wavelength, and loading coils for achieving in-phase feeding are inserted between the radiator elements. This type of antenna is also termed the *loading antenna* and is often adopted as a vehicular antenna [37]. Figure 3.48 shows the structure of a vehicular antenna developed by using a quarter-wavelength radiator element for the lower element and a half-wavelength radiator element for the upper element. Since the lower element acts as a half-wavelength dipole antenna, due to the image formed by the ground plane, this antenna acts

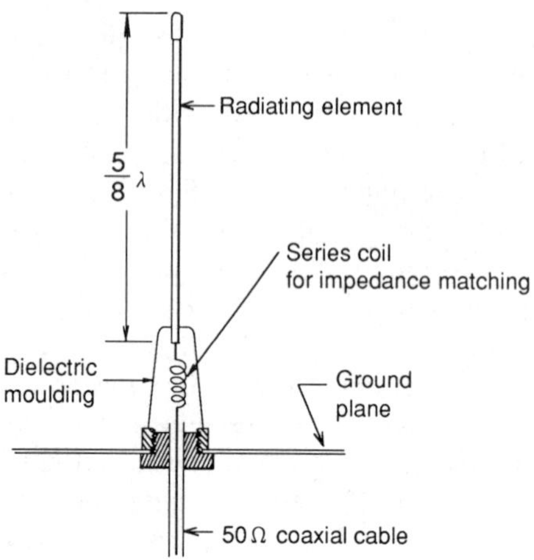

Figure 3.46 Typical structure of 5/8-wavelength whip antenna.

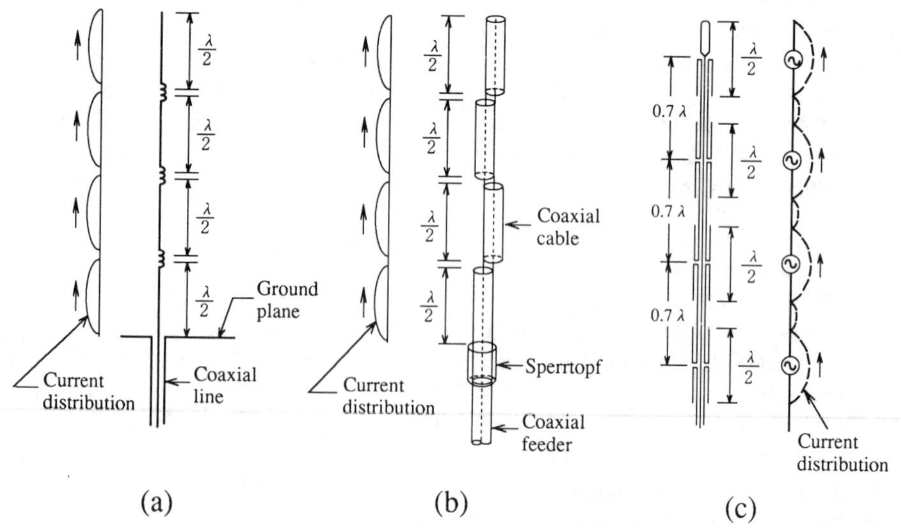

Figure 3.47 Structures of collinear dipole array: (a) Franklin antenna; (b) series-fed transposed coaxial collinear antenna; (c) series-fed symmetrical coaxial collinear antenna. (From [49], © 1991 IEEE, and [51], © 1961 McGraw-Hill.)

as a two-element collinear array antenna, where each element is a half-wavelength dipole antenna. Hence, the directional gain of this antenna is theoretically 3 dB higher than that of a simple half-wavelength dipole antenna. In practice, the length of radiator elements is selected in the range from 1/4 to 5/8 wavelength. This type of antenna also is used as an on-glass antenna [48]. The antennas shown in Figures 3.47(b,c) are commonly known as *coaxial collinear antennas* [49–51]. These antennas are used as vehicular antennas for amateur radio use. Since, in the case of each antenna shown in Figure 3.47(b or c), the in-phase feeding depends on the length of radiator elements and the distance between radiator elements, the bandwidth is generally narrow. Thus, these antennas are not suitable for a broadband mobile radio communications system: however, they can be used in a narrow-band system, such as a time division multiple access/time division duplex (TDMA/TDD) digital mobile radio communication system.

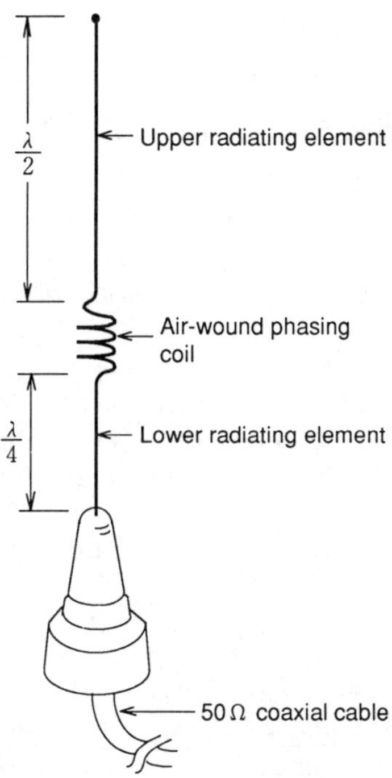

Figure 3.48 High-grain whip antenna for the AMPS system.

Microstrip antennas and IFAs are well known as typical low-profile antennas. The basic structure of an IFA is shown in Figure 3.49(a). This antenna can be configured by bending a quarter-wavelength monopole element mounted on a ground plane into an L-shape and by feeding at a point offset from the mounting point. Impedance matching to the feed line can be easily obtained by choosing the feeding point. The frequency bandwidth increases in proportion to the antenna height H. This antenna has been studied as a missile antenna [52], and it is also called a *blade antenna*. In land mobile communication, a car cab antenna has been developed in which the antenna element is printed on a dielectric substrate; a parasitic element is used for wide-band operation [53]. Replacing the wire-type radiating element with a planar element as shown in Figure 3.49(b) yields wide-band resonance

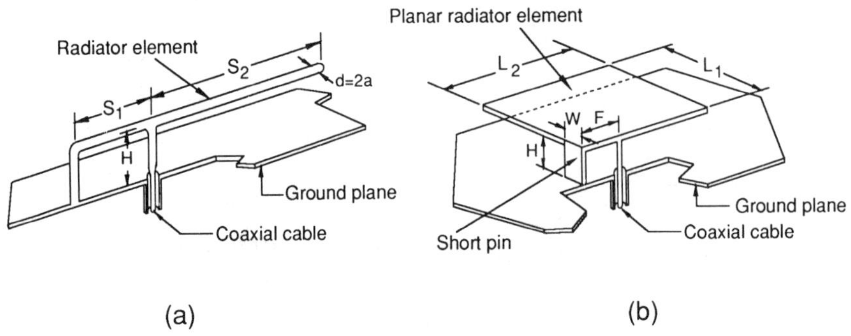

Figure 3.49 Inverted-F antenna: (a) wire inverted-F antenna; (b) planar inverted-F antenna.

characteristics. Moreover, the size can be reduced and resonance is obtained, when the peripheral length of the planar element is about a half wavelength. Planar IFAs are suitable for portable telephones and have been adopted as the internal antennas for a transportable phone and a handheld-type portable telephone [54]. The fundamental characteristics of planar IFAs have been analyzed by T. Taga, and details have been published [55].

The structure of microstrip antennas is shown in Figure 3.50(a). They are constructed by printing conductors on dielectric substrates. They are also called *patch antennas*. Many kinds of radiator shapes are possible, such as rectangular patch, circular patch, triangular patch, and ring patch. This antenna is derived from microstrip resonators, and it uses the radiation loss of the resonators in a positive manner. Therefore, the bandwidth of this antenna is basically narrow. A larger bandwidth can be obtained by increasing the electrical thickness of the

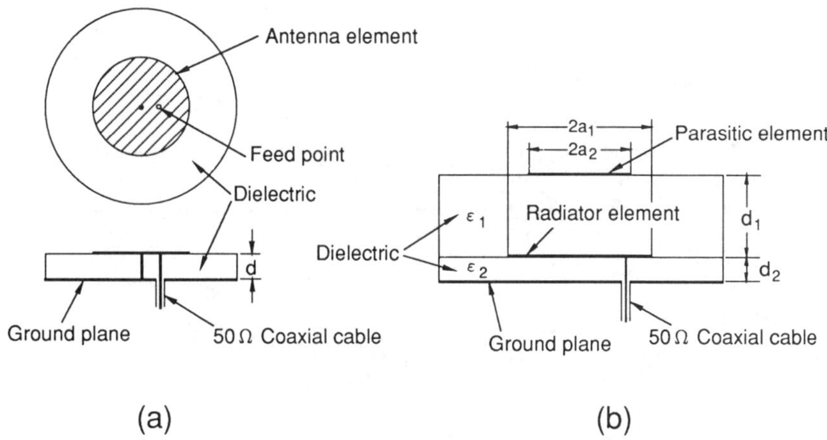

Figure 3.50 Microstrip antenna: (a) basic structure of circular microstrip antenna; (b) broadband microstrip antenna with stacked parasitic element.

substrate or by choosing a substrate with a lower dielectric constant. However, for broadband operation, a stacked microstrip antenna is more effective. A parasitic patch is added to load the double-layered structure, as shown in Figure 3.50(b). This type was developed as a car cab antenna for land mobile radio communication [56].

3.5 DEVELOPMENTS IN MOBILE PHONE ANTENNA CAR INSTALLATIONS

The automobile phone service was the first mobile phone system to be developed. In the initial stage, the system was far less sophisticated than the present cellular system, which at the moment is extremely popular. Nonetheless, vehicular-mount antennas have been researched since the 1950s. This section introduces vehicular-mount antennas, and other special designs, such as a diversity antenna, a cabin-mounted antenna, a dual-frequency antenna, and a system compatible antenna, are described in more detail.

3.5.1 Design Considerations

A typical vehicular-mount antenna is a rod antenna such as the monopole and sleeve dipole. There are some installation variations for the car body, as illustrated in Figure 3.51 [57]. A cabin-mounted antenna has been developed, but it is not

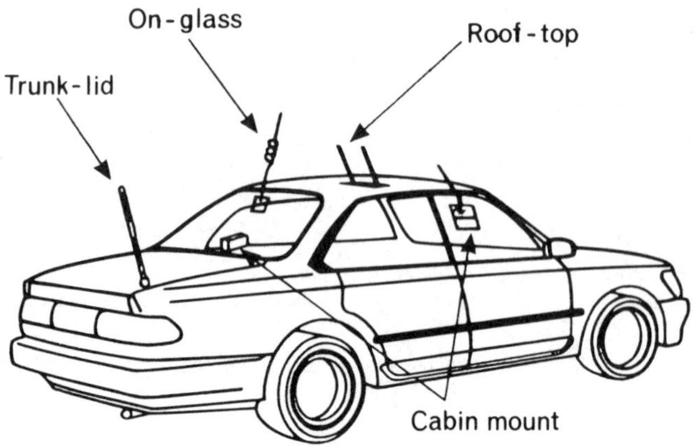

Figure 3.51 Installation example of vehicular-mount antenna. (After [57].)

used in most public mobile communication systems. Other major items for consideration with regard to antenna design are shown in Figure 3.4 of Section 3.1. A high installation location for the outside antenna is preferable. The cable connecting the antenna and radio equipment should pass through the car body, which requires the body to be penetrated. To avoid this, an on-glass-mount antenna was developed. Since this antenna couples electromagnetically through the glass window, there is no need to make a hole or to pass a cable inside the passenger compartment. A cabin-mounted antenna has the advantage of connecting the antenna to the radio equipment without modifying the car body, but its disadvantage is that the radiation pattern may severely deteriorate due to the car body itself. When designing the cabin-mounted antenna, it is of utmost importance to minimize the deterioration of the radiation pattern. It is equally important to be able to estimate the radiation pattern of the outer antenna when it is installed on or within the car. This problem has been thoroughly investigated, and useful data for designing practical antennas is available [58].

3.5.2 Antenna Evolution

Various types of vehicular-mount antennas have been developed. A typical example is the rod-type antenna; some designs of this category are shown in Figure 3.52. The 5/8-wavelength antenna has greater gain and is less affected by the car than the quarter-wavelength monopole. For these reasons, the former is more frequently used in cellular systems. Of the rod-type antennas, the electromagnetically coupled antenna, the on-glass antenna, is unique. Since it uses electromagnetic coupling,

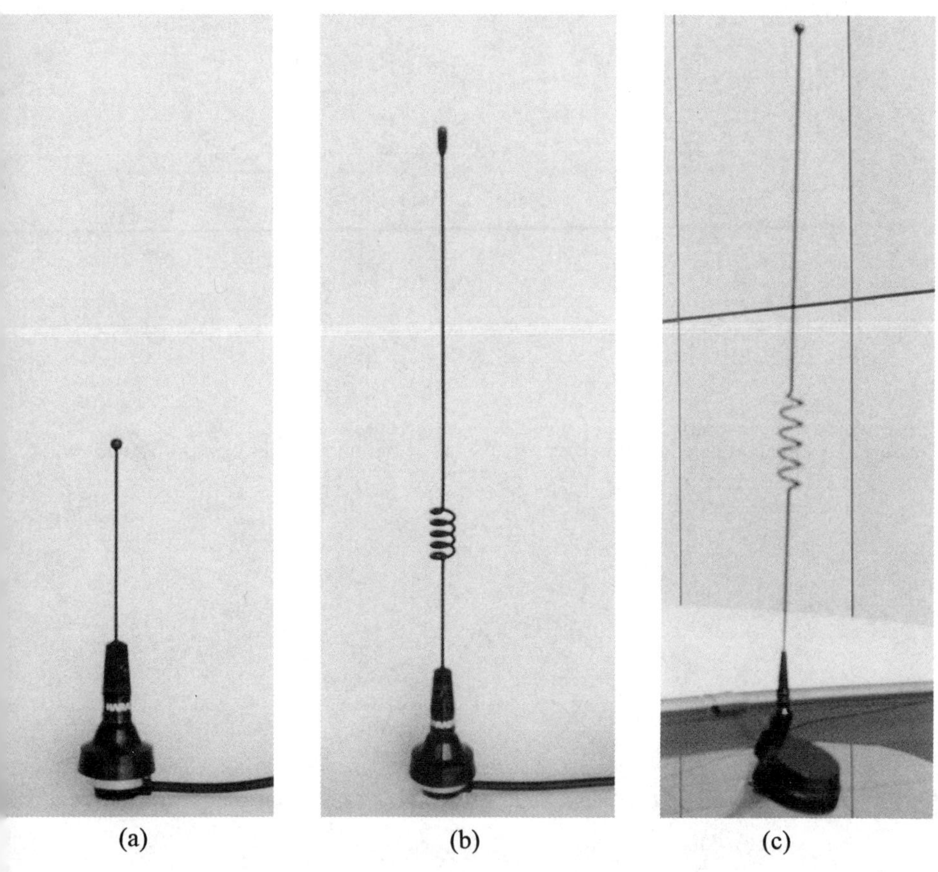

(a) (b) (c)

Figure 3.52 Rod-type vehicular-mount antenna: (a) 1/4-wavelength whip; (b) 5/8-wavelength whip; (c) on-glass antenna. (Courtesy of Harada Industry Co., Ltd.)

it is easy to install without modifying the car body. It is very popular around the world as a cellular antenna.

Turning to the cabin-mounted antenna, Ebine developed the dual type shown in Figure 3.53. These two unidirectional antennas are combined with a hybrid circuit to achieve an omnidirectional radiation pattern without the deterioration usually caused by the car body [57,59]. Ebine also demonstrated that the antennas could function as a diversity antenna if the two output ports of the hybrid circuit were used as diversity branches. The measured radiation pattern of the antenna is shown in Figure 3.54. It is understood that the omnidirectional pattern is obtained for its envelope. Although there are ripples in the ±90-deg direction, they might be filled due to the multipath wave propagation experienced in urban areas.

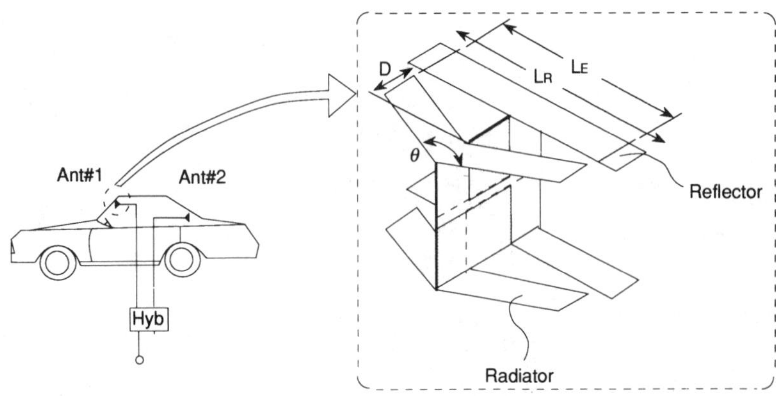

Figure 3.53 Configuration of a cabin-mounted antenna and geometry of its element. (After [59].)

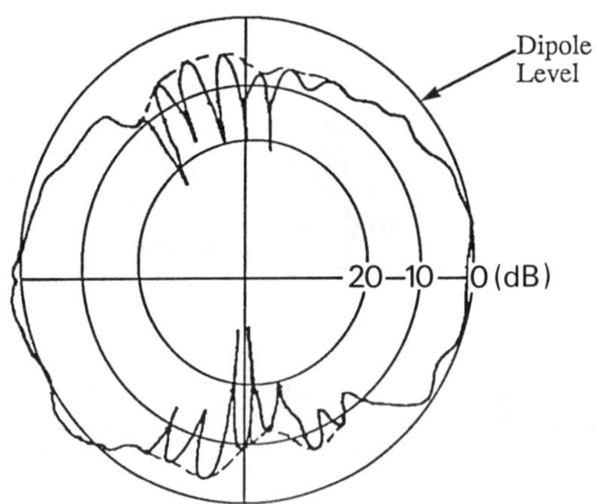

Figure 3.54 Radiation pattern of a cabin-mounted antenna when installed in a cabin.

Although development of the vehicular antenna seems to have reached maturity, no doubt there will be continuous technological development in this area accompanying the advance of communication systems as a whole. Figure 3.55 shows the historical progress of vehicular antennas. This figure includes the diversity antenna developed for NTT's Large-Capacity System, a dual-frequency antenna,

Year
1978
1986
(1987)
(1990)
1993
1995
Systems
Cellular phone system for metropolitan areas
High-capacity cellular phone system
High-capacity cellular phone system (adaptor antenna for portable telephone)
Integrated mobile communication system using terrestrial and satellite systems
Cellular phone system for urban areas
(Mobile satellite system experiment using Engineering Test Satellite -V)
Digital cellular phone system / High-capacity cellular phone system
Requirements
Horizontal gain OdBd
Easy installation on vehicles
Diversity
Use of portable telephone in a vehicle
Wide bandwidth operation in 800MHz freq. band (7%→17%)
Two-frequency operation (800MHz /1.5GHz)
Integration of terrestrial and satellite mobile antennas
High gain mobile satellite antennas (12 ~ 14dBi)
Developed antenna
Terrestrial system
Trunk-lid type antenna
Trunk-lid type
Roof-top type
Indoor type
Front
Back
Diversity Antennas
On-glass antenna
Trunk-lid type diversity antenna for 800MHz/1.5GHz frequency bands
Compatible antenna for cellular and mobile satellite system
Satellite system
Quadrifilar helical antenna
ETS-V
Development of compact feed configuration
Bifilar helical antenna
60% size reduction compared with quadrifilar helical antenna
Integration of a diversity antenna used in the terrestrial system
High gain electronically beam-steering antenna

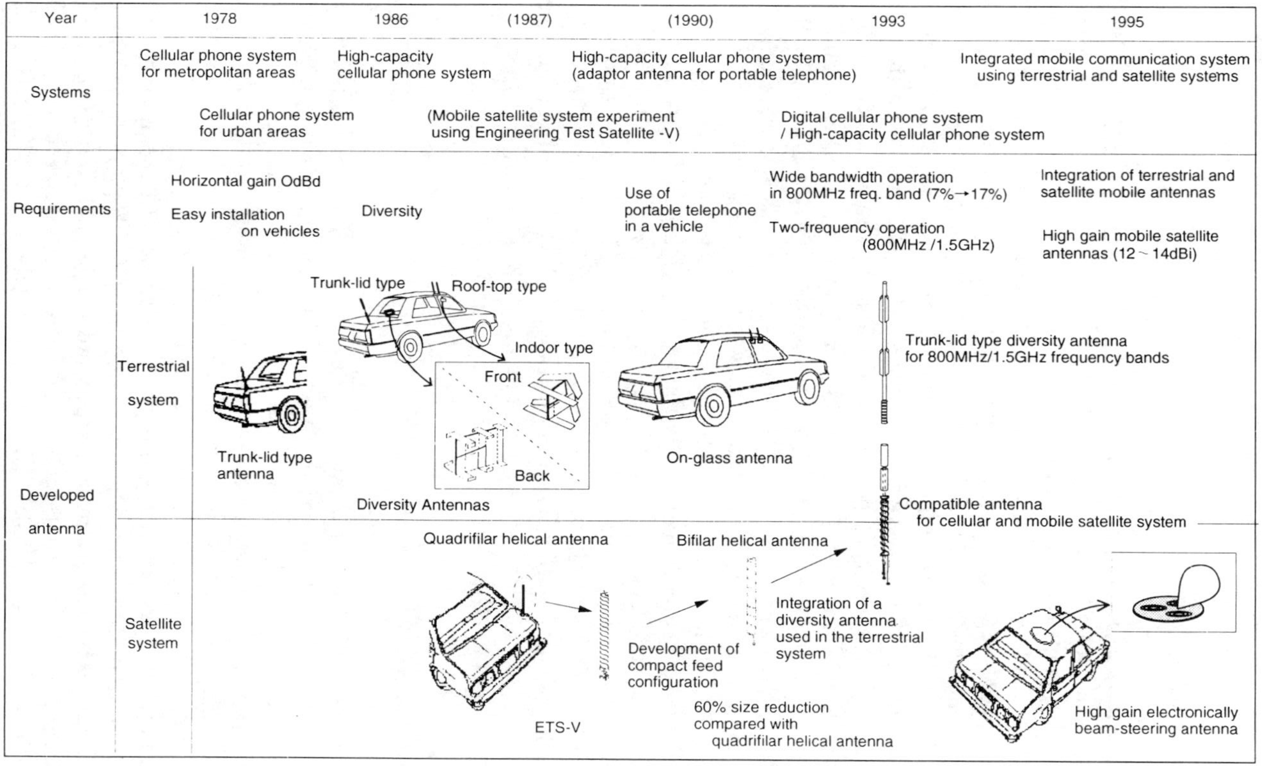

Figure 3.55 Historical progress of vehicular antennas.

and a system-compatible antenna. These promise to make important contributions to system performance and are described in more detail in the following sections.

3.5.3 Diversity Antenna

Diversity antennas were studied in the early development stages of the mobile system as a means of overcoming fading. However, they are not used in any commercial systems outside of Japan. NTT's Large-Capacity System has adopted a diversity antenna system for both mobile and base station antennas.

There are three types of diversity antennas for vehicular mounting, as shown in Figure 3.56. Figure 3.56(a) shows the conventional horizontally spaced diversity antenna. The space between the two antennas is about a half wavelength, and the antenna elements are half-wave sleeve dipoles. This is effective for reducing fading, but requires two discrete elements. Figure 3.56(b) shows the vertically spaced diversity antenna, which looks like a single antenna. Ebine et al. found that vertically spaced diversity antennas [60] were almost as effective as horizontally spaced ones. This was shown in the basic experiment wherein two half-wavelength dipoles were arranged vertically. Ebine subsequently developed the trunk-lid diversity antenna as shown in Figure 3.57 [60]. The correlation coefficient drops to 0.5 A value suitable for both urban and suburban areas. Figure 3.56(c) depicts the diversity antenna installed inside the cabin described previously. In this case, the two terminals for diversity reception are the outputs of the hybrid circuit joining the two antennas. The correlation between two terminals decreases due to the difference of the radiation pattern, as shown in Figure 3.54. The measured correlation coefficients of this antenna are shown in Figure 3.58. A sufficiently low correlation coefficient is obtained [57].

3.5.4 Dual-Frequency Antenna

To satisfy the rapid growth of mobile telephones, it is often necessary to add a new frequency band to the existing system. For example, if the 800-MHz band becomes overly crowded, a new frequency band, such as 1.5 GHz, is needed to increase the number of channels. In this case, a mobile station usually requires two antennas for each frequency band. If a single antenna can operate in dual-frequency bands, the system could be expanded easily.

Figure 3.59 shows the trunk-lid-type dual-frequency band antenna, which can be operated in both the 800-MHz band and the 1.5-GHz band [61]. By adding a parasitic element, whose length corresponds to one-half the wavelength of the higher frequency, to the dipole antenna, dual-frequency operation can be achieved.

A dual-frequency antenna for cabin-mounted use was also developed and is shown in Figure 3.60 [62]. One original antenna used in the 800-MHz band is the

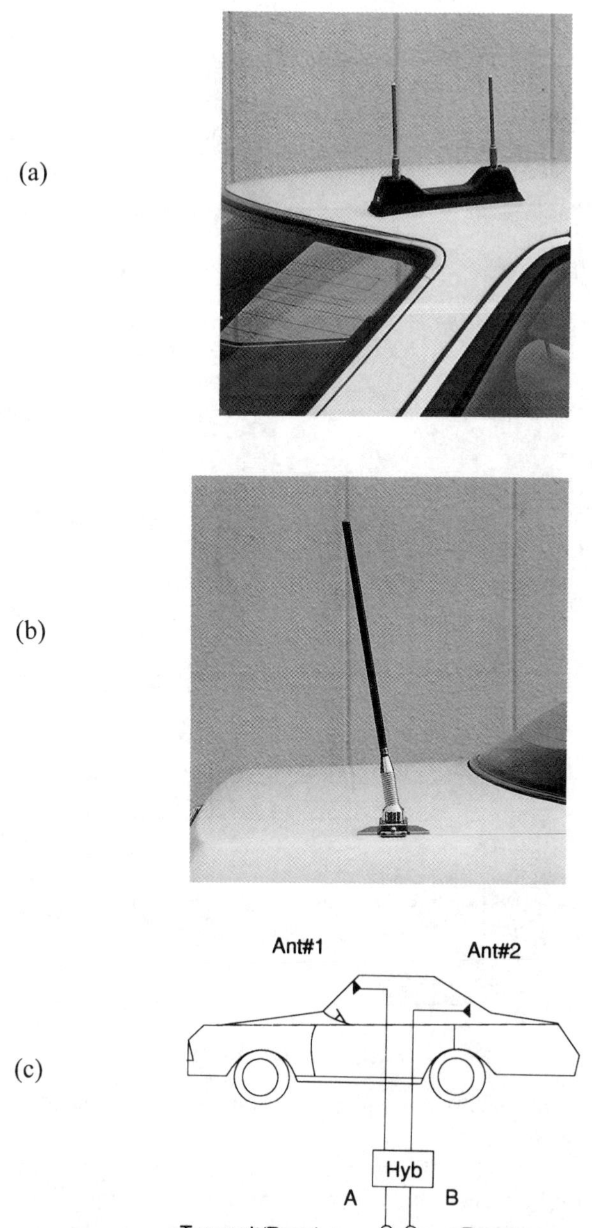

Figure 3.56 Types of diversity antennas: (a) horizontally spaced; (b) vertically spaced; (c) pattern diversity antenna. (Courtesy of Nippon Telegraph and Telephone Corp.)

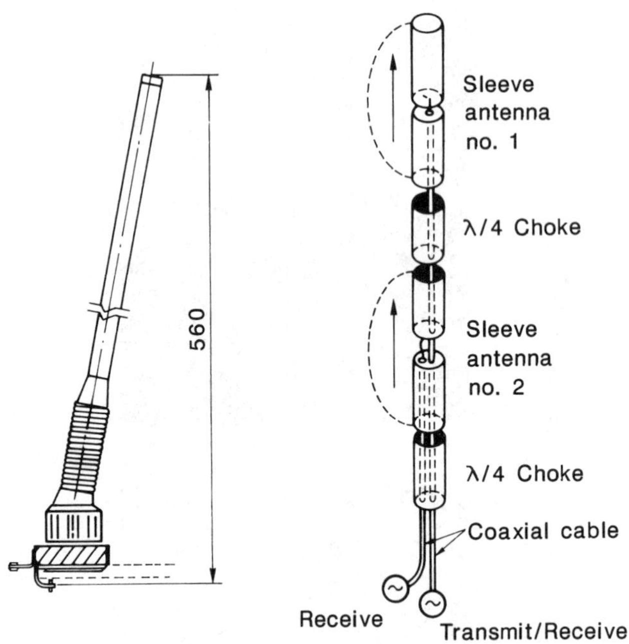

Figure 3.57 Trunk-lid-type diversity antenna. (From [60], © 1990 IEEE.)

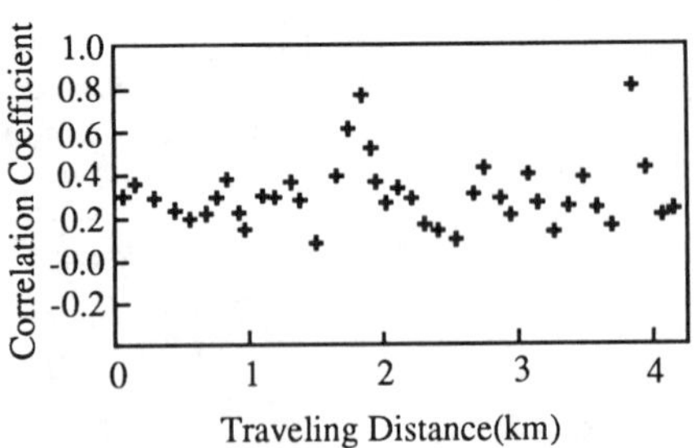

Figure 3.58 Correlation coefficient of a cabin-mounted diversity antenna. (From [57], © 1988 IEEE.)

capacity-loaded printed dipole, which is the same as the one mentioned in Section 3.5.2. To make dual-frequency operation possible, a parasitic element is added near the exciter of the original antenna. Outer dimensions of the antenna are approximately 90 mm in width, 70 mm in depth, and 60 mm in height. This can be attached at the back of the rear view mirror, as shown in Figure 3.61.

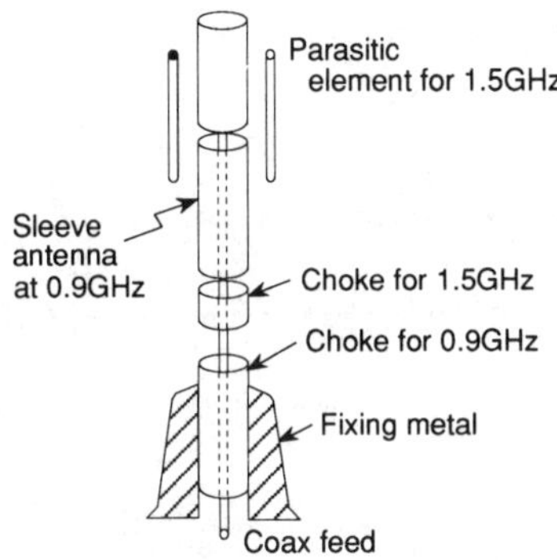

Figure 3.59 Trunk-lid-type dual-frequency antenna. (After [61].)

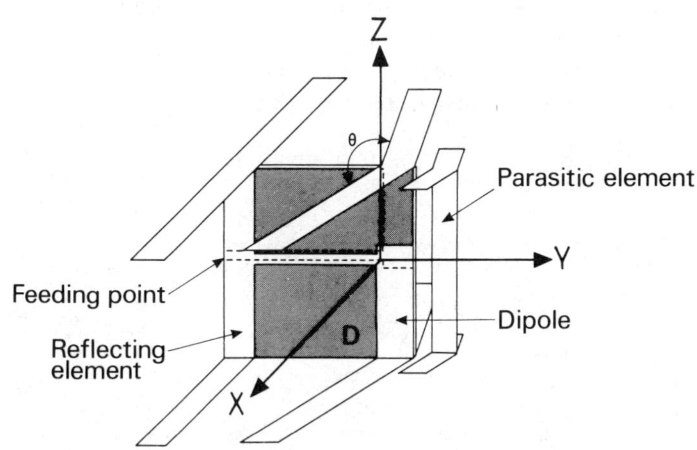

Figure 3.60 Cabin-mount-type dual-frequency antenna. (From [62], © 1991 IEEE.)

3.5.5 System-Compatible Antennas

When expanding the mobile telephone service coverage nationwide, it is inefficient to cover the service area using only one land mobile system, such as the cellular system. A satellite system is far more efficient in covering a wide area containing few mobile stations. Usually, the land mobile system and the satellite system use different frequencies and different polarization states. Moreover, the antenna pat-

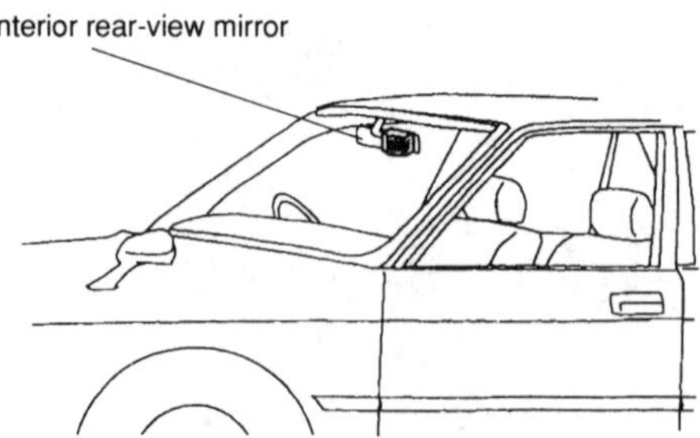

Figure 3.61 Installation example of cabin-mounted antenna. (From [62], © 1991 IEEE.)

tern required for individual mobile stations in each system may differ. An example of the requirements of a dual-system mobile station antenna are shown in Table 3.4. Although it is difficult to satisfy the requirements of both systems simultaneously, a compatible antenna able to achieve this would be very effective. Figure 3.62 shows an example of a system-compatible antenna [63] which can satisfy all the requirements shown in Table 3.4. Motorola announced a portable telephone system, the *Iridium System*, which uses low-earth-orbit satellites. This would require

Table 3.4
Antenna Requirement for Each System

	Land Mobile	*Satellite Mobile*
Frequency	900 MHz	S-band (2.5/2.6 GHz)
Polarization	Linear (V-pol)	Circular
Radiation pattern	Omnidirectional in horizontal plane and maximum exists in its plane	Omnidirectional in horizontal plane and maximum exists at some elevation angle (nearly equal to 45 deg)
Gain	0 dB or more	5 dBd
Diversity	Necessary	None

an antenna compatible with the land mobile system, and the mobile satellite system would be a more challenging undertaking, perhaps, than that shown in Figure 3.62. This indicates the necessity of a system-compatible antenna to realize highly effective dual systems.

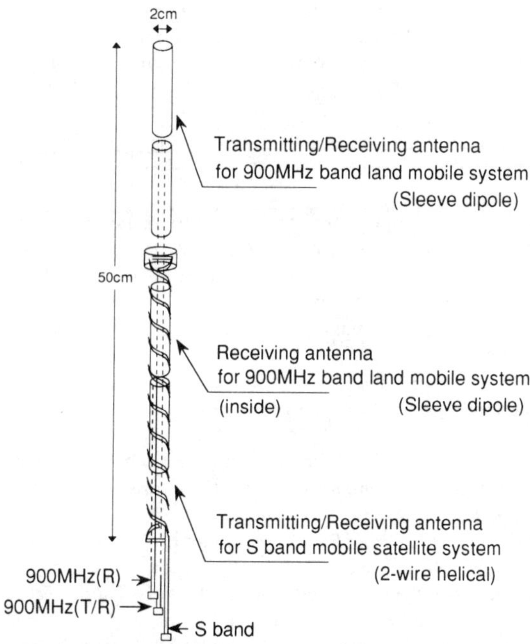

Figure 3.62 System-compatible antenna for land mobile and mobile satellite (diversity reception for land mobile system). (After [63].)

REFERENCES

[1] Kagoshima, K., "Key Factor and Technique for Antenna Design," *J. IEICE*, Vol. 71, No. 6, 1988, pp. 607–609.

[2] Taga, T., "Analysis for Mean Effective Gain of Mobile Antennas and in Land Mobile Radio Environments," IEEE Trans. VT, Vol. 39, No. 2, 1990, pp. 117–131.

[3] Ebine, Y., "Antenna Characteristics in Vertically Gaussian Distributed Mobile Radio Propagation," *IEEE AP-S'90*, Dallas, Texas, 7–11 May 1990, pp. 1800–1803.

[4] Sato, K., K. Matsumoto, K. Fujimoto, and K. Hirasawa, "Characteristics of a Planar Inverted-F Antennas on a Rectangular Conducting Body," *IEICE Trans.*, Vol. J71-B, No. 11, 1988, pp. 1237–1243.

[5] Analoui, M., and Y. Kagawa, "Surface Patch Analysis of a Built-in Planar Inverted-F Antenna," *Proc. 1992 Int. Symp. on Antennas and Propagation*, Sapporo, Sept. 22–25, 1992, pp. 661–664.

[6] Reineix, A., and B. Jecko, "Analysis of Microstrip Patch Antennas Using Finite Difference Time Domain Method," *IEEE Trans. AP*, Vol. 37, No. 11, 1989, pp. 1361–1368.

[7] Wu, C., K. L. Wu, Z. Q. Bi, and J. Litva, "Accurate Characterization of Planar Printed Antennas Using Finite-Difference Time-Domain Method," *IEEE Trans. AP*, Vol. 40, No. 5, 1992, pp. 526–534.

[8] Kagoshima, K., K. Tsunekawa, and A. Ando, "Analysis of a Planar Inverted-F Antenna Fed by Electromagnetic Coupling," *IEEE AP-S'92*, Chicago, July 18–25, 1992, pp. 1702–1705.

[9] Mittra, R., ed., "Computer Techniques for Electromagnetics," Pergamon Press, 1973.

[10] Stutzaman, W. L., and G. A. Thiele, "Antenna Theory and Design," John Wiley, 1981.

[11] Okumura, Y., and A. Akeyama, "Radio Wave Propagation in Mobile Communications," Ch. 2 of *Basic Technology of Mobile Communications*, Y. Okumura, and M. Shinji, eds., IEICE press (Japanese), 1986, pp. 24–59.

[12] Okumura, Y., E. Ohmori, T. Kohno, and K. Fukuda, "Field Strength and Its Variability in VHF and UHF Land Mobile Radio Service," *Rev. Elec. Comm. Lab.*, Vol. 16, No. 9–10, 1968, pp. 825–873.

[13] Special issue on mobile radio propagation, "Coverage Prediction for Mobile Radio System Operating in the 800/900 MHz Frequency Range," *IEEE VT.*, Vol. 37, No. 1, 1988, pp. 3–72.

[14] Kozono, S., and T. Takeuchi, "Recent Propagation Studies on Land Mobile Radio in Japan," *IEICE Trans.*, Vol. E-74, No. 6, 1991, pp. 1538–1546.

[15] Special issue on "Digital Mobile Communications," *IEICE Trans.*, Vol. J73-B-II, No. 11, 1990, pp. 569–804.

[16] Mishima, H., Y. Ebine, and K. Watanabe, "Base Station Antennas and Multiplexer System for Land Mobile Telephone System," *Elect. Com. Lab. Tech. J.*, Vol. 26, No. 7, 1977, pp. 2011–2036.

[17] Nakajima, N., H. Mishima, and Y. Yamada, "Mobile Communications Antennas," Ch. 10 of *Basic Technology of Mobile Communications*, Y. Okumura, and M. Shinji, eds., IEICE press (Japanese), 1986, pp. 239–260.

[18] Yamada, Y., Y. Ebine, and K. Tsunekawa, "Base and Mobile Station Antenna for Land Mobile Radio Systems," *IEICE Trans.*, Vol. E74, No. 6, 1991, pp. 1547–1555.

[19] Sakamoto, M., "System Configuration and Control," Ch. 8 of *Basic Technology of Mobile Communications*, Y. Okumura, and M. Shinji, eds., IEICE press (Japanese), 1986, pp. 188–217.

[20] Mishima, H., Y. Ebine, and K. Watanabe, "Base Station Antennas and Multiplexer System for Land Mobile Telephone System," *Elect. Com. Lab. Tech. J.*, Vol. 26, No. 7, 1977, pp. 2011–2036.

[21] Yamada, Y., T. Nara, S. Kameo, Y. Chatani, and H. Abe, "A Variable Beamwidth Corner Reflector Antenna," *IECE Nat. Conv. Record*, No. 694, 1986.

[22] Kraus, J. D., *Antennas*, 2nd edition, McGraw-Hill, 1988, pp. 549–558.

[23] Suzuki, T., and K. Kagoshima, "Corner Reflector Antenna With the Same Beamwidth in Two Frequency Bands," *IEICE Trans.*, Vol. J75-B-II, No. 12, 1992, pp. 950–956.

[24] Nara, T., Y. Ebine, and N. Nakajima, "Beam Tilting Effect of Base Station Antenna," *IECE Nat. Conv. Record*, No. S5-15, 1985.

[25] Elliott, R. S., Ch. 4, Sec. 4.4 in *Antenna Theory and Design*, Prentice-Hall, 1981, p. 128.

[26] Kijima, M., and Y. Yamada, "Relationship Between Array Excitation Distribution and Radiation Pattern Ripple Depth," *IEICE Trans.*, Vol. J73-B-II, No. 12, 1990, pp. 860–868.

[27] Elliott, R. S., Ch. 5 in *Antenna Theory and Design*, Prentice-Hall, 1981.

[28] Kijima, M., and Y. Yamada, "Beam Narrowing Method for Radiation Pattern With Suppressing Some Sidelobes," *IEICE Tech. Rep. on Antennas and Propagation*, A·P91–125, 1992.

[29] Kijima, M., and Y. Yamada, "Determining Excitation Coefficients of Dual-Frequency Shaped Beam Linear Array Antenna for Mobile Base Station," *IEEE AP-S'91*, London, Ontario, 24–28 June 1991, pp. 932–935.

[30] Clark, R. H., "A Statistical Theory of Mobile-Radio Receptions," *B.S.T.J.*, Vol. 47, No. 6, 1968, pp. 957–1000.

[31] Yamada, Y., K. Kagoshima, and K. Tsunekawa, "Diversity Antennas for Base and Mobile Stations in Land Mobile Communication Systems," *IEICE Trans.*, Vol. E74, No. 10, 1991, pp. 3202–3209.

[32] Nara, T., Y. Ebine, and Y. Yamada, "Characteristics of Polarization Diversity Base Station Antenna," *IECE Nat. Conv. Rec.*, No. 2362, 1986.

[33] Ebine, Y., T. Takahashi, and Y. Yamada, "A Study of Vertical Space Diversity for a Land Mobile Radio," *IEICE Trans.*, Vol. J73-B-II, No. 6, 1990, pp. 286–292.
[34] Gardiner, J. G., and R. E. Fudge, "Aerials and Base Station Design," Ch. 4 of *Land Mobile Radio Systems*, R. J. Holbeche, ed., IEE Telecommunications Series 14, Peter Peregrinus, 1985, pp. 45–71.
[35] "Public Land Mobile Telephone Systems," *CCIR SG-8*, Report 742-3.
[36] Horn, D. W., "Vehicle-Caused Pattern Distortion at 800 MHz," *Proc. 33rd IEEE Vehicular Technology Conf.*, Toronto, Ontario, May 1983, pp. 25–27.
[37] Losee, M., "Antennas: Squelching Problems," *Cellular Business*, August 1985, pp. 20–23.
[38] Horikawa, I., "Major Techniques for High-Capacity Land Mobile Communication System—Antenna and Propagation," *NTT Int. Symp., Tokyo, Japan*, July 1983, pp. 81–93.
[39] Lee, W. C. Y., and R. H. Brandt, "Elevation Angle of Mobile Radio Signal Arrival," *Joint IEEE Comm. Soc.-Veh. Tech. Group Special Trans. Mobile Radio Communications*, Nov. 1973, pp. 1194–1197.
[40] Watanabe, K., H. Mishima, and Y. Ebine, "Measurement of Elevation Angle of Land Mobile Radio Signal Arrival," *IECE Trans. Japan*, Vol. J60-B, No. 11, Nov. 1977, pp. 880–887.
[41] Ikegami, F., and S. Yoshida, "Analysis of Multipath Propagation Structure in Urban Mobile Radio Environments," *IEEE Trans. Antennas and Propagation*, Vol. AP-28, No. 4, July 1980, pp. 531–537.
[42] Taga, T., "Analysis for Mean Effective Gain of Mobile Antennas in Land Mobile Radio Environments," *IEEE Trans. Vehicular Technology*, Vol. VT-39, No. 2, May 1990, pp. 117–131.
[43] Rustako, A. J., Jr., Y. S. Yeh, and R. R. Murray, "Performance of Feedback and Switch Space Diversity 900 MHz FM Mobile Radio Systems With Rayleigh Fading," *IEEE Trans. Communications*, Vol. COM-21, No. 11, Nov. 1973, pp. 1257–1268.
[44] Adachi, F., and K. Ohno, "Experimental Evaluation of Postdetection Diversity Reception of Narrow-Band Digital FM Signals in Rayleigh Fading," *IEEE Trans. Vehicular Technology*, Vol. VT-38, No. 4, Nov. 1989, pp. 216–221.
[45] Kraus, J. D., Ch. 9 in *Antennas*, New York: McGraw-Hill, 1950, p. 242.
[46] Kraus, J. D., Ch. 5 in *Antennas*, New York: McGraw-Hill, 1950, p. 141.
[47] Hahn, R. F., and J. G. Fikioris, "Impedance and Radiation Pattern of Antennas Above Flat Discs," *IEEE Trans. Antennas and Propagation*, Vol. AP-21, No. 1, Jan. 1973, pp. 97–100.
[48] Horn, D. W., "Cellular 'On-Glass' Antenna Technology," *Proc. 34th IEEE Vehicular Technology Conf.*, Pittsburgh, PA, 21–23 May 1984, pp. 65–68.
[49] Sakitani, A., and S. Egashira, "Analysis of Coaxial Collinear Antenna: Recurrence Formula of Voltages and Admittances at Connections," *IEEE Trans. Antennas and Propagation*, Vol. AP-39, No. 1, Jan. 1991, pp. 15–20.
[50] Judasz, T. J., W. L. Ecklund, and B. B. Balsley, "The Coaxial Collinear Antenna: Current Distribution From the Cylindrical Antenna Equation," *IEEE Trans. Antennas and Propagation*, Vol. AP-35, No. 3, March 1987, pp. 327–331.
[51] Jasik, H. ed., *Antenna Engineering Handbook*, pp. 22–28, New York: McGraw-Hill, 1961.
[52] King, R. W. P., C. W. Harrison, Jr., and D. H. Denton, Jr., "Transmission-Line Missile Antennas," *IRE Trans. Antennas and Propagation*, Vol. 8, No. 1, Jan. 1960, pp. 88–90.
[53] Mishima, H., and T. Taga, "Mobile Antennas and Duplexer for 800 MHz Band Mobile Telephone System," *IEEE AP-S Int. Symp. Digest*, Quebec, June 1980, pp. 508–511.
[54] Seki, S., N. Kanmuri, and A. Sasaki, "Detachable Unit Service in 800 MHz Band Cellular Radiotelephone System," *IEEE Communications Magazine*, Vol. 24, No. 2, Feb. 1986, pp. 47–52.
[55] Hirasawa, K., and M. Haneishi, eds., Ch. 5 in *Analysis, Design, and Measurement of Small and Low-Profile Antennas*, Norwood, MA: Artech House, 1991.
[56] Mishima, H., and T. Taga, "Antenna and Duplexer for New Mobile Radio Unit," *Review of the ECL, NTT*, Vol. 30, No. 2, March 1982, pp. 359–370.

[57] Ebine, Y., and Y. Yamada, "Vehicular-Mounted Diversity Antennas for Land Mobile Radio," *IEEE VTC'88*, Philadelphia, 15–17 June 1988, pp. 326–333.

[58] Nishikawa, K., and Y. Asano, "Vertical Radiation Pattern of Trunk Mount Antennas for Mobile Radio Communications," *IEICE Trans.*, Vol. E74, No. 10, 1991, pp. 3227–3232.

[59] Ebine, Y., N. Shimada, and K. Kosaka, "Print Dipole With Capacity Plate," *IECE Nat. Conv. Rec.*, No. 641, 1986.

[60] Ebine, Y., and Y. Yamada, "A Vehicular-Mounted Vertical Space Diversity Antenna for a Land Mobile Radio," *IEEE Trans.*, Vol. VT-40, No. 2, 1991, pp. 420–425.

[61] Ebine, Y., and K. Kagoshima, "Multi-Frequency Dipole Antenna With Closed-Spaced Parasitic Elements," *IEICE Trans.*, Vol. J71-B, No. 11, 1988, pp. 1252–1258.

[62] Kagoshima, K., T. Takahashi, and Y. Ebine, "Dual Frequency Low Profile Printed Dipole Yagi-Array for a Cabin Mounted Vehicular Antenna," *IEEE AP-S'91*, London, Ontario, 24–28 June 1991, pp. 962–965.

[63] Terada, N., and K. Kagoshima, "Compatible Mobile Antenna for Mobile Satellite and Cellular Communication Systems," *IEICE Tech. Rep. on Antennas and Propagation*, A·P91-65, 1991.

Chapter 4

Land Mobile Antenna Systems II: Pagers, Portable Phones, and Safety

R. Mumford, Q. Balzano, and T. Taga

4.1 PRACTICAL REQUIREMENTS OF AND CONSTRAINTS ON PAGER ANTENNA DESIGN

Radio paging is a cost-effective solution for locating staff, alerting personnel, and transmitting one-way messages or data. In Europe, the industry has developed from single onsite systems, the first of these being an inductive loop system installed in St. Thomas's Hospital in England by Multitone Electronics in 1956. Transmissions were in the 30- to 50-kHz range, and each receiver was tuned to its own unique frequency. Nationwide area systems in Europe started by the Netherlands PTT in 1964 were followed up by public operator licenses being granted to private consortia. A pan-European paging system with access from any country is now planned and specified.

Three main types of pagers are currently available: the audible-tone alerter, the numeric message displaying up to 20 digits, and the alphanumeric displaying about 100 characters and storing up to 5,000-character messages. Figure 4.1 shows a typical VHF tone alert pager with a strip antenna, and Figure 4.2 shows a high-frequency numeric pager with a ferrite antenna. Paging receivers have been substantially reduced in size and greatly increased in features in the last 30 years. This has had the effect of reducing possible integral antenna size and bringing RF noisy digital circuitry (e.g., gate arrays and microprocessors) much closer to the antenna and the RF front-end circuits.

Paging receivers are used in private onsite situations, usually with one low-power transmitter radiating 1W to 5W of RF power, and in large wide-area systems where there are multiple transmitters of 100W to 200W, each covering a 5- to 20-mile radius. It is important in the latter case for the paging receiver to be as sensitive as possible in order to economize on the number of transmitters in a large system. The onsite pager, which may be operating, for example, on a warehouse site, a brewery, hotel, or supermarket, must be sensitive enough to operate in antenna shadow

Figure 4.1 VHF tone alert pager. (Courtesy Multitone Electronics, Ltd.)

regions and between buildings. The pager's own internal antenna is arguably the most important factor in defining sensitivity, since it is much smaller than a portable two-way radio and is worn on the belt or in the shirt or jacket top pocket.

Frequency bands allocated to paging vary according to country across the world and are in the total range of 25 to 910 MHz. Traditionally in Europe and the United States, wide-area systems, such as British Telecom's service (covering nearly all the populated areas of the United Kingdom) and large North American radio common carrier subscriber networks, operated on frequencies in the VHF band 147 to 174 MHz. In Japan, frequencies ranged up to 280 MHz. However the attraction of greater penetration and propagation in buildings has been one factor contributing to UHF frequencies in the 450- to 512-MHz band being used for citywide networks such as the Cityruf system in Germany. Various small integral antenna designs are therefore required for this very wide frequency range.

The designer's aim is to optimize the antenna performance in the presence of a wide variety of problems such as the following.

- The small size of the unit;
- The proximity of the circuit components;
- The RF noise generated by the rest of the receiver circuitry;
- The ease of receiver alignment and the maintenance of the alignment when the pager is worn on the body;

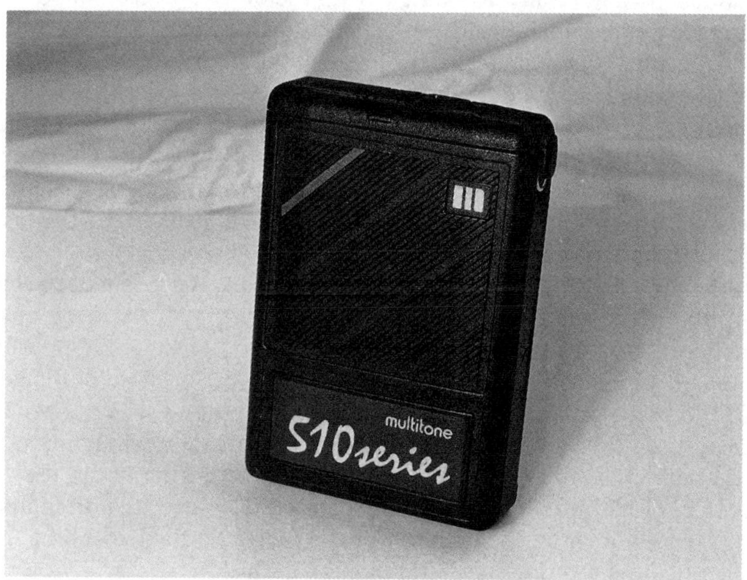

Figure 4.2 HF numeric pager. (Courtesy Multitone Electronics, Ltd.)

- The effect of the proximity of the human body on antenna performance;
- The unit is frequently transported at widely varying speeds;
- Use inside a building or vehicle is commonplace.

It is evident that in order to deal with the first two of these problems, the aerial development must include the pager chassis as an inseparable part of the antenna. If possible, its metal parts should be used to aid the performance. To avoid loss of calls when the transmitter direction is in a pager antenna pattern null, any such nulls must be made as shallow as possible. A somewhat obvious solution to this problem would be to use two or more antennas arranged so that their nulls do not coincide.

The latter three problems in the above list all stem from the nature of an electromagnetic field near a reflective or absorptive surface. Fades caused by absorption cannot be counteracted as the power is dissipated within the surface of the object. With the reflective nulls, however, the situation is different. We will now discuss some of these issues in more detail.

4.1.1 Effect of the Human Body on Antenna

Over the frequency range under consideration (20 to 1,000 MHz), the human body has a conductivity and relative permittivity on the order of 1.5 S/m and 75, respectively, and acts predominantly as a reflector [1] with varying degrees of efficiency. Figure 4.3 illustrates the effect of an object composed of simulated body tissue on a uniform electromagnetic field in two dimensions. The low impedance, presented to the wave by the body, reduces the electric field close by, while increasing the magnetic field. At a distance of a quarter wavelength away from the body towards the transmitter, a high impedance is presented to the approaching wavefront which enhances the electric field while reducing the magnetic field. This effect is periodic in nature, resulting in the situation illustrated in Figure 4.3.

On the far side of the body from the transmitter, there is a deep null caused by absorption of power by the body. This, of course, affects both electric and magnetic fields. It is evident from the above that for good onbody performance a magnetic antenna must be used. It is also evident that because of the conductive nature of the body any magnetic field close to the body will be tangential to the body. This is useful because it means that if a magnetic loop antenna is used and positioned such that the axis of the loop is tangential to the body, the maximum magnetic field available will always be intercepted, whatever the orientation of the body [2]. Figure 4.4 shows a plan view of the magnetic field around a cylindrical lossy dielectric object representing the human body.

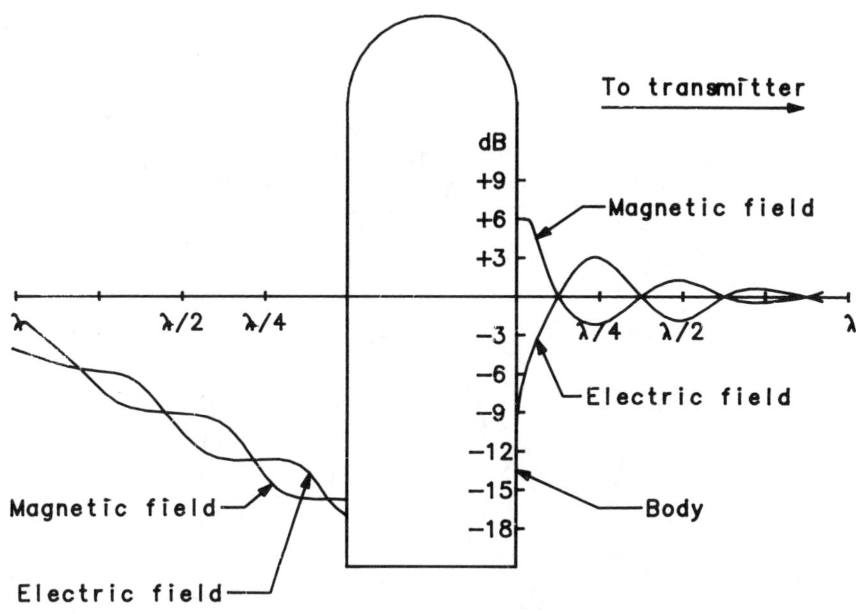

Figure 4.3 Effects of the human body on an electromagnetic field.

Magnetic Antennas

The magnetic field enhancement of approximately 6 dB (at 150 MHz, 3 dB at 450 MHz, and 0 dB at 900 MHz) close to the body has led to the use of predominantly magnetic antennas, which naturally perform best on the body. These antennas are based mainly on the electrically small loop [3,4], and the performance of the latter when the loop is body-mounted, compared to the free-space behavior, is shown in the azimuth radiation pattern plots of Figure 4.5. The free-space pattern has two finite nulls in the 90- and 270-deg positions. When body-mounted, the resulting pattern shape depends on the body location and loop orientation. Simple small-loop theory tends to assume constant current around the loop with no phase change. This in turn would predict infinitely deep nulls. In practice, the loop size is large enough to produce a nonuniform current distribution around the loop, which has the effect of filling in the two nulls. For electrically larger loops, the notch depth is decreased, and if the circumference approaches half a wavelength, the null disappears altogether and the antenna exhibits electric rather than magnetic characteristics, with the associated poorer onbody response. A typical one-turn air loop mounted on a pager chassis has a null depth of about 15 to 20 dB (depending on

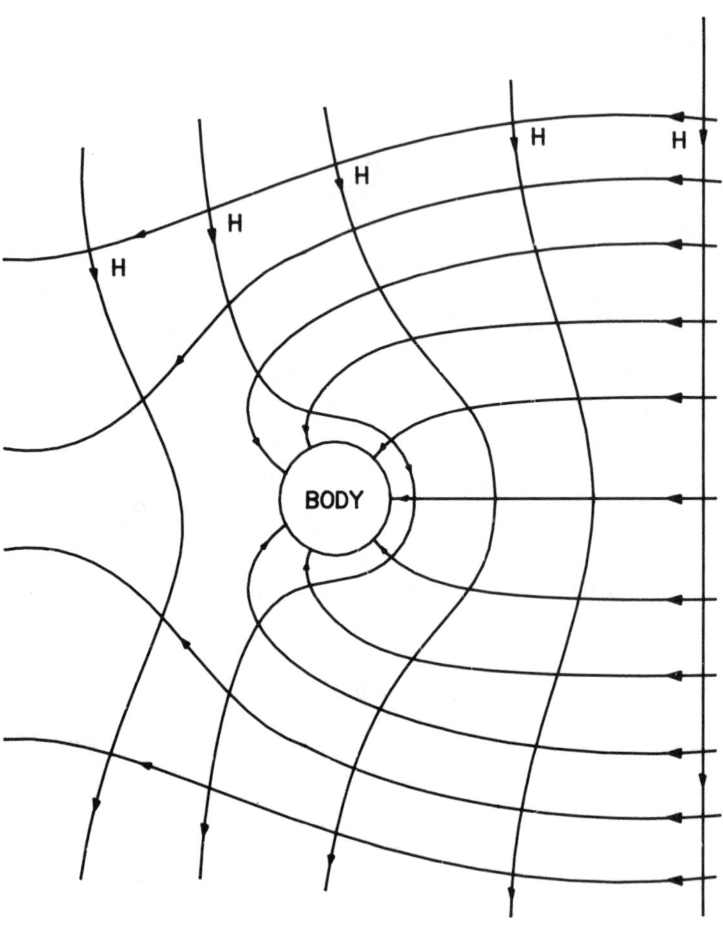

Figure 4.4 Plan view of magnetic field around the human body.

frequency) while maintaining a predominantly magnetic effect, and is thus very effective on the body.

An improvement in performance may be obtained by integrating metallic parts of the receiver into the antenna. However, the antenna cannot be reoriented such that it is wound around the maximum pager cross-sectional area, since this has the effect of rotating the radiation pattern by 90 deg, and creating a null in the forward direction, which is clearly not acceptable.

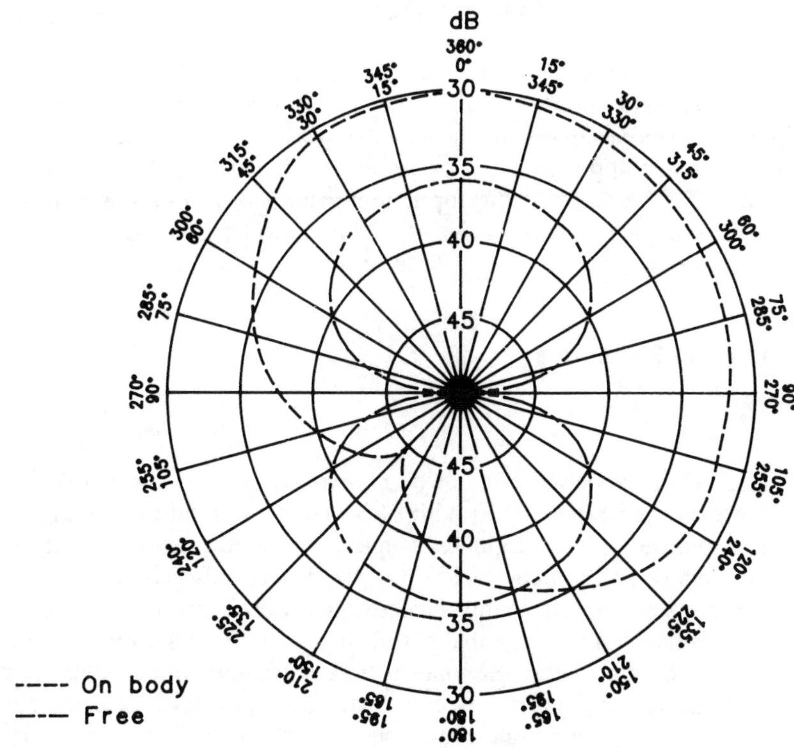

Figure 4.5 Radiation pattern of typical magnetic antenna expressed as ratio of electric field (μV/m) to antenna output voltage (μV).

Electric Antennas in Free Space

Electric field antennas have not been used in pagers in the VHF frequency band, principally because of the deep null in the electric field close to the human body. However, in a free-space situation, there are several advantages to be gained by using an electric field antenna, especially if it is used in conjunction with a magnetic antenna.

Experiments have proved that the most efficient electric field antenna that is compatible with a pager chassis is the top-loaded monopole consisting of a short length of wire loaded with a widely spaced coil. Typically, four turns, wound conformally to the extreme dimensions of the chassis, are found to be resonant in the pager bands. The azimuth pattern performance of this antenna is compared with magnetic antenna performance in Figure 4.6. These patterns show a significantly superior free-space performance on the part of the electric field antenna, as well as a total absence of nulls in this azimuth plane. The loaded monopole radiation resistance is typically five times that of the magnetic antennas for typical pager dimensions and accounts for the superior performance of the latter of some 2 to 6 dB. For some applications, a pager will be required to function when not body-mounted. It is found that the top loading coil acts as an antenna in its own right when the chassis is placed on its side or face, which is a useful design compromise when the pager is lying on a desk or table.

4.1.2 Aspects of Manufacture

Noise

Thermal noise is unavoidable, but can be minimized by conventional design procedures; however, noise generated within the receiver circuitry requires careful consideration. The noise may originate from local oscillator-derived signals, limiter signals, data signals, and any combination of the latter. This may be minimized by judicious layout, screening, and interfering signal level control. It is also evident that antenna immunity to this problem will depend very much on the geometry and near-field response of the antenna itself. Experiments have established that the loop antenna is much more susceptible to internally generated interference than the ferrite loop antenna, due to the fact that the near fields of the latter are condensed closer to the ferrite core.

Antenna Tuning and Matching

Antenna tuning and matching can be one of the most critical processes in manufacturing a radio pager. The main requirements are:

- The physical movements required to accurately tune and match the antenna should not be too small.
- When the antenna is correctly tuned and matched, the presence of nearby objects and particularly the human body must not be significant.

Both the above constraints require the antenna Q factor to be designed to less than a given value. A design guideline based on the angle of rotation of a commercially available trimmer capacitor is

$$Q < \frac{9f}{f_2 - f_1} \qquad (4.1$$

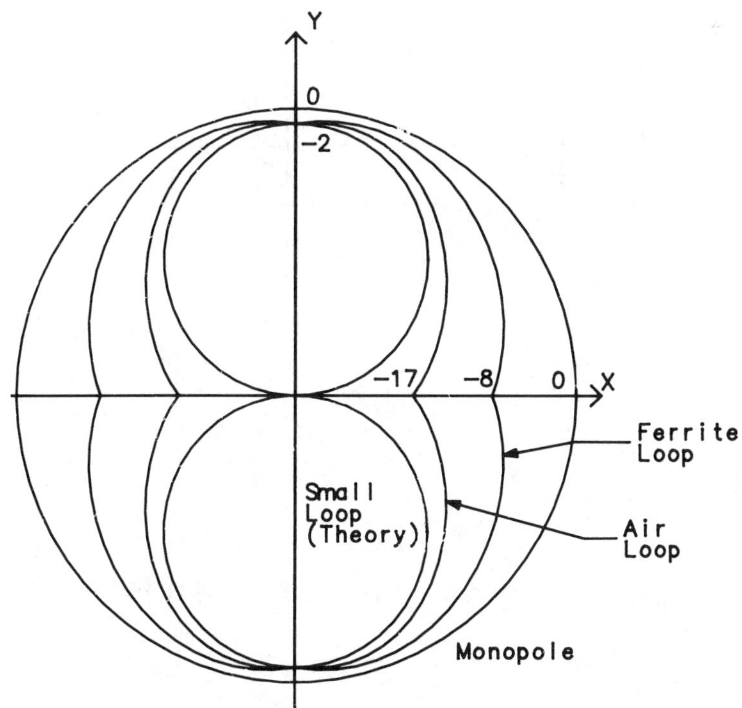

Figure 4.6 Top-loaded monopole performance compared with magnetic antennas.

where f is the nominal frequency of the pager and f_2 and f_1 the upper and lower limits of the band. For instance, Q should be less than 67 for a 20-MHz band centered on 150 MHz to ensure sufficient angular physical movement while tuning up. As regards body proximity effects, induced losses have the effect of reducing sensitivity, but the detuning effect can be reduced by limiting Q so that

$$Q < C/\Delta C \tag{4.2}$$

where C is the tuning capacitance and ΔC is the change of capacitance induced into the antenna by the proximity of the human body. For the air-cored loop antennas of Figure 4.7, typical values are $C = 16$ pf, $\Delta C = 0.2$ pf; hence, $Q < 80$, while for the ferrite-cored loops $C = 8$ pf and $Q < 40$. For more complicated antenna circuits (i.e., having capacitive or inductive matching taps), these capacity values will differ in magnitude.

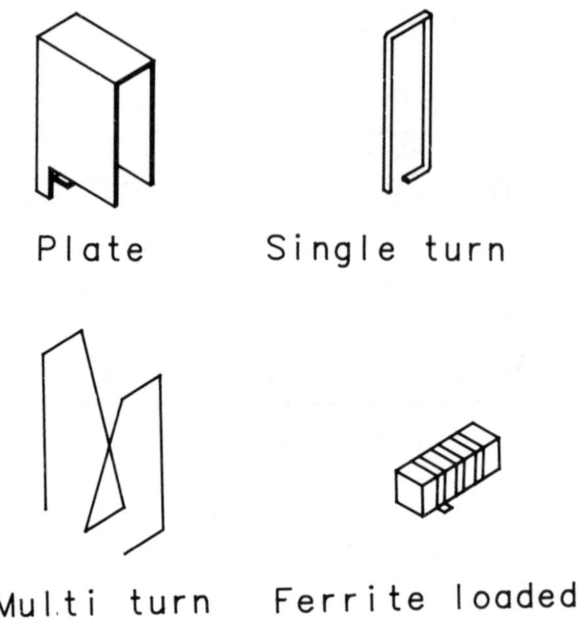

Figure 4.7 Loop antennas.

4.1.3 Measurement of Pager Antenna Performance

The accurate measurement of pager antenna performance requires a certain amount of care in decoupling the device under test from any conductors in the immediate vicinity. Such conductors (e.g., cables) could have better antenna performance than the pager under test.

The method often used in testing pager performance is to detect incoming signals at the threshold of the pager sensitivity. If this parameter is known by bench measurements, then the antenna equivalent area or its gain can be determined if the incident electric and magnetic fields are also known. This methodology has the clear advantage of using the pager acoustic or mechanical alerting devices to indicate the presence of a signal strong enough to be detected by the receiver. In this condition, no metal wires need to be connected to the pager, whose antenna performance can be severely affected by the presence of such conductors. Another method consists of packing a low-power (few milliwatts) transmitter in the form factor of the paging receiver and using it to test antenna performance. Although this procedure can be used to establish the effects of the human body on the angular coverage of the pager, the real performance of the antenna cannot be measured until the radio receiver with all its printed circuit (PC) boards and components is placed in the package. In order to collect meaningful results, the testing fixture for the pager antenna should resemble the final package of the product in the geometry of the conducting surfaces.

Let us now quantify the considerations of the method to detect pager antenna performance. From basic theory [5], the following relationship exists between the gain above isotopic and the equivalent area of an antenna with no match loss:

$$G = 4\pi A/\lambda^2 \tag{4.3}$$

where λ is the wavelength of the incident electromagnetic energy. The power density of the incident wave is simply given by

$$P_d = |E|^2/120\pi \tag{4.4}$$

in watts per square meter or watts per square centimeter, depending on the units of $|E|$, the electric field. The power received by an antenna of equivalent area A is then

$$P_r = \frac{|E|^2\lambda^2 G}{480\pi^2} \quad \text{watts} \tag{4.5}$$

The power received by the antenna is delivered to a load, representing the input impedance of the receiver. For the convenience of the design effort, the receiver input impedance is set very often at (50 + j0)Ω. Using the voltage delivered by the antenna to a 50Ω load, V is

$$V = |E|\,\lambda\sqrt{\frac{G}{94.7}} \quad \text{volts} \tag{4.6}$$

Equation (4.6) simply states the obvious result that a constant cross-section antenna maintains its performance as the frequency decreases. It is often convenient to measure the incident electric field in microvolts/meter (μV/m) and express the pager sensitivity in the same units; in the process it is also convenient to measure the received voltage in terms of dBμV (decibels relative to microvolts across a 50Ω load). Equation (4.6) expressed in decibels becomes

$$V_{\text{dB}\mu\text{V}} = E_{\text{dB}\mu\text{V/m}} + 20\log\lambda + G_{\text{dBi}} - 19.8 \tag{4.7}$$

In (4.7), the wavelength λ is measured in meters and the gain G in decibels above isotropic. Equation (4.7) relates all the important characteristics of the paging receiver. In a good antenna range or an anechoic chamber, the incident E field can be calibrated to within ± 0.25 dB; the received RF power necessary to trigger a pager can be measured within 0.1 dB with commercially available instruments. Using (4.7), the absolute gain of the pager antenna can be measured with an uncertainty of ± 0.35 dB, which is acceptable for most commercial applications.

Let us now briefly discuss the characterization of pager antenna performance. Clearly, from (4.7) it is possible to measure the gain $G(\theta, \phi)$ for any given angular direction. For directional antennas pointing to specific transmitters or targets, the performance is normally evaluated in terms of maximum gain and sidelobe level below the peak of the beam. This criterion is strictly a way to evaluate the energy transfer from one antenna to another or from one antenna to itself by means of an echo from a target. For pager antennas, the use of the maximum gain criterion only may not be the best for the application, because if there are wide gaps in the coverage of pager antenna, the user may not receive some messages incident from the blind directions. A better evaluation method than the maximum gain criterion is to average the gain of the pager antenna over the 360-deg angular sector at the horizon of the user. This hypothesis is correct in most cases. Given the slow angular variation of the gain of small antennas, the receive characteristics remain practically constant in the elevation sector ± 10 deg around a direction at the horizon. Using (4.7), the performance parameter can be obtained directly as a gain average or by averaging the square of the received voltages in microvolts.

Table 4.1 summarizes the measured performance of pagers from a variety of manufacturers at frequency bands allocated to paging. There is some spread of values in each frequency band, depending on antenna size and packaging of the receiver. The values given in the second column in Table 4.1 represent an eight-point average of the gain of the antenna when the pager is worn at the belt level of a human-equivalent phantom [6]. In the third column, the maximum measured value of the gain is tabulated. In the last column, the maximum spread of gain value over the eight-point sample is shown at a given frequency.

At the very low frequencies (30 to 80 MHz), the performance of the antennas is uniformly poor. The paging device is probably just coupling into the currents

Table 4.1
Measured Pager Antenna Performance (Tone-Only Signals)

Frequency Band	*Average Gain (dBi)*	*Maximum Gain (dBi)*	*Gain Spread (dB)*	*Receiver Sensitivity (dBm)*
30–80 MHz	−32	−30	5	−129
VHF	−23	−20	10	−129
UHF	−19	−14	18	−128
900 MHz	−9.0	−2.5	30	−127

excited on the simulated human, which is the true antenna at these long wavelengths. The only positive quality of these antennas is that they provide almost uniform coverage in function of an azimuthal angular reference. The gain values of pagers show a substantial improvement in the 150- to 280-MHz band. The average values given in the table reflect a somewhat uniform coverage in the azimuthal plane, with variations in gain at the horizon of no more than 5 dB. At 450 MHz, the peak value of pager gain approximates 0 dBi, but the coverage ceases to be uniform. Deep nulls, on the order of −10 dB below the maximum, appear in the direction opposite to where the pager is located on the phantom. The human equivalent is a column of salt water 30 cm in diameter and 180 cm high. At 450 MHz, the phantom casts a penumbra at the backside of the incident wave. Finally, at 800 to 900 MHz, the shadow effect from the phantom becomes very clear. Deep nulls (20 to 30 dB) below the peak gain value are found in an angular region 20 to 30 deg opposite the location of the antenna.

The values given in Table 4.1 are representative of the expected performance on an average human, if such a person exists. The designer should expect some variation in average and maximum value if a specific individual is used in the measurement. Experiments have shown that belt-level results were so dependent on the individuals wearing the pager that the human phantom, however crude, was a welcome standard setter [6].

4.1.4 Pocket Pager Size Constraints

The antenna size (it must fit in a small pocket) constitutes the essential problem for this class of radio receivers and is often the only obstacle to good performance of the equipment. All the comments relative to loss versus antenna size presented in the previous sections can be repeated here with stronger emphasis. In addition, the orientation of the pager with respect to the incident field is no longer predictable as in the case of belt-worn receivers. With these constraints in mind, let us review the few options available to the designer.

To minimize matching losses, the frequency of operation of the pager should be as close as possible to the frequency of resonance of the antenna. The use of self-resonant antennas is strongly recommended. It is the designer's good fortune that most pocket-sized equipment is requested for the newer available paging channels, which are at higher frequencies than older ones.

Finally, there are obvious exceptions to the considerations of this section. A pen-sized pager has one dimension (length), which gives the designer an obvious advantage. The design of pocket-sized antennas becomes problematic if the form factor of the pager is similar to that of a credit card, a small cigarette lighter, a watch, a lipstick container, or other items normally found in the pocket or the purse, and the frequency of operation is relatively low.

4.1.5 Concluding Remarks

This section has been descriptive and qualitative rather than analytical and quantitative, due to the impossibility of clearly defining the antenna for a very small, pocket-sized pager. As already pointed out, the components or the PC board of the device will be practically part of the antenna because of the tight packing requirements. In addition, the presence of other objects in the pocket or in the purse can radically affect the performance of the paging device. While from a purely experimental point of view there seems to be little relief for the design effort, analytical methods are very much on the verge of providing the tools to solve some of the problems presented in this section. Moment methods, finite element methods, or other analytical techniques that can solve Maxwell's equations for very complicated boundary conditions represent practically the only chance to properly design these antennas. The use of a modern high-speed computer to simulate deterministically the pager package performance as an antenna and, statistically, the influence of nearby external objects is the only tool on the horizon that can rationalize the design of these devices. Certainly, the availability of high-frequency, low-loss ferrites, low-loss dielectric materials, and high-temperature superconductors will help the performance level of pocket-sized pagers. However, the predictability of the performance of these devices will be based essentially on analytical techniques and early product definition, which, properly coupled, are the basis of any successful engineering design effort.

4.2 PAGER TYPES AND PERFORMANCE

4.2.1 Design Considerations

The salient design factors for a pager antenna are as follows. First, the paging service in Japan uses the VHF band, and since the pager must be less than 10 cm long, the antenna is electrically small for the service frequency wavelength. The pager's antenna, which is built into the body, must thus be an electrically small antenna. The radiation resistance of such an antenna is smaller than the ohmic loss of the antenna element [7–9]; therefore, the efficiency of the pager's antenna is low. Second, when the antenna is built into the pager, it must be in close proximity to metal parts or a battery; this alters the impedance and the gain of the antenna [10]. Third, pagers are carried in shirt pockets or hung on the belt. Thus, the antenna characteristic is affected by the human body [11–15]. For good pager sensitivity, the antenna must have a high efficiency that is not unduly reduced in the vicinity of a human body or metal parts. Small-loop antennas are known to retain their gain near the human body, and are one of the most promising antennas for pagers. The characteristics of a small-loop antenna and basic design formulas are now summarized.

Antenna Characteristics Near the Human Body

The influence of the human body on the electric and magnetic components of an electromagnetic wave is a substantial effect [11]. The electric component is reduced significantly near the human body, whereas the magnetic component increases. Small-loop antennas are strongly affected by the magnetic component. A loop antenna can be regarded as a magnetic dipole normal to the loop area [16], as shown in Figure 4.8. Therefore, if the loop antenna is built into a pager so that it is perpendicular to the human body, the magnetic dipole is parallel to the human body and the sensitivity of the pager is not reduced. The magnetic component in the high-frequency band can be captured very efficiently if the antenna is reinforced with ferrite material [17,18]. Unfortunately, the loss of ferrite is excessive in the VHF band [19].

Efficiency

The efficiency η of a small-loop antenna is [2,4]:

$$\eta = \frac{R_r}{(R_r + R_{\text{loss}})} \tag{4.8}$$

where R_r = radiation resistance and R_{loss} = ohmic loss of the loop element. There

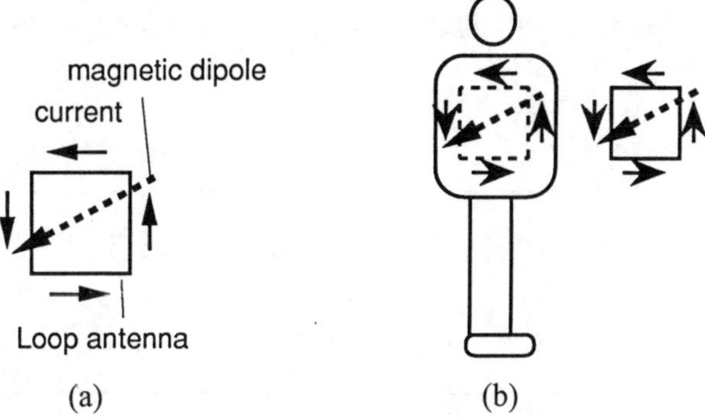

Figure 4.8 The response of a small-loop antenna in the vicinity of a human body: (a) action of small-loop antenna; (b) response in the vicinity of a human body.

are two methods of increasing the efficiency: one is to increase R_r and the other is to decrease R_{loss}. A small-rectangular-loop antenna, whose longer side is l_a and whose shorter side is l_b, has a radiation resistance R_r and an ohmic loss R_{loss} [2] given by

$$\begin{aligned} R_r &= 320\ \pi^4 \left(\frac{l_a^2 l_b^2}{\lambda^4} \right) \\ R_{\text{loss}} &= \frac{1}{\pi a} (l_a + l_b) \sqrt{\frac{\pi \mu f}{\sigma}} \end{aligned} \tag{4.9}$$

where a is the wire radius of the loop element, μ is the permeability, f is the frequency, λ is the wavelength, and σ is the conductivity of the element.

R_r can be effectively increased by increasing the loop area or the number of loop turns [7]. The ohmic loss can be reduced by thickening the diameter of the element, by using platelike elements to increase the element surface, or by using low-loss material for the antenna element. In commercial pager antennas, the ohmic loss is reduced by using plate-shaped elements.

Matching Circuit

The input resistance of a small-loop antenna is very low (less than 1Ω), and a matching circuit is required to connect the antenna to the receiving circuit. Thus, the loss of the matching circuit affects pager sensitivity significantly. Pagers frequently employ L-shaped matching circuits with two capacitors. The principle of the matching circuit is shown in Figure 4.9(a) [20], and in general the circuit consists

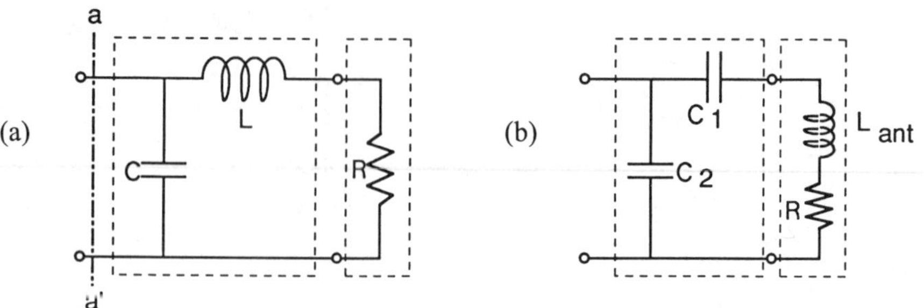

Figure 4.9 Matching circuit for: (a) load R; (b) small-loop antenna. (From [20] © 1964 Ohm Co., Ltd.)

of an inductor and a capacitor. The admittance from *a-a′* toward the right-hand side in Figure 4.9(a) is expressed as

$$Y = \left(\frac{R + \frac{X_1}{Q_L}}{\left(R + \frac{X_1}{Q_L}\right)^2 + X_1^2} + \frac{Q_C}{X_2(1 + Q_C^2)} \right) + \mathrm{j}\left(\frac{Q_C^2}{X_2(1 + Q_C^2)} - \frac{X_1}{\left(R + \frac{X_1}{Q_L}\right)^2 + X_1^2} \right)$$

$$X_1 = \omega L = 2\pi f L \qquad (4.10)$$

$$X_2 = \frac{1}{\omega C} = \frac{1}{2\pi f C}$$

where R is the load resistance, L is the inductance, and C is the capacitance shown in Figure 4.9(a). Q_L and Q_C are Q factors of the inductance L and the capacitance C, respectively. X_1/Q_L and X_2/Q_C are the loss of the inductance and the capacitance, respectively. Generally, the loss of an inductance is larger than that of a capacitor. Impedance matching is realized if inductance L tunes with C, and then the real part of the right-hand side of (4.10) is the required input resistance of the receiving circuit.

For impedance matching of a small-loop antenna to a receiving circuit, the low-loss matching circuit can be constructed with just capacitors C_1 and C_2 because of the large inductive reactance L_{ant} of the loop antenna (see Figure 4.9(b)); ωL in Figure 4.9(a) is then equivalent to the series combination of ωL_{ant} and $1/\omega C_1$ in Figure 4.9(b).

Another matching arrangement with low loss can be designed by unifying the antenna and the matching circuit [21]. This circuit is shown in Figure 4.10(a). Since there is only one capacitor in the circuit, the loss of the matching section is much lower than that in the first example. In this case, impedance matching between the antenna and the circuit is carried out by capacitance tuning at the paging system frequency and by selecting a tapping point on the loop element so that the input impedance of the antenna equals the input impedance of the circuit. The equivalent circuit of this circuit is shown in Figure 4.10(b). The input admittance Y_{in} is expressed as (4.11), given the approximation of a small-loop antenna, ωL, $1/\omega C \gg R_r + R_{\mathrm{loss}}$.

$$Y_{\mathrm{in}} = \frac{\omega^2 (R_r + R_{\mathrm{loss}})(L_1 + M)^2}{\left[\omega^2 (L_1 + M)^2 - \omega L_1 \left(\omega L_T - \frac{1}{\omega C}\right)\right]^2} + \mathrm{j}\frac{\left(\omega L_T - \frac{1}{\omega C}\right)}{\omega^2 (L_1 + M)^2 - \omega L_1 \left(\omega L_T - \frac{1}{\omega C}\right)}$$

$$L_T = L_1 + L_2 + 2M \qquad (4.11)$$

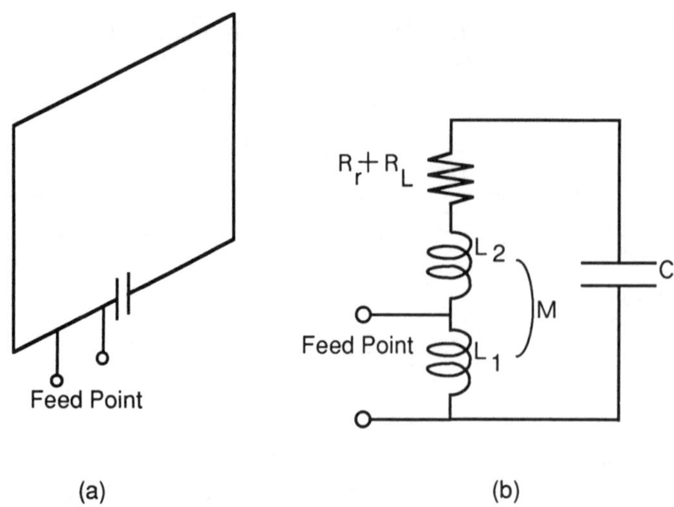

Figure 4.10 Matching circuit unifying the antenna: (a) construction figure; (b) equivalent circuit. (From [11], © 1973 Matsushita Electric Industrial Co., Ltd.)

where L_1 and L_2 are the inductances and C is the loading capacitance shown in Figure 4.10(b). M is the mutual inductance existing between the inductances L_1 and L_2. At resonance, the input admittance $Y_{\text{in/res}}$ is expressed as (4.12), where $\omega L_T - 1/\omega C = 0$.

$$Y_{\text{in/res}} = \frac{(R_r + R_{\text{loss}})}{\omega^2 (L_1 + M)^2} \tag{4.12}$$

L_1 and M are decided by the tap position.

Directivity and Polarization

When a pager is carried, it is turned in all directions. Thus, the pager antenna must be omnidirectional, and the vertical and horizontal polarization characteristics of the antenna must be carefully designed. A commercial pager can have a built-in antenna that is both vertically and horizontally polarized. The antenna design specifies unbalanced current flow on the loop element and circuit boards and that the circuit board functions as part of the antenna [2]. Figure 4.11 illustrates balanced and unbalanced systems of the loop antenna, which is placed on the side of the pager. The behavior of the antenna in this location is partly balanced and partly

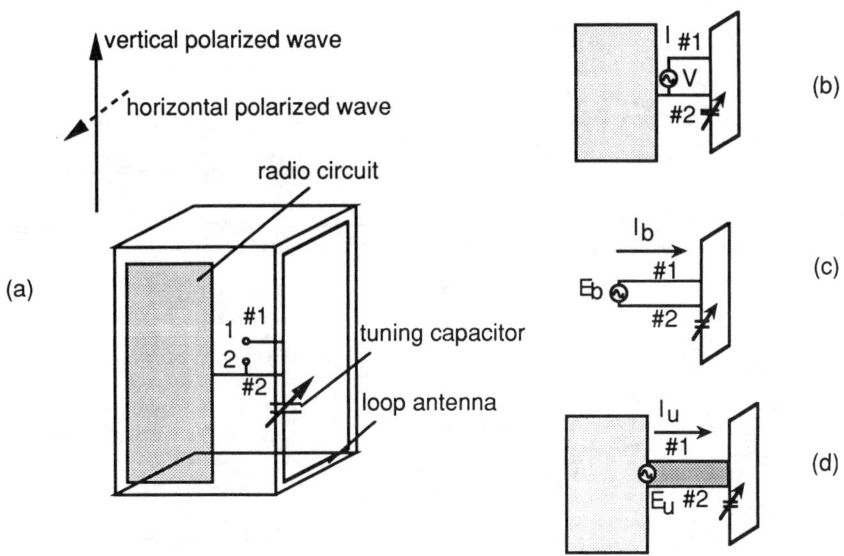

Figure 4.11 The behavior of a small built-in loop antenna: (a) small-loop antenna built into the pager; (b) original; (c) balanced mode; (d) unbalanced mode. (From [11], © 1973 Matsushita Electric Industrial Co., Ltd. Reprinted with permission.)

unbalanced, as shown in Figure 4.11. In the balanced mode, a balanced current flows on the loop element and the antenna receives the vertical polarized wave (shown in Figure 4.11(a,c)). In the unbalanced mode, the loop antenna and the ground of the circuit board function as an unsynthesized dipole antenna, and the antenna receives the horizontal polarized wave (shown in Figure 4.11(a,d)). Figure 4.12 illustrates the receiving pattern of this combination antenna in free space. The polarization of the loop antenna is the same as the polarization of the incident wave in case A. The polarization of the loop antenna is perpendicular to the incident wave in case C; the receiving level in case C is not unduly reduced compared to that in case A (the receiving level is 5 dB less than in case A). This is why this pager antenna responds as a combination antenna. On the other hand, when the loop antenna is built into the pager so that it is perpendicular to the polarization of the incident wave in case B, the receiving level has two nulls that are 15 dB less than they would be if the loop antenna were parallel to the polarization of the incident wave. Figure 4.13 illustrates the receiving pattern of the combination antenna in the vicinity of a human body. The directional gain increases about 5 dB from the free-space value.

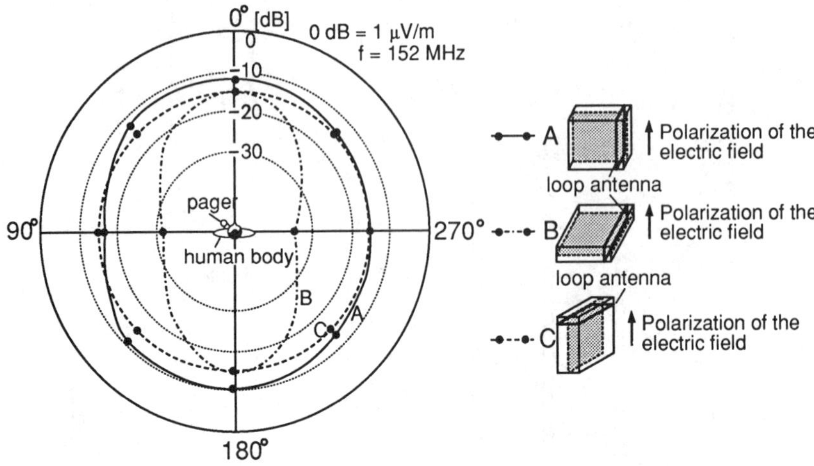

Figure 4.12 Receiving pattern of a pager in free space. (From [11], © 1973 Matsushita Electric Industrial Co., Ltd.)

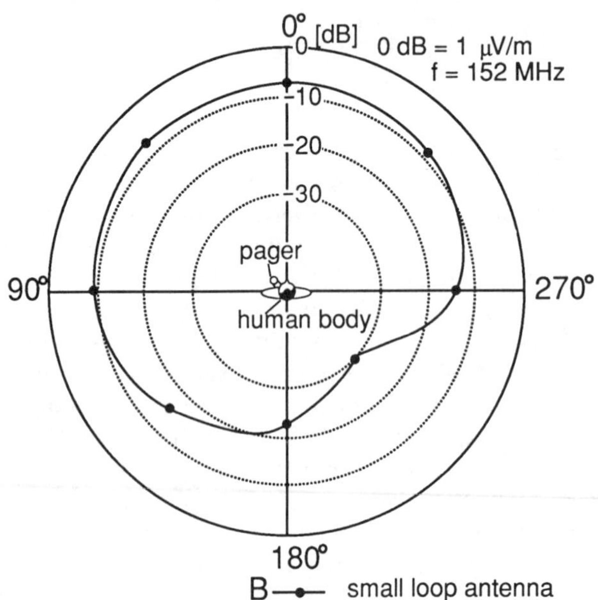

Figure 4.13 Horizontal receiving pattern of the pager in the vicinity of the human body. (From [11], © 1973 Matsushita Electric Industrial Co., Ltd.)

4.2.2 Pocket-Sized Equipment

The first Japanese paging system operated in the 150-MHz band and was designed by NTT [22]. The initial service was rather simple, and only sounds and flashing LEDs were used. Pager size was about 120 × 60 × 20 mm. After the number of users started to rapidly increase, the frequency band was changed from the 150- to the 250-MHz band, and a digital signaling system was introduced to increase customer capacity. There are now many kinds of services [23], such as LCD screens, display alphanumerics, Kanji, and illustrations. Pagers now come in many shapes. Figure 4.14 illustrates some typical current pagers. The service areas were first centered in each prefecture of Japan. So if users went to another prefecture (area), they could not receive any service. To solve this problem, the Wide-Range Paging Service was made available, employing a pager that can receive plural frequencies.

From the beginning of the paging service, ferrite-loaded loop antennas were employed. After the introduction of the 250-MHz frequency band, air-spaced loop antennas were used because there was no ferrite material that had low loss in this band. We will describe some examples of current commercial pagers and their antennas.

The first example is the tone-only pager with a size of 67 × 42 × 13 mm. It must achieve its maximum gain when it is carried in the breast pocket of a shirt (shown in Figure 4.15(a)). Therefore, the loop antenna is perpendicular to the human body and vertical so that the antenna is parallel to the polarization of the incident wave (shown in Figure 4.15(b)). The loop size is 60 × 10 mm, and the element diameter is 1 mm. Figure 4.15(c) illustrates the receiving patterns of the antenna built into the pager in free space and in the vicinity of a human body. The sensitivity is increased with proximity to the human body.

The second example is the pen-type pager, whose size is 130 × 24 × 13 mm. It is also designed to be carried in the breast pocket of a shirt (shown in Figure 4.16(a)). The antenna is also a rectangular-loop antenna, and its size is 45 × 8 mm. A plate-shaped element is used that is 2 mm wide and 1 mm thick. This antenna lies along the long side of the pager and its length is half that of the pager (shown in Figure 4.16(b)). The antenna is kept away from the circuit board, the battery, and the display, because these reduce antenna efficiency. Figure 4.16(c) illustrates the receiving patterns of the pager in free space and in the pocket. Close to the human body, the sensitivity increases by about 2 dB, so the small-loop antenna characteristic is quite discernible.

The third example is the tone-only type of pager of the Post Office Code Standardization Advisory Group (POCSAG) system, which has a loop antenna in the long side of the pager, like the NTT system (shown in Figure 4.17(a)). The POCSAG system uses the 150-MHz band. The receiving patterns of the antenna are shown in Figure 4.17(b). The sensitivity increases by about 5 dB if placed in the shirt pocket.

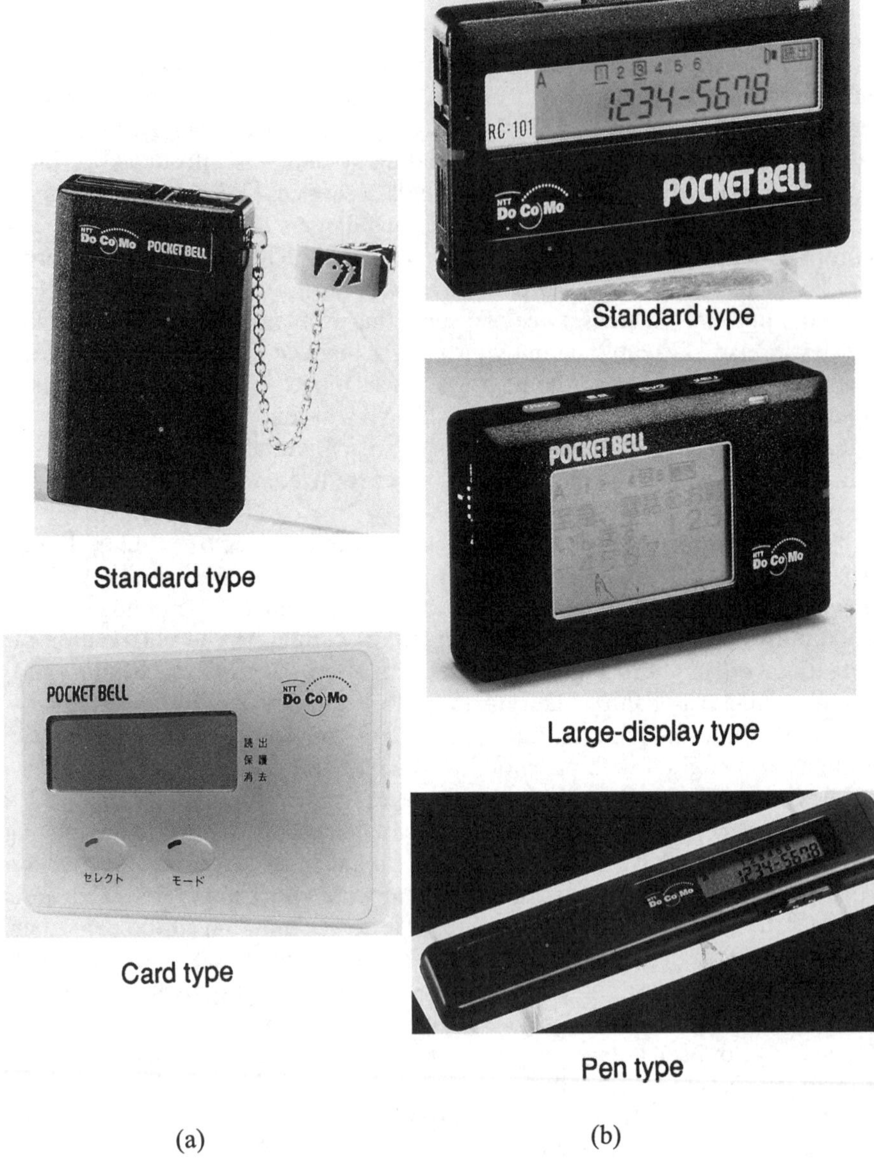

Figure 4.14 Samples of pagers in Japan: (a) tone-only pager; (b) display pager.

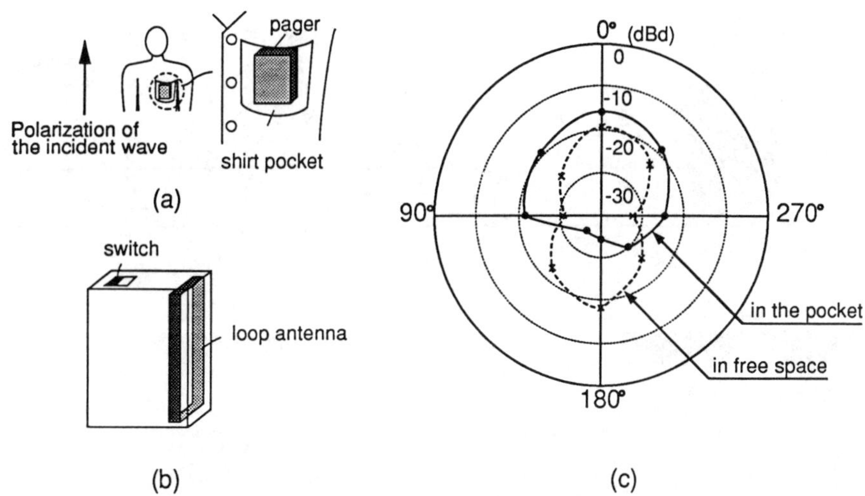

Figure 4.15 Tone-only type of pager: (a) intended usage; (b) built-in antenna; (c) receiving pattern.

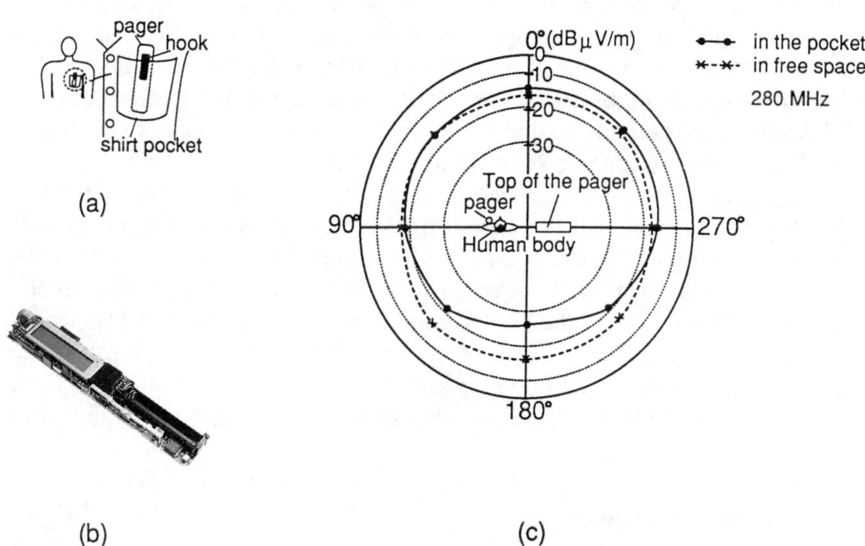

Figure 4.16 Pen type of pager: (a) intended usage; (b) built-in antenna; (c) receiving pattern. ((b,c) are courtesy Matsushita Communication Industrial Co., Ltd.)

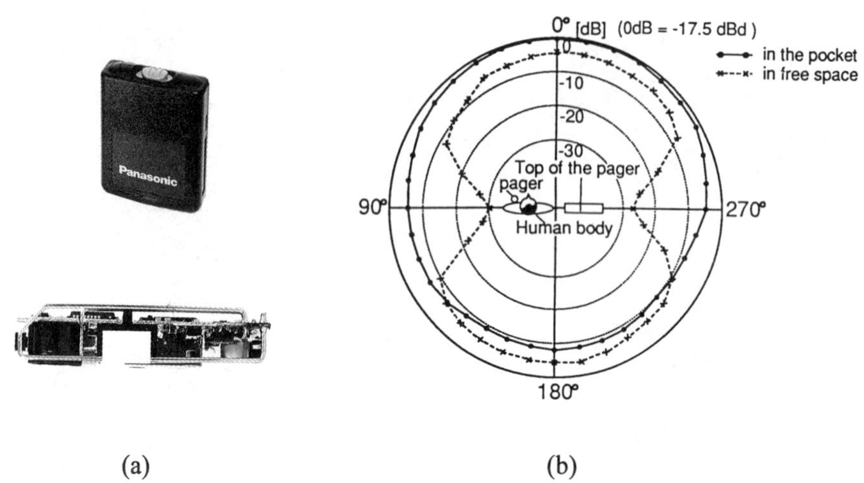

Figure 4.17 Tone-only type of pager (in POCSAG): (a) built-in antenna; (b) receiving pattern. ((a,b) are courtesy Matsushita Communication Industrial Co., Ltd.)

The fourth example is the Kanji display model of the POCSAG system. It is designed to achieve maximum gain when it is waist-mounted. The loop antenna is thus arranged along the short side of the pager housing so that the antenna is parallel to the polarization of the incident wave (shown in Figure 4.18(a)). The antenna size is about 50 × 15 mm (shown in Figure 4.18(b)). Figure 4.18(c) illustrates the receiving patterns. In this case, the sensitivity is increased by about 5 dB when the pager is carried on a waist belt.

Figure 4.19(a) illustrates how pager antenna gain can be increased by increasing the number of turns. In this example, the number of turns is two, loop sizes are 49 × 9.5 and 37 × 9.5 mm, and the element diameter is 1 mm; these loop antennas are to be built into a pager, which is 45 × 60 × 15.5 mm. As the number of turns increases, both the effective aperture of the antenna and the gain increase. The receiving patterns of this antenna are shown in Figure 4.19(b). The sensitivity in proximity to the human body is about 3 dB more than that of the Kanji display model (whose loop size is almost the same), so the pager has high sensitivity.

4.2.3 Card-Sized Equipment

A card-type pager was shown in Figure 4.14. The thickness of the pager is only 5 mm, which is half that of the tone-only type. Thus, it is difficult to achieve an antenna aperture that yields satisfactory pager sensitivity. To solve this problem,

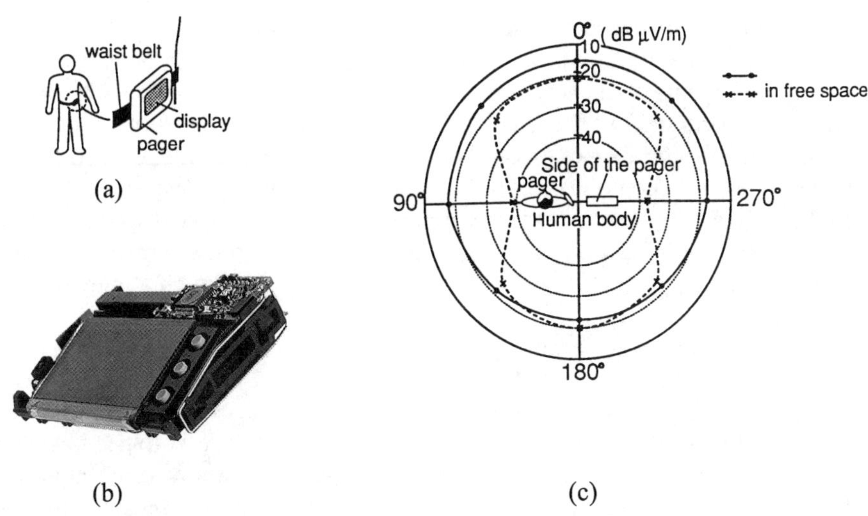

Figure 4.18 Kanji display type of pager: (a) intended usage; (b) built-in antenna; (c) receiving pattern. ((b,c) are courtesy Matsushita Communication Industrial Co., Ltd.)

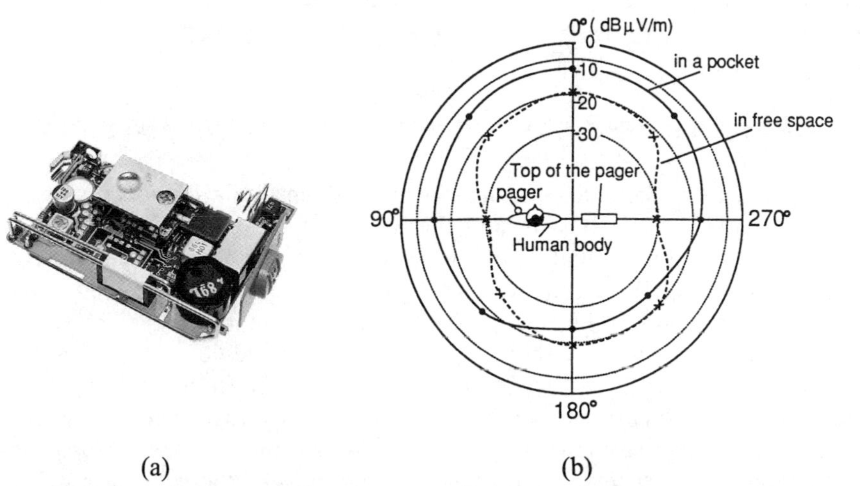

Figure 4.19 Two-turn loop antenna for a pager: (a) built-in antenna; (b) receiving pattern. (Courtesy Matsushita Communication Industrial Co., Ltd.)

the antenna is constructed as shown in Figure 4.20(a) (i.e., the pager body is made of two large metal plates). One edge of both side plates is connected to the input circuit and some of the other edges are shorted, so the pager becomes a loop antenna. Using the two large side plates allows the loop aperture to be maximized while reducing the element loss. The receiving pattern of this antenna is shown in Figure 4.20(b). The polarization direction of this antenna runs from the feed point to the shorted point [24]. If the shorted point is moved along the edge of the pager, the polarization direction is changed because the currents on the antenna (metal plate) are changed. Figure 4.21 illustrates the polarization characteristic when the shorted point is moved. This characteristic can be used to increase pager sensitivity by matching the polarization of the antenna to that of the incident wave.

Another method of controlling the polarization direction is to use electronic circuits [24]. This is carried out by impressing voltages on points A and B and shorting the points C and D (see Figure 4.21). When the voltages have the same amplitudes and phases, the receiving pattern is maximum at $\phi = 90$ and 270 deg (shown in Figure 4.22(a)). On the other hand, if the phases are opposite, the receiving pattern is changed and the maximum directions are 0 and 180 deg (shown in Figure 4.22(b)). The reason is as follows. When the phases are the same, the current on the antenna flows mainly between A and D and between B and C. Few currents flow between A and B or D and C. When the phases are in opposition, the current flows mainly between A and B, and few currents flow between A and D or B and C.

This section has described the small antennas of existing pagers. Public demand is for even smaller pagers that have higher performance. It is necessary, therefore, to develop very small but high-performance antennas.

4.3 DESIGN TECHNIQUES FOR PORTABLE PHONE ANTENNAS

The trend of the portable telephone technology in the last few years has been to dramatically decrease the size and the weight of the unit. In the initial offering (1984), the portable cellular phone was about 35 in^3 in volume (including batteries) and weighed about 30 oz (0.85 kg). Today (1991), the most recent cellular portables have a volume of 11.6 in^3 and weigh less than 7.7 oz (0.22 kg). These dramatic weight and volume improvements have necessitated a rapid evolution of the antennas used for the phones. The design efforts have been to maintain approximately the same antenna performance in terms of gain, coverage, and bandwidth in the face of rapidly decreasing size requirements. We will limit our discussion to antennas in the 800- to 1,000-MHz band, which covers most of the cellular systems in use at this time. These antennas can be used at higher frequencies by simply reducing their size by frequency scaling techniques.

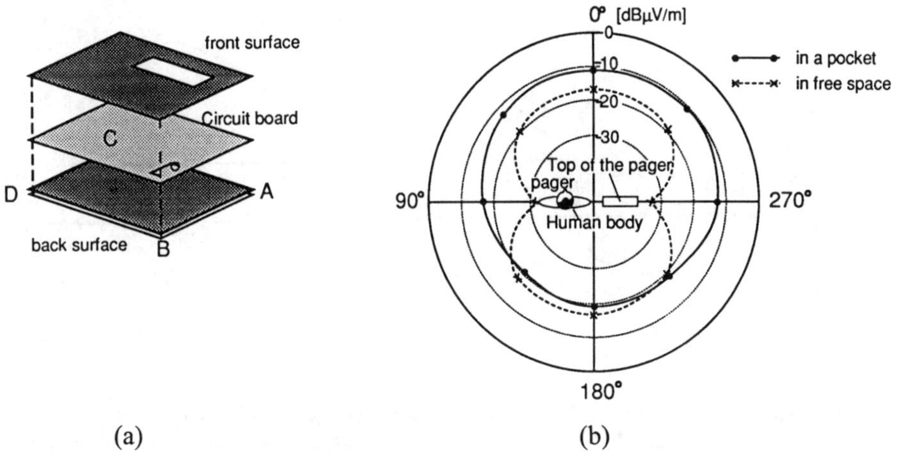

Figure 4.20 Antenna of the card type of pager: (a) construction of the antenna; (b) receiving pattern. ((b) is courtesy Kokusai Electric Co., Ltd.)

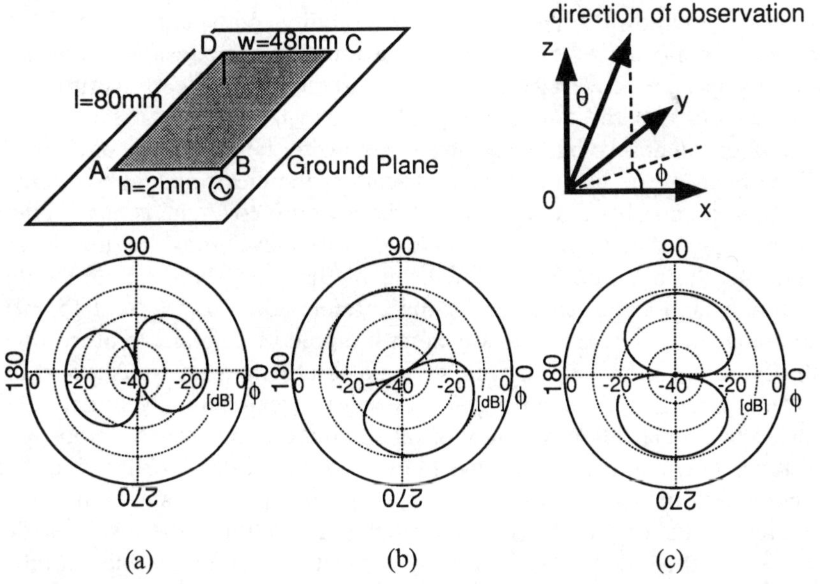

Figure 4.21 The mechanical method of polarization control [24]: (a) feed point is B, shorted point is A; (b) feed point is B, shorted point is C; (c) feed point is B, shorted point is D.

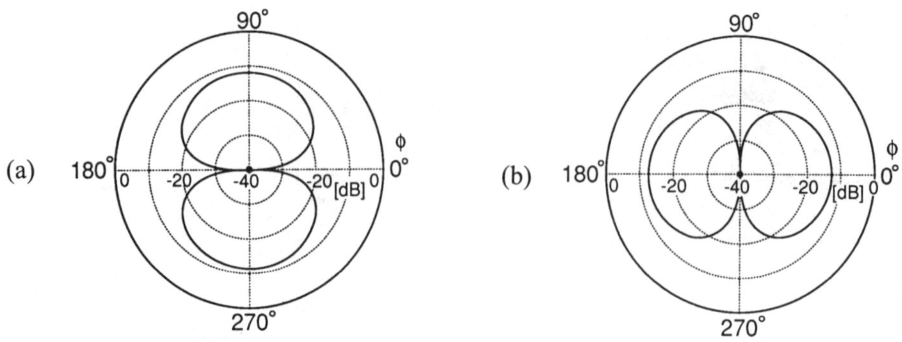

Figure 4.22 The electrical method of polarization control [24]: (a) polarization characterization with same amplitude and phase at A and B; (b) polarization characterization with same amplitude and opposite phase at A and B.

4.3.1 Design Considerations

The main factors affecting the design of portable cellular phone antennas are the relatively large bandwidth (~10%), the desirability of a small size, and the need to provide a uniform coverage over the azimuthal angle, with a gain of about 0 dbi or higher. The coverage and size requirements are clearly contradictory in view of the fact that the antenna is held close to the head of the user, which is approximately an absorbing dielectric ellipsoid with the major axis one wavelength long. The head of the user absorbs and scatters the electromagnetic energy emanating from the nearby antenna, so that the azimuth coverage is never uniform if the head is close to the radiator. A simple way to achieve uniform azimuth coverage is to elevate the antenna above the head of the user. This solution requires a nonradiating supporting structure for the antenna about 6 in (about 15 cm) long. The antenna, in this case a half-wavelength dipole or similar radiator, also has a length of about 6 in (about 15 cm), giving a total length for the support and the radiator of about 1 ft (about 30 cm). This size is unacceptable, given that the largest dimension of cellular phones is no more than about 25 cm, which is approximately the distance between the mouth and the ears of an adult. A collapsible structure of 30-cm length presents serious mechanical problems in the small and constantly decreasing volume of portable cellular phones. In addition, the result of dropping the radio with the antenna in the extended position may be an unacceptable degradation of performance.

Alternative structures would require separating the radio from the antenna, which would be located on a hat, a headband, or some type of head-mounted support. This solution requires a link between the radio and the antennas; in addition, the users may not accept any head-mounted solution of the coverage problem.

So far, only relatively small antennas have found acceptance in the rapidly growing cellular phone market. In the following sections, we will discuss the following antenna types: (1) the sleeve dipole, (2) the helical antenna, (3) the quarter-wavelength whip, and (4) the dipole helical antenna combination. These antennas represent most of the radiators used in cellular phone technology. The quarter-wavelength whip has not found favor with the cellular phone market, so it will be only briefly discussed in the section dedicated to the helical antenna.

4.3.2 Antenna Types

Sleeve Dipole

A sleeve dipole operating in the 800- to 900-MHz band is shown in Figure 4.23. It is essentially a half-wave dipole fed from one end by a coaxial line. The structure has cylindrical symmetry, so in free space it has excellent radiation pattern uniformity in azimuth.

The radiating structure is an asymmetric dipole made of conductors of different diameters and slightly different lengths. The thinner radiator is normally the inner conductor of the coaxial line feeding the antenna. This conductor must have an appropriate length to achieve good antenna match in the band of operation. The conductor of larger diameter has the critical function in proper antenna operation and must be designed with some care. The large-diameter conductor must provide effective choking of the RF currents at its own open end and also one-half of the radiating dipole. This conductor is shorted to the braid of the coaxial line feeding the sleeve.

To provide good RF current choking, it is essential that the sleeve part of the antenna have as large a diameter as possible compatible with the aesthetic acceptability of the antenna. The choke works most effectively if the sleeve transmission line formed by the braid of the coaxial line and the inner surface of the sleeve is resonant. In these conditions, the impedance presented by this line is

$$Z = \mathrm{j}Z_o \tan(kl) \sim \infty \qquad \mathrm{k}l = \pi/2 \tag{4.13}$$

$$Z_0 = \frac{60}{\sqrt{\epsilon_r}} \ln \frac{b}{a} \tag{4.14}$$

which also equals the characteristic impedance of sleeve line, where l is the length of the sleeve line, k is the propagation constant of the line, and ϵ_r is the relative dielectric constant of the insert. The sleeve line is normally dielectrically loaded.

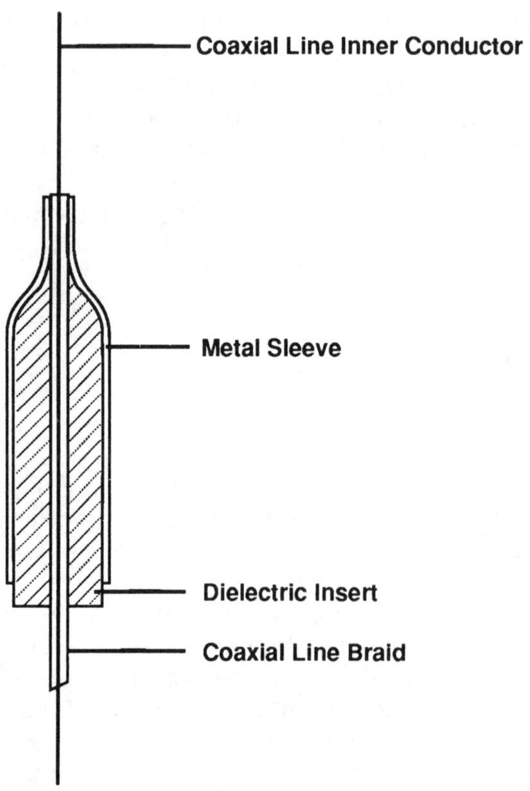

Figure 4.23 Sleeve dipole antenna cutaway.

A low-loss, low-dielectric constant cylinder is needed to ensure the concentricity of line conductors, which is essential to the proper performance of the antenna. The dielectric material should have low ϵ_r to achieve maximum bandwidth of the antenna. In (4.14), a is the feeding coaxial line braid outer diameter and b is the metal sleeve inner diameter.

The outer surface of the metal sleeve is part of the radiating dipole, so its resonant length is slightly less than a quarter wavelength in free space, depending on its outer diameter. The differential length required for the outer and inner surfaces of the sleeve is made up by shorting the sleeve to the braid of the feeding coaxial line, as shown in Figure 4.23.

If the impedance of the sleeve line is reduced sufficiently, then RF currents will propagate down the feeding coaxial line and the radiator is no longer only the half-wavelength dipole. The radio case also becomes part of the radiating structure.

Its pattern performance depends on the length of the exposed section of the feeding coaxial line and the size and shape of the metal surfaces of the radio. Rapid pattern deteriorations are seen within ±5% of the frequency given by (4.13) for relatively thin chokes (about 1-cm radius).

Radiation From Sleeve Dipoles

The radiation from half-wavelength dipoles has been the subject of a large body of literature [25–33], familiar to most readers. The sleeve dipole has a diameter discontinuity at the shoulder of the sleeve and at the feed point of the antenna. Sleeve dipoles with diameter discontinuity are dealt with in [34], which gives the radiation impedance of the antenna.

The bandwidth limitation of this antenna is dictated more by pattern performance than by impedance variation. If the choke is operated at a frequency of about ±5% away from the resonance of the sleeve line, RF currents will flow on the outer surface of the coaxial line feeding the antenna. These currents will excite the radio case, which radiates in phase opposition to the sleeve antenna, as shown in Figure 4.24. The pattern of the entire radiating structure has a minimum in the directions orthogonal to the axis of the antenna. This is normally the direction of the maximum desired coverage.

Figure 4.25 shows the radiation pattern of a sleeve antenna when the choke is near resonance, and so minimal RF currents excite the radio case. Note that the pattern of the antenna is practically that of a center-fed half-wavelength dipole. To check that the radio case has small or no RF currents flowing on it, the radio case was scanned with an isotropic RF-transparent, physically small (2.5-mm diameter) E-field probe (see Section 4.5.3). Figure 4.26 depicts the results of the scan, which show that the metal parts of the radio case have minimal RF currents or charges.

If the frequency of operation of the antenna is changed by +5% of the frequency of resonance of the choke, pattern lobing becomes clearly apparent, as shown in Figure 4.27, and similar pattern changes occur in Figure 4.28 for a 5% reduction in frequency. Clearly, these patterns are not as suitable for land mobile communication as that in Figure 4.25. A pattern notch in the direction of the horizon represents an effective coverage loss in most propagation environments. In addition, RF currents conducted on the radio case are partly absorbed by the hand of the user, causing an additional ohmic loss.

The gain of sleeve dipoles at the resonance frequency of the sleeve choke is almost the theoretical value for the half-wavelength dipole with absolute gain equal to 1.61 or 2.1 dBi. Ohmic losses in the thin braided coaxial line feeding the dipole (normally 0.05-in diameter or less) and in the thick dielectric molding used to protect the antenna reduce the absolute free-space gain of this dipole to 1 to 1.5 dBi.

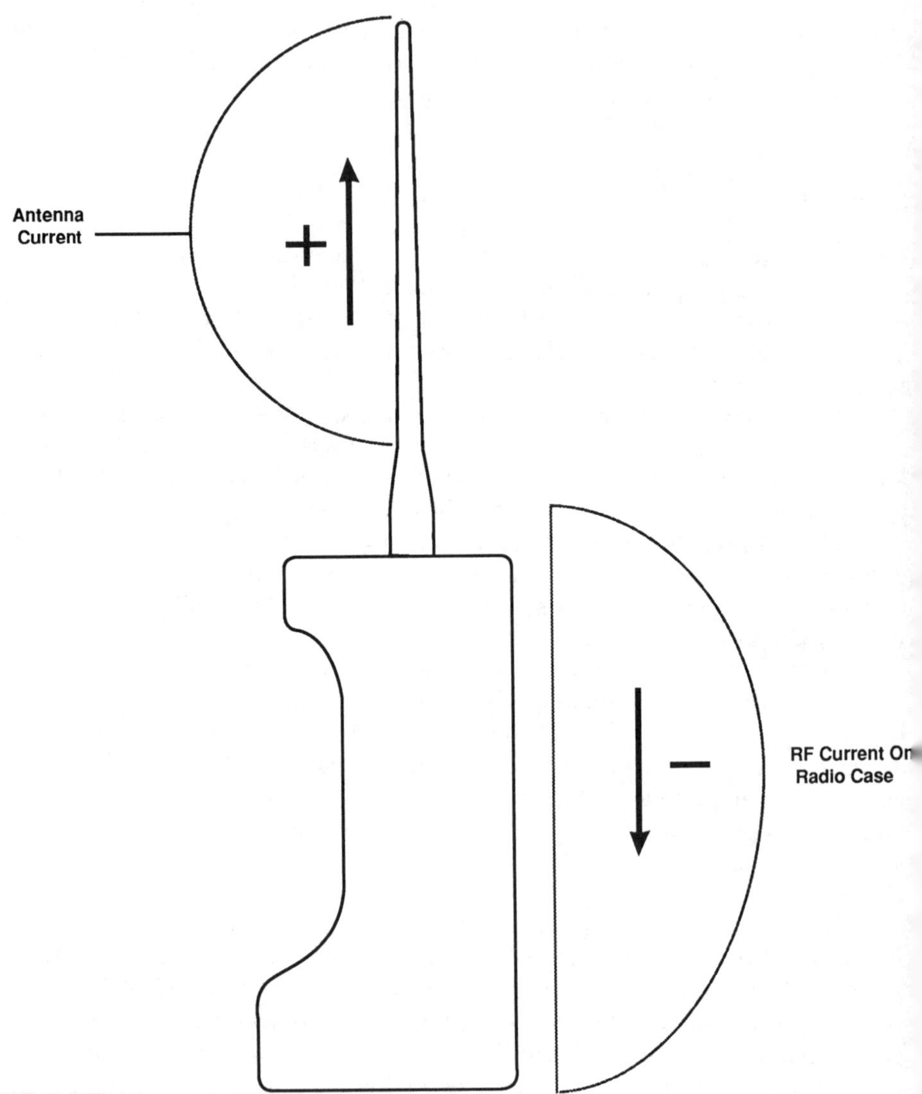

Figure 4.24 RF currents on a radio with sleeve dipole antenna.

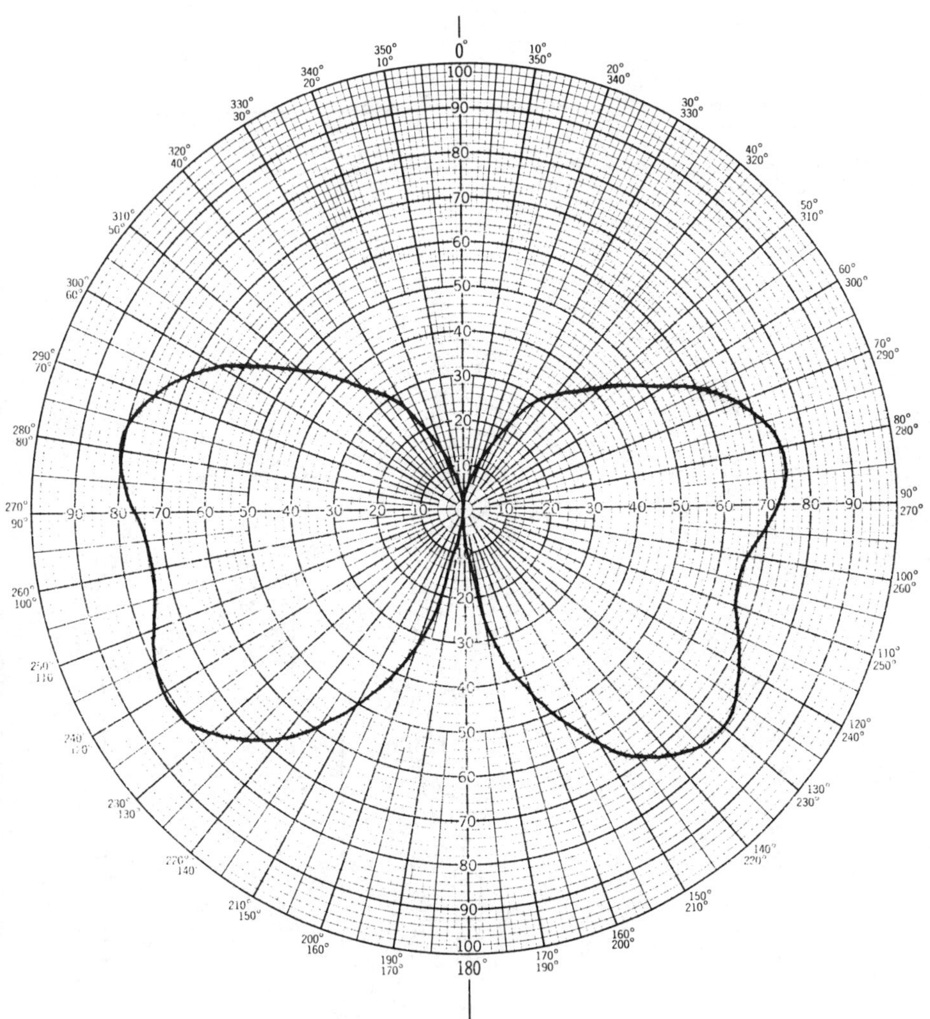

Figure 4.25 Sleeve dipole antenna elevation pattern at center frequency ($f = f_c$).

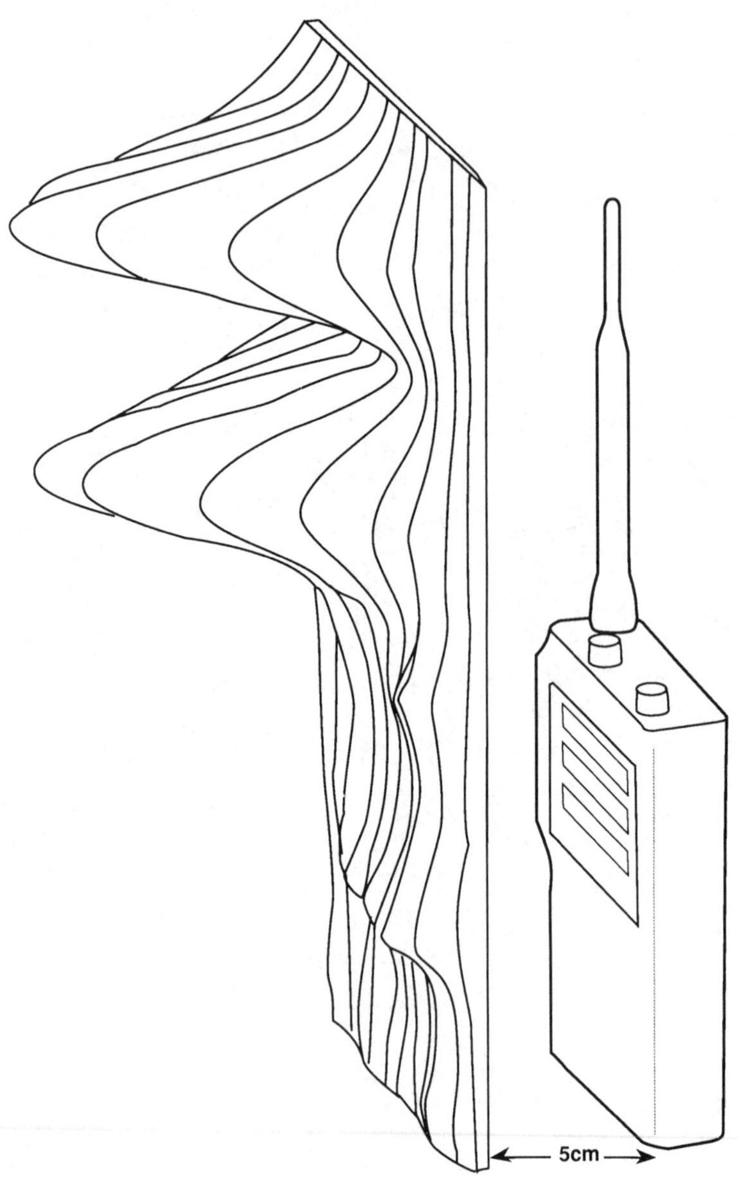

Figure 4.26 RF field near a radio with sleeve dipole at 840 MHz.

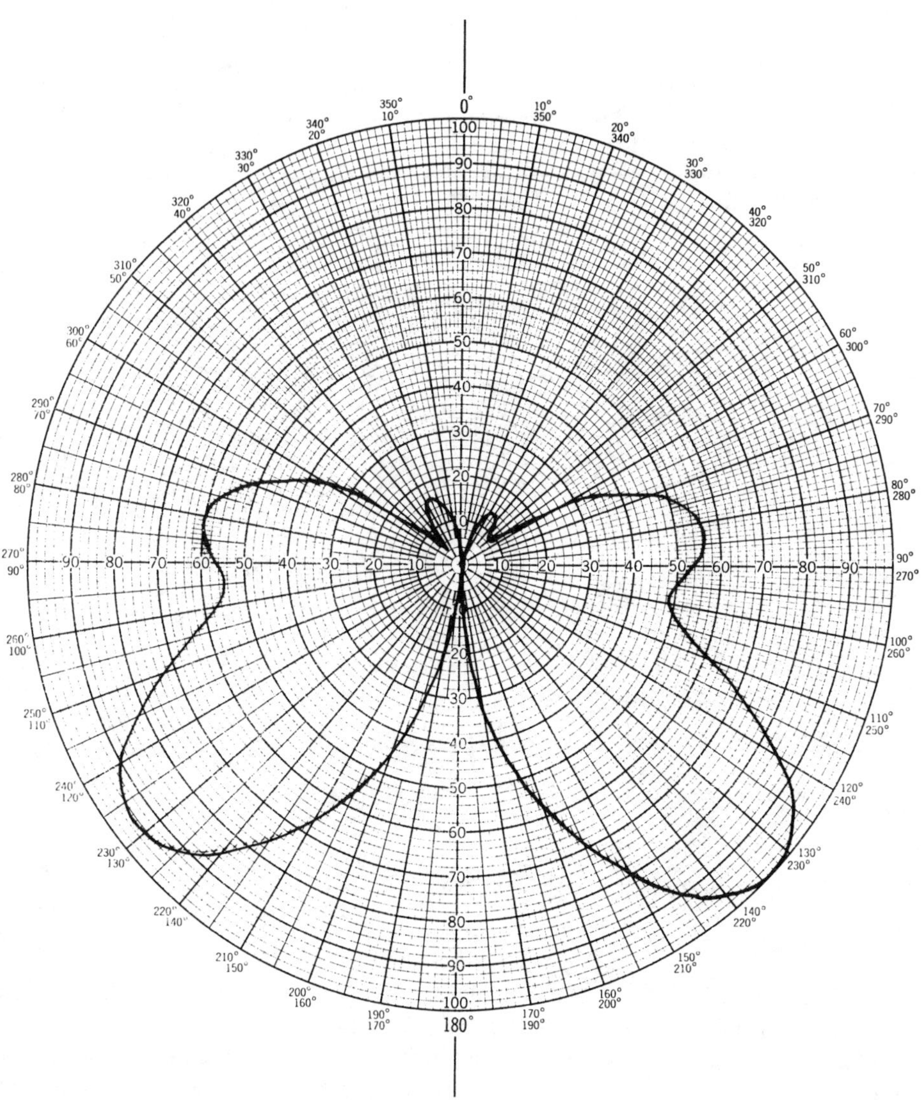

Figure 4.27 Sleeve dipole antenna elevation pattern ($f = f_c + 5\% f_c$).

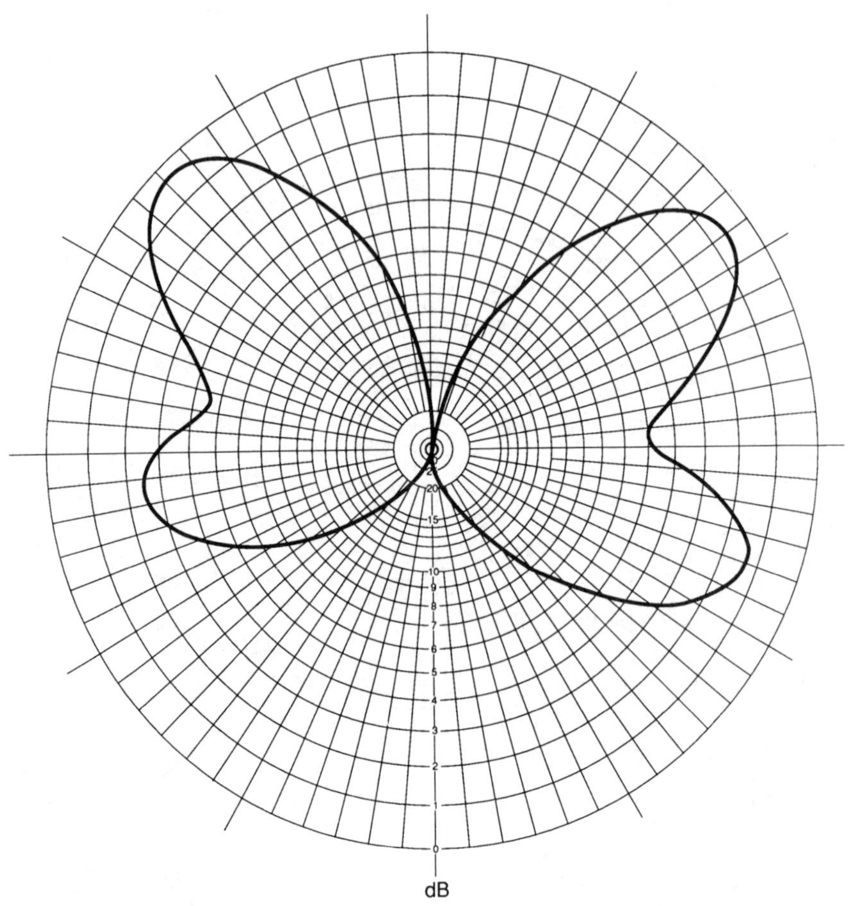

Figure 4.28 Sleeve dipole antenna elevation pattern ($f = f_c - 5\% f_c$).

End-fed half-wave dipoles are also used for portable phone systems. They normally consist of a half-wavelength wire antenna fed by a quarter-wavelength transformer, which is necessary to match the high radiation resistance of the antenna to the 50Ω output impedance of the RF final amplifier in the radio. These antennas, depending on their construction and the bandwidth of the quarter-wavelength transformer, have performance characteristics similar to those just described for the sleeve dipole.

Helical Antennas

Helical antennas are used in portable communication radios at the low-frequency bands (30 to 150 MHz) to reduce the size of the radiator to comfortable lengths. With the proper selection of parameters, the normal-mode helical dipole is an efficient radiating structure, with pattern and gain performance similar to those of the half-wavelength dipole. The bandwidth of an efficient and short helical dipole is narrower than that of the half-wavelength antenna because of the higher Q factor, which depends on the number and the diameter of the turns of the antenna [35].

The radiation from the normal-mode helix, also called *omnidirectional*, has been the object of various works [35–38]. The reader can review the mathematical theory of the radiation of the helical antenna in the available literature. Here, we will summarize the limitations of the helical antenna in its applications to the two-way portable communication equipment, especially 800- to 900-MHz cellular phones.

The resonant quarter-wavelength helical antenna excites strong RF currents on the portable radio case, which is an integral part of the radiating system. The phenomenon can be easily detected by scanning the near field of the portable radio (about 5-cm distance or less). A picture of such a scan of a UHF portable radio is shown in Figure 4.29. Note the flow of the currents on the metal parts of the radio case. These RF currents are partly absorbed into the hand of the user, which must be considered as a lossy dielectric material wrapped around the radio case.

In most applications, the helical antenna is used to reduce by a factor of three or more the length of a resonant thin-wire antenna operating at the same frequency. The monopole helixes normally used in the portable two-way communication technology have a physical length of about 1/12 wavelength and an electrical length of about a quarter wavelength. The physical length can be reduced well below 1/12 wavelength, but the radiation performance loss is normally intolerable. The following considerations hold for approximately 1/12-wavelength-long resonant helical antennas; the losses are much higher for shorter antennas.

Gain measurements at the UHF band (450 MHz) have shown (see Section 4.5) that a substantial gain loss with respect to the radio in free space is recorded when the radio is handheld. The RF currents on the radio case are conducted through the hand and dissipated through the arm of the user. In addition, some of the fields emanating from the helical antenna are directly coupled into the hand of the user, thus providing an additional loss mechanism. A gain loss of about 3 dB is commonly detected with the radio case held by a medium-sized person with dry hands.

The radiating structure helix plus radio case has enough bandwidth (about 50 MHz) to cover a substantial portion of the UHF two-way portable communication spectrum. With the gain losses due to RF absorption by the user, one finds that one helical antenna can cover the entire two-way UHF land mobile band

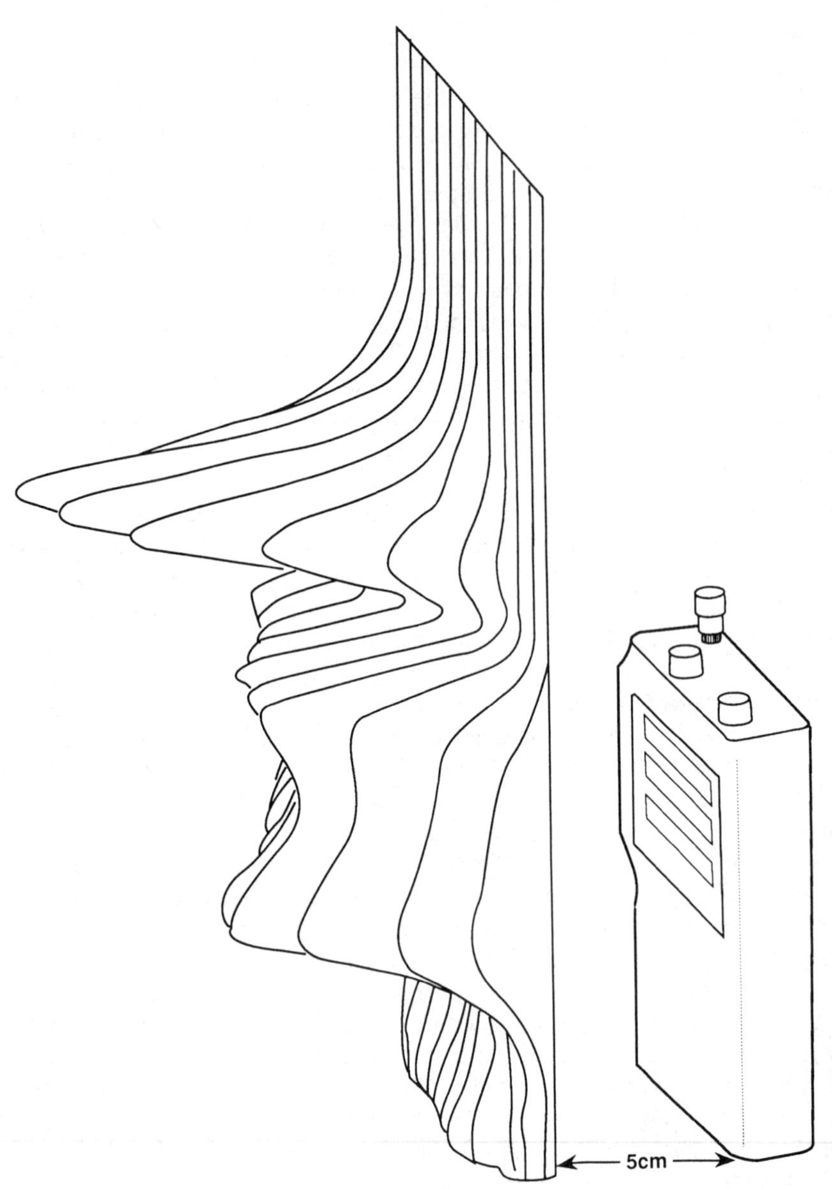

Figure 4.29 RF field near a radio with helical antenna at 450 MHz.

(about 100 MHz), although the same helix exhibits a Q factor of about 60 if tested over a large conducting ground plane or in a helical dipole configuration.

The application of the helical antenna at the 800- to 900-MHz band has an additional disadvantage with respect to its use at UHF unless the radio case is about one-quarter wavelength long (about 9 cm). If the radio case is substantially longer than a quarter wavelength, the currents on part of the radio case are in phase opposition with that of the helix, producing pattern lobing with a gain loss at the horizon, if the radio case is substantially longer than a quarter wavelength. Early models of cellular portables had a radio case about 20 cm long, which was producing patterns of the type shown in Figure 4.30. Finally, the antenna is so small that it hardly radiates in the angular sector shadowed by the head of the user. A substantial pattern coverage gap is caused by the helix being held very close (2 to 4 cm) to the head of the user. From pattern measurements, the helical antenna on a portable phone near the head of a human shows an average gain loss of about 12 dB with respect to a half-wavelength free-space dipole. Pattern and RF losses cause the helical antenna to be hardly suitable for cellular radio application at 800 to 900 MHz, except for the small size of the antenna, which makes it very attractive for pocket-sized radios.

Most of the considerations regarding the helical antennas hold true for the quarter-wavelength whips. They radiate somewhat more efficiently than the foreshortened helical antennas, but do not overcome any of the pattern shape problems just discussed. Since there is no true advantage in using the quarter-wavelength whips over the helical antennas, the latter ones have found portable phone market acceptability. Practically no portable cellular phone model uses the quarter-wavelength monopole antennas, which have found acceptance in the mobile cellular technology.

4.3.3 Antenna Diversity

The poor performance of the helical antenna for pocket-sized portable phones at 800 to 900 MHz has forced designers to offer an alternative radiator to the short helix. The antennas currently used are two helixes: a primary fixed helix, approximately 2 cm long (quarter-wave electrical length) and a secondary helical resonator approximately 10 cm long (half-wave electrical length), which can be collapsed inside the radio case. When extended, the secondary helix becomes the dominant radiator because it is free from the losses caused by the hand of the user holding the radio case (Figure 4.31).

The secondary helix is end-fed by the primary antenna and exhibits excellent free-space performance, depending on its own length and the ohmic losses in the metal wire forming the helix. At the top of the secondary antenna there is a 2-cm-

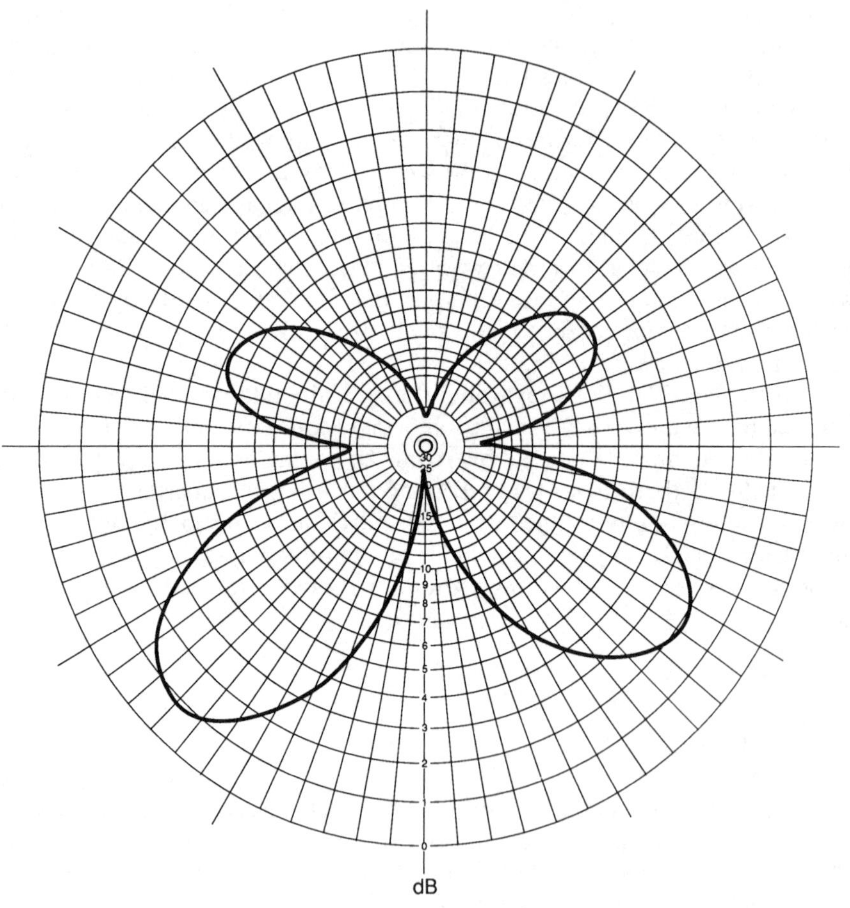

Figure 4.30 Elevation pattern of helical antenna on portable cellular phone.

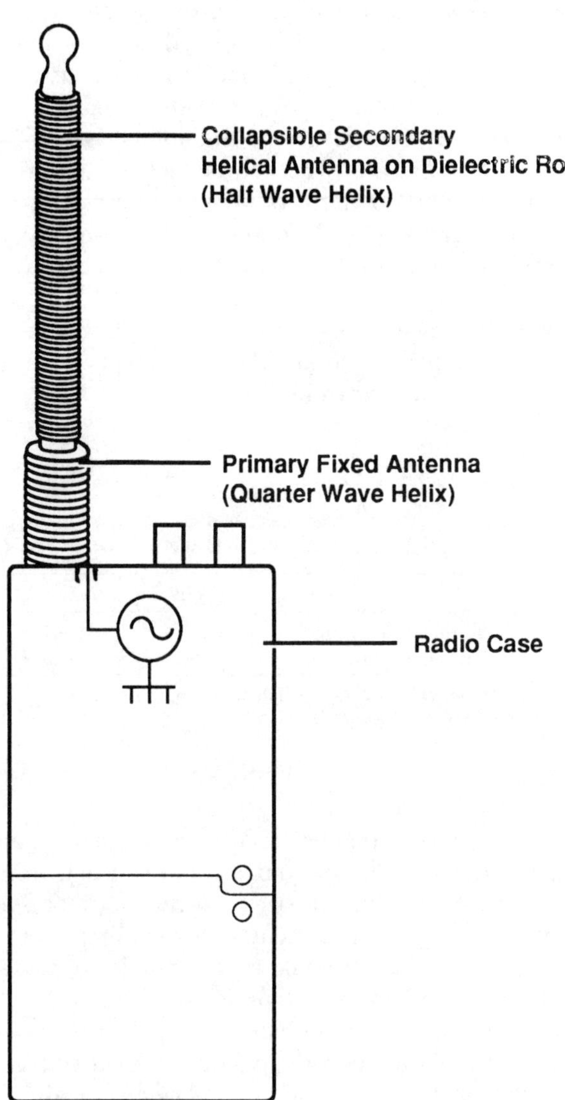

Figure 4.31 Gain diversity helical antenna system.

long dielectric rod to decouple the two helixes, when the secondary antenna is retracted, so minimal conducted RF is injected into the radio case through this path. This arrangement has provided some improvement to the communication range achievable by pocket phones with the helical antennas.

A summary of the performance of the antenna diversity arrangement just described is given in Table 4.2. The gain values were recorded in an anechoic chamber by taking elevation patterns of the radio case in free space and handheld at the head of a user like a normal household telephone. As is obvious from the results in the last column of Table 4.2, the performance of this radio telephone is far from optimal, although the pocket cellular phone has become extremely popular, which proves that engineering perfection is not always a condition for market success. Certainly, the next generation of pocket phones should provide some improvement on the results shown in Table 4.2.

Table 4.2
Recorded Free Space and Handheld Gain
Loss of Diversity Helical Antenna*

	Free Space	*Handheld*
Secondary extended	−4	−7
Secondary retracted	−6	−13

*Gain referenced to a half-wavelength dipole.

4.3.4 Conclusions

The size reduction of cellular phones has driven designers to antenna choices that using traditional portable radiators (dipoles and helices), have gain performance far from satisfactory. Considering the size of current pocket cellular phones (roughly 12 × 6 × 3 cm), there is no fundamental physical limitation for why the performance summarized in Table 4.2 should be acceptable. A new set of basic radiator choices should be made and optimized for this application. The traditional antennas are all axially symmetric, so they radiate uniformly at the horizon. The necessity of this choice as the radiators become physically small with respect to the size of the head of the user should be questioned. Coverage in the directions shadowed by the head of the user is practically impossible and produces only RF ohmic losses in the head and neck of the subscriber to the phone service. The situation can be alleviated by using basic radiators with some form of directivity. In addition, the antenna should be decoupled from the RF absorbing hand of the user. With careful design and selection of the size and location of the radiator, it is possible to improve the current performance of pocket cellular phone antennas, even if their size is further reduced.

4.4 PORTABLE PHONE ANTENNA SYSTEMS

4.4.1 System Design Aspects

The two chief design characteristics of any portable radio telephone antennas are that it must be mounted on the housing of a portable telephone and that during operation the set will be held by a human, who may randomly point the set in any direction. Because the antenna is forced into close proximity with the housing, the antenna current is induced not only into the antenna element, but also into the conductive housing. This current dispersion changes the shape of the original radiation pattern. Radiation efficiency is further degraded by the antenna's forced proximity to the human body, since the antenna is necessarily used near an operator. The polarization of radiation pattern is also changed by the changes in antenna direction caused by operator movements and habits. These difficult design constraints are complicated by the need to develop very small antenna elements to meet the demand for compact, portable equipment; it is common knowledge that radiation efficiency and bandwidth degrade as antenna element size is decreased. In addition to these design constraints, which are commonly found in portable radio telephone applications, system-based considerations such as the security of the specified frequency bandwidth and the need to support antenna diversity reception must also be taken into account. Figure 4.32 shows the interaction and impact of these technical subjects on the design of antennas mounted on portable radio telephone sets.

The principal subject that must be investigated is the effective gain of antennas in a multipath radio propagation environment. The key design task is to study these subjects synthetically and to maximize the effective gain in the desired frequency range as much as possible. High effective gain allows the size and weight to be reduced, while the usage time is increased. Not only does it permit the portable radio telephone to be used in areas of low electrical field strength, but it also allows transmitting power to be reduced. This reduction in transmitting power is extremely effective in reducing battery capacity (i.e., the weight of the portable radio telephone) or extending the usage time of a full battery. Therefore, the effective gain of antennas to be mounted on portable radio telephones must be maximized to develop small and high-performance portable telephone sets.

Simple Method for Evaluating Mean Effective Gain of Antennas in Multipath Propagation Environments

As described in Chapter 2, the mean effective gain of mobile antennas in a multipath propagation environment is determined by the joint contribution of the antenna pattern and the characteristics of the waves randomly arriving from the environ-

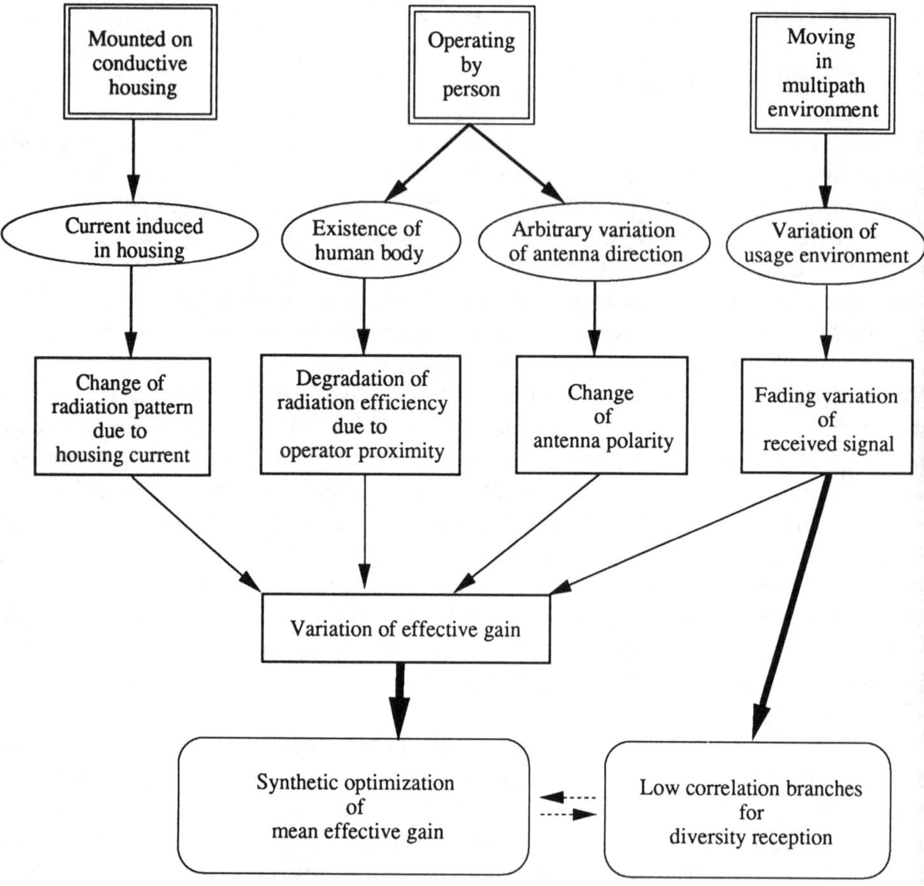

Figure 4.32 Technical subjects for designing antennas for portable telephone equipment.

ment. The method of analyzing the mean effective gain also has been clarified However, the statistical distribution of the arriving waves, which is needed in the analytical method, has not yet been sufficiently clarified. Furthermore, the three-dimensional radiation pattern is needed, but it is not easy in practice to measure such patterns. Hence, given the present situation, a simplified method [39], derived from the analytical method described in Chapter 2, must be used to design antennas. In this simplified method, it is first assumed that the incident waves arriving at the mobile antenna are concentrated in the horizontal plane, and that the angular density function of incident waves follows the distribution function.

$$P_\theta = P_\phi = \frac{1}{2\pi}\delta\left(\theta - \frac{\pi}{2}\right) \tag{4.15}$$

where P_θ and P_ϕ are the θ and ϕ components of the angular density functions of arriving plane waves, respectively. This function indicates that the mean effective gain is represented by the following equation.

$$G_e = \frac{1}{2\pi}\int_0^{2\pi} \frac{\text{XPR}}{1 + \text{XPR}}\left[G_\theta\left(\frac{\pi}{2}, \phi\right) + \frac{1}{\text{XPR}} G_\phi\left(\frac{\pi}{2}, \phi\right)\right] d\phi \tag{4.16}$$

where XPR is the average cross-polarization power ratio, and G_θ and G_ϕ are the θ and ϕ components of the antenna power gain pattern, respectively. Based on previous research and outdoor experiments, the XPR value used in the above equation (4.16) is assumed to be 9 dB [40,41]. This XPR value evaluates the lowest contribution of the ϕ component of the antenna power gain pattern to the effective gain. The mean effective gain of the target antenna with respect to a reference antenna is estimated by the following equation.

$$G_p = \frac{G_e \text{ (testing mobile antennas)}}{G_e \text{ (reference antennas)}} \tag{4.17}$$

As described in Chapter 2, G_p presents the mean effective gain with respect to an isotropic antenna if the reference effective gain is the mean effective gain of a 55-deg inclined half-wavelength dipole antenna incremented by 3 dB. However, to evaluate the gain performance more easily, the mean effective gain of a vertically oriented half-wavelength dipole antenna (2.15 dBi) is normally used as the reference mean effective gain because it can be assumed that incoming plane waves are concentrated in the horizontal direction. It should be noted that, in general, this estimation value is not sufficiently accurate to assess gain performance in real-world propagation environments, because the mean effective gain of vertically oriented half-wavelength dipole antennas changes with propagation conditions and should be assumed to be lower than 2.15 dBi [42].

Since the mean effective gain evaluated by (4.17) is equal to the ratio of the average strengths of the radiation patterns of the target and reference antennas in the horizontal plane, it is also called the *pattern averaging gain* (PAG) [39]. PAG allows the practical evaluation of antennas to be simplified as follows.

- The performance of a target antenna can be easily evaluated by pattern measurement in a radio anechoic chamber.

- The gain degradation due to variations of radiation pattern can be easily evaluated irrespective of whether the variations are caused by the conductive housing or by human operation.
- The degradation of radiation efficiency due to the variation of input impedance caused by either proximity to a human body or the absorption of radiation power by the human body can be evaluated similarly.
- The frequency dependency of the effective gain can be evaluated by changing the measurement frequency.

Effect of Conductive Housing and Inclination of Portable Radio Unit

The effect of conductive housings on antenna radiation pattern was discussed in Chapter 2. The radiation pattern of antennas mounted on a conductive housing can be calculated by applying the moment method to a wire-grid model that approximates the configuration of the antenna element and conductive housing. It is commonly known that the radiation pattern of an antenna mounted on a portable telephone housing varies according to the form and size of the housing and the configuration of the antenna element [39,43]. In this section we shall consider the PAG characteristics of such a portable mobile antenna, which has a varied radiation pattern and is inclined in elevation. We then develop a guiding principle for designing antennas mounted on portable radio telephones.

Figure 4.33 shows antenna configurations investigated in this section. Figures 4.33(a–c) show planar inverted-F antennas (PIFA) mounted in various orientations on a conductive rectangular housing, and Figure 4.33(d) shows a half-wavelength sleeve dipole antenna mounted on a conductive rectangular housing. Figure 4.34 shows wire-grid models for housing-mounted PIFAs. When the microphone, speaker, and the dialing panel are mounted on the front of the housing, the antenna element is mounted on the back or side of the housing (type A), as shown in Figure 4.34(a), while Figure 4.34(b) shows the case in which the antenna element is mounted on the top of the housing (type B). The radiation pattern for each case was calculated by moment analysis, and the calculated and experimental results in the 900-MHz band are shown in Figures 4.35 and 4.36. The calculated results of both the E_θ and E_ϕ components closely agree with experimental results; thus, this wire-grid model analysis is effective in calculating these antenna patterns. PAG as a function of set inclination can be calculated by entering into the equation the calculated horizontal plane radiation pattern for the case where the housing is inclined from the zenith at angle α. Figures 4.37 and 4.38 show the PAG characteristics calculated from the wire-grid models of the cases in Figures 4.33(a) and 4.33(b,c), respectively. They show that the effective gain characteristics depend on housing dimension. This analytical method can be used effectively to optimize the antenna configuration in terms of mean effective gain; moreover, the optimum

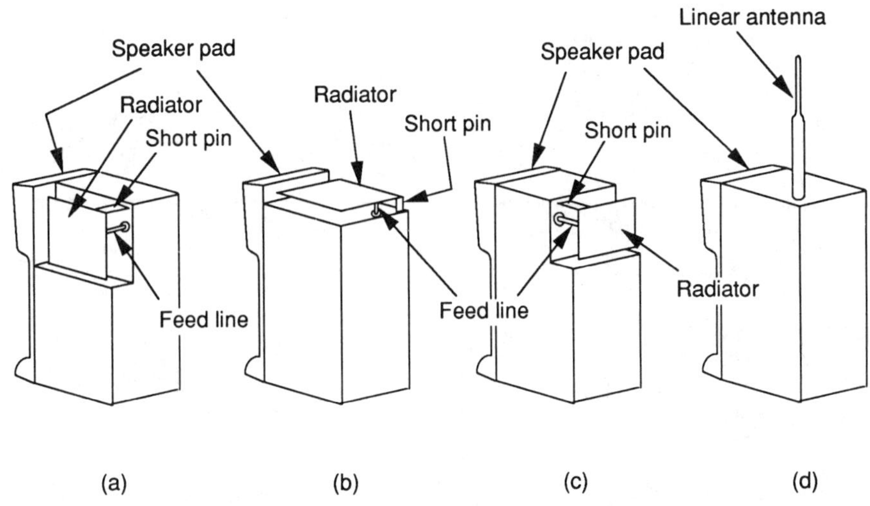

Figure 4.33 Antenna configuration for portable transceiver units: (a) type A; (b) type B; (c) type C; (d) type D. (From [39], © 1987 IEEE.)

antenna configuration for a particular operating system, such as an outdoor mobile communication system or an indoor system. Figure 4.39 shows the measured distribution of the inclination angle during speech [44]. It can be considered that the typical inclination angle is 60 deg. Figure 4.40 compares the calculated PAG values for the antenna configurations shown in Figure 4.33 against the experimental values and mean effective gain, as measured in a suburban area with the random field measurement (RFM) method [45]. This comparison shows that side-mounted PIFAs have the smallest variation in effective gain against the inclination angle of the housing, as well as the highest gain of the configurations examined. This antenna configuration has been adopted for the portable telephone unit of the 800-MHz band, high-capacity land mobile radio communication system (NTT system) in Japan [46].

Evaluation of Gain Degradation Due to Operator Proximity and Its Influence on Design

The effective gain of a portable radio antenna depends on antenna radiation performance, which is affected by operator proximity. The effect of the operator holding the portable radio unit on antenna patterns can be experimentally inves-

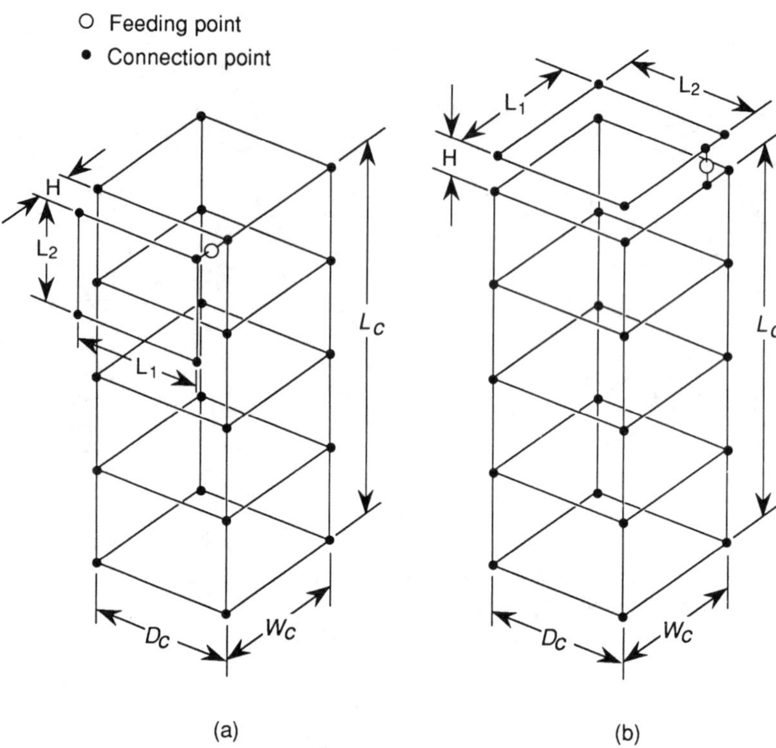

Figure 4.34 Wire-grid models of portable radio unit with PIFA element: (a) type A, (b) type B. (From [39], © 1987 IEEE.)

tigated by using the PAG estimation method. First, the radiation patterns in the horizontal plane for both the vertically and horizontally polarized components of the antenna mounted on the portable radio unit should be measured in a radio anechoic chamber, the inclination angle should be 60 deg, and no human body should be present. Second, the radiation patterns should be similarly measured while the unit is held by an operator in the speaking position at an inclination angle of 60 deg in the same radio anechoic chamber. These measurements should be made using a small oscillator placed within the portable unit as the radio power supply source, instead of the more common technique, which is to employ a coaxial cable to feed or detect the radio frequency power. This is because the antenna radiation pattern is disturbed by the radiation emitted by the antenna current leaking onto the surface of the coaxial cable. For each type of antenna configuration shown in Figures 4.33(a–c), the PIFA radiation patterns were measured when (1) the radio units were held to the operator's head and (2) the radio housing was

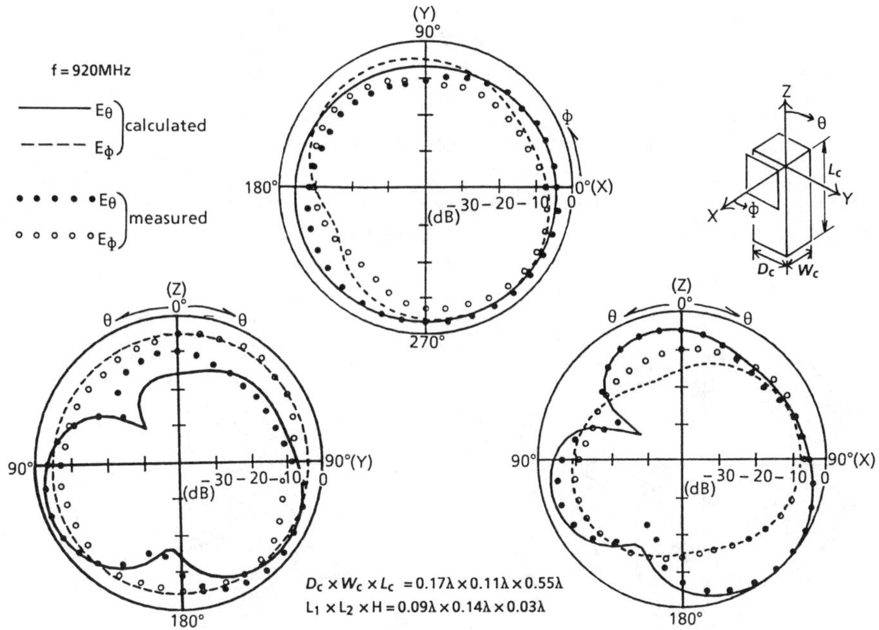

Figure 4.35 Radiation patterns of PIFA mounted on the side of the housing (type A). (D = 0.17λ, W = 0.11λ, L = 0.55λ, l = 0.14λ, w = 0.09λ, h = 0.03λ, f = 920 MHz). (From [39], © 1987 IEEE.)

inclined at 60 deg without an operator. The measured results are shown in Figure 4.41, where 0 dBd represents the maximum radiation strength of a half-wavelength dipole antenna. In this experiment, the operator's head was separated from the radio housing by a 0.06λ thick foam spacer in place of the speaker pad and dials. Not surprisingly, it can be seen that the operator's presence caused pattern changes and degraded the vertically and horizontally polarized components.

Applying these measured radiation patterns to the PAG estimation method (i.e., to (4.16)) yields a gain degradation value for the operator effect, which is shown in Table 4.3. The XPR is assumed to be 9 dB. These results show that the operator-induced gain degradation is considerably lower with type A and C configurations than for type B.

Figure 4.42 shows the frequency dependency of the effective gain of a PIFA mounted on the side of the radio case (type A). The solid lines are the characteristics estimated by the PAG method, and the open triangles and dotted circles are the experimental values measured by the RFM method in a suburban area. The exper-

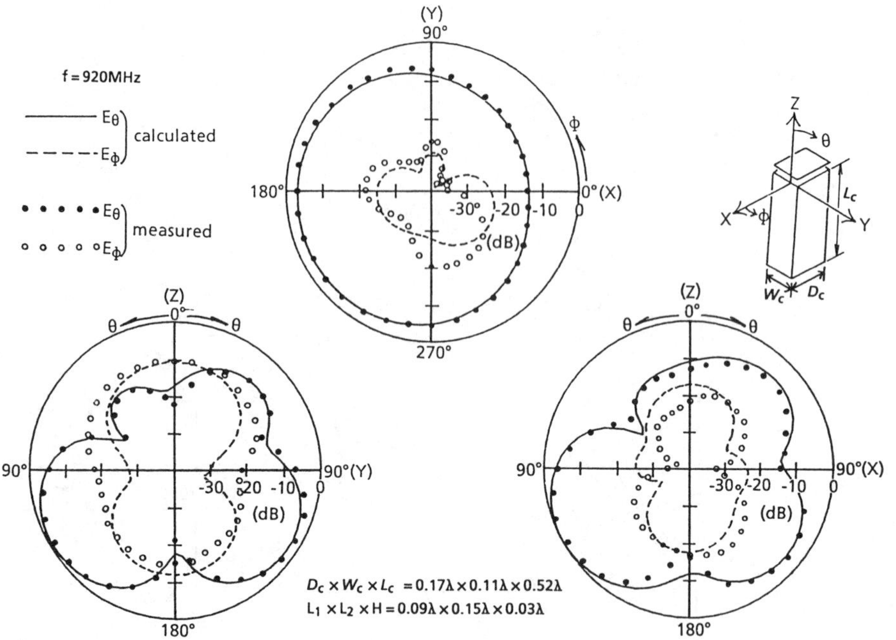

Figure 4.36 Radiation patterns of PIFA mounted on the top of the housing (type B): ($D = 0.17\lambda$, $W = 0.11\lambda$, $L = 0.52\lambda$, $l = 0.15\lambda$, $w = 0.09\lambda$, $h = 0.03\lambda$, $f = 920$ MHz). (From [39], © 1987 IEEE.)

imental values are lower than the PAG values because the distribution of arriving waves in the measurement environment is spread in elevation. However, the overall agreement is quite good, which shows that the PAG estimation method is effective in determining approximately the effective gain characteristics of mobile antennas. For the gain degradation shown in Figure 4.42, operator proximity reduces the gain by about 2.5 dB and is caused by three factors: VSWR degradation, pattern changing, and the absorption and scattering of radiated radio waves by the human body.

Designing for Antenna Diversity

The correlation coefficient between antenna diversity branches can be investigated theoretically by calculating the complex radiation patterns of the antennas that form each branch in antenna diversity and applying them to the analytical method

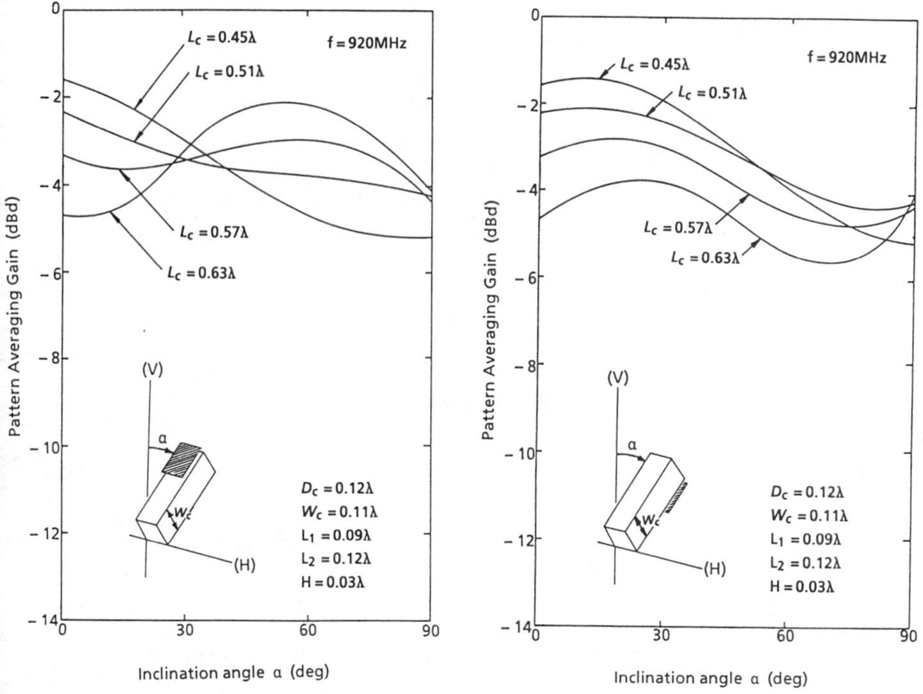

Figure 4.37 Gain characteristics of the PIFA mounted on the side of the housing (type A): ($D = 0.12\lambda$, $W = 0.11\lambda$, $l = 0.12\lambda$, $w = 0.09\lambda$, $h = 0.03\lambda$, $f = 920$ MHz). (From [39], © 1987 IEEE.)

described in Chapter 2. Figure 4.43 shows a space diversity configuration using two parallel half-wavelength dipole antennas. The dipole antennas are inclined from the vertical (Z) direction around the Y-axis. It is also assumed that this antenna diversity branch is used in a postdetection diversity reception system (i.e., each branch is connected to a receiver whose input impedance is 50Ω). Because of the mutual coupling between the two antennas, the correlation coefficient depends on the antenna spacing, as shown in Figure 4.44. The XPR is the cross-polarization power ratio, m_V and m_H are, respectively, the mean elevation angle of the VP and the HP wave distributions observed from the horizontal direction, and σ_V and σ_H are, respectively, the standard deviation of the VP and HP wave distributions, as described in Chapter 2. When the inclination angle α is 0 deg, noncorrelation is achieved when the antenna spacing is larger than 0.3λ. When $\alpha = 90$ deg, the correlation can be strongly increased, as shown in Figure 4.45. This is because the overlap of the complex radiation patterns of the antenna branches increases with the inclination angle.

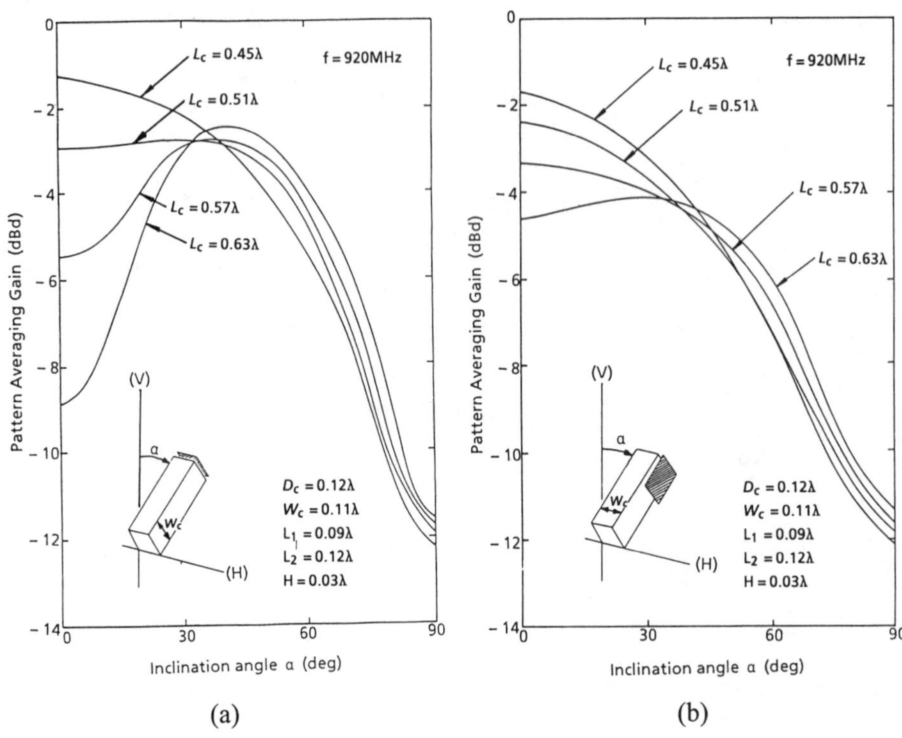

Figure 4.38 Gain characteristics of the PIFA on the inclined housing: (a) type B; (b) type C. (D = 0.12λ, $W = 0.11\lambda$, $l = 0.12\lambda$, $w = 0.09\lambda$, $h = 0.03\lambda$, $f = 920$ MHz). (From [39], © 1987 IEEE.)

The characteristics of the MEG of the antennas can be calculated similarly by using the analytical method described in Chapter 2. Figure 4.46 shows the MEG characteristics versus the antenna spacing at the antenna inclination angles α of 0 and 90 deg. MEG decreases rapidly when the antenna spacing is smaller than 0.2λ because the radiation efficiency is decreased due to the mutual coupling loss. Figure 4.46 shows that the optimum antenna spacing for this type of space diversity with two parallel half-wavelength dipole antennas lies in the range 0.3λ to 0.4λ. As described above, the optimum configuration of antenna diversity can be designed by considering the characteristics of both the correlation coefficient and MEG.

In practice, the antenna diversity elements are mounted on the portable radio housing. In this case, if we calculate only the radiation pattern of the antennas by using the wire-grid model and the moment method, and input the pattern so determined into the method described in Chapter 2, the MEG characteristics are

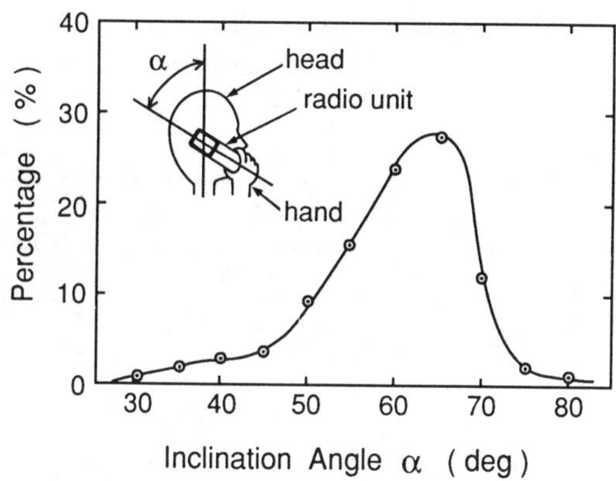

Figure 4.39 Inclination angle distribution of portable radio units.

The correlation characteristics are obtained for the portable diversity antennas. This approach can also be used to evaluate the effect of the operator. However, as described earlier, the statistical distribution of arriving waves has not yet been sufficiently studied, even though it is needed in the analytical method. Furthermore, the analytical method requires knowledge of the three-dimensional radiation pattern, but measuring such patterns is not easy in practice. Hence, in the design of antenna diversity systems, it is considered effective to apply a simplified method derived from the analytical method described in Chapter 2 to evaluate the correlation characteristics of antenna diversity. We must make an assumption similar to that used in the PAG estimation method in Section 4.4.1, whereby the angular density function of incident waves is assumed to be modeled by (4.15). This results in the following simplified formula for evaluating the correlation coefficient.

$$\rho_e = \frac{\int_0^{2\pi} [\mathrm{XPR}\cdot \overline{E}_{\theta 1}\overline{E}^*_{\theta 2} + \overline{E}_{\phi 1}\overline{E}^*_{\phi 2}]e^{-j\beta x}\,d\phi}{\int_0^{2\pi} [\mathrm{XPR}\cdot \overline{E}_{\theta 1}\overline{E}^*_{\theta 1} + \overline{E}_{\phi 1}\overline{E}^*_{\phi 1}]\,d\phi \times \int_0^{2\pi} [\mathrm{XPR}\cdot \overline{E}_{\theta 2}\overline{E}^*_{\theta 2} + \overline{E}_{\phi 2}\overline{E}^*_{\phi 2}]\,d\phi} \tag{4.18}$$

where $\overline{E}_{\theta n}$ and $\overline{E}_{\phi n}$ $(n = 1, 2)$ are the complex envelopes of the θ and ϕ components of the electric field pattern in the horizontal plane, respectively. The contribution of the ϕ components of the electric field pattern should be rated lower, and thus it seems to be reasonable to assume that the XPR value used in (4.18) is 9 dB [40,41].

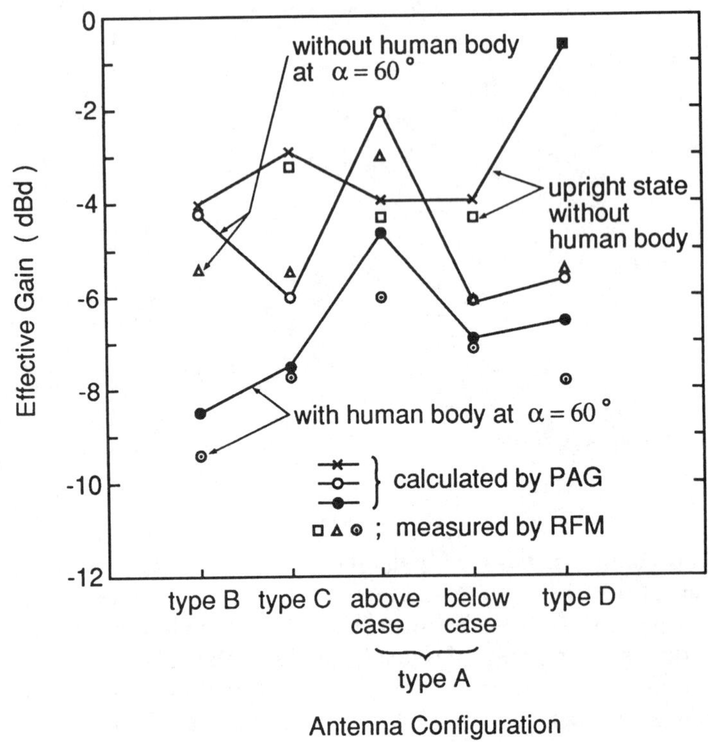

Figure 4.40 Comparison of PAG value and measured mean effective gain with respect to antenna configuration for type A to type D (inclination angle is 60 deg. f = 920 MHz).

4.4.2 Handheld Systems

Antennas for Nondiversity Systems

Many kinds of handheld portable telephone units have been developed for the 900-MHz band land mobile communication systems. In the Japanese system MCS-L1 [47], which does not adopt diversity reception, the mobile stations use the 870 to 885-MHz band for receiving and the 925 to 940-MHz band for transmitting. Thus, to cover the whole frequency band, 870 to 940 MHz, the antennas for the mobile stations must have a relative bandwidth of about 7.8%. The most popular antenna for handheld units is the whip antenna, which not only has relatively broad bandwidth characteristics, but can also retract into the unit. Furthermore, the whip antenna has three other advantages: simple element structure, small element volume, and easy installation. However, antenna currents are impressed not only o

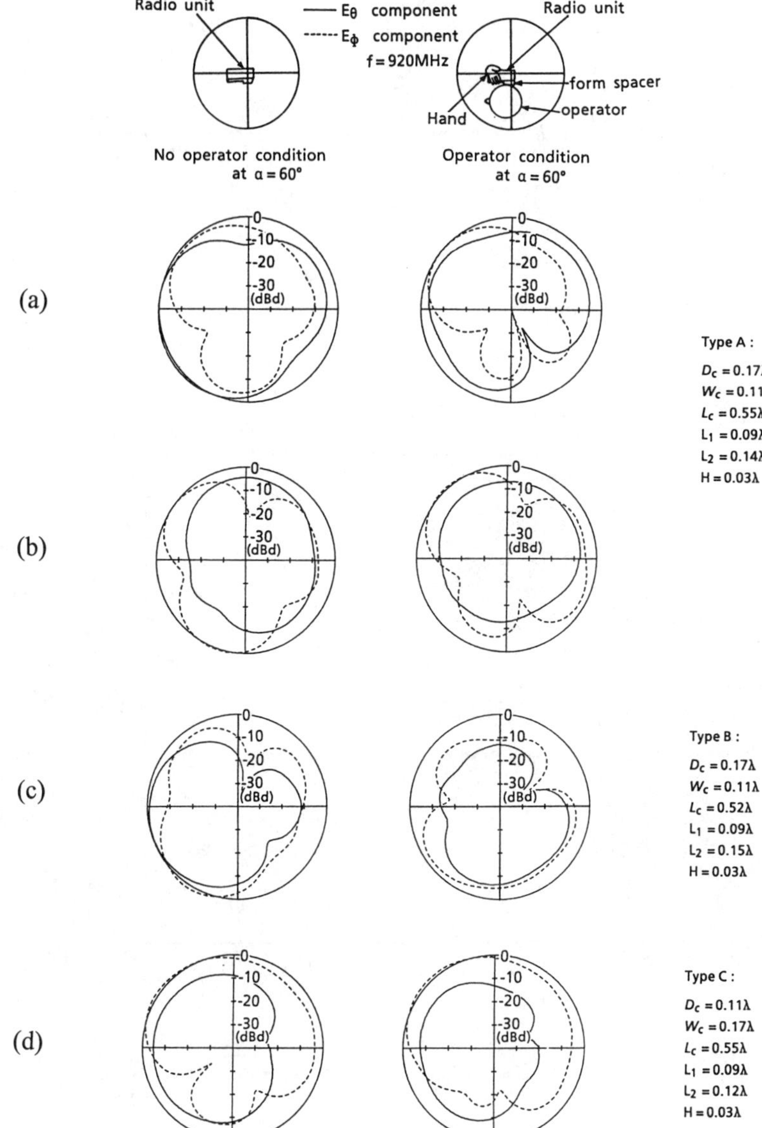

Figure 4.41 Degradation of radiation patterns due to operator proximity: (a) upper-side-mounted PIFA (type A); (b) lower-side-mounted PIFA (type A); (c) top-mounted PIFA (type B); (d) back-mounted PIFA (type C) (type A and B dimensions as in Figures 4.35 and 4.36; type C: $D = 0.11\lambda$, $W = 0.17\lambda$, $L = 0.55\lambda$, $l = 0.12\lambda$, $w = 0.09\lambda$, $h = 0.03\lambda$). (From [39], © 1987 IEEE.)

Table 4.3
Evaluation of the Operator Effect by the PAG Method

Type of Radio Unit	*Type A* Above Radio Housing	*Type A* Below Radio Housing	*Type B*	*Type C*
No-operator condition at $\alpha = 60$ deg (dBd)	−2.0	−6.1	−4.2	−6.0
Operator condition at $\alpha = 60$ deg (dBd)	−4.6	−6.9	−8.5	−7.6
Operator effect (dB)	2.6	0.8	4.3	1.6

Source: [39]. © 1987 IEEE. Reprinted with permission.

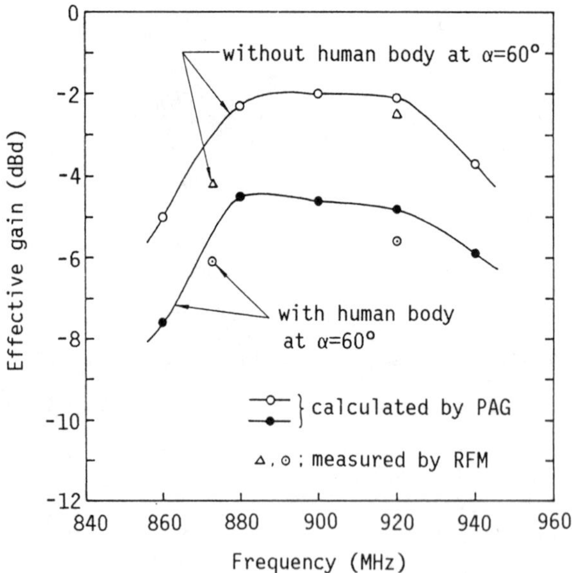

Figure 4.42 Mean effective gain characteristics of the PIFA above the inclined housing (type A).

the antenna element, but also on the conductive housing. This is why the radiation pattern depends on the size of the housing. Figure 4.47 shows the calculated radiation patterns of quarter-wavelength whip antennas mounted on a conductive housing 160 mm long, 30 mm wide, and 55 mm deep. In the case of the quarter-

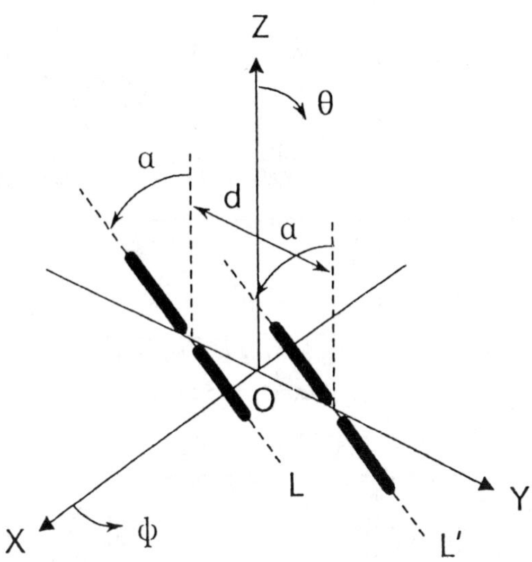

Figure 4.43 Space diversity configuration of parallel dipole antennas and the coordinate system.

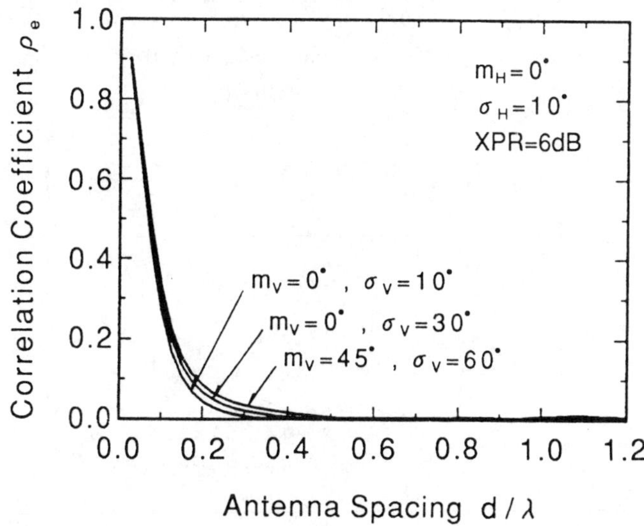

Figure 4.44 Correlation characteristics at antenna inclination angle α of 0 deg.

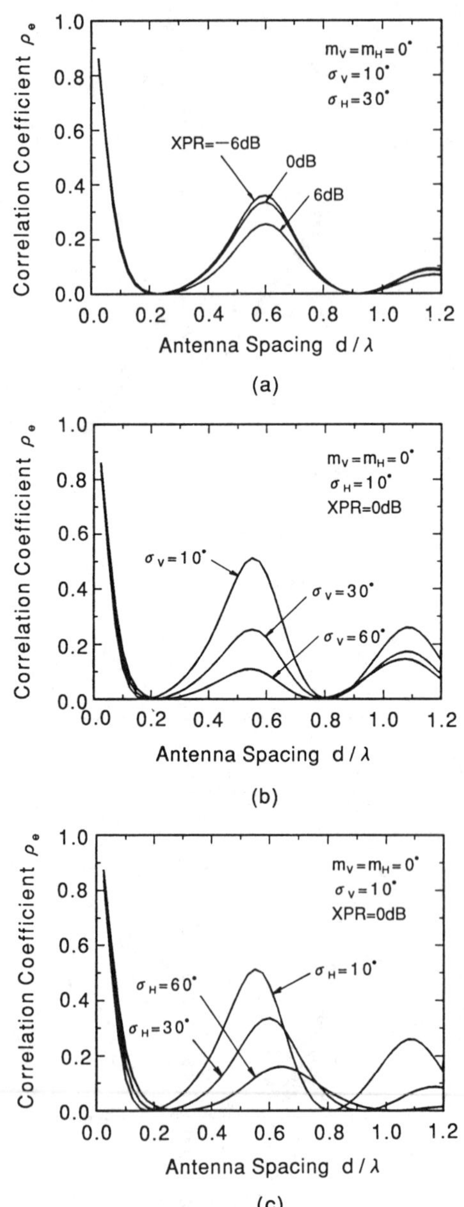

Figure 4.45 Correlation characteristics at antenna inclination angle α of 90 deg.

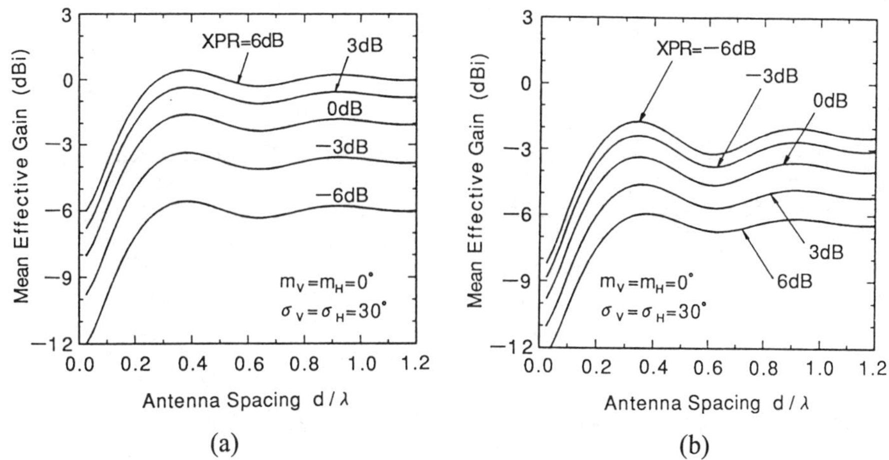

Figure 4.46 MEG characteristics of parallel dipole antennas: (a) $\alpha = 0$ deg; (b) $\alpha = 90$ deg.

wavelength whip antenna, the currents in the antenna element and the housing are not in phase. This results in a radiation pattern that has four vertical radiation lobes, and the degradation of the directivity in the horizontal plane is considerable. Figure 4.48 shows the PAG characteristics, calculated at 900 MHz, for various-length whip antennas mounted on a conductive housing in the upright condition. It shows that the effective gain depends mainly on the length of the whip antenna element and the length of the conductive housing. Thus, in this type of handheld unit, we can optimize the antenna performance by selecting the optimum lengths. Figure 4.49 shows a commercial handheld unit that uses a whip antenna. Its length, width, and depth are 178 mm, 40 mm, and 73 mm, respectively; that is, its volume is about 520 cc. The length of the antenna element is 160 mm in use, and this corresponds to about one-half wavelength. According to the PAG evaluation in Figure 4.48, this handheld unit was designed to yield the highest effective gain in the upright position. It can be expected that the operator would use this handheld unit in this position to improve the speech quality. The effective gain of this antenna, measured in a typical mobile radio environment, is about −1 dBd in the upright position, and this measured result corresponds to the result given in Figure 4.48, in which 0 dBd means the effective gain of a vertically oriented half-wavelength dipole antenna.

1983 saw the development of a handheld telephone unit that employed a built-in antenna. The unit was designed to be used in cars and as a portable radio unit [48]. This handheld type of detachable transceiver is pictured in Figure 4.50. The antenna element is a PIFA [39]. Its length, width, and depth are 180 mm, 38 mm, and 70 mm, respectively; that is, its volume is about 480 cc. The PIFA element is

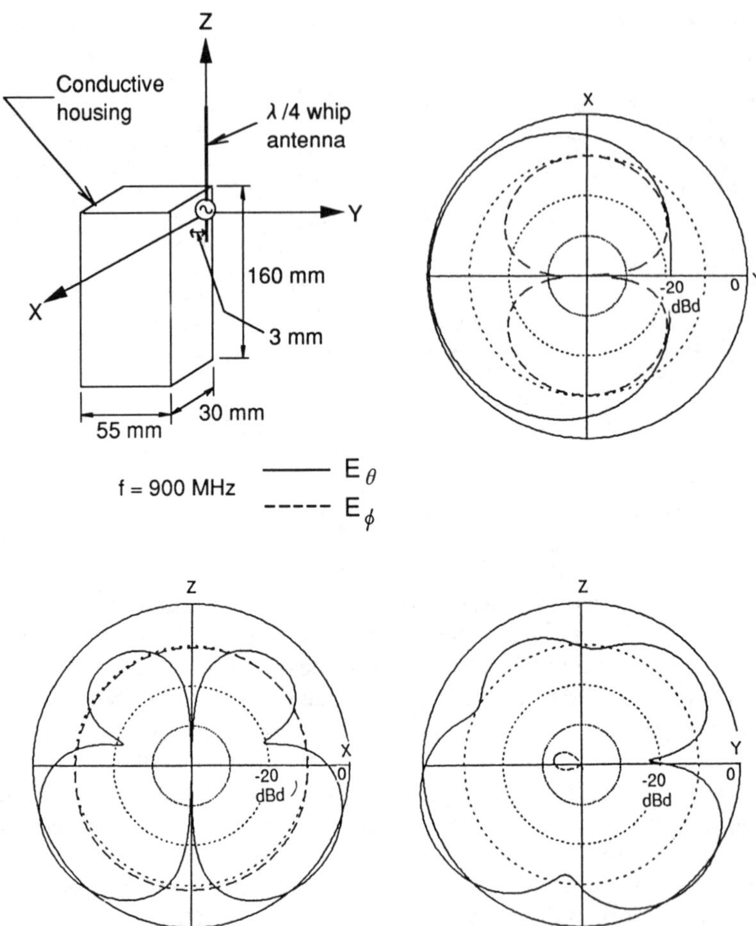

Figure 4.47 Calculated radiation patterns of quarter-wavelength whip antennas mounted on conductive housing 160 mm long, 30 mm wide, and 55 mm deep.

mounted on the upper quarter of the back of the housing. This antenna configuration corresponds to the type C antenna configuration shown in Figure 4.33(c). Table 4.4 shows the experimental results of the effective gain measured in a suburban mobile environment for the antenna configurations shown in Figure 4.33. In the upright condition, the effective gain of this PIFA configuration is 2.6 dB lower than that of the half-wavelength sleeve dipole configuration, but in the speaking condition their effective gains are almost the same. The effect of operator

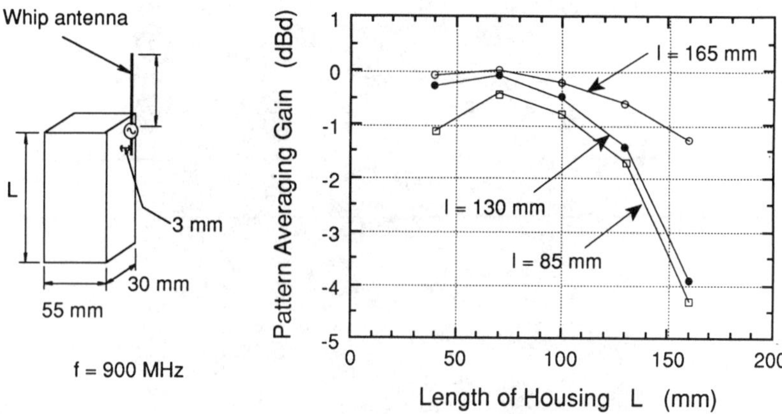

Figure 4.48 PAG characteristics of whip antenna mounted on conductive housing 30 mm wide and 55 mm deep.

proximity is also 2.1 dB, which is smaller than that of the half-wavelength sleeve dipole antenna configuration. This is because the antenna element is separated from the operator's head in the speaking position, and the operator's hand is also kept away from the antenna element. According to the measured effective gain in the speaking position, the type A configuration yields the best performance against inclination of the unit. However, the type C configuration is adopted because it radiates the least amount of power to the human body. This handheld unit is considered to be an epoch-making unit because it was the first to adopt a built-in antenna configuration.

Antennas for Diversity Systems

The Japanese system MCS-L2 [47] was put into commercial service in Tokyo in April 1988. Its radio channel spacing was narrowed from 25 to 12.5 kHz and diversity reception was adopted [46]. In this system, also called the *high-capacity land mobile telephone system*, since postdetection selection diversity reception is adopted in both base and mobile stations, the optimum antenna diversity configuration was required.

One important characteristic of the diversity antennas for the MCS-L2 system was compactness. The handheld unit placed into commercial service in February 1989 is pictured in Figure 4.51. The diversity antenna is composed of a retractable whip antenna and a built-in PIFA element. The length, width, and depth of this handheld unit are 174 mm, 42 mm, and 68 mm, respectively; that is, its volume is

Figure 4.49 Handheld transceiver unit using half-wavelength whip antenna.

about 500 cc. Figure 4.52 shows a schematic view of the diversity antenna. The length of the whip antenna element is 145 mm in use, which is a little shorter than one-half wavelength. This whip antenna is used for both transmitting and receiving, and the PIFA element is used only for diversity reception. The effective gain in the upright condition without the operator is -2.6 dBd for the whip antenna and -6.8 dBd for the PIFA, and the correlation coefficient between the two antennas is 0.2. However, the effective gain with the operator present is -6.8 dBd for the whip antenna and -8.4 dBd for the PIFA, and the correlation coefficient between the two antennas is 0.3, which is sufficiently low for diversity reception.

In March 1991, a series of extremely miniaturized pocket-sized handheld units, whose volume is less than 150 cc and whose weight is less than 230g, was developed

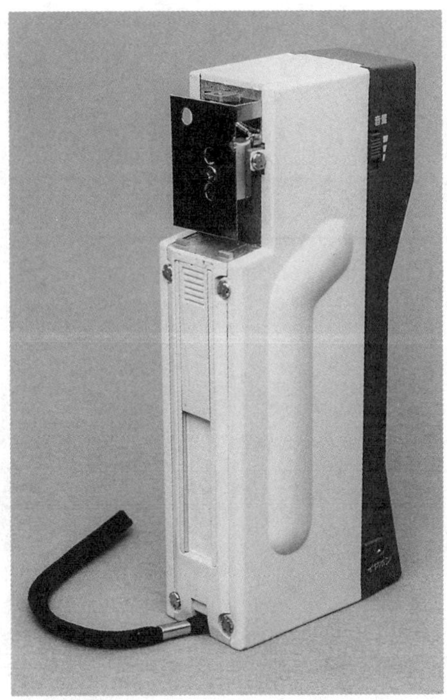

Figure 4.50 Handheld type of detachable transceiver unit using built-in PIFA element.

Table 4.4
Measured Mean Effective Gain of Antennas Mounted on Handheld Transreceiver Unit in Multipath Propagation Environment

Antenna Configuration	*No Body Contact (dBd)* $\alpha = 0$ *deg*	*No Body Contact (dBd)* $\alpha = 60$ *deg*	*Speaking Position* $\alpha = 60$ *deg (dBd)*	*Operator Effect (dB)*
PIFA mounted on lateral side (Type A):				
Upper side	−4.2	−2.6	−5.9	3.3
Lower side	−4.2	−6.0	−7.0	1.0
PIFA mounted on top side (Type B)	−4.5	−5.4	−9.5	4.1
PIFA mounted on back side (Type C)	−3.2	−5.5	−7.6	2.1
$\lambda/2$ sleeve dipole antenna (Type D)	−0.6	−5.3	−7.9	2.6

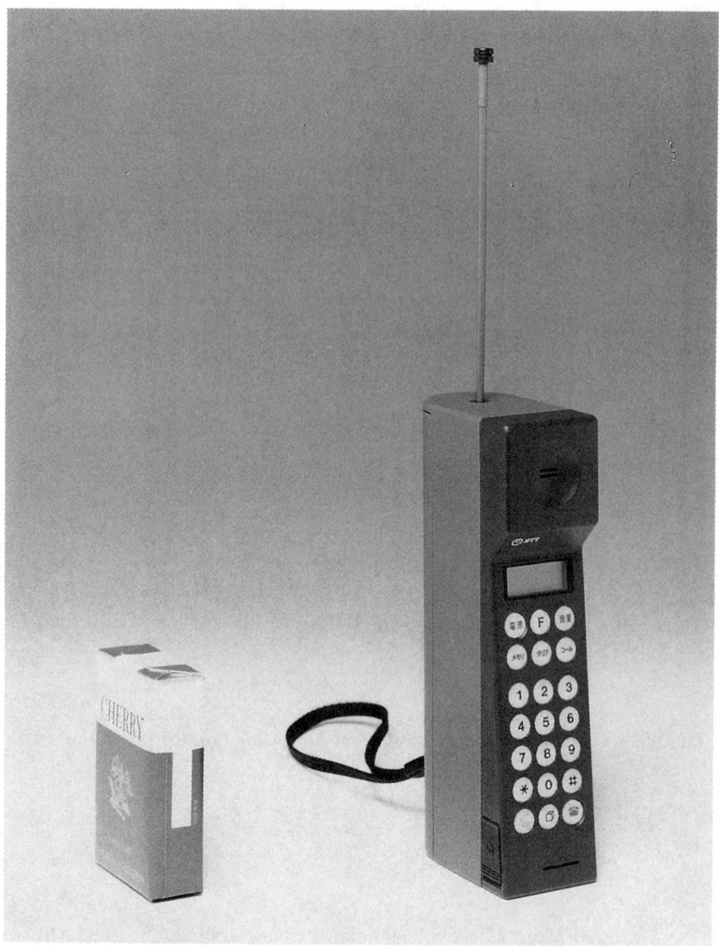

Figure 4.51 Handheld radio telephone unit using diversity antenna system.

by NTT [49]. Figure 4.53 shows their appearance. Every unit uses a diversity antenna system composed of a retractable whip antenna and a built-in PIFA element. The miniaturization of antenna elements has been the key technology in developing such compactness. A 3/8-wavelength whip antenna is adopted for the transmitting and receiving antennas. The PIFA element, used only as a receiving antenna, was miniaturized by putting the receiver RF filter into the cavity of the PIFA element. The PIFA element was also reduced in size because the capacitance between the planar element of the PIFA and the ground plane is increased by the proximity of the conductive case of the RF filter, which is at the ground voltage

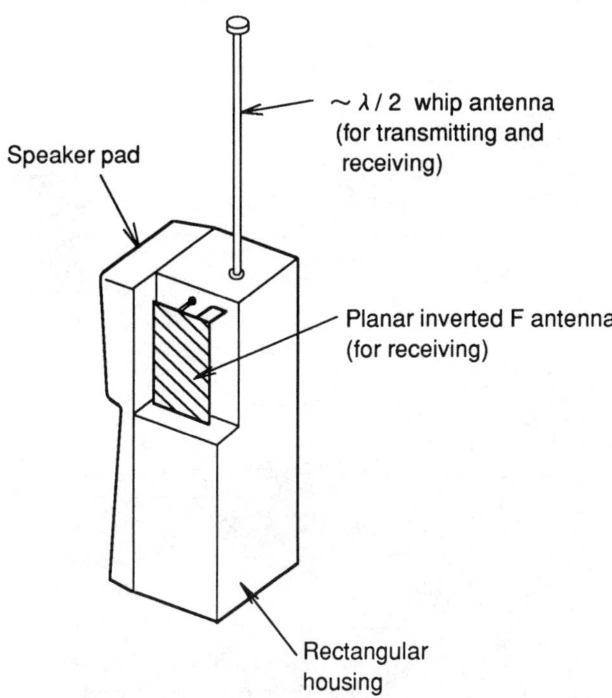

Figure 4.52 Schematic view of diversity antenna configuration.

level. With this integration, the volume of the antenna and filter was reduced to one-fifth that of former models, as shown in Figure 4.54. Figure 4.55 shows the radiation pattern of the 3/8-wavelength whip antenna in the upright condition without the operator. The PAG estimation values of the whip and PIFA are about -2 and -6 dBd, respectively, in the upright condition without the operator. The received signal envelope correlation coefficient of the two antennas is very important in achieving high diversity gain. This antenna configuration yields a correlation coefficient of less than 0.5 under a Rayleigh multipath environment.

4.4.3 Transportable Systems

Antennas for Nondiversity Systems

A transportable telephone unit called the *shoulder-phone* was developed in 1983 as a detachable mobile radio unit for the 800-MHz band cellular radio telephone

Figure 4.53 External shape of pocket-sized radio telephones.

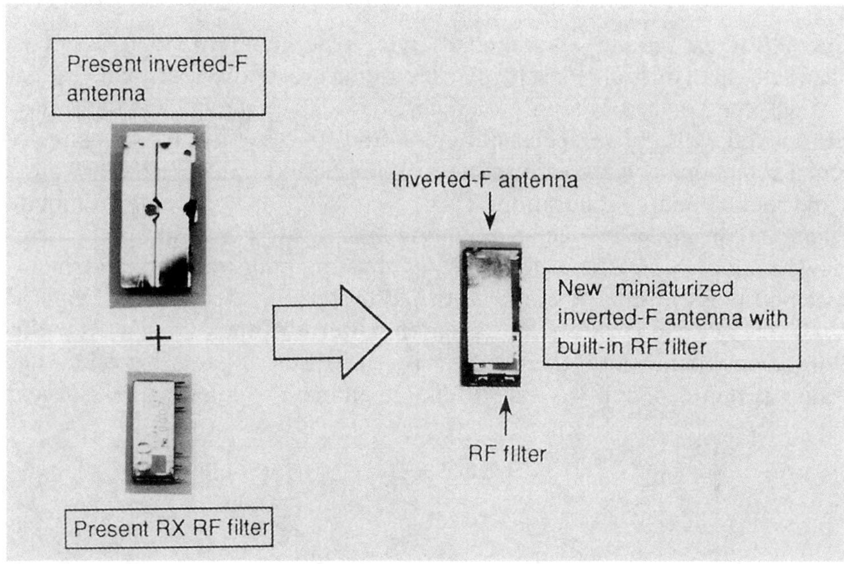

Figure 4.54 Miniaturized planar inverted-F antenna with RF filter.

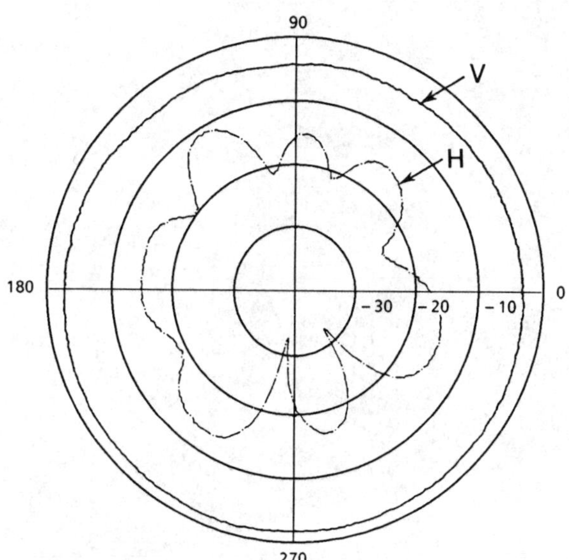

Figure 4.55 Radiation pattern of the 3/8-wavelength whip antenna in the pocket-sized radio telephone.

system [48]. It was put into commercial service in September 1985. This detachable mobile radio unit consists of a vehicular-mounting unit and a detachable transceiver unit, which can be used as a portable telephone. When the detachable transceiver unit is docked with the vehicular-mounting unit, one can use it as a conventional vehicular telephone. Figure 4.56 shows a photograph of the detachable transceiver unit and its antenna configuration. This type of unit must be easy to mount into and demount from the vehicular-mounting unit; thus, a built-in antenna configuration was adopted. A PIFA element is used as the built-in antenna element, and its specified bandwidth (7.8%) was satisfied by making the element height larger than 0.04λ. Figure 4.57 shows the radiation patterns of the antenna. The effective gain of the antenna without the operator is -0.5 dBd [50], and this value is almost the same as that of a half-wavelength dipole antenna. The effective gain with the

Figure 4.56 Transportable type of detachable transceiver unit and built-in PIFA element.

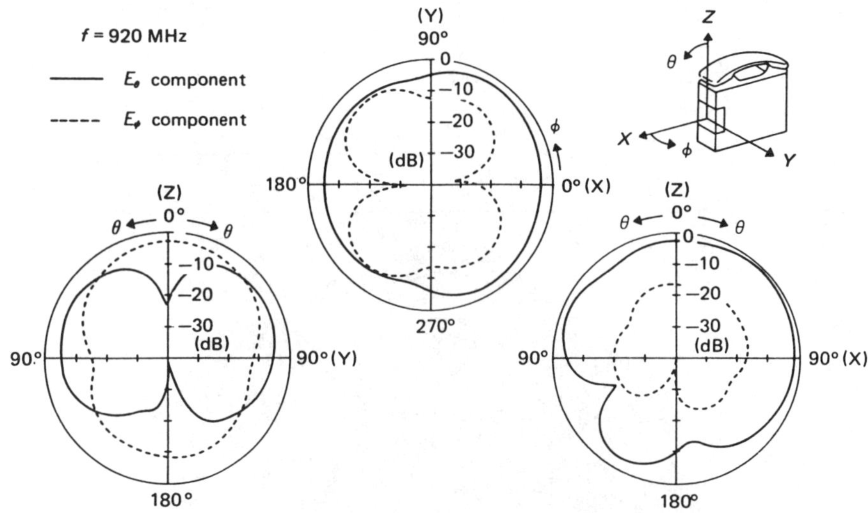

Figure 4.57 Radiation patterns of PIFA element on transportable type of detachable transceiver unit.

operator present is −3.5 dBd, and the gain degradation due to human proximity is about 3 dB.

Another type of transportable unit that uses a half-wavelength sleeve dipole antenna or a whip antenna has also been developed as a commercial transportable unit. Figure 4.58 shows the appearance of the unit, produced by Matsushita Communication Industrial Corporation, and the antenna. This unit uses a half-wavelength monopole (whip) antenna. Figure 4.59 shows the radiation pattern of the antenna without the operator. Figure 4.60 shows the PAG estimation result for the whip antenna, and it is found that the effective gain of this antenna is considered to be −2 to 4 dBd. This type unit is common in the North American (AMPS) system.

Antennas for Diversity Systems

A transportable unit with antenna diversity reception was developed for NTT's high-capacity land mobile telephone system and was put into commercial service in May 1988 [51]. Figure 4.61 shows a photograph of the transportable unit. This unit consists of a transceiver unit, a handset, and an antenna-battery unit, which contains built-in diversity antennas. The set is used as a portable telephone by stacking them together. The adoption of this configuration allows the transceiver

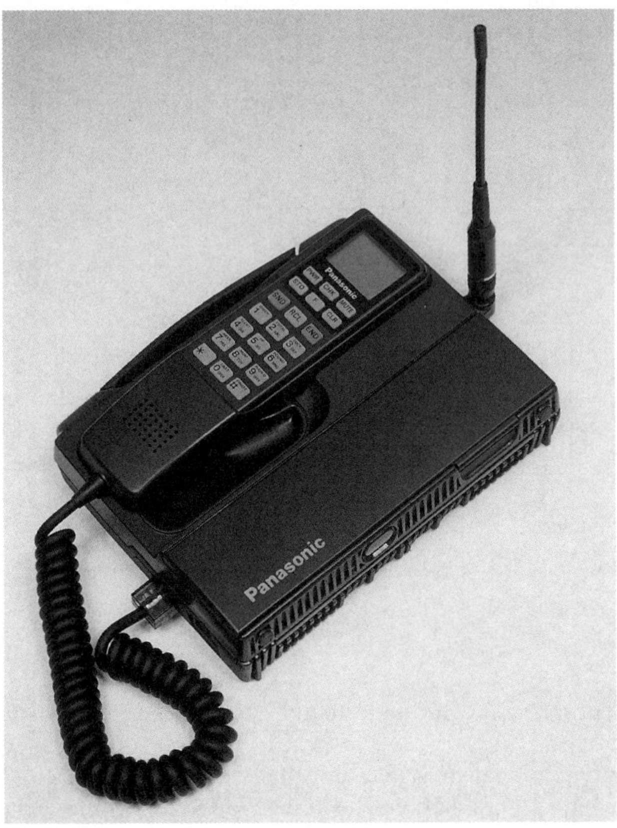

Figure 4.58 Transportable telephone unit using conventional whip antenna. (Courtesy Matsushita Communication Industrial Co., Ltd.)

unit to be used as a vehicular telephone in conjunction with the vehicle's handset, antenna, and battery. As a result of this design concept, the antenna elements are built into the battery unit. The antenna location is an important parameter for optimizing antenna gain, since the conductive housing on which the PIFA element is mounted also acts as an antenna, because of the induced currents present on the housing. Figure 4.62 shows four schematic views of the antenna configurations possible in the antenna-battery unit. In type S1, the two PIFA elements are built into either end of a flat antenna-battery unit. In type S2, the elements are built in parallel in the side of the antenna-battery unit. In type S3, the built-in elements are mounted in parallel on the wide side of an L-shaped antenna-battery unit. In

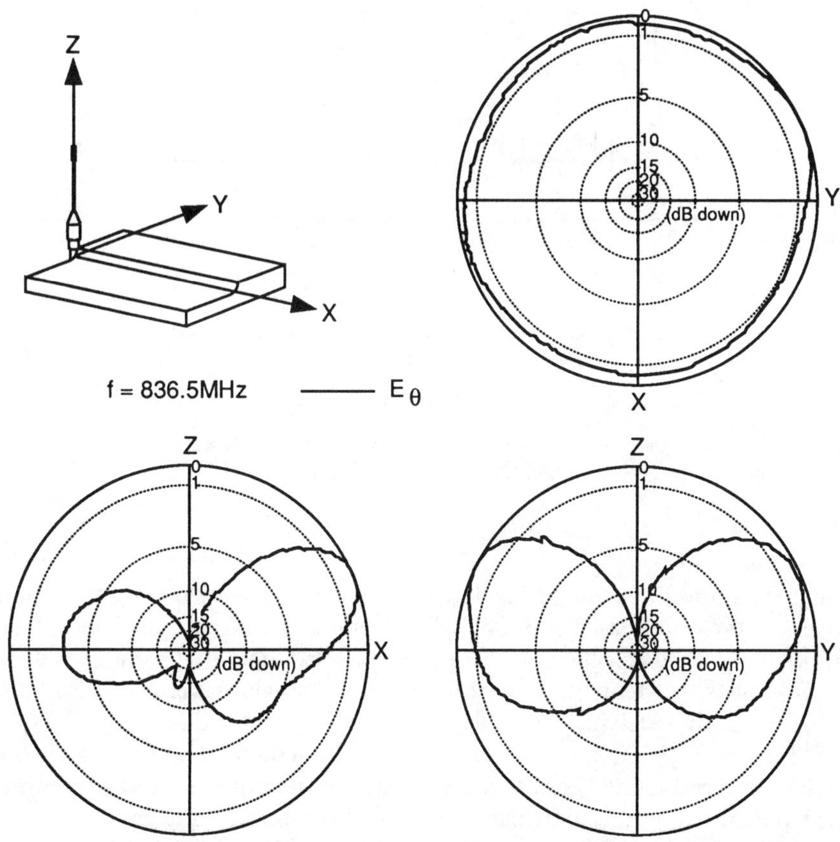

gure 4.59 Radiation pattern of conventional half-wavelength sleeve dipole antenna mounted on the transportable telephone unit. (Courtesy Matsushita Communication Industrial Co., Ltd.)

pe S4, the elements are built in parallel on the end of a differently configured -shaped antenna-battery unit. Figure 4.63 shows the PAG estimation results for ese four antenna configurations in the carrying position. In the carrying position, e portable unit is slung over the operator's shoulder and held close to the waist. ypes S2 and S3 have two different gains, depending on whether the operator's ody contacts the antenna-mounted side or the opposite side. The PAG for type l is −3.2 dBd and is relatively high compared to those of the others. The PAG r type S3 is −5.0 dBd when the operator's body is close to the side on which e PIFA elements are mounted, but −2.3 dBd when the operator's body is close the opposite side. Type S1 was adopted for NTT's commercial transportable lephones because of its high effective gain in its carrying position, the good

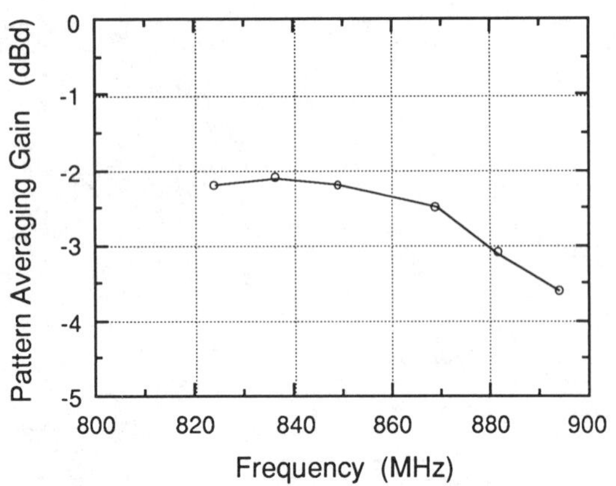

Figure 4.60 PAG characteristics of conventional whip antenna mounted on the transportable telepho unit. (Courtesy Matsushita Communication Industrial Co., Ltd.)

antenna separation needed to maximize the space diversity effect, and the con pactness of the antenna-battery unit. The commercial model, depicted in Figu 4.64, occupies 1,500 cc. The two PIFA elements are separated from each other 0.6λ. One of these elements is used for transmitting and receiving (TR), and t other is used for receiving (R) only. The relative bandwidth for VSWR ≦ 2 10.1% for the TR-element and 6.3% for the R-element. Figure 4.65 shows t radiation pattern of the TR-antenna without the operator. Outdoor experimen carried out in Tokyo urban areas using the down-link frequency of the curre mobile telephone system measured the effective gain of both the TR-antenna a the R-antenna and the correlation coefficient between the antennas; the resu are shown in Table 4.5. The correlation coefficient is low enough to obtain hi diversity gain.

The type S3 antenna configuration was adopted by Nippon Idou Tsush Corporation (IDO) to create a transportable unit; built-in PIFA elements we adopted. Figure 4.66 shows a photograph of the unit, which is used for mob telephone services. Figure 4.67 shows the schematic view of the antenna confi uration of the IDO unit. The design of this unit ensures that the operator's bo is not close to the antennas; thus, the effective gain as well as the PAG value improved, as indicated by the open square shown in Figure 4.63.

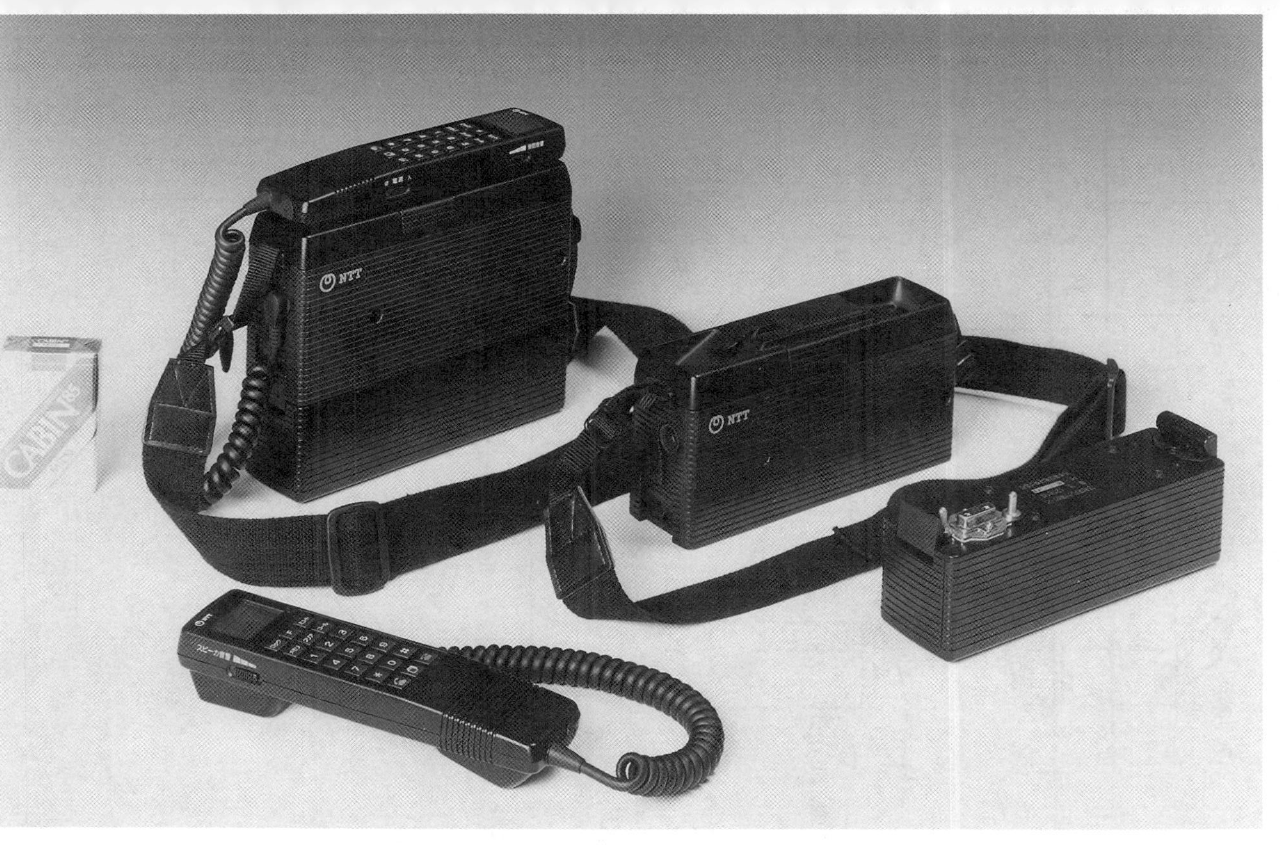

Figure 4.61 NTTs transportable telephone units with diversity antenna system.

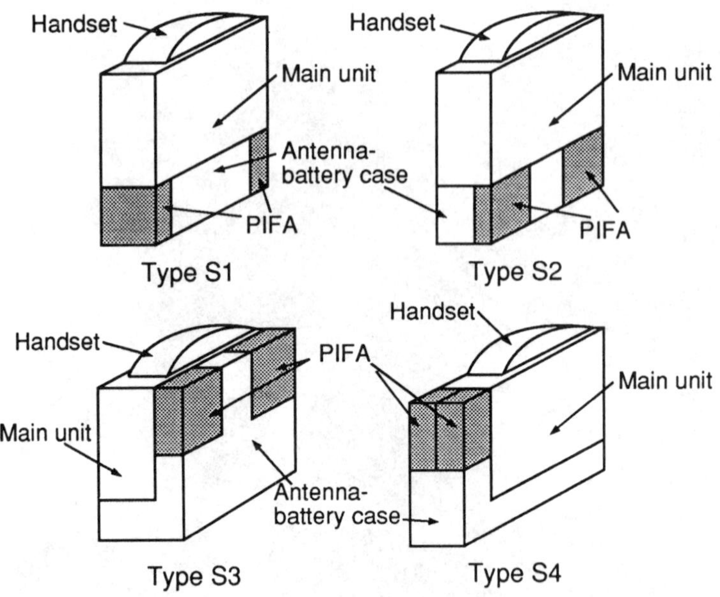

Figure 4.62 Schematic views of diversity antenna configurations using PIFA elements for transportable telephone units.

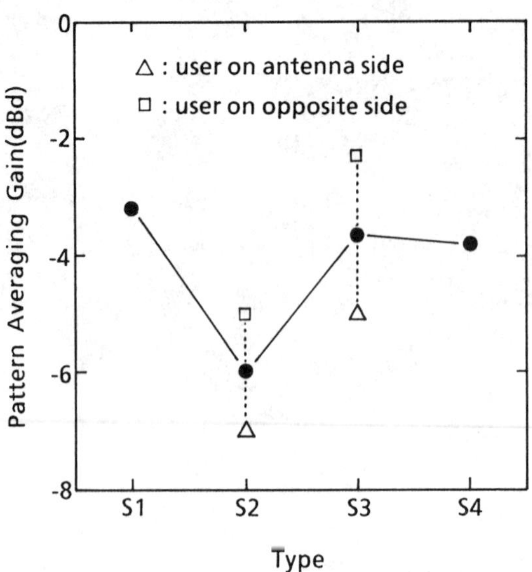

Figure 4.63 PAG estimation results for the diversity antenna element in carrying position.

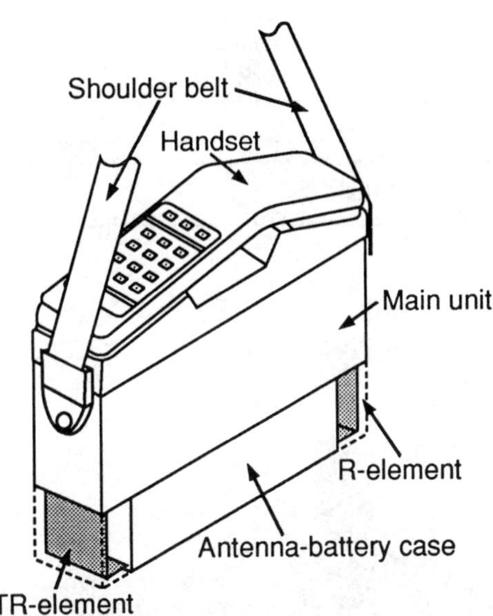

gure 4.64 Schematic views of antenna configuration in NTT's transportable telephone unit.

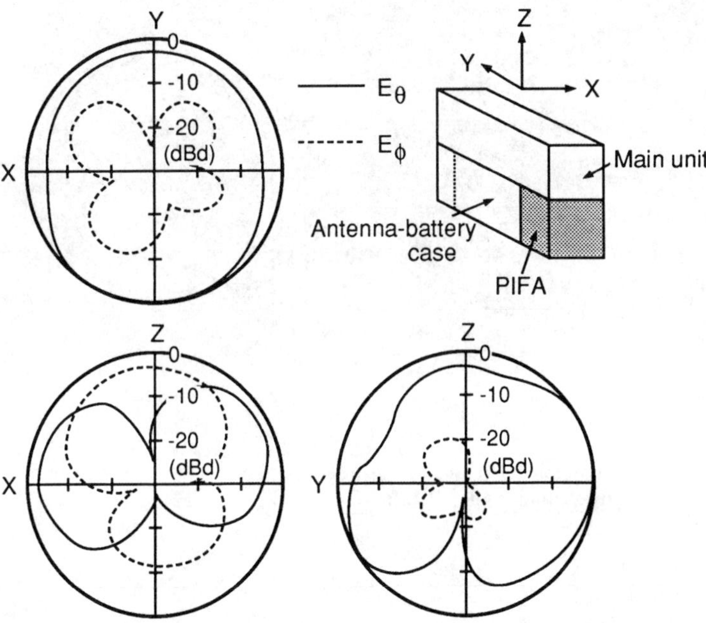

gure 4.65 Radiation patterns of the built-in TR-antenna in NTT's transportable telephone unit.

Table 4.5
Performance Test Results of the Built-in PIFA of NTT's Transportable Telephone Unit in Tokyo Urban Area

Test	*Result*
Mean effective gain of TR-element	−2.9 dBd
Mean effective gain of R-element	−5.4 dBd
Correlation coefficient between TR- and R-elements	0.27

Figure 4.66 IDO's transportable telephone unit using PIFA elements mounted on lateral side of batte unit. (Courtesy Nippon Idou Tsushin Co.)

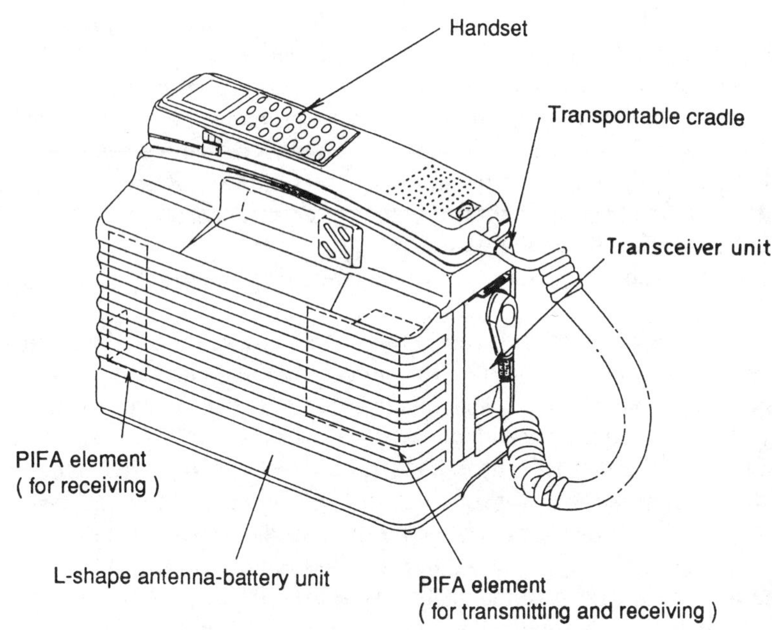

gure 4.67 Schematic view of antenna configuration in IDO's transportable telephone unit. (Courtesy Nippon Idou Tsushin Co.)

.5 SAFETY ASPECTS OF PORTABLE AND MOBILE COMMUNICATION ANTENNA DESIGN

.5.1 Exposure to RF Energy and the Safety of the User of RF Communication Equipment

Iobile and portable two-way wireless communication radios emit RF energy, which as low as a fraction of a watt in the case of some portable phones and as high s 130W for certain mobile stations. In normal use, handheld radios have the otential for causing higher exposure than mobiles because, although portable quipment rarely emits more than 7W of RF power, a person's vital organs (e.g., ıe head) are in the immediate vicinity of the RF source. Mobile radio antennas an be spaced by a few feet from car passengers, so the exposure is normally educed by distance. In addition, in metal-body cars, the vehicle's metal surfaces orm an effective RF shield for the car passengers if the antenna is mounted in ertain positions, as will be discussed in Section 4.5.5.

Let us now define what is relevant in an RF exposure. Until recently (1982) the exposure to RF electromagnetic fields (EMF) in the band of 100 KHz t 100 GHz was quantified in terms of the incident power density measured in watts square meter or milliwatts. This method of measurement, borrowed from photom etry, proved to be grossly unsatisfactory at some RF frequencies of the electro magnetic (EM) spectrum. Two phenomena contribute to differentiate the absorp tion of RF versus infrared or visible light: coherent absorption and the structur of the EM fields in the vicinity of sources.

Coherent absorption is a typical resonance phenomenon. A human expose to an EM wave E-polarized in the direction of the axis of the body supports relatively strong coherent RF current if the wavelength of the incident field is abou five times or two-and-a-half times the height of the human, depending on th grounding conditions. If the body is well grounded, the current in the human i similar to that of a quarter-wavelength dipole over a ground. If the body is poorl grounded, the coherent RF current resembles the one of a resonant half-wav dipole. In both situations, a strong current is excited [1] and the peaks of th absorption of RF energy occur where the human anatomy has the smallest cros section (e.g., the ankles, the neck, and the inguinal area) [52].

The near-field exposure makes the use of the term *power density* practicall meaningless. Clearly, the relevant exposure from low-power (7W or less) portabl devices happens within a few centimeters' distance from the antenna, where som high EM energy density values may be found. At this short distance, the radiatin source has a finite angular extent as seen from the observation point, so it i impossible to define a direction of power flow (see Figure 4.68). Very close to a source, the direction of propagation of the RF energy may not coincide with th direction of the Poynting vector ($=1/2\ \mathbf{E} \times \mathbf{H}^*$), which is normally associated wit power flow in the far field of RF sources. In these conditions, we cannot simpl define *power flow*, which makes any attempt to measure it practically impossible

Another consideration, which may help us to define the exposure paramete of relevance, is the observation that only the EM energy actually absorbed by th human subject can cause a bioeffect. Given the large amount of reflection an scattering that a human body can cause in the near field of a source of comparabl size, it makes more sense to measure the RF power absorbed by a human tha the power incident. The question that arises at this point is: How does a huma absorb EM energy in the RF band?

Since the human body has practically irrelevant amounts of magnetic mate rials, it does not directly absorb magnetic energy. The human body is mainl composed of water, electrolytes, and complex molecules with a large net dipol moment [53] With these constituents, the human body extracts energy from th RF E-field by ionic motion and oscillation of polar molecules. At the present tim other mechanisms of RF coupling with living tissue (e.g., Larmor precession c nuclei) are not considered relevant to the effects discussed in this section. Th

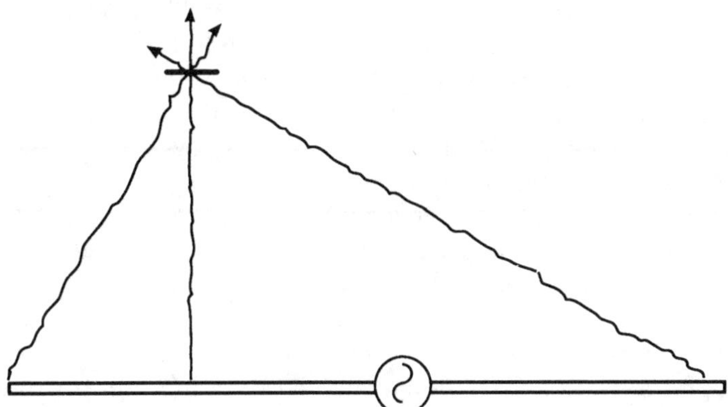

Figure 4.68 Energy multipath near a half-wave length dipole.

absorption of energy from the H-field is through the concatenation phenomenon of Faraday ($\nabla \times \mathbf{E} = -\mu\, \partial \mathbf{H}/\partial t$).

4.5.2 Specific Absorption Rate of Electromagnetic Energy

From the consideration in the previous section, the evaluation of the exposure of a human in the near field of RF sources, like portable and mobile radios, can be accomplished by measuring the E-field inside the body. If we denote with the symbol σ the conductivity of human tissue due to both conductive and lossy displacement currents (σ is measured in siemens/m), then, by definition, the ohmic loss per unit volume is given by the dot product of the current density **J** and the conjugate of the electric field ($\mathbf{E}^*$). The power absorbed per unit volume of tissue is

$$P_v = \frac{1}{2}\mathbf{J}\cdot\mathbf{E}^* = \frac{1}{2}\sigma\,|E|^2 \qquad \text{W/m}^3 \tag{4.19}$$

If the absorption of RF per unit mass is desired rather than the absorption per unit volume, we need only to introduce in (4.19) the density p (in kilograms/cubic meter) of the tissue, thus obtaining

$$P_g = \frac{1}{2}\frac{\sigma}{p}|E|^2 \qquad \text{W/kg} \tag{4.20}$$

The significance of (4.20) should be clear to the reader. P_g is called *specific absorption rate* (SAR) and gives the ratio between the infinitesimal amount of RF power dW absorbed in the infinitesimal mass dm of tissue surrounding a specific point: $P_g = \mathrm{d}W/\mathrm{d}m$. If the RF power P absorbed by a specific organ is desired, then the result is given by the integral relation: $P = \int_M P_g \, \mathrm{d}m$. where M is the mass of the organ of interest, which can be an entire human body.

From the definition of SAR, it should be clear that the phenomenon of RF power absorption by the human body is very complex. The conductivity is a function of frequency and tissue constitution [54]: $\sigma = \sigma(x,y,z,f)$, where x,y,z are the three-dimensional coordinates of an arbitrary reference system connected with the human body. The tissue density also changes as a function of its water content and its structure, so $p = p(x,y,z)$. Equations (4.19) and (4.20) are valid for time-harmonic fields. The incident field may be time-varying (e.g., a pulse train or several fields of different frequencies are incident). In these conditions, it is possible to apply (4.20) to the spectral content of the absorbed field $\mathbf{E}(t)$. If the signal has a broadband spectrum, care must be taken to use the appropriate value of σ at different frequencies. In [55], tables are given of the conductivity versus frequency of various tissues.

The time integral of the SAR is called *specific absorption* (SA) and represents the ratio of the infinitesimal amount of EM energy (dE) absorbed by an infinitesimal volume of tissue and its mass (dm): SA $= \mathrm{d}E/\mathrm{d}m$. The concepts of SA and SAR have been borrowed from radiation medicine, where *dose* and *dose rate* have been used for many years in the treatment of tumors with x-rays.

Clearly, using the SAR methodology in assessing the exposure of a human to incident EM energy totally bypasses the issue of near or far fields from sources. It is still desirable to relate the SAR distribution of an exposure to the structure of the incident EM fields, but it is not strictly necessary. Let us now review the methods currently available for measuring SAR.

4.5.3 Measurement of SAR and Safety Criteria

One method of measuring SAR is to directly apply (4.19) or (4.20), which requires the local probing of the E-field in tissue. The construction of small E-field probes, which are accurate, sensitive, mechanically sturdy, and transparent to RF energy, except for the sensing device, has been difficult. The initial efforts to evaluate SAR made use of temperature measurements.

All the energy absorbed by living tissue from EMFs ends up increasing the status of mechanical agitation of the molecules and ions of the cells forming the tissue. If such mechanical motion is truly random, then we can say that all the EM energy is absorbed as heat and the only final outcome of the exposure is an increase of the temperature of the exposed tissue. At the present time (1991), this issue is being hotly contested. There are researchers who believe that the mechanical motion imparted by certain periodic EMFs to long molecular structures in the living cell are not random, but coherent, at least for several periods of the incident wave [56,57]. In these conditions, before thermalization (random motion), the EM energy conceivably could trigger some biochemical reactions (e.g., the activation of an enzyme).

The matter of the thermalization or the coherent absorption of EM energy by living tissue will not be settled for some time. From a simple experimental point of view, we can consider the energy coherently absorbed by tissue to be negligible if any, and that practically all the RF energy is instantly thermalized. If we neglect the effect of heat exchange mechanisms during a short but intense RF exposure, the following equation holds, expressing the instant thermalization of the absorbed EM energy:

$$\mathrm{SAR}\,\Delta t = \mathrm{c}\gamma\Delta T \tag{4.21}$$

where Δt is the duration of the exposure, c is Joule's constant (c = 4.185 J/cal), γ is the thermal capacity of the exposed tissue (in calories/grams/degrees Celsius) and ΔT is the temperature increase due to the RF exposure.

Various researchers [58,59] have developed human tissue equivalent models, so it is possible to perform SAR measurements using phantoms or dummy humans. Using computerized axial tomography (CAT) or magnetic resonance imaging (MRI) results, it is possible to achieve human-equivalent phantoms as sophisticated [60] as necessary to study RF absorption by specific organs, even at relatively high frequencies (6 GHz).

Phantom models can be built, so they can be disassembled repeatedly after exposure and scanned by an infrared camera for temperature differential measurements [61]. The temperature rise in the models can also be detected by using small temperature sensors [62,63]. Thermometric measurements have the main drawback that they require a substantial amount of RF power to be accurate. A more sensitive method requires the employment of calibrated E-field probes, which are RF-transparent, except for the sensors, so they do not disturb the field to be measured, are small with respect to wavelength (in the phantom material) so as to measure accurately rapid spacial changes of SAR, and are mechanically sturdy, so they can be inserted into the simulated tissue. Such probes have been developed [64,65] and have been used to collect some of the data presented in the next sections.

Now that we can characterize the RF exposure of a human to a substantial degree of accuracy through SAR measurements, we are left with the fundamental question "What are the safe levels of SAR?" in order to decide whether the incident EMFs are safe or not. The answer goes beyond engineering and physics; it requires the confluence of knowledge from a variety of disciplines: pathology, biology, health physics, epidemiology, and veterinary medicine (animals are exposed during experiments to establish threshold levels of adverse bioeffects).

Today (1991), most western countries (including Japan and the United States) have adopted or are in the process of adopting safety regulations for the exposure of humans to RF energy, based on the American National Standards Institute (ANSI) standard C95.1-1982 [66]. The standard states that the exposure of humans in the 3-KHz to 100-GHz band over any 6-min period must be limited to a body-averaged SAR of 0.4 mW/g, with peaks no higher than 8 mW/g; the peaks are averaged over any 1g of tissue shaped like a cube. The ANSI C95.1-1982 standard is currently being revised. In the proposed new standard, two classes of exposure limits are recommended in the 30-MHz to 15-GHz band:

1. Controlled exposure limits: 0.4 mW/g (averaged over the body); 8-mW/g peak (averaged over 1g of tissue); averaging time: 6 min.
2. Uncontrolled exposure limits: 0.08 mW/g (averaged over the body); 1.6-mW/g peak (averaged over 1g of tissue); averaging time: 30 min.

The proposed new standard recognizes two groups of exposed individuals. In the controlled exposures, the human subjects are aware of the presence of RF energy and have control of the sources of the fields. In the uncontrolled exposure conditions, the individuals have no knowledge of the presence of the fields and no control of the RF sources.

To help the reader understand the exposure limits just mentioned, we will analyze the significance of 0.4-W/kg SAR for an adult human weighing 70 kg. For a continuous exposure lasting 6 min or more, the maximum RF power absorbed by the human is limited to 0.4 W/kg $\times$ 70 kg = 28W; the maximum total permissible energy absorption is 28W $\times$ 6 $\times$ 60 sec = 10,080J over the same 6-min period. The power of 28W is approximately one-quarter of the resting metabolic rate of an adult [67]. The maximum permitted exposure does not significantly increase the core temperature of the human subject [68] in comfortable temperature and humidity conditions.

4.5.4 Exposure from Mobile Radios

The exposure of humans by mobile radios has been investigated by a variety of researchers [69–72]. All the available data show that there are concentrations of E-field energy or E-field *hot spots* located in and around the vehicles with RF transmitters operating at VHF and lower bands. The E-field hot spots are practically absent in vehicles with center-of-the-roof-mounted antennas operating at 450-MHz and higher bands. The same phenomena were detected with center-of-the-trunk-mounted antennas. The hot spots have caused some concern because of the possibility of sharp, localized high values of SAR. These concerns turned out to be unfounded for the reasons given below.

Experimental results [70] indicate that the vehicle passengers are significantly shielded by the metal surfaces of the car body. Measured radiation patterns of antennas mounted on cars show little dependence on the number of passengers in the car or their location in the cabin. Using a quarter-wave antenna at 150 MHz mounted on the front fender at about a 50-cm distance from the driver, radiation pattern levels with or without the driver show that a person in the front seat on the side of the antenna absorbs less than 1% of the transmitted power. For a 100W mobile radio, such an exposure is about 1W of RF power absorbed.

At high frequencies (450-MHz and higher bands), the presence or absence of passengers makes no difference in the radiated level from mobile antennas. Stuffing the cabin of the car with RF-absorbing material or humans makes no measurable difference in the power levels radiated from mobile radios with center-of-the-roof-mounted antennas. The result points out that the total absorption of RF from the passengers is much less than 1% of the power radiated. The main factors in determining the RF emission from a mobile antenna are its location on the vehicle (often the center of the roof or the trunk lid) and the size and shape of the cab. Measurements and computations show that the dimensions of the vehicle roof are an essential factor for the antenna efficiency and general pattern shape. The details of the pattern are determined by the height of the antenna above ground [73,74].

At low frequencies (150-MHz and lower bands), with wavelengths of the order of magnitude of a vehicle size, practically the entire metal body of the car participates in the radiation process by supporting the RF currents emanating from the antenna feed point. These currents terminate in electric charges where the conducting external surfaces of the car come to an abrupt end. These changes account for the strong E-fields (1 to 2 V/cm) detected in the wheel wells and at the bottom of vehicles, with the low-frequency (<150 MHz) 100W transmitters.

Strong electric fields (about 1 V/cm) have been also detected inside the vehicle near the top of the steering wheel, at the top of the foot pedals and the transmission hump. These fields are probably due to RF currents that penetrate the vehicle and terminate at sharp metal edges. The E-fields do not give rise to SAR values larger than 0.2 mW/g in human-equivalent phantoms, a fact that points out that the fields are connected with static charge accumulation rather than with radiating currents, as pointed out in [75].

High field intensities are found in the immediate vicinity of the antennas of mobile transmitters. At 150 MHz, measurements have shown that E-fields of 3.8 V/cm exist at a 20-cm distance from roof- or trunk-mounted antennas with 100W transmitted power. Exposure to these fields causes SAR values exceeding the ANSI limits. The strong E-fields just mentioned decay almost linearly with distance from the antenna. Measurements show that for 100W radiated at 150

MHz, the E-field is 1.4 V/cm at the edge of the roof or the trunk at about 60 cm from the antenna. A partial body exposure to this field is slightly above the ANSI limits. The exposure is within the ANSI limits for a 50W or lower power transmitter.

From the above considerations, based on measured data, it is clear that 100W mobile radio equipment emits strong EMFs. However, if the antennas are properly located at places that do not have easy access, these strong fields are attenuated by space propagation before incidence at places where people may be around the vehicle. Mobile radio operators and passengers are shielded from RF energy by the metal surfaces of the vehicle, if the antenna is located on the roof or on the trunk lid. Roof gutter and fender mounts are not recommended for high-power mobile radios (7W or more) because of the potential for overexposure of a bystander. Finally, for vehicles with plastic bodies, great care should be placed to locate the mobile radio antenna and its grounding surfaces, if any, at a sufficient distance from the driver and the passengers so as to avoid potential overexposure.

4.5.5 Exposure of Portable Radio Operators

Because of the relatively low power of portable transmitters (usually less than 6W), the exposure of concern is only in the very close proximity to the antenna, where the head of the user is located during the normal operation of the device. Phantoms have been used extensively to determine the peak and average SAR in users of portable radios. Typical phantoms are shown in Figures 4.69 to 4.71. The full-body phantom is used to measure the overall amount of RF power coupled between the radio with the simulated human. The measurement is performed by contour plots of the radiation pattern of the radio in the presence and the absence of the simulated human. The difference between the total powers radiated by the portable device represents, as a first approximation, the total power absorbed by the simulated human. Some differential mismatch losses are neglected in this process, but these are not large, as detailed in [76]. The method yields a worst-case approximation.

The flat phantom shown in Figure 4.70 is used to simulate with a simple model the tightest coupling conditions between the radio antenna and the body. A belt-worn radio with a remote press-to-talk switch produces an exposure of this type. Also visible in Figure 4.70 is the implantable E-field probe, which is immersed in a liquid simulating soft muscle tissue.

The head-shaped phantom of Figure 4.71 is used to perform SAR measurements in the human brain and skull when exposed to portable radio telephones. During the exposure measurements, the implantable E-field probe penetrates the simulated skull from the direction opposite the one where the radio is placed near this phantom. A sophisticated robotic system should be used to locate precisely the E-field sensor inside the simulated head, since the probe tip cannot be seen from the outside. The correct measurement of SAR in the near field of RF sources

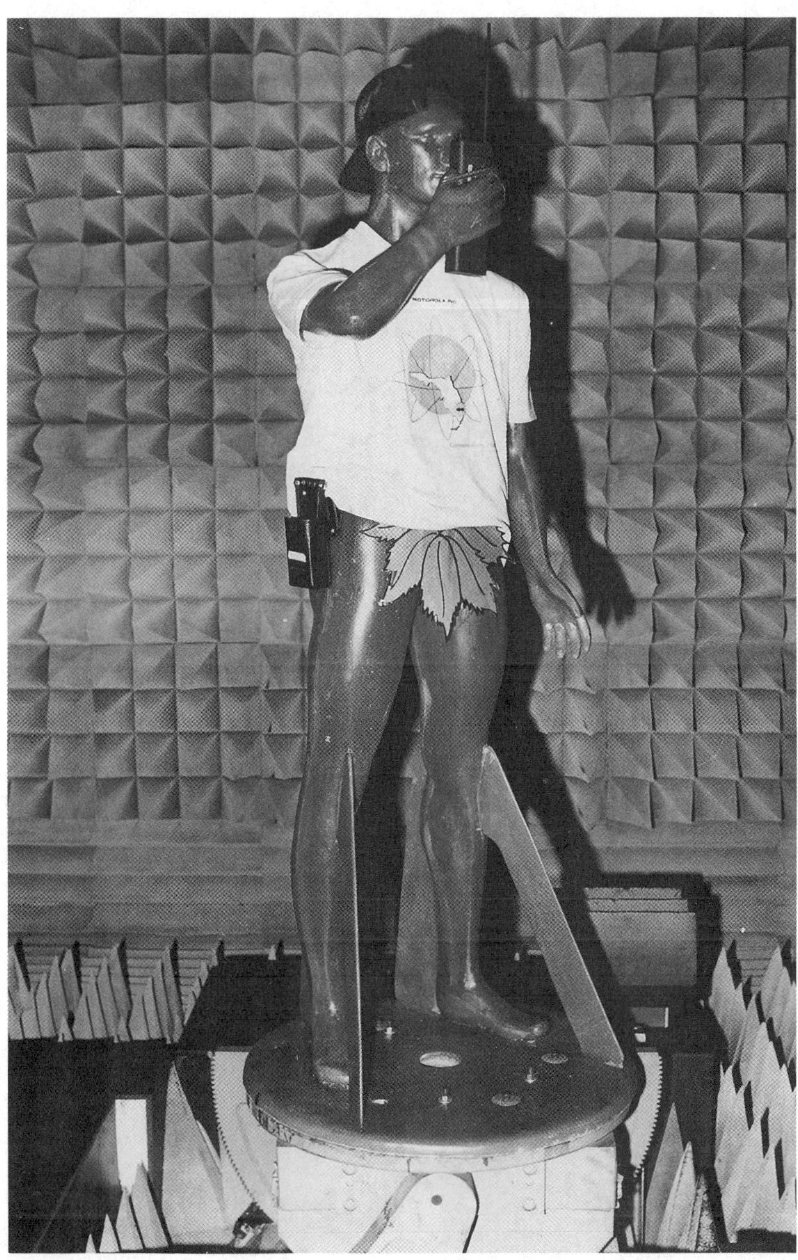

Figure 4.69 Full human body phantom.

Figure 4.70 Box phantom.

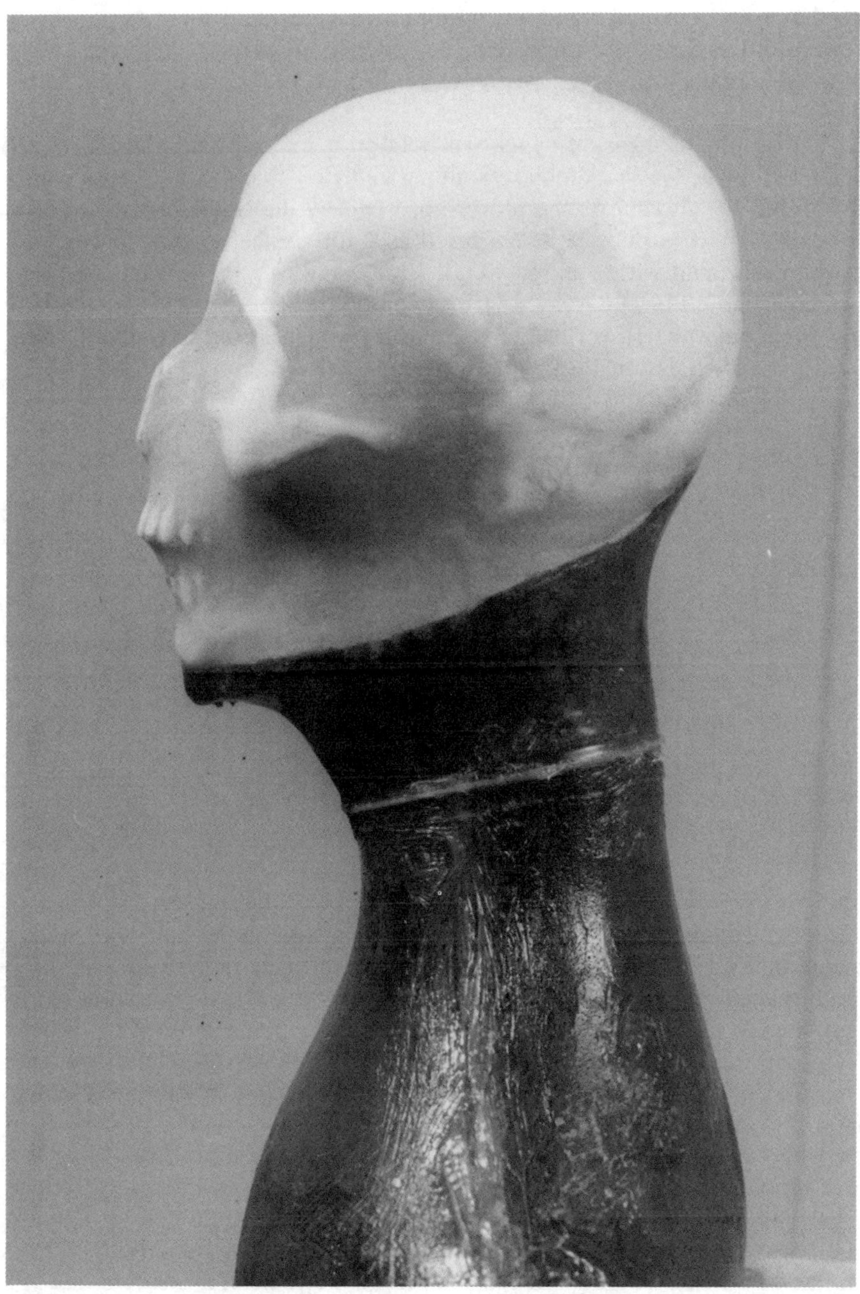

Figure 4.71 Human head phantom.

using flat or head-shaped phantoms requires good mechanical positioning accuracy (± 1 mm) of the E-field sensor. The measurements are difficult: a relative accuracy of $\pm 10\%$ is obtained at the cost of great attention to detail and substantial dedicated time.

The results of measurements of whole-body exposure conducted at various frequencies using 6.4W radios are summarized in Table 4.6 for various antenna types. Table 4.6 also shows the power absorbed by the simulated head of a user. The head total absorption was measured as a difference of total power radiated with the radio held with a fully extended arm, and at 5.0 cm from the face with the radio at mouth level. Cross-polarized components of the field are about 10 dB below the main radiated vertical polarization. In the case of the isolated radio, the cross-polarized radiation (horizontal polarization) is about 20 dB below the vertically polarized component.

Table 4.6
RF Absorption of Portable Radio User 6.4W Transmitter Power

Frequency (MHz)	*Antenna Type*	*Location*	*Power Absorbed (W)*
30	Collapsible	Body and head	4
30	Helix	Body and head	5
150	Helix	Body	3.5
		Head	0.5
450	Dipole	Body	3.0
		Head	0.7
450	Helix	Body	3.5
		Head	0.5
800–900	Sleeve Dipole	Body	1.5
		Head	1.0

Some comments are in order about the results at 30 and 800 MHz. The measurements at 30 MHz have been extremely difficult to perform. The isolated radio has poor radiation performance (especially with the helical antenna, which is only 40 cm long), making the measurements of total power emitted by the isolated radio somewhat imprecise. Only when the phantom operator holds the radio is there efficient RF emission. This phenomenon points out that the user is actually part of the antenna of the radio. The radio excites RF currents on the body of the user, who has the physical dimension of a good antenna. In the 800-MHz band, a sleeve dipole was used in the measurements, with the specific aim to reduce the RF currents on the radio case. The limited role of the radio case as a source of EM energy explains the lower losses through the hand, arm, and body of the user than those at the lower frequencies. In all frequency bands, measurements of SAR in the hands and the arms of a simulated human have not detected a peak value greater than 0.5 mW/g.

4.5.6 Peak SAR Values

In the 30-MHz band, it has not been possible to perform SAR measurement either with a small implantable E-field probe or a temperature sensing device with 0.005°C sensitivity. The radio used in the measurement was a 6.4W portable device.

From the absence of measurable temperature increase with a 0.005°C sensitive thermometer, it is possible to conclude that the peak SAR value in a user of a 6.4W 30-MHz radio is no more than 0.3 mW/g. From Table 4.6, the average body SAR for a 70-kg human is 57 to 71 μW/g.

In the VHF frequency band (about 150 MHz), tests were conducted using 6.4W portable radios with helical antennas. Collapsible quarter-wavelength telescopic antennas have been used in the past at this frequency band, but they have been abandoned because of poor mechanical performance in drop tests. The exposure of users of VHF radios with extended collapsible antennas is expected to be lower than the values presented herein for the same power radiated and the same distance between the antenna and the user. The collapsible antennas are longer by a factor of three than the helical antennas commonly used at this frequency band. If all other conditions are the same, the energy density incident on the user is smaller in the case of the longer antenna.

Figure 4.72 shows the value and the location of the peak SAR value detected in a head phantom with a distance of 5 cm between the simulated lips and the radio case. The peak SAR is about 0.5 mW/g, located at 2.5 cm above the eyebrow on the side of the antenna. Anatomically, the peak SAR location is on the subcutaneous fat tissue or at the surface of the skull. The average brain exposure is 0.1 mW/g. In Figure 4.72, the average head SAR value is obtained by dividing the power-absorbed value in Table 4.6 by 4.5 kg (approximate average weight of the head and neck of a 70-kg adult). The average body SAR is obtained by dividing the sum of the power absorbed by the body and the head by 70 kg. Radios operating in the 430- to 512-MHz band are normally equipped with a quarter-wavelength whip or a helical antenna. The results are shown in Figures 4.73 and 4.74; measurements were performed at 450 MHz with a 6.4W radio.

In both cases, the peak exposure is at the surface of the eye that is closer to the antenna. It is interesting to note that the whip antenna, although physically longer, produces the larger peak exposure. The cause of this phenomenon can be traced to the impedance (ratio $|E|/|H|$) of the fields incident on the eye. The fields generated by the whip have a relatively low impedance and so more readily penetrate the tissues rich in water like the eye.

If the portable radio is brought closer to the face so that the radio case is practically touching the nose of the user (about 2-cm distance from the mouth), the peak SAR at the surface of the eye is less than 4 mW/g for both antenna types. The RF penetrating the eye decays following the approximately exponential falloff shown in Figure 4.75. Within 2 cm from the surface, the SAR is 50% lower than the peak value.

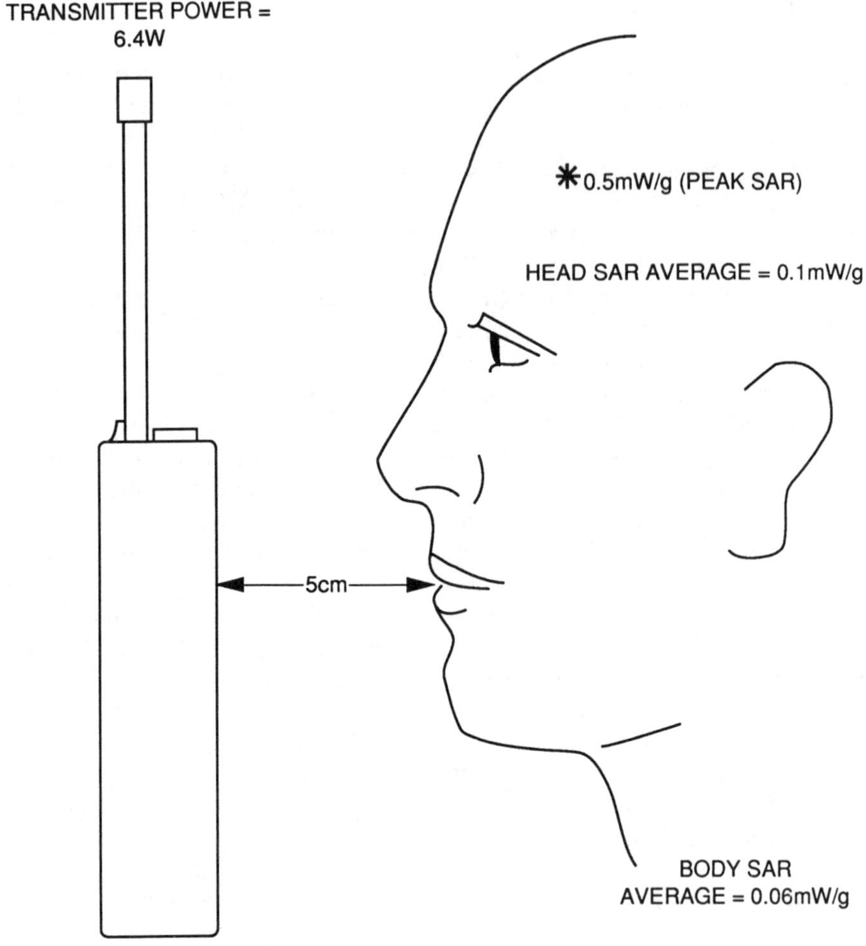

Figure 4.72 SAR levels in the head and body of portable-transmitter users at 150 MHz (helical antenna).

In the 800- to 900-MHz band, portable radios are equipped with sleeve dipoles, so the antenna is decoupled from the radio case. This arrangement substantially reduces the absorption of RF by the hand and arm, but increases the total amount of power absorbed by the head of the user, as can be seen from the values in Table 4.6. With this antenna, more than 50% of the available RF power is radiated for the purpose of communication. At the lower frequencies, more than 50% of the available RF power is absorbed by the body of the user. Figure 4.76 summarizes the results of the exposure measurements performed with a 6.4W radio

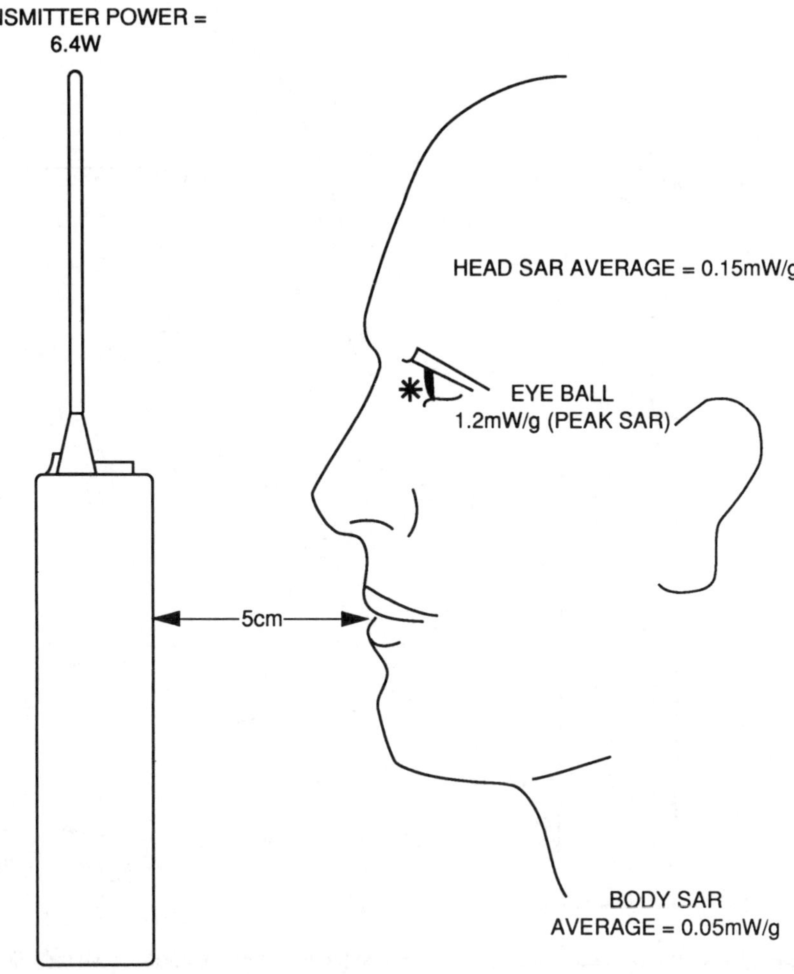

Figure 4.73 SAR levels in the head and body of portable-transmitter users at 450 MHz (whip antenna).

at 840 MHz. The peak SAR is located on the frontal lobe on the side of the antenna in correspondence with the feed points of sleeve dipole. The peak SAR is detected at the skull-brain interface.

If the radio case is held very close to the face (2-cm distance from the mouth), the peak SAR is found at the surface of the eye that is closer to the antenna; this peak value is about 3 mW/g. The penetration of RF into the eye is given in Figure 4.77. Comparing Figures 4.75 and 4.77, it can be seen that the RF energy deposition

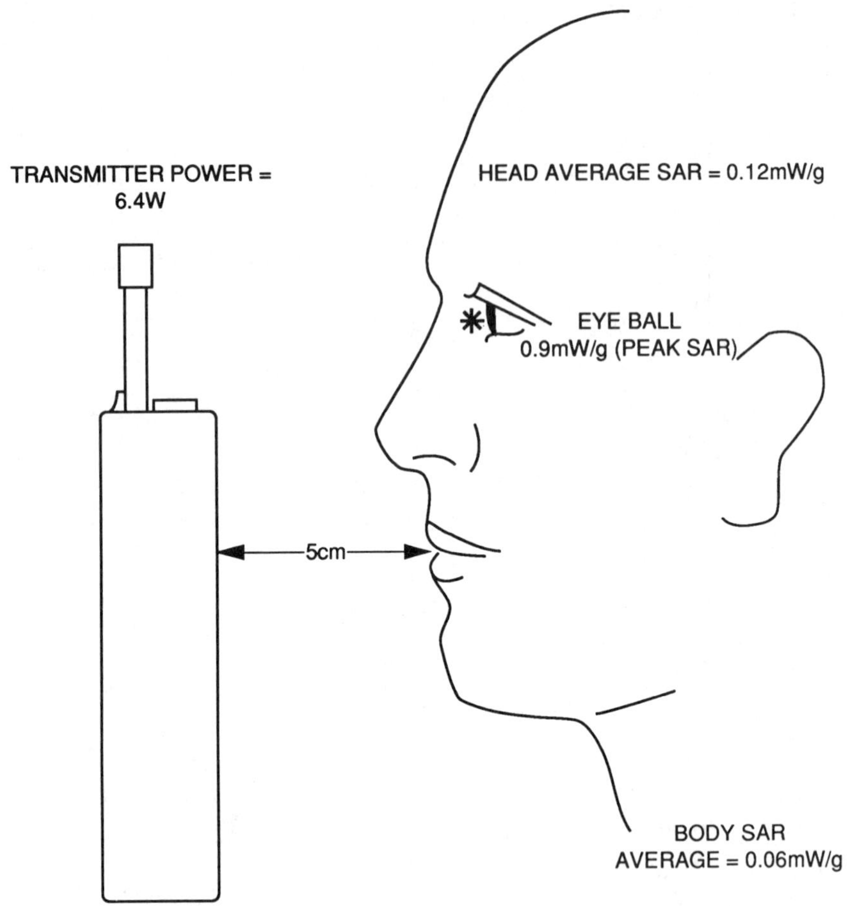

Figure 4.74 SAR levels in the head and body of portable-transmitter users at 450 MHz (helical antenna).

in the eye in the 800- to 900-MHz band with a sleeve dipole is much lower than at 450 MHz with a whip antenna for the same exposure conditions.

The quarter-wave whip antenna is seldom used in the 800- to 900-MHz band because of its poor coverage properties for handheld radios. Measurements show that the quarter-wavelength whip [77,78] in close proximity (radio held at 2 cm from the mouth) to the face produces a peak SAR of about 7 mW/g in the eye that is closer to the antenna. The power radiated is 6.4W, as in the previous cases.

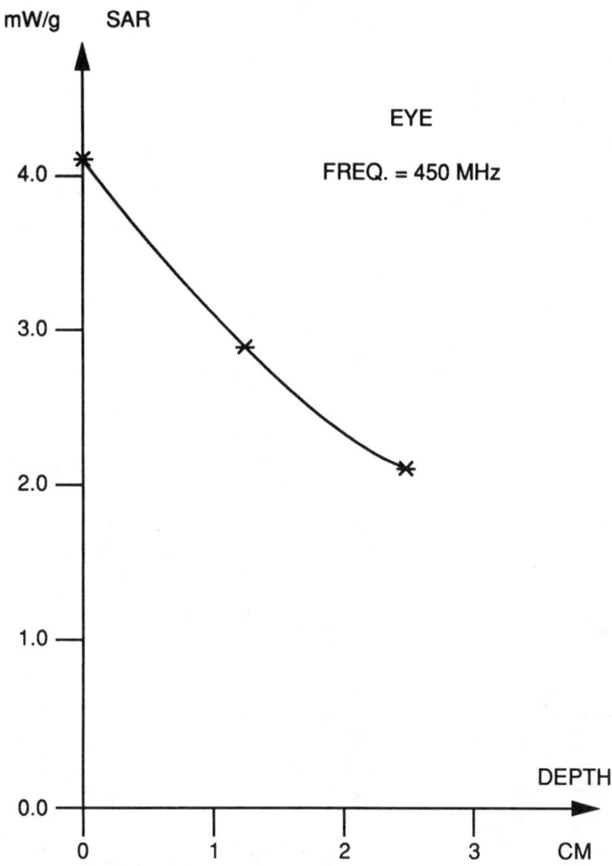

Figure 4.75 SAR versus depth in the eye.

4.5.7 Conclusions

From the results presented in the previous sections, the reader should conclude that dispatch-type portable radios expose the user at instantaneous SAR values below the limits of the ANSI C95.1-1982 safety standard. Time averaging further reduces the overall exposure of the user of portable radios. Dispatch channels are normally shared by 50 to 100 subscribers, who transmit messages of total average length of 20 sec [79]. Considering that the transmitter is on approximately half of the 20-sec average message and that a user may place or receive at most a couple of calls during any 6-min period, the time averaging reduction factor for the RF

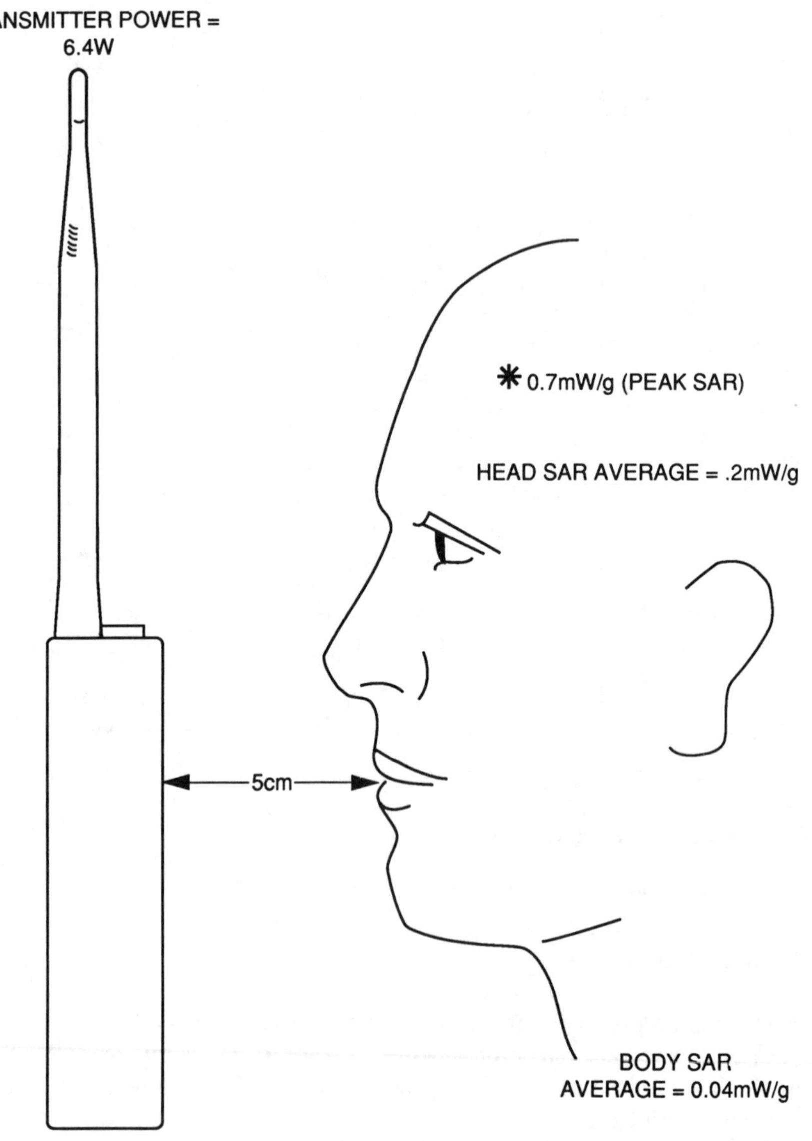

Figure 4.76 SAR levels in the head and body of portable-transmitter users at 900 MHz (sleeve antenna).

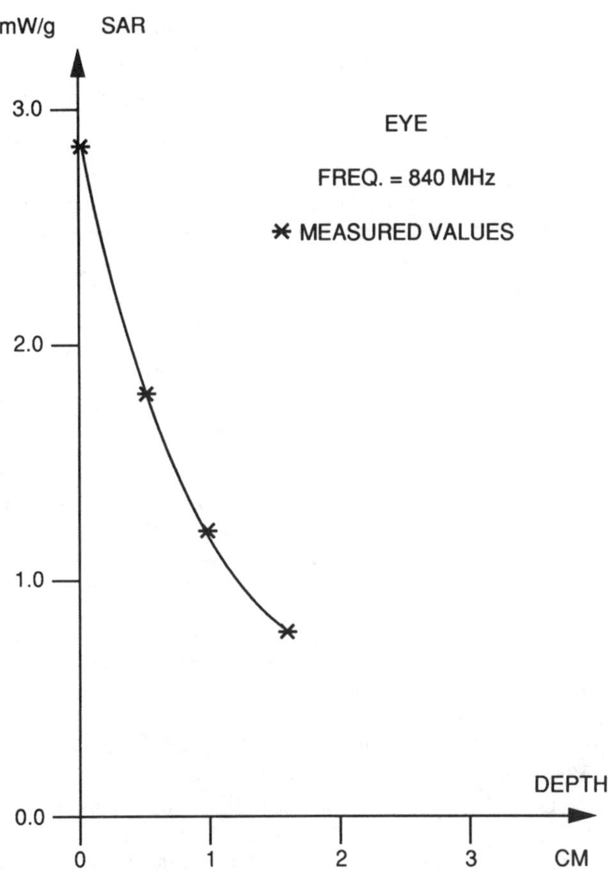

Figure 4.77 SAR versus depth in the eye.

exposure is about 18. With time averaging, it is clear that the users of portable transmitters are exposed substantially below the ANSI C95.1-1982 limits. Time averaging does not apply to cellular phone users. The portable cellular phones have a power output of about 0.6W maximum.

Measurements [80] show that the peak SAR values from miniature cellular phones are less than 0.6 mW/g, which is substantially lower than the present and possible future human exposure limits of ANSI. The peak SAR value from the exposure to portable cellular phones is found in the temporal area of the brain, near the base of the antenna. The emitted RF power of the portable unit ensures that the exposure of the user is below the ANSI limits even if the averaging time is extended to 30 min, which would represent a long telephone conversation.

In conclusion, the currently available portable two-way communicatio radios, with RF power transmission limited to 7W or less, expose their users to time-averaged peak and average SAR levels well below widely accepted standard for the safety of humans exposed to RF energy.

REFERENCES

[1] Gandhi, O. P., "Conditions of Strongest Electromagnetic Power Deposition in Man and Animals," *IEEE Trans. MTT*, Vol. MTT-23, No. 12, Dec. 1975, pp. 1021–1029.

[2] Fujimoto, K., A. Henderson, K. Hirasawa, and J. R. James, *Small Antennas*, Letchworth England: Research Studies Press, 1987, pp. 75–116.

[3] Balzano, Q., and K. Siwiak, "The Near Field of Annular Antennas," *IEEE Trans. Vehicula Technology*, Vol. VT-36, No. 4, Nov. 1987, pp. 173–183.

[4] Smith, G. S., "Radiation Efficiency of Electrically Small Multiturn Loop Antennas," *IEEE Trans Ant. Propag.*, Vol. AP-20, No. 5, Sept. 1972, pp. 656–657.

[5] Kraus, J. D., *Antennas*, 2nd edition, New York: McGraw-Hill, 1988, pp. 46–49.

[6] Johnson, C. C., and A. W. Guy, "Nonionizing Electromagnetic Wave Effects In Biological Mate rials and Systems," *Proc. IEEE*, Vol. 60, June 1972, pp. 642–718; IEC 489-6, 2d ed., Appendi H, "Methods of Measurement for Radio Equipment Used in Mobile Services," 1987.

[7] Wheeler, H. A., "Small Antennas," *IEEE Trans. Ant. Propag.*, Vol. AP-23, No. 4, July 197 pp. 462–469.

[8] Fujimoto, K., "Small Antennas," *J. IECE Japan*, Vol. 60, No. 4, April 1977, pp. 391–397 (Japanese).

[9] Fujimoto, K., "A Review of Research on Small Antennas," *J. IEICE Japan*, Vol. 70, No. 8, Au 1987, pp. 830–838 (in Japanese).

[10] Newman, E. H., "Small Antenna Location Synthesis Using Characteristic Modes," *IEEE Tran Ant. Propag.*, Vol. AP-27, No. 4, July 1979, pp. 530–531.

[11] Ito, H., H. Haruki, and K. Fujimoto, "A Small-Loop Antenna for Pocket-Size VHF Radi Equipment," *National Technical Review*, Vol. 19, No. 2, April 1973, pp. 145–154 (in Japanese)

[12] King, H. D., et al., "Effects of Human Body on a Dipole Antenna at 450 and 900 MHz," *IEE Trans. Ant. Propag.*, Vol. AP-25, No. 3, May 1977, pp. 376–379.

[13] Ebinuma, T., et al., "The Properties of Dipole Antenna Located Near the Human Body in th UHF Band," *Trans. IECE Japan*, Vol. J62-B, No. 9, Sept. 1979, pp. 1066–1067 (in Japanese).

[14] Krupka, Z., "The Effect of the Human Body on Radiation Properties of Small-Sized Comm nication Systems," *IEEE Trans. Ant. Propag.*, Vol. AP-16, No. 2, March 1968, pp. 154–163.

[15] King, H. E., "Characteristics of Body-Mounted Antennas for Personal Radiosets," *IEEE Tran Ant. Propag.*, Vol. AP-23, No. 2, March 1975, pp. 242–244.

[16] Kraus, J. D., *Antennas*, 2d ed., New York: McGraw-Hill, 1988, pp. 241–242.

[17] IECE Japan, *Antenna Engineering Handbook*, Ohm, 1980, pp. 474–477 (in Japanese).

[18] Devore, R., and P. Bohley, "The Electrically Small Magnetically Loaded Multiturn Lo Antenna," *IEEE Trans. Ant. Propag.*, Vol. AP-25, No. 4, July 1977, pp. 496–504.

[19] Sato, S., and Y. Naito, "Dipole Antenna Covered With Ferrite Sleeve," *Trans. IECE Japan*, Vo J58-B, No. 6, June 1975, pp. 285–292 (in Japanese).

[20] Uda, S., *Radio Engineering, Transmission Volume*, Maruzen, 1964, pp. 79–82 (in Japanese).

[21] Haruki, H., H. Ito, and Y. Hiroi, "A Small Loop Antenna for Pocket-Size VHF, UHF Radi Equipment," *Paper of Technical Group, IECE Japan*, Vol. AP74-29, Aug. 1974, pp. 7–12 (Japanese).

22] Wada, M., and Y. Shimanuki, "New Developments in NTT's Radio Paging System," *NTT Review*, Vol. 2, No. 4, July 1990, pp. 129–137.

23] Wada, M., "NTT Paging System," *NTT Review*, Vol. 1, No. 1, May 1989, pp. 45–53.

24] Ishii, N., and K. Itho, "A Consideration on the Numerical Method for a Card-Sized Thin Planar Antenna," *Paper of Technical Group, IEICE Japan*, Vol. AP91-36, June 1991, pp. 9–14 (in Japanese).

25] Hallen, E., "Transmitting and Receiving Qualities of Antennas," *Nova Acta Reg. Soc. Sc. Ups.*, Vol. 11, No. 4, Series 4, Uppsala, Sweden, Nov. 1938, pp. 2–44.

26] Schelkunoff, S. A., and H. T. Friis, *Antennas, Theory and Practices*, London: Chapman and Hall, 1952.

27] King, R.W.P., *The Theory of Linear Antennas*, Cambridge, MA: Harvard University Press, 1956.

28] Neff, N. P., et al., "A Trigonometric Approximation of the Current in the Solution to Hallen's Equation," *IEEE Trans. Ant. Propag.*, Vol. AP-17, No. 6, Nov. 1969, pp. 804–805.

29] King, R. W. P., and C. W. Harrison, *Antennas and Waves: A Modern Approach*, Cambridge, MA: MIT Press, 1969.

30] Jordan, E. C., and K. G. Balmain, *Electromagnetic Waves and Radiating Systems*, 2nd edition, Englewood Cliffs, N.J.: Prentice-Hall, 1968, pp. 333–338.

31] Kraus, J. D., *Antennas*, New York: McGraw-Hill, 1950, pp. 232–238.

32] King, R.W.P., and C. W. Harrison, "The Distribution of Current Along a Symmetrical Center-Drive Antenna," *Proc. IRE*, Vol. 31, Oct. 1943, pp. 548–567.

33] Balzano, Q., et al., "The Near Field of Dipole Antennas," *IEEE Trans. Vehicular Technology*, Vol. VT-30, No. 4, Nov. 1981, pp. 161–181.

34] Jasik, H., ed., *Antenna Engineering Handbook*, 1st edition, New York: McGraw-Hill, 1961, Sect. 3.5, pp. 3017–3018.

35] Balzano, Q., et al., "The Near Field of Omnidirectional Helical Antennas," *IEEE Trans. Vehicular Technology*, Vol. VT-31, No. 4. Nov. 1982, pp. 173–185.

36] Cha, A. G., "Wave Propagation on Helical Antennas," *IEEE Trans. Ant. Propag.*, Vol. AP-20, No. 5, Sept. 1972, pp. 556–560.

37] Collin, R. E., Ch. 9 in *Field Theory of Guided Waves*, New York: McGraw-Hill, 1960, Sec. 9.8.

38] Fujimoto, K., A. Henderson, K. Hirasawa, and J. R. James, *Small Antennas*, Letchworth, England: Research Study Press, 1987, pp. 59–71.

39] Taga, T., and K. Tsunekawa, "Performance Analysis of a Built-In Planar Inverted F Antenna for 800 MHz Band Portable Radio Units," *IEEE Trans. Selected Areas in Communication*, Vol. SAC-5, No. 5, June 1987, pp. 921–929.

40] Takegawa, I., and T. Taga, "Effective Gain Estimation of Portable Antenna," *Conv. Rec. IECE of Japan* (in Japanese), March 1984, p. 2454.

41] Lee, W.C.Y., and Y. S. Yeh, "Polarization Diversity System for Mobile Radio," *IEEE Trans. Communications*, Vol. COM-20, No. 5, Oct. 1972, pp. 912–923.

42] Taga, T., "Analysis for Mean Effective Gain of Mobile Antennas in Land Mobile Radio Environments," *IEEE Trans. Vehicular Technology*, Vol. VT-39, No. 2, May 1990, pp. 117–131.

43] Fujimoto, K., A. Henderson, K. Hirasawa, and J. R. James, *Small Antennas*, Letchworth, England: Research Studies Press LTD., 1987, pp. 135–151, 270–293.

44] Taga, T., and K. Tsunekawa, "A Built-In Antenna for 800 MHz Band Portable Radio Units," *Proc. Int. Symp. Ant. Propag.*, Kyoto, Japan., 20–22 Aug. 1985, pp. 425–428.

45] Andersen, J. B., and F. Hansen, "Antennas for VHF/UHF Personal Radio: A Theoretical and Experimental Study of Characteristics and Performance," *IEEE Trans. Vehicular Technology*, Vol. VT-26, No. 4, Nov. 1977, pp. 349–357.

46] Mitsuishi, T., "Automobile and Portable Telephones in Japan," *NTT Review, NTT, Japan*, Vol. 1, No. 1, May 1989, pp. 30–39.

[47] "Public Land Mobile Telephone Systems," *CCIR SG-8*, Report 742-3.
[48] Kobayashi, K., S. Nishiki, T. Taga, and A. Sasaki, "Detachable Mobile Radio Units for the 80 MHz Land Mobile Radio System," *Proc. 34th IEEE Vehicular Technology Conf.*, Pittsburgh, 21 23 May 1984, pp. 6–11.
[49] Shimizu, I., S. Urabe, K. Hirade, K. Nagata, and S. Yuki, "A New Pocket-Size Cellular Telephon for NTT High-Capacity Land Mobile Communication System," *Proc. 41st IEEE Vehicular Tech nology Conf.*, St. Louis, Missouri, 19–22 May 1991, pp. 114–119.
[50] Taga, T., K. Tsunekawa, and A. Sasaki, "Antennas for Detachable Mobile Radio Units," *Revie of the ECL, NTT, Japan*, Vol. 35, No. 1, Jan. 1987, pp. 59–65.
[51] Tsunekawa, K., "Diversity Antennas for Portable Telephones," *Proc. 39th IEEE Vehicular Tech nology Conf.*, San Francisco, 29 April–3 May 1989, pp. 50–56.
[52] Gandhi, O. P., "State of Knowledge for Electromagnetic Absorbed Doses in Man and Animals, *Proc. IEEE*, Vol. 68, Jan. 1980, pp. 24–32.
[53] Pethig, R., *Dielectric and Electronic Properties of Biological Materials*, New York: John Wile 1979.
[54] Johnson, C. C., and A. W. Guy, "Nonionizing Electromagnetic Wave Effects in Biological Mat rials and Systems," *Proc. IEEE*, Vol. 60, June 1972, pp. 692–718.
[55] Ibid., pp. 694–695.
[56] Adey, W. R., "Tissue Interaction With Nonionizing Electromagnetic Fields," *Physiologic Reviews*, Vol. 61, No. 2, April 1961, pp. 435–507.
[57] Taylor, L. S., "The Mechanisms of Athermal Biological Effects," *Bioelectromagnetics*, 1981, Vc 2, No. 3, pp. 241–246.
[58] Guy, A. W., private communication.
[59] Hartsgrove, G. W., "New Formulas for Tissue Equivalent Materials Used in Electromagne Absorption Studies," University of Ottawa, Dept. of Electrical Engineering, 770 King Edwa Avenue, Ottawa, Ontario, Canada.
[60] Grandolfo, M., et al., "Magnetic Resonance Imaging: Calculation of Rates of Energy Absorpti by a Human Torso Model," *Bioelectromagnetics*, Vol. 11, No. 2, 1990, pp. 117–128.
[61] Guy, A. W., and C. K. Chou, "Impact of the RF Radiation Controversy on Cellular Mobi Systems," *Convergence 84*, 22–24 Oct. 1984, pp. II-7–II-17.
[62] Olson, R. G., and R. R. Bowman, "Simple Non-perturbing Temperature Probe for Microwav RF Dosimetry," *Bioelectromagnetics*, 1989 No. 2, pp. 209–213.
[63] Balzano, Q., et al., "Heating of Biological Tissue in the Induction Field of VHF Portable Radios *IEEE Trans. Vehicular Technology*, Vol. VT-27, No. 2, May 1978, pp. 51–56.
[64] Bassen, H., "Internal Dosimetry and External Microwave Field Measurements Using Miniatu Electric Field Probes," *Proc. Symp. Biological Effects*, FDA, Rockville, MD, 16–18 Feb. 197 pp. 136–151.
[65] Batchman, T. E., and G. Gimpelson, "An Implantable Electric Field Probe of Submillimet Dimensions," *IEEE Trans. Microwave Theory and Techniques*, Vol. MTT-31, No. 9, Sept. 198 pp. 745–751.
[66] American National Standards Institute, ANSI C95.1-1982, "Safety Levels With Respect to Huma Exposure to Radio Frequency Electromagnetic Fields, 300 KHz–100 GHz," Copyright IEEE, Inc 345 East 47th Street, New York, NY 10017.
[67] Polk, C., and E. Postow, eds., Ch. 3 in *CRC Handbook of Biological Effects of Electromagne Fields*, CRC Press, Inc., Boca Raton, FL, 2nd printing, 1987, Sec. IIB.
[68] Ibid., Sec. IIC
[69] Shafer, J. F., "Field Strength Levels in Vehicles Resulting From Communication Transmitters National Institute of Justice, Report No. 200-83.
[70] Guy, A. W., and C. K. Chou, "Specific Absorption Rate of Energy in Man Models Exposed

Cellular UHF Mobile Antenna Fields," *IEEE Trans. Microwave Theory and Techniques*, Vol. MTT-34, No. 6, June 1986, pp. 671–680.

[71] Balzano, Q., "The Near Field of Portable and Mobile Transmitters and the Exposure of the Users," report submitted to the U.S. Federal Communication Commission in response to the Notice of Inquiry, General Docket No. 79-144, 1979. This report is available from Q. Balzano.

[72] Lambdin, D. L., "An Investigation of Energy Densities in the Vicinity of Vehicles With Mobile Communication Equipment and Near a Handheld Walkie Talkie," U.S. EPA, Ref. No. ORP/EAD 79-2, March 1979.

[73] Davidson, A. C., "Mobile Antenna Gain at 900 MHz," *IEEE Trans. Vehicular Technology*, Vol. VT-24, No. 4, Nov. 1976, pp. 54–58.

[74] Davidson, A. L., and W. J. Turney, "Mobile Antenna Gain at 900 MHz," *IEEE Trans. Vehicular Technology*, Vol. VT-26, No. 4, Nov. 1977, pp. 345–348.

[75] Balzano, Q., et al., "Energy Deposition in Biological Tissue Near Portable Radio Transmitters at VHF and UHF," *Conf. Record of the 27th Conf. IEEE Vehicular Technology Group*, Orlando, FL, 16–18 March 1977, pp. 25–39.

[76] Balzano, Q., et al., "Investigation of the Impedance Variation and the Radiation Loss in Portable Radios," *IEEE Symp. Ant. Propag.*, Urbana, IL, 2–4 June 1975, pp. 89–92.

[77] Balzano, Q., et al., "Energy Deposition in Simulated Human Operators of 800 MHz Portable Transmitters," *IEEE Trans. Vehicular Technology*, Vol. VT-27, No. 4, Nov. 1978, pp. 174–181.

[78] Cleveland, R. F., and T. W. Athey, "Specific Absorption Rate in Models of the Human Head Exposed to Hand-held UHF Portable Radios," *Bioelectromagnetics*, Vol. 10, No. 2, 1989, pp. 173–186.

[79] Cohen, P., et al., "Traffic Characterization and Classification of Users of Land Mobile Communication Channels," *IEEE Trans. Vehicular Technology*, Vol. VT-33, Nov. 1984, pp. 276–284.

[80] Balzano, Q., and O. Garay, "Dyna TAC Exposure Measurements," internal Motorola report, 15 May 1984, obtainable from Q. Balzano.

Chapter 5

Land Mobile Antenna Systems III: Cars, Trains, Buses

H. K. Lindenmeier, L. Reiter, J. Hopf, K. Fujimoto, and K. Hirasawa

5.1 ANTENNA SYSTEMS FOR BROADCAST RECEPTION IN CARS

t has been common practice for several decades to fit commercial vehicles and private cars with radios to receive national broadcast programs. The performance of these radios has now become more critical in view of the demand for high-fidelity audio, the increasing number of local radio stations, and, not least, the ecent escalation in the electronic complexity of automobile information systems, which include the reception of Television, navigation, and traffic control data.

The typical problems in the design of modern car radio antennas are: (1) matching the antenna to the amplitude modulation (AM) receiver in the frequency ange of 526.5 to 1606.5 kHz and to the TV receiver for wide-frequency bands overing both VHF and UHF bands; (2) reduction of multipath fading in FM and TV broadcast reception; and (3) avoiding the reception of noise generated by the engine and electronic circuits installed in the car. In the AM frequency bands, xact matching is impossible, since the antennas commonly used are electrically ery small and thus have extremely small radiation resistance and very large reactance. This problem has been solved by placing an amplifier with low gain (about dB) close to the antenna terminals. The amplifier has a high input impedance and an output impedance matched to the receiver in order to reduce transmission osses due to mismatch. The same concept is applied to the design of antennas for AM broadcast reception that are printed on window glass.

Ignition noise from the engine enters the antenna by both direct radiation and coupling to the vehicle body. Digital electronic circuits in a car may also act as a source of interference. The current in digital circuits contains many harmonic frequency components in both the FM and TV broadcast bands. When either electromagnetic or electrostatic coupling exists between digital circuits and antenna or receiver circuits, digital currents may be induced on the antenna element, thus appearing in the FM or TV receiver circuits and interfering with broadcast reception. Thus, the place to mount an antenna on the body of a car must be selected carefully in order to avoid noise interference. When a monopole element is used, the best place to mount it is often considered to be the rear fender. For an antenna printed on the window glass, the chosen location depends on the need to maintain unobstructed vision and the nature of the ambient noise interference.

Section 5.1 places emphasis on the FM reception and the improvements obtained by diversity techniques in the presence of multipath fading. Section 5.2 illustrates the additional complexity required to include the reception of TV and further complements the technical details given in Section 5.1. The practical examples in both sections are taken from systems currently being manufactured in Germany and Japan.

5.1.1 Introduction

The well-known standard whip antenna used for car radios since the early 1950s is inadequate nowadays. The electrical performance of new types of car antennas such as active window antennas, has been proved to be at least equivalent to the whip. Those modern antennas make use of improved antenna structures; they profit from the progress in semiconductor technology and the dramatically reduced price of active elements; and they take advantage of much research done with respect to low noise and highly linear amplifier circuitry.

A multitude of car antennas for AM and FM reception with different principles of operation have been developed and implemented mainly by German and Japanese car producers. Section 5.1.4 refers to antennas that have been developed at the Institute for High-Frequency Techniques, University of the Bundeswehr Munich, Germany, in cooperation with car radio and antenna producers and that are in mass production with several German car producers.

The AM reception quality mainly depends on the propagation path between the transmitting broadcast station and the mobile car radio. In the morning hours and the early afternoon, only a few stations with comparatively small service area can be received due to the surface wave propagation along the ground. Receiving distortions are caused mainly by motor ignition pulses and manmade noise.

In the evening hours and in the night, especially, the AM radio waves are reflected and diffracted at the ionosphere. The service area of the radio station is therefore large, and a multitude of waves superimpose at the receiving antenna with different amplitudes and phases, leading to signal fadings. With a signal level exceeding the receiver noise level, this effect does not reduce the reception quality because of the automatic gain control (AGC) control circuits in the radio. In this case, the main distortions are caused by cochannel interference due to the long-range propagation conditions in the night.

Newly developed active AM antennas are discussed in Section 5.1.4. Due to the long AM wavelength antenna (space), diversity is not an appropriate means to improve the AM reception quality in cars. With FM reception, however, the FM distortions well known from the reception with the whip antenna still occur with the new active window antennas; these distortions result from the field situation in the FM range and not from an inferiority of the antennas. The advantage of the new antennas is of a different type, because unlike the whip, the window antennas completely meet the requirements as far as mechanical and other automotive aspects are concerned. There is no corrosion, no need for cleaning, no danger of breaking, no handicap with respect to styling, and no noise due to air turbulence.

For a considerable improvement of the quality of reception in the FM range, however, more complex techniques are required, such as multiantenna arrangements applied to scanning diversity systems. The overall costs are moderate, since only one receiver is required, and very compact antenna designs with up to four antennas in a single window have been developed. Such narrowly spaced window antennas with high efficiency within the diversity system can only be realized if the basic principle of active antennas is applied. Multiantenna diversity systems have been in mass production since 1989 with a German car producer.

Neither modern active window antennas, nor compact multiantenna arrangements, nor the extremely fast distortion detector implemented in the diversity processor could have been developed without a deep knowledge of the basic field situation in the FM range and the distortions associated with it. Therefore, some of the basic results obtained from the research on multipath propagation in combination with high-power FM broadcasting will be presented below as a precondition for a better understanding of the antennas and their systems.

5.1.2 Signal Analysis with FM Multipath Propagation

In the FM range, the transmitting antenna is normally arranged on the top of a mountain or a tower in order to get a large service area. The RF carrier wave from the transmitter is diffracted, reflected, and scattered at mountains, buildings, and obstacles such as wire fences. Therefore, many propagation paths exist between the transmitter and the receiving antenna.

In contrast to this, the receiving antenna for home reception is typically attached at a height of approximately 10m above ground and normally has a directional antenna pattern. With this arrangement, disturbing multipath reception can be avoided and the sound is of high-fidelity quality, assuming enough signal strength of at least 54 dBμV [1]. With mobile reception, however, the receiving antenna is fixed on the vehicle at a height of approximately 1m. In this case, a multitude of waves with different amplitudes and phases superimpose at the location of the receiving antenna. In order to investigate the distortions due to multipath propagation, it is appropriate to create a model in order to simulate typical distortions.

FM Multipath Reception Model

To create the multipath reception model, it is first assumed that the transmitting antenna is arranged on a mountain (Figure 5.1). The transmitting signal $\mathbf{V}_T$ is radiated from an antenna with an omnidirectional pattern. This RF carrier wave is reflected at n other mountains. Therefore, the reflected and the direct waves superimpose at the receiving antenna, which has an omnidirectional antenna pattern, too. Therefore, the receiving voltage $\mathbf{V}_R$ is a combination of $n + 1$ signals having different transmission coefficients and delay times due to the different length of their propagation paths. The transmission coefficients A_i correspond to the amplitudes of the different waves at the receiving antenna. In practice, the absolute delay time is irrelevant, since only the differences in delay time τ_i between the waves are of importance. In the following, the wave with the shortest propagation time between transmitter and receiving antenna is assumed to be the reference wave with an amplitude A_0. This definition has the advantage that only positive delay times are considered, and receiving situations with no direct view to the transmitter are regarded, too. This kind of transmission can be described by a two-

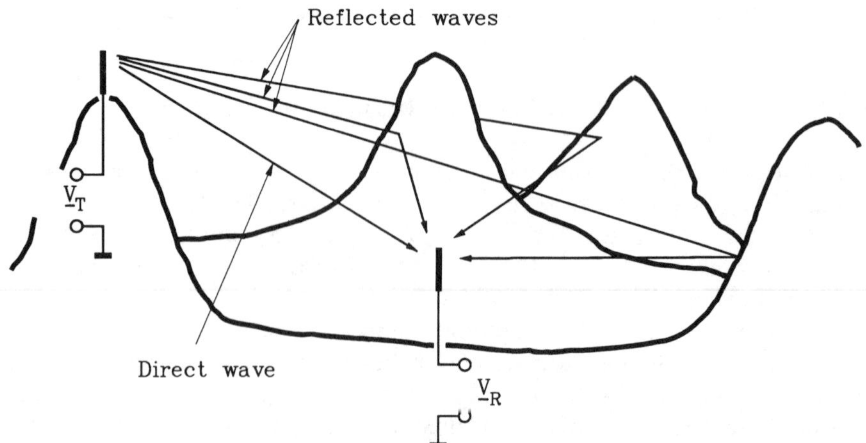

Figure 5.1 Wave propagation between transmitter and receiver with one direct and n reflected waves.

port [2] with a transmission function $\mathbf{S}(f)$, which is described by (5.1), and the pulse response $s(t)$, which is determined by (5.2):

$$\mathbf{S}(f) = \mathbf{V}_R/\mathbf{V}_T = A_0 + \sum_{i=1}^{n} A_i \exp(-j\omega_c \tau_i) \qquad (5.1)$$

$$s(t) = A_0\, \delta(t) + \sum_{i=1}^{n} A_i \delta(t - \tau_i) \qquad (5.2)$$

where $\delta(t)$ is a Dirac pulse at time t and f_c is the carrier frequency.

In the case of stationary reception, this two-port is only frequency-dependent. The transmission function $\mathbf{S}(f)$ has the character of a nonrecursive delay network with a structure as shown in Figure 5.2. Each branch of the network delivers a contribution to the receiver input signal with an amplitude A_i and with a difference in delay time τ_i related to the direct wave. The pulse response represents the transfer characteristics $\mathbf{S}(f)$ in a simple way, which is displayed in Figure 5.3 for a measured receiving situation in a mountainous area. Up to four echo waves have been registered with different amplitudes A_i and different time-delay differences τ_i.

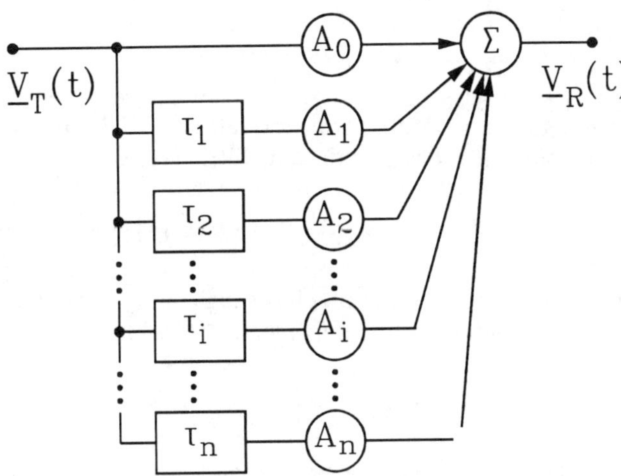

Figure 5.2 Structure of a nonrecursive time-delay filter.

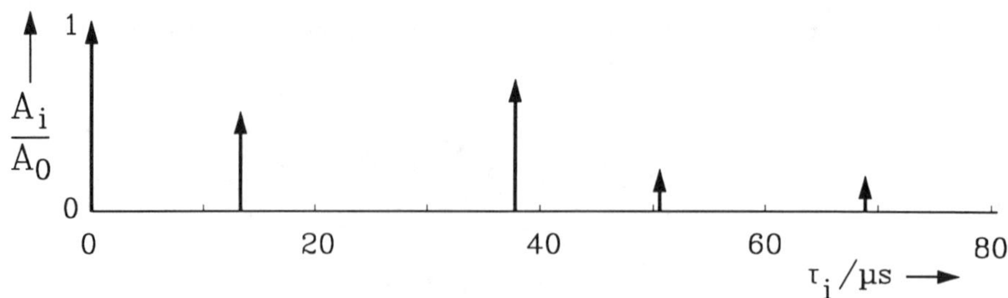

Figure 5.3 Measured pulse response with multipath propagation.

Analysis of FM Multipath Distortions Caused by Large Delay Times

Multipath reception disturbs the FM modulation content, and in order to compute the audio frequency (AF) distortions, the multipath reception model displayed in Figure 5.1 has been applied. The received erroneous signal can be computed as a function of the channel parameters A_i and τ_i and of the AF modulation signal. Therefore, the resulting receiver voltage $\mathbf{V}_R(t)$ in a standing car obtained from the superposition of n waves can be calculated by (5.3):

$$\mathbf{V}_R(t) = \mathbf{V}_T(t) * \mathbf{S}(t) \tag{5.3}$$

which is obtained from the Fourier transformation of (5.1). In the case of FM, the complex transmission signal can be described by

$$\mathbf{V}_T(t) = |\mathbf{V}_T(t)| \exp[-j\omega_c t + j\phi(t)] \tag{5.4}$$

with $\phi(t)$ representing the modulation phase angle and f_c the RF carrier frequency In the case of a sinusoidal modulation signal, $\phi(t)$ is

$$\phi(t) = -\eta \cdot \cos(\omega_{AF} t) \tag{5.5}$$

with $\eta = \Delta f / f_{AF}$ being the modulation index and Δf the maximum frequency peak deviation. Equation (5.3) applied to (5.4) delivers the receiver input voltage $\mathbf{V}_R(t)$

$$\mathbf{V}_R(t) = A_0 \cdot |\mathbf{V}_T(t)| \exp[-j\omega_c t + j\phi(t)] + \sum_{i=1}^{n} A_i |\mathbf{V}_T(t)| \exp[-j\omega_c(t - \tau_i) + j\phi(t - \tau_i)] \tag{5.6}$$

The computation of the modulation phase $\alpha(t)$ of $\mathbf{V}_R(t)$ is described by

$$\begin{aligned}\alpha(t) &= \arctan \frac{\operatorname{Im}\{\mathbf{V}_R(t)\}}{\operatorname{Re}\{\mathbf{V}_R(t)\}} \\ &= \arctan \frac{A_0 \sin \phi(t) + \sum_{i=1}^{n} \sin[-\omega_c \tau_i + \phi(t - \tau_i)]}{A_0 \cos \phi(t) + \sum_{i=1}^{n} A_i \cos[-\omega_c \tau_i + \phi(t - \tau_i)]}\end{aligned} \tag{5.7}$$

and the following differentiation gives the instantaneous angular frequency deviation $\Omega(t)$

$$\Omega(t) = \frac{\mathrm{d}}{\mathrm{d}t}\alpha(t) \tag{5.8}$$

Assuming an ideal FM demodulator with the transfer characteristic

$$\mathbf{V}_{AF}(t) = \mathrm{k}_{Dem} \cdot \Omega(t) \tag{5.9}$$

with k_{Dem} being the demodulator constant, and the amplitude of the AF signal is given by

$$\mathbf{V}_{AF}(t) = \mathrm{k}_{Dem} \frac{\sum_{i=0}^{n} \sum_{j=0}^{n} A_i A_j \cos \sigma_{ij}(t) \cdot \frac{\mathrm{d}}{\mathrm{d}t}\phi(t - \tau_i)}{\sum_{i=0}^{n} \sum_{j=0}^{n} A_i A_j \cos \sigma_{ij}(t)} \tag{5.10}$$

with

$$\sigma_{ij}(t) = \omega_c (\tau_j - \tau_i) + \phi(t - \tau_i) - \phi(t - \tau_j) \tag{5.11}$$

Since A_i cannot be assumed to be small for any i, all terms in (5.10) have to be considered. Figure 5.4 shows the distorted AF output signal $\mathbf{V}_{AF}$ (bold curve) with an ideal FM demodulator, assuming two echo waves compared to an undistorted AF signal. In this example, the wave parameters are $A_1/A_0 = 0.6$, $A_2/A_0 = 0.4$, $\tau_1 = 20\ \mu s$, $\tau_2 = 70\ \mu s$, and $f_{AF} = 1$ kHz with a peak deviation $\Delta f = 75$ kHz. The IF bandwidth was assumed to be 160 kHz. These results have been obtained by numerical evaluation of (5.10), and the distortion spikes are evident.

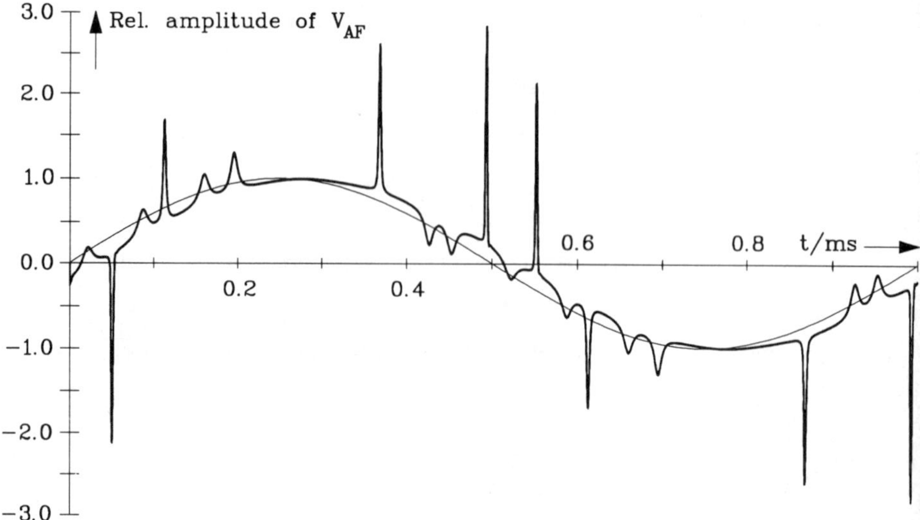

Figure 5.4 Distorted AF signal at the receiver output with three incident waves (bold curve) and with one wave (thin curve).

Distortion Factor

The frequency spectrum resulting from the distorted AF signal of Figure 5.4 at the demodulator output is distributed within a wide frequency range. The distortion factor—which was weighed with a 15-kHz low-pass filter—is displayed in Figure 5.5 as a function of the time-delay difference between a direct and one echo wave, with an AF frequency f_{AF} = 1 kHz and with Δf = 25 kHz peak deviation. The distortion factor reaches maximum values up to about 30% in the case of small differences between the amplitudes of the incident waves and depends on the number of distortion spikes in the AF signal. From Figure 5.5 it can be learned that differences in delay time greater than 3 μs cause considerable distortion factors. For a receiving situation with more than one echo wave, the distortion factor can reach values up to 100%.

Although these results are helpful for an understanding of the nature of multipath distortions, the receiving situations found in practice are much more complex than the above assumed superposition of only two or three relevant waves. In fact, in a driven car, the receiver voltage $\mathbf{V}_R(t)$ varies rapidly within a wide range due to the reflection, diffraction, and scattering at local obstacles of the waves within a wave bundle of equal delay time. The superposition of a multitude of wave bundles causes a rapid variation of the factor A_i/A_0, generating the well-known sporadic distortions that can be recognized in a multipath distorted area.

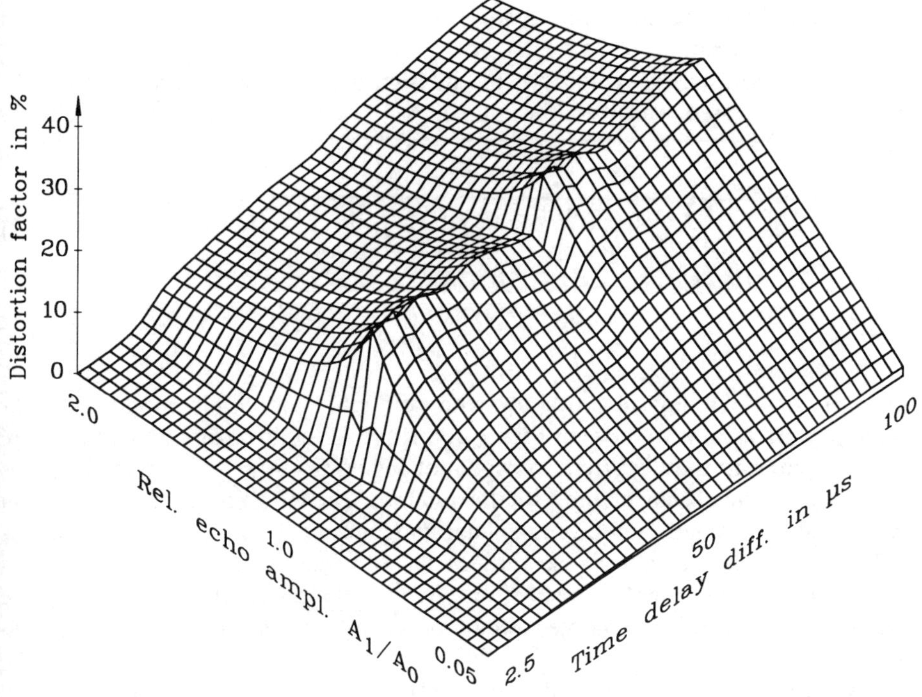

Figure 5.5 Distortion factor with two incident waves (Δf = 25 kHz, f_{AF} = 1 kHz) versus time-delay difference τ and normalized echo amplitude A_1/A_0.

Additional signals from a cochannel station lead to significant AF distortions, too. An intermodulation interference can be regarded in the same way if the frequency of the intermodulation product of two undesired signals appears in the received RF channel. If the receiver selectivity is lacking, and with a dense coverage of FM stations, the probability increases that an undesired frequency spectrum overlaps with the desired signal spectrum. If a distortion detector is available that recognizes these undesired signals, then cochannel, adjacent channel, and intermodulation interference can be reduced with antenna diversity.

FM-AM Conversion With FM Multipath Reception

Since multipath reception with great differences in delay time is always correlated with FM-AM conversion, this amplitude modulation can be used for the detection of multipath distortion. The sole use of this criterion, however, results in a poor distortion recognition performance, because in a mobile receiving system, the received signal level varies within a wide range of 20 to 40 dB with the variation of the receiving location.

In the following, this amplitude modulation due to great differences in delay time will be computed. The absolute value of $|\mathbf{V}_R(t)|$ of (5.6) describes the amplitude modulation:

$$
\begin{aligned}
|\mathbf{V}_R(t)| &= \sqrt{\mathrm{Re}^2\,[\mathbf{V}_R\,(t)] + \mathrm{Im}^2\,[\mathbf{V}_R\,(t)]} \\
&= \sqrt{\sum_{i=0}^{n}\sum_{j=0}^{n} A_i \cdot A_j \cos \sigma_{ij}(t)}
\end{aligned}
\tag{5.12}
$$

with $\sigma_{ij}(t)$ of (5.1). Figure 5.6 shows the computed amplitude modulation for parameters identical with those of Figure 5.4. If the normalized amplitude of Figure 5.6 is compared to the distorted AF signal of Figure 5.4, the AF distortion spikes are seen to be strongly correlated with the minima in the RF level. This phenomenon can be easily explained by the vector diagram (Figure 5.7), where only two waves of different amplitudes and phase shift δ_A are considered. Both complex vectors rotate with different angular speed if an FM modulation is applied, but only the relative movement of the complex vector of wave 1 in relation to wave 0 is considered.

Assuming great frequency deviation and a difference in delay time τ in the range of 3 to 100 μs, the phase angle between both waves can reach the factor 2π or more. With a periodic modulation signal, several complete rotations of the complex vector of wave 1 related to wave 0 will occur, and the resulting complex

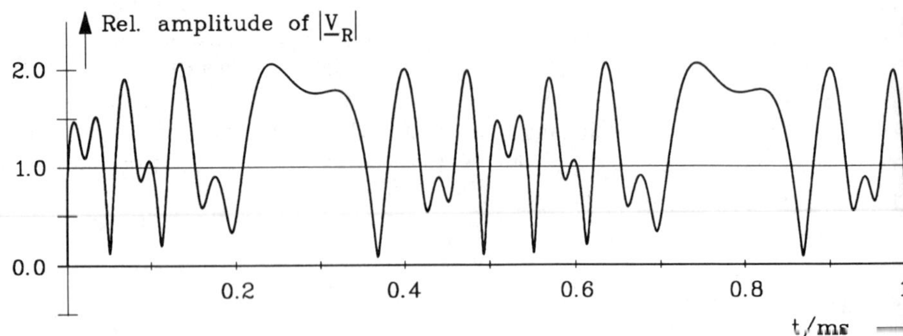

Figure 5.6 Normalized RF amplitude modulation at the receiver input caused by multipath reception (bold curve) and with only one wave (thin curve).

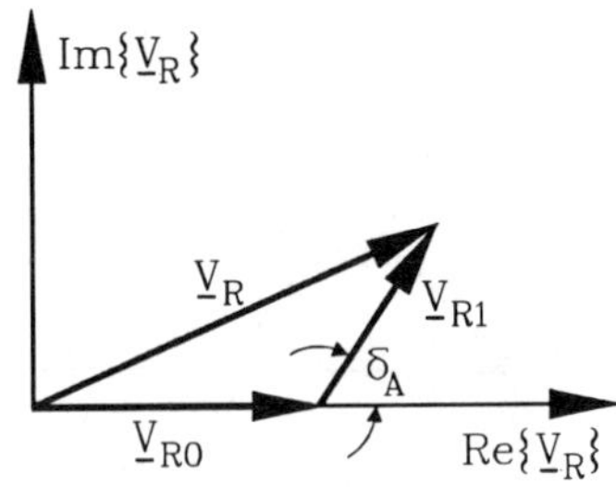

Figure 5.7 Vector diagram for a direct and one reflected wave.

vector will pass several minima. The instantaneous distortion frequency deviation is represented by (5.8). Any sufficiently great difference in angular velocity between both complex vectors generates distortion spikes. Therefore, with the example of a constant difference in angular velocity, the differential of the resulting complex vector of the waves is increased in the range where both waves are of opposite phase. If the absolute values of both complex vectors are equal, spikes of infinite height are obtained, and the resulting vector representing the instantaneous RF level becomes zero. A similar phenomenon exists with the superposition of more than two waves.

These investigations have been made with a sinusoidal AF signal. It is obvious that a regular modulation content from a broadcasting station covers a wide AF frequency band. Therefore, the distortion factor in practice is greater than with a single AF frequency example considered here. In addition, with a stereo multiplex signal, the channel separation decreases rapidly with multipath reception.

5.1.3 FM Distortions Due to Multipath Reception

Multipath Reception in a Standing Car

In order to learn the statistics of the distortions with FM radio and to validate the multipath model, numerous field measurements have been made. In cooperation with the Bavarian Broadcasting Corporation, a special test signal was radiated with 100-kW ERP, with horizontal polarization during the night from the transmitting station on the Wendelstein in the Bavarian Alps (height 1,850m). With a vertically or circularly polarized wave, similar results can be expected. This fact has been proved in tests.

In Figure 5.8, the radiated test signal is displayed showing a sinusoidal AF modulation signal with embedded 1-kHz pulses of 4-μs time duration. The sinu-

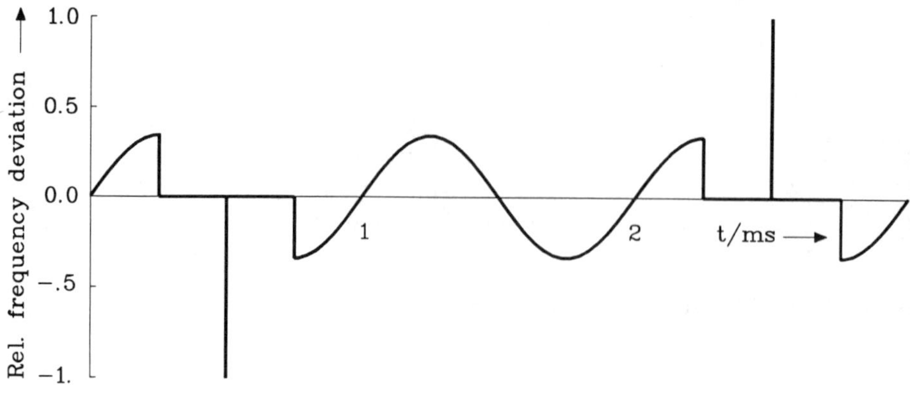

Figure 5.8 Test signal for registration of disturbing echo waves.

soidal part of the test signal serves for evaluation of the distortion factor in a multipath area. With the pulses in the test signal, the number of incident waves, their normalized amplitudes, and the difference in delay time between them could be determined. Waves with a difference in time delay less than 5 μs were registered as one wave. The test signal was radiated with the maximum frequency deviation of 75 kHz.

The number of measured echo waves is presented in Figure 5.9. This diagram shows that with a probability of approximately 50%, three waves form a distortion factor greater than 20%. In mountain areas, the number of echo waves (shadowed) is greater than in suburban areas. The probability distribution of the time delay between superimposed waves with a distortion factor being greater than 20% is presented in Figure 5.10 for a mountainous area and a suburban area. The delay time difference in mountain areas is greater than in flat areas, and in some cases time delays up to 100 μs have been measured. In suburban areas, a time delay greater than 30 μs rarely occurs. The wave amplitudes belonging to a set of echo waves related to the total amplitude of the resulting carrier is presented in Figure 5.11. With distortion factors greater than 20%, normalized amplitudes of the echo waves up to 0.6 have been registered. In Figure 5.12, the probability distribution of the distortion factor is displayed. The greater distortion factor in mountain areas is a result of the above discussed effects.

These measurements have shown for the first time that the distortions due to multipath propagation do not depend primarily on the average RF carrier, but are mainly caused by the differences in delay time between the superimposed waves.

Therefore, a diversity system that operates exclusively on the basis of an evaluation of the RF level is less suitable for improving mobile FM reception.

Mobile Reception With Multipath Propagation

With the above applied measurement of the pulse response, it is demonstrated that the wave parameters A_i and τ_i vary within a wide range and the measurements show that the amplitudes of the waves are Rayleigh distributed [3]. Therefore, the

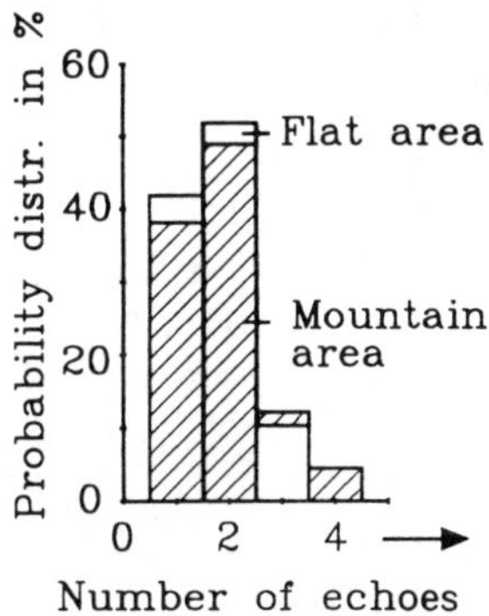

Figure 5.9 Probability distribution of the number of echo waves for an audio distortion factor >20%.

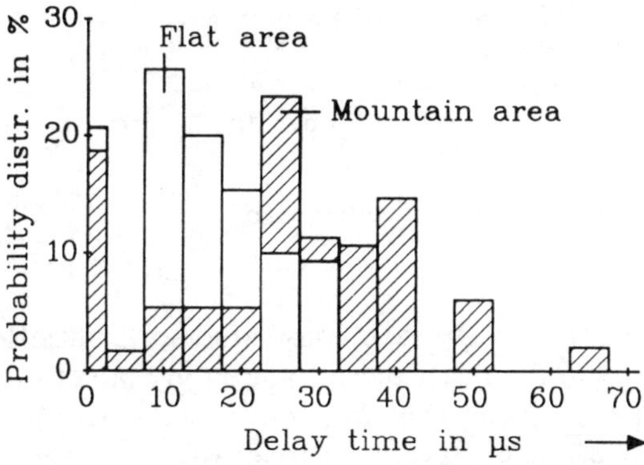

Figure 5.10 Probability distribution of delay times.

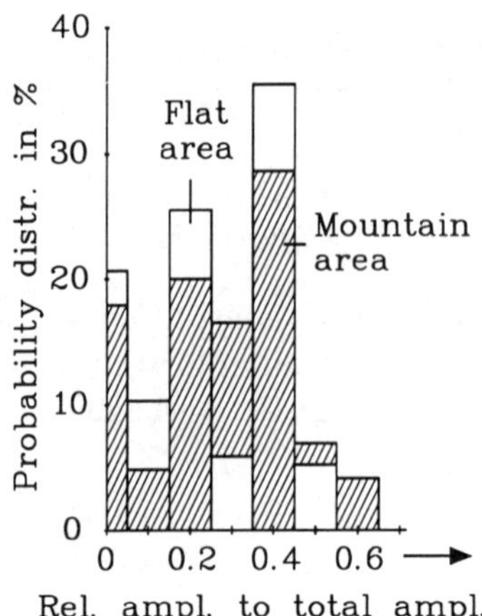

Figure 5.11 Probability distribution of the wave amplitudes related to the total amplitude for a distortion factor >20% in flat and mountain areas.

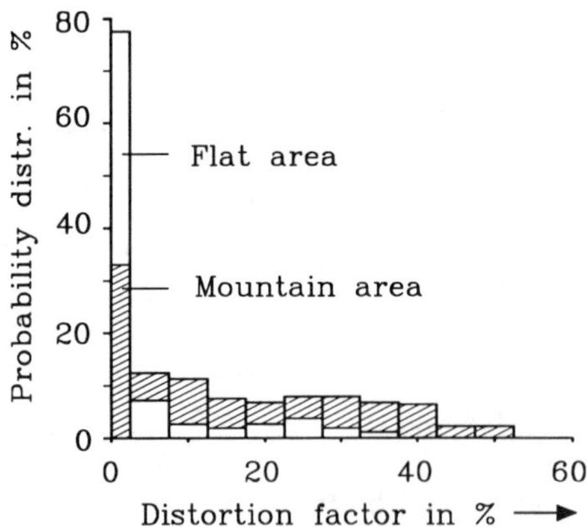

Figure 5.12 Probability distribution of the distortion factor.

assumption is justified that each wave considered so far has to be regarded as a bundle of waves. Due to the restricted resolution in time of the above method, a new measurement procedure was necessary to prove this phenomenon.

In cooperation with the Bavarian Broadcasting Corporation, an extremely frequency-stabilized RF carrier was radiated from a chain of transmitting stations located along the northern rim of the Alps. The standard transmitting power was

100-kW ERP per station. Doppler shift measurements were performed [4] and the absolute value of the azimuth angles $|\varphi|$ of incidence of the waves in a driven car can be found by

$$|\varphi| = \arccos\left(\frac{f_d}{f_c}\cdot\frac{c}{v}\right) \tag{5.13}$$

with f_c representing the carrier frequency, f_d the Doppler frequency shift, v the car speed, and c the velocity of light.

With $f_c = 98.5$ MHz and $v = 100$ km/h, the maximum Doppler shift f_d is found to be ± 9.12 Hz, representing the width of the received frequency spectrum. The azimuthal distribution of the incident waves can be read from the intensity distribution of the received spectrum. With a measuring interval of 2.5 sec, this corresponds to a covered distance of 71m and an angle resolution of 4 deg could be realized. Measurements in flat areas (Figure 5.13(a)) show a priority direction for incident waves. In contrast to this, in mountainous areas (Figure 5.13(b)), waves from nearly all azimuthal directions are incident at the location of reception. Due to the applied Doppler method, it is not possible to distinguish between positive and negative angles φ of incidence. The arrows in Figure 5.13 are displayed in a linear scale representing the amplitude distribution over the absolute value of the angle φ of incident waves.

In order to take these effects into account, the multipath reception model has to be extended (Figure 5.14). Therefore, the waves in Figure 5.1 have to be

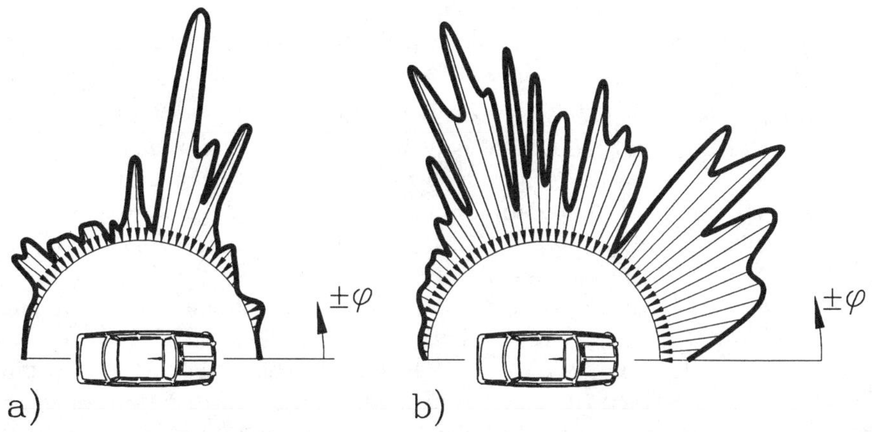

Figure 5.13 (a) Azimuthal distribution of incident waves in a flat area; (b) azimuthal distribution of incident waves in a mountainous area.

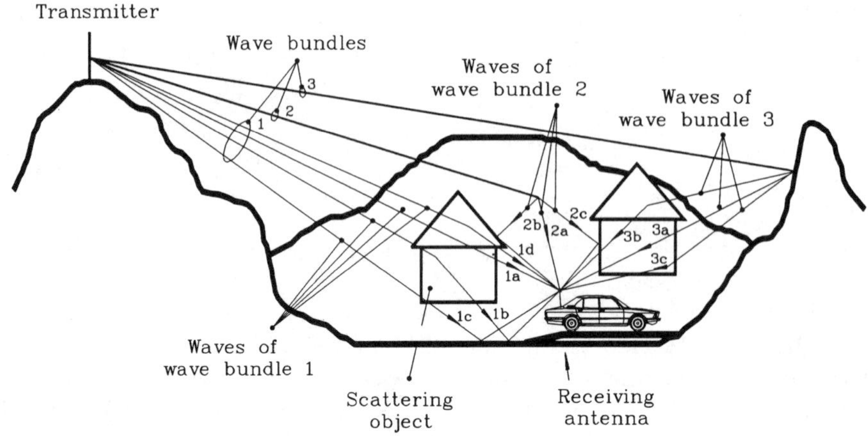

Figure 5.14 Multipath propagation with three wave bundles.

considered as wave bundles, which are reflected, diffracted, and scattered at local obstacles near the receiving antenna. The delay-time differences τ_{ij} describe the time delays of the ith wave bundle due to the jth of m local scattering objects. This effect leads to a pulse response as shown in Figure 5.15 which can be described by

$$s(t) = \sum_{i=0}^{n} \sum_{j=0}^{m} A_{ij} \, \delta(t - \tau_{ij}) \tag{5.14}$$

In practice, it is advisable to distinguish between small and great differences in time delay.

Small Differences in Time Delay (τ smaller than 3 μs). The received signal in a standing car is frequency modulated and the erroneous amplitude modulation is negligibly small, causing no recognizable modulation distortions. However, in a driven car, an amplitude modulation is produced due to local scattering objects near the receiving antenna. Figure 5.16 shows computed examples of this amplitude modulation with different antenna pattern. If the RF level exceeds the noise level of the receiver, the receiving system is not disturbed in spite of a level minimum. During the time that the RF level is below the noise level of the receiver, noise can be heard. These minima cannot be avoided by measures at the transmitter side. With a single-antenna system, the abundance of distortions can only be reduced by optimizing the receiver sensitivity. The excitation of the antenna may be considered to be caused by a singular wave bundle, and the differences in delay time between the waves within the wave bundle are negligible.

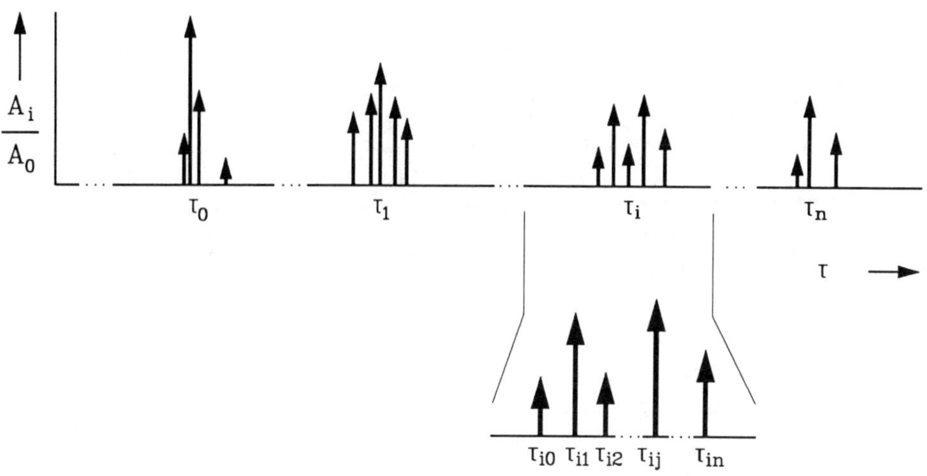

Figure 5.15 Pulse response in a multipath distorted area.

Great Differences in Time Delay (τ greater than 3 μs between wave bundles). Further investigations have shown that at different locations noise is present in spite of the fact that the superimposed RF carrier exceeds the noise level of the receiver. This causes a space-dependent level requirement at the receiver input for minimum reception quality (Figure 5.16). Scattering objects at a distance of more than 1 km to the receiving antenna form a multitude of wave bundles with great differences in delay time between the superimposed bundles. After FM demodulation of the disturbed RF signal, the resulting AF distortions generate a significant impression of distortion in the human ear. The reason for this is that the frequency modulated signal is subject to linear amplitude and phase distortions which are produced on the propagation path.

Influence of the Antenna Characteristics

Very often the main specification relevant for describing the performance of an FM car antenna is considered to be the directional diagram. In order to measure the diagram, the car is rotated on a turntable. However, this artificial situation of only one wave being incident at the location of reception in no way represents the physical behavior of FM car radio reception in practice. In fact, as shown above, at any location of reception, there is a multitude of coherent waves of complex amplitudes $\mathbf{A}_1$ to $\mathbf{A}_n$ incident from various directions in azimuth and elevation.

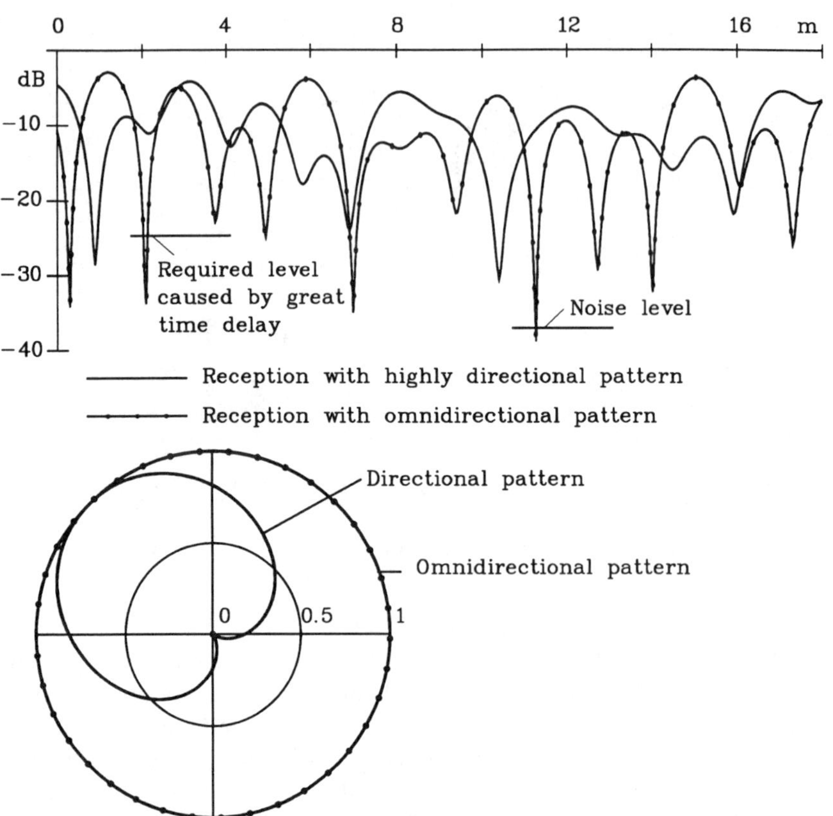

Figure 5.16 Signal level variation with an omnidirectional antenna pattern (—•—•—•—•—) and with a directional pattern (————————).

With $\mathbf{C}(\varphi,\theta)$ representing the complex antenna factor and $\mathbf{A}_i \cdot \mathbf{C}(\varphi_i, \theta_i)$ representing the contribution of wave i to the complex voltage amplitude at the receiver input, it becomes obvious that the total complex amplitude at the receiver in a standing car is

$$\mathbf{V}_{\mathrm{R}} = \sum_{i=1}^{n} \mathbf{A}_i \cdot \mathbf{C}(\varphi_i, \theta_i) \tag{5.15}$$

and can cancel out even if the directional diagram is omnidirectional ($\mathbf{C}(\varphi,\theta) = 1$). In a moving car, the set of complex wave amplitudes varies very rapidly in magnitude, phase, angle of incidence, and number of waves to be considered, so that the receiver input voltage is

$$\mathbf{V}_R(t) = \sum_{i=1}^{n} \mathbf{A}_i \cdot \mathbf{C}[\varphi_i(t), \theta_i(t)] \tag{5.16}$$

and varies within a wide range.

In a Rayleigh field, the computed signal received with an omnidirectional antenna is represented in Figure 5.16 by the curve with dots versus the driven distance. The solid curve is obtained if a directional antenna pattern is deployed, thus generating less deep signal minima and therefore causing fewer distortions. Since it is obvious that a directional antenna cannot satisfy the requirements for a mobile receiving antenna, the receiving problem can only be solved by means of a time-variable antenna system such as an antenna diversity system, which is described in Section 5.1.5.

5.1.4 Single-Antenna Reception for AM and FM

Antennas for car radios are required for reception of the AM and FM bands. AM (long-, medium-, and short-wave frequency band) covers the frequency range from 150 kHz to approximately 6 MHz. In Europe and North America, the FM band is located within 87.5 and 108 MHz. In the AM band, the polarization is vertical. In the FM range in Europe, most stations are horizontally polarized, whereas the United States, for example, vertically or circularly polarized stations broadcast, too.

One of the main problems with mobile reception results from the fact that the maximum antenna output voltage delivered to the receiver input is many orders of magnitude higher than the noise level. Taking the FM range and a receiver with 150Ω input impedance, for example, the maximum output voltage was found to be up to 125 dB above the noise level of the system, which is typically at -5 dBμV, if a receiver bandwidth of 120 kHz is taken as a basis. In fact, in the very vicinity of strong transmitting stations, for example, in downtown New York, output signals of 1V (effective value) at the receiver input have been registered. On the other hand, a signal with a mean level of only 25 dBμV has been found to be still satisfactory [5]. In the AM range, the basic situation is similar.

Consequently, all components of a mobile receiving system containing active and therefore nonlinear elements are endangered by intermodulation and cross-modulation distortion. Receiver and active antennas therefore have to meet very hard requirements, not only with respect to sensitivity, but also with respect to large signal behavior.

Passive Whip Antennas

Since the early 1950s, a passive whip about 1m long has been in use for radio reception in cars. The antenna is mostly mounted on the fenders or, for example, in France, on the roof of the car. The received signal is delivered to the receiver by means of a coaxial cable with a low value of capacitance per unit length (about 30 pF/m) and therefore with a high value of the nominal impedance of about 150Ω. The standard length of the cable is about 1.5m, which fits an antenna mounted in the front of a car. If the antenna is mounted in the rear, an additional cable of up to 6m length is required.

In contrast to the FM range, no matching of the antenna impedance with the nominal impedance of the cable is attempted in the AM range, since there is no chance of arriving at a broadband solution. This results from an antenna impedance with an extremely low radiator bandwidth of only 2.8×10^{-7}, due to an antenna impedance of approximately $(0.0044 - j15600)\Omega$ (for example [6], for a 1m whip antenna of 1 MHz).

AM Range. Modern receivers in the AM range use a capacitive nonresonant input stage with an input capacity of typically $C_i = 20$ pF. In Figure 5.17, the equivalent circuit for a signal-to-noise evaluation is displayed. The antenna is represented by its impedance $\mathbf{Z}_a = R_a + jX_a$ (with X_a resulting from C_a and with negligible value of R_a) and by its open voltage source $E \cdot h_{eff}$, with h_{eff} being the effective height of the antenna and E representing the electric field strength of the incident wave. The coaxial cable is described by the capacitive load C_p, because the cable is very short in comparison to the wavelength. The input stage of the receiver is represented by the input capacitance C_i and by the equivalent series noise source $\overline{u_{en}^2} = 4kT_0BR_{en}$, where Boltzmann's constant is k and $T_0 = 290$K is the ambient temperature. R_{en} is the equivalent noise resistance of $\overline{u_{en}^2}$, which is typically 200Ω. From

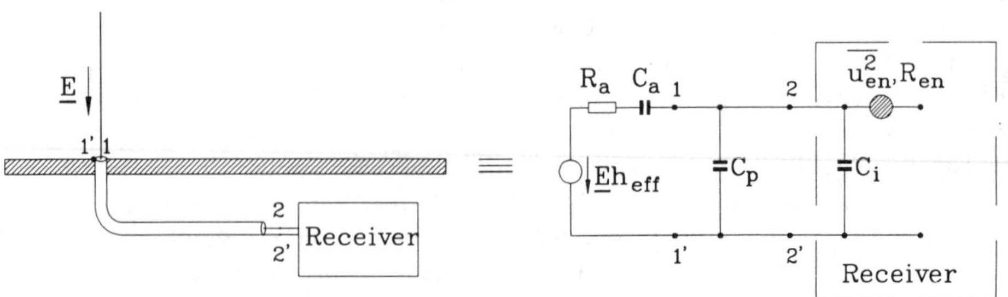

Figure 5.17 Passive whip antenna and receiver and the equivalent circuit in the AM range.

Figure 5.17, the field strength sensitivity E_e, which results in a signal-to-noise ratio of 1, can be derived as

$$E_e = \sqrt{\overline{u_{\text{en}}^2}} \cdot \frac{1 + \dfrac{C_i + C_p}{C_a}}{h_{\text{eff}}} \tag{5.17}$$

which shows the reduction in sensitivity by an additional capacitive load C_p. Consequently, an extension cable of another 5m to 6m, required for antennas mounted in the rear, reduces the sensitivity for another 8 to 10 dB. The active antennas discussed below profit from (5.17) by avoiding C_p. Therefore, an equivalent sensitivity can be obtained with smaller values of h_{eff} and/or C_a.

FM Range. In the FM range, the antenna impedance matches the nominal impedance of the cable with a VSWR value of about 2. The mainly vertically oriented whip, however, does not fit the horizontally polarized FM waves almost exclusively broadcast in Europe, since a rod antenna vertically mounted above ground delivers no output signal with horizontally polarized waves. As a result, the excitation of any antenna mounted on a point of symmetry on a car (e.g., in the middle of the roof or in the middle of the trunk) is small. Therefore, the combination of the car and the whip form the antenna arrangement with its special characteristic. It is obvious from this that the horizontal antenna pattern cannot be expected to be omnidirectional, as can be seen from Figure 5.18.

Figure 5.18(a,b) refer to a whip antenna and Figure 5.18(c,d) refer to a screen antenna. All the diagrams have been measured on a turntable. With the pattern displayed in Figure 5.18(a,c), the excitation was an ideal plane wave radiated from a transmitting antenna positioned nearby. A comparison between the diagrams of the whip and the screen antenna shows that there is no fundamental difference. Both diagrams are not at all omnidirectional, but have minima and sidelobes.

The pattern of Figure 5.18(b,d) has been recorded by receiving an FM station broadcasting from about 100 km from a high mountain. In spite of the fact that the location of the turntable was in a rural area and in spite of the optical view to the transmitter, the pattern shapes of Figure 5.18(b,d) are not similar to those measured with the ideal plane wave. This results from the inevitable multipath propagation. In consequence, no conclusions should be derived from the pattern measured with an ideal plane wave. As explained below, only statistical evaluations of the time-dependent signal level and of the time-dependent distortions are adequate for comparing the performance of different car antennas. It is well known that with vertical and circular polarization, multipath propagation reception occurs

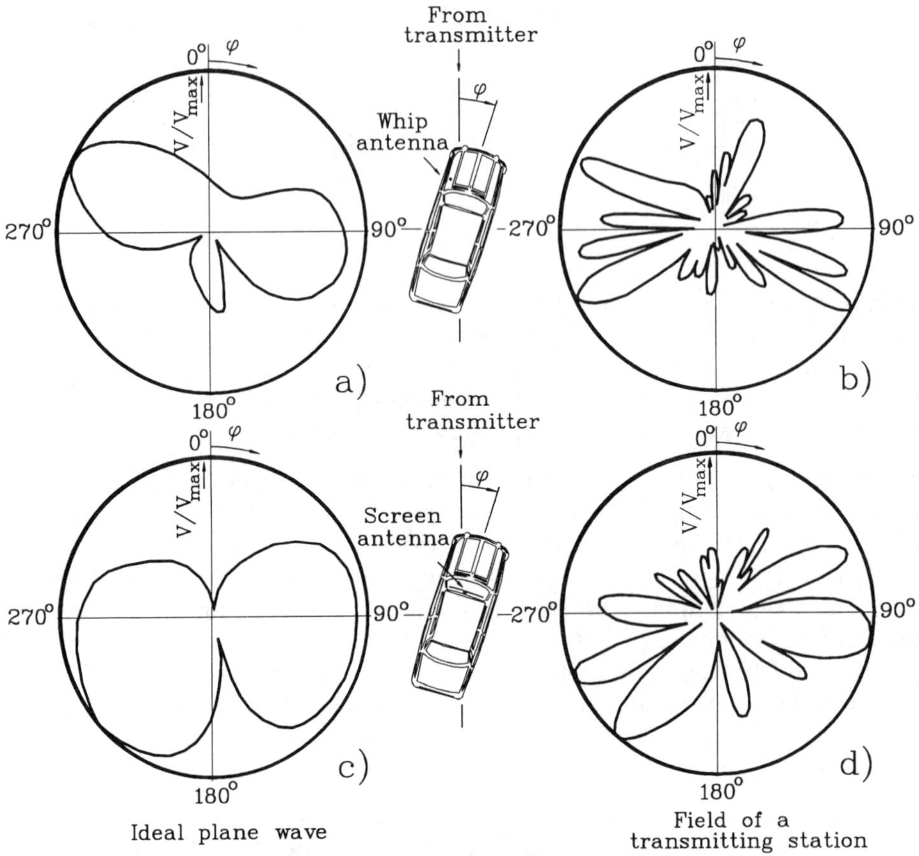

Figure 5.18 Antenna pattern obtained with a whip (a,b) and with a screen antenna (c,d), measured with an ideal plane wave (a,c) and with a high-power broadcasting station (b,d).

in a similar way. Therefore, the shape of the antenna diagram is of minor importance, even with those polarizations.

From the beginning of mobile reception there was a strong demand for antennas smaller than the 1m-long whip. In Figure 5.19, the measured reduction of sensitivity is displayed versus antenna length for a passive whip having a 1.5m-long coaxial cable for both the AM and FM bands. We see that other solutions are required to arrive at smaller antennas, since no decrease in sensitivity can be tolerated.

Passive Screen Antennas

Passive screen antennas meet the mechanical automotive requirements excellently. Two antenna types have been introduced in the market, the antenna structures of which are displayed in Figure 5.20. Neither is still in production, due to an intolerably reduced sensitivity. The antenna structure of Figure 5.20(b), for example, comes out to between −5 and −6 dB compared to a standard whip antenna in the AM and FM range. The unsatisfactory performance of those passive screen antennas, however, has been misinterpreted in the past: the shape of the diagram is not more directional than that of the whip antenna. The antenna directional

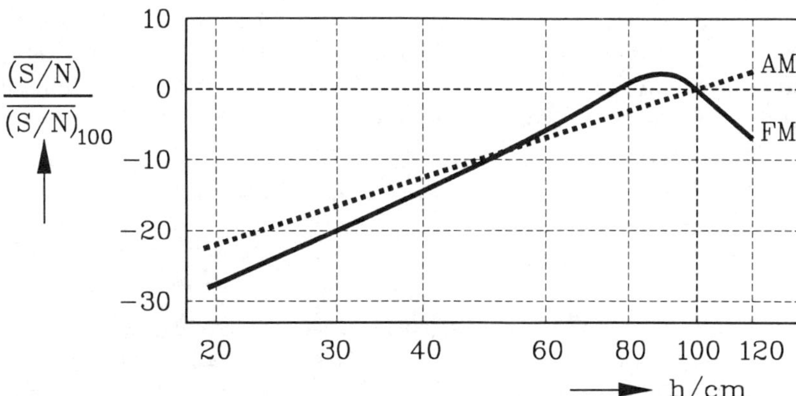

Figure 5.19 Related field strength sensitivity in the AM and FM bands versus the height of a passive whip antenna.

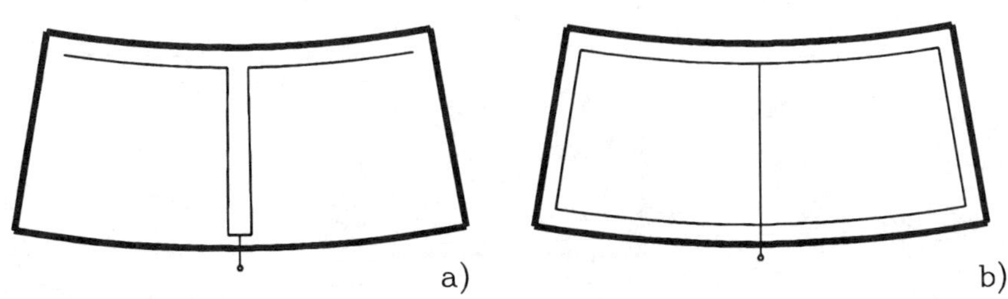

Figure 5.20 Antenna structures applied for passive windscreen antennas.

diagram is not responsible for the degradation; rather, the reduced average field strength sensitivity is resulting in deeper absolute signal values during the minima of reception.

Fundamentals of Active Receiving Antennas

Following the basic definition, an antenna has to be understood as a two-port that transforms the free-space wave to the transverse electromagnetic (TEM) wave delivered to the coaxial output cable (Figure 5.21). Consequently, if the antenna two-port includes active elements, it is an active two-port and an active antenna, respectively. As displayed in Figure 5.22, an active antenna consists of a passive and an active part. In general, low-loss reactive elements are included in addition to the antenna element in the passive part. These provide an optimum combination of the antenna element and the active part with respect to sensitivity and linearity.

The ratio between signal strength and the total intensity of all kinds of distortions is the basic limitation of the maximum transferable information content in a receiving system. Considering the respective values of post-demodulation delivered power within the receiver, this ratio can be expressed as

$$\frac{P_d}{P_{\text{dist}}} = \frac{P_d/P_{\text{exn}}}{1 + \dfrac{P_{\text{ren}}}{P_{\text{exn}}} \cdot \left(1 + \dfrac{P_{\text{int}}}{P_{\text{ren}}} + \dfrac{P_{\text{crm}}}{P_{\text{ren}}}\right)} \tag{5.18}$$

where P_d is the power of the desired signal, P_{dist} is the total power distortion resulting from the received external noise (P_{exn}), from the internal noise of the receiving system (P_{ren}), from intermodulation (P_{int}), and from cross-modulation (P_{crm}). The respective field strength vectors are displayed in Figure 5.22, with E_d referring to P_d and E_{exn} to P_{exn}. The intermodulation resulting in P_{int} at the frequency f_{int3} is

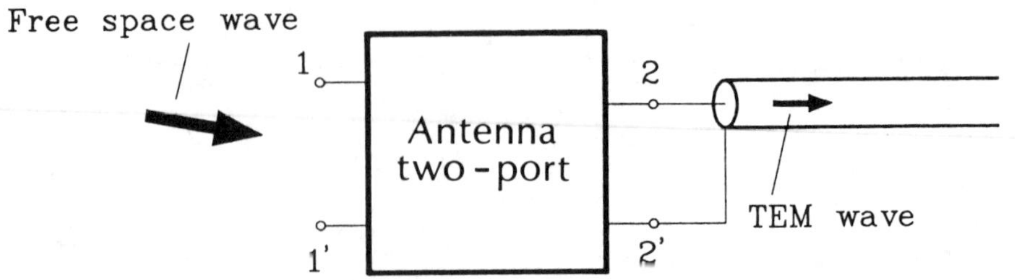

Figure 5.21 Basic definition of an antenna: an antenna is a two-port transforming the free-space wave into the TEM wave or vice versa.

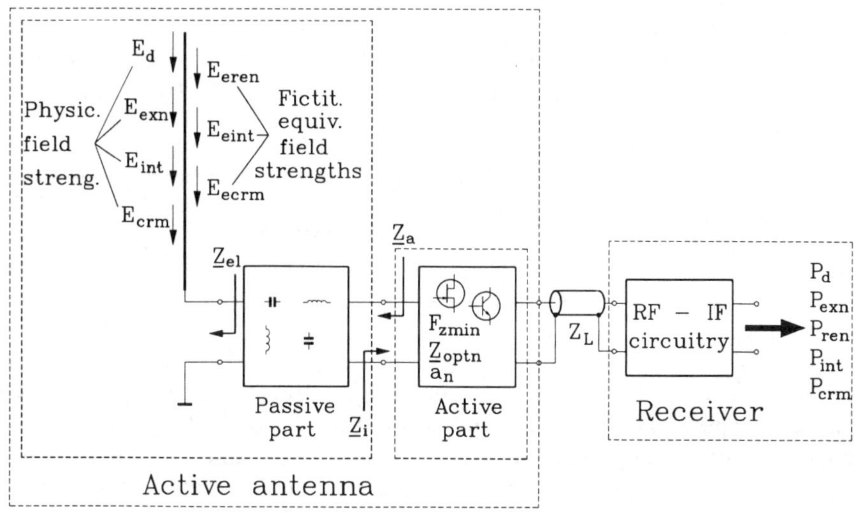

Figure 5.22 Physical and fictitious field strengths and power in a receiving system.

produced by at least two strong undesired signals with field strengths E_{int} at the respective frequencies f_{int1} and f_{int2}.

If a third-order intermodulation product is taken, for example, the frequency f_{int3} is equal to $2 \cdot f_{int1} \pm f_{int2}$ or $2 \cdot f_{int2} \pm f_{int1}$. Third-order intermodulation is the dominating nonlinear effect with FM. With AM, cross-modulation is of major importance, where P_{crm} may be produced by a strong signal E_{crm}. In contrast to those field strengths that are existing physically, E_{eren}, E_{eint}, and E_{ecrm} are equivalent and therefore fictitious values. E_{eren}, for example, is the fictitious equivalent field intensity, which would lead to the same value of P_{ren} if the receiving system is considered to be noiseless, as with a noisy receiving system. In a similar way, E_{eint} describes P_{int} at f_{int3} and E_{ecrm} corresponds to P_{crm}.

With a passive antenna, P_{ren}, P_{int}, and P_{crm} exclusively result from the receiver due to its finite linearity and its noise sources. With an active antenna being part of the receiving system, the antenna contributes to P_{ren}, P_{int}, and P_{crm}, too. With these kinds of distortions, an exact analysis is required in order to determine if the relevant distortion is produced in the active antenna or in the receiver.

In recent years, important improvements with respect to the linearity of broadband antenna amplifiers have been achieved. Consequently, with the respective circuitries applied, the contribution to P_{ren}, P_{int}, and P_{crm} with modern active antennas is negligible. This will be considered in more detail later on.

Active antennas profit from the optimum combination between passive and active antenna components, avoiding the restrictions imposed by cables. With respect to this, power matching (5.19) and noise matching (5.20) techniques have to be considered:

$$\frac{P_{\max}}{P} = 1 + \alpha_p \frac{|\mathbf{Z}_a - \mathbf{Z}_{\text{opt}p}|^2}{R_a \cdot R_{\text{opt}p}} \tag{5.19}$$

$$\frac{F_z}{F_{z\min}} = 1 + \alpha_n \frac{|\mathbf{Z}_a - \mathbf{Z}_{\text{opt}n}|^2}{R_a \cdot R_{\text{opt}n}} \tag{5.20}$$

where $\mathbf{Z}_{\text{opt}p} = \mathbf{Z}_i^*$ is the impedance required for power matching, which is the conjugate complex value of the input impedance $\mathbf{Z}_i$. With $\mathbf{Z}_a = R_a + jX_a$ the impedance of the passive antenna part is equal to $\mathbf{Z}_{\text{opt}p}$ and P becomes equal to $P_{\max}$, which is the available power of the source. $\alpha_p = 0.25$ represents the coefficient for power mismatch. The structures of (5.19) and (5.20) differ merely in the coefficients α_p and α_n. $\mathbf{Z}_{\text{opt}n}$ is the impedance required for noise matching providing optimum signal-to-noise ratio S/N. With $\mathbf{Z}_a = \mathbf{Z}_{\text{opt}n}$, the (additional) noise figure F_z becomes a minimum and is $F_{z\min}$. α_n is the weighing factor for noise mismatch with values between 0.25 and 0.5, depending on the correlation coefficient between the equivalent noise voltage and current sources at the input of the active circuitry. In practice, only lossy structures in front of the active element provide $\mathbf{Z}_{\text{opt}n} = \mathbf{Z}_{\text{opt}p}$. Therefore, with optimum antenna design, noise matching is not obtained when power matching is applied and vice versa.

With passive antennas where the received signals are led to the receiver via an output line, power matching is required at the input of the receiver in order to be independent of the length of the cable. This results in a considerably high noise mismatch. Active antennas profit from noise matching, which can be realized with no compromise, since no power match is required. Consequently, the considerably reduced noise figure can be used to minimize the antenna element, and car antennas can be designed to be smaller than a passive 1m whip antenna.

Optimization of the Antenna Element With Active Antennas

It is desirable that car antennas be as inconspicuous as possible. On the other hand, they have to meet the overall performance of the whip in the AM and FM bands, and therefore the field strength sensitivity obtained with it as well. All passive window antenna structures up to now do not satisfy the sensitivity requirement, whereas active antennas perform even better than the standard whip antenna.

In the AM range, the antenna amplifier is realized by a field-effect transistor (FET) input stage in common drain circuit and bipolar output stages, often in a complementary push-pull class A-B circuit (Figure 5.23). The FET provides the antenna amplifier with a small input capacitance of typically 10 pF and with low equivalent noise sources. The bipolar output stage and the output transformer are used to provide the antenna with the required voltage gain in combination with an excellent large-signal performance, since the large-signal quality is limited by the supply voltage.

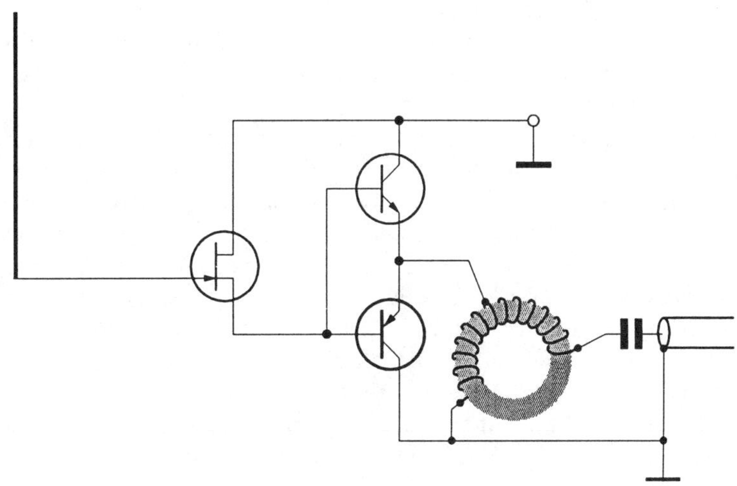

Figure 5.23 Basic circuit for highly linear active antennas applied in the AM range.

For good AM reception, antenna elements are open-end structures with an antenna impedance of capacitive character, and the equivalent circuit of Figure 5.17 and (5.17) is valid. For the active antenna, $C_p = 0$ and C_i is only approximately 10 pF, while with FETs R_{en} is on the order of 200Ω. Consequently, the required sensitivity can be obtained with a smaller value of h_{eff} and/or C_a. Taking the front window antenna, for example, discussed later on, the related effective height h_{eff} of the structure used is approximately 20 dB below that of a 1m passive whip. The antenna capacitance, however, is comparatively high with values of 55 pF, so the obtained signal-to-noise ratio is roughly equal to the passive whip mounted in the front of the car, but is considerably better than with a whip in the rear.

Due to the small car size in comparison to the wavelength in the AM band, the field is homogeneous. The obtained sensitivity in the AM range can easily be measured in a stationary car and can even be judged subjectively by listening to weak stations and immediately switching from the antenna under test to the reference antenna. For further details with respect to active antenna solutions for AM reception, see [7].

In the FM range, statistical methods are required for antenna optimization due to the inhomogeneous field resulting from the multipath propagation situation. With mobile measuring equipment consisting of a commercial receiver, a scanner, and a controlling computer, the output signals of the antenna structure (device under test) $V_{dut}(t)$ and of a reference antenna $V_{ref}(t)$ in dBμV are recorded for a set of frequencies spread over the FM band and for all polarizations for which the antenna is to be used. Those data are the basis for an evaluation of the mean logarithmic signal level V_{mdut} and V_{mref} (for example, in dBμV) and of the probability distribution of the signal levels of the respective antennas at the respective frequencies and polarizations.

The typical shapes of those curves can be taken from Figure 5.24, where they refer to a passive whip (curve with dots) and an active windscreen antenna (bold curve). In Figure 5.24(a), the respective receiver input signals versus time in a car driven at about 5 m/s are displayed. The shapes of both curves are of same character, with sharp minima and broad maxima and with a level variation of typically 20 to 30 dB. The *exceeding probability* evaluated from a longer section of the same test drive leads to the diagram of Figure 5.24(b), with an equivalent shape for both antennas which proves the identity of the dynamic behavior with respect to signal level. With the active antenna (bold curve), the mean signal level is about 2 dB higher than the passive whip antenna. With the passive structure of this windshield antenna, the same curve, but 5 dB down, could be obtained without signal amplification.

With active car antennas, the passive antenna structure and the transformation network have to be designed in such a way that a noise match is obtained at terminals

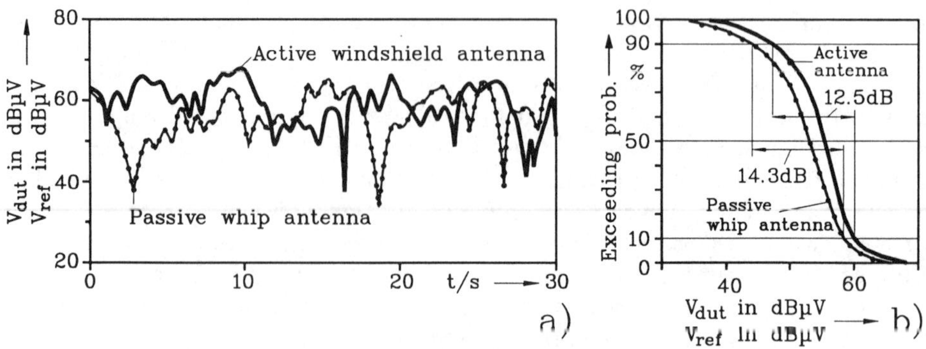

Figure 5.24 Signal level versus time (a) and the respective exceeding probability function (b) measured with the active windshield antenna of Figure 5.29 and with a passive whip antenna.

2 and 2′ in Figure 5.25 for the transistor being connected to these terminals. For this, the frequency response of the impedance of the antenna element $\mathbf{Z}_a$ is measured. The optimum shape of the transformed antenna impedance curve at terminals 2 and 2′ to cover the FM range is a loop around $\mathbf{Z}_{\text{opt}n}$. The circle of constant noise mismatch depends on the impedance bandwidth of the antenna structure, on the band to be covered and on the degree of the transformation network.

The available power with window antenna structures in general is lower than that of a 1m whip antenna, which means that the passive antenna structure does not meet the performance of a whip. Due to the low minimum noise figure of the active antenna amplifiers (typically 4 dB), which is typically 6 dB better than the respective value of car radios (typically 10 dB), an increase of signal-to-noise ratio is obtained. Consequently, antenna structures with available power p_{rel} of -6 dB compared to the whip can be used in active antennas without decrease in sensitivity.

This, however, is only true if the dynamic behavior of the antennas to be compared in a driven car is equivalent. Therefore the exceeding probability has to be considered also. From several respective measurements, the conclusion can be drawn that the typical shape of the exceeding probability function (Figure 5.24(b)) is formed mainly by the random Rayleigh field and only rarely by the individual antenna structure. Consequently, the shape of the exceeding probability function for most antenna structures tested up to now was quite similar to that of the 1m whip antenna.

The results obtained with such an optimization are now to be discussed. The structure under test is shown in Figure 5.26, with l being the total length of the antenna wire embedded between the layers of the laminated glass of a front window. For each of the spot frequencies (horizontal polarization), a resonant peak of excitation is obtained (Figure 5.27). The optimum length of the wire for the FM

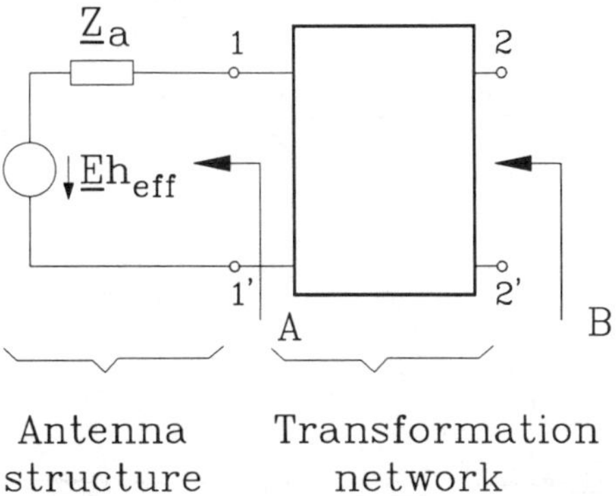

Figure 5.25 Noise match and power match.

range comes out to be about $l = 175$ cm, leading to an average value of p_{rel} of about -3 dB. This value is good enough to arrive at an excellent performance as an active antenna.

Antenna Amplifiers for Active Receiving Antennas

In the past, active antennas met little acceptance, and their development has been retarded by the fear that the problem of nonlinear distortions in a broadband active circuit might not be solvable. In the meantime, good improvements have been achieved with amplifiers with respect to third-order intermodulation distortion, which is the dominating nonlinear effect with FM reception. Various car manu-

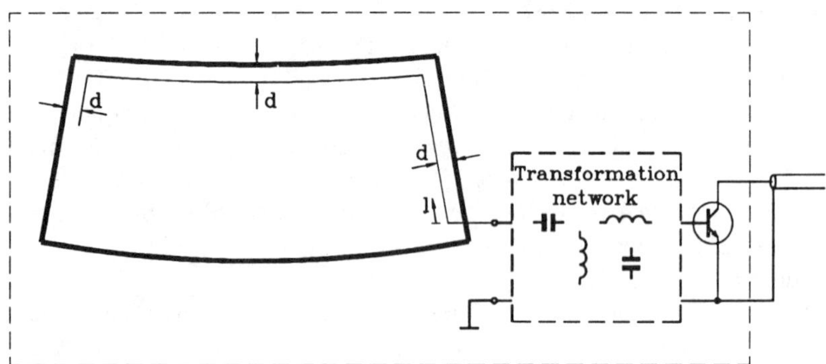

Figure 5.26 Example of an active windscreen antenna.

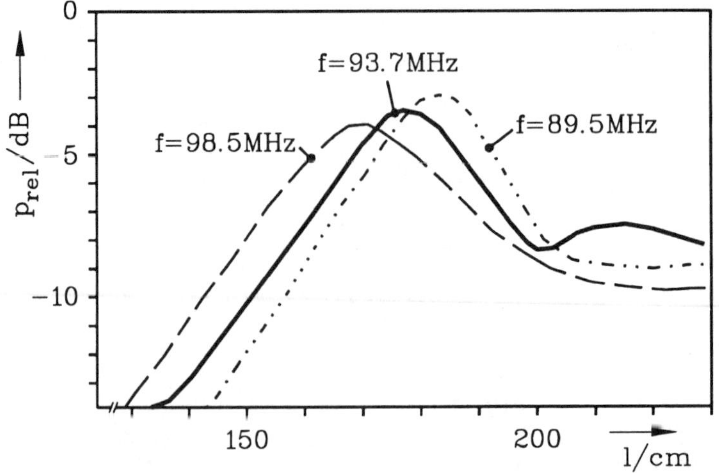

Figure 5.27 Antenna performance of the antenna structure of Figure 5.26 versus the length l of the antenna wire.

facturers have taken active car antennas into series production since 1982. The performance of active antennas nowadays shows that nonlinearity must no longer be considered a problem in practice. This progress has been obtained by advanced circuitry and by better active elements. The suppression of nonlinear effects today is of even greater importance than some years ago due to the increased number of FM stations. Extreme demands arise in large cities where, typically, a multitude of FM stations are concentrated on radio towers. Consequently, extremely high antenna output signals delivered to 150Ω of up to 105 dBμV have been recorded in Berlin and up to 120 dBμV in New York near the Empire State Building.

The respective performance of different types of FM antenna amplifiers is displayed in Figure 5.28. In this diagram, the level of the third-order intermodulation product (two-signal method) at frequencies $2 \cdot f_1 \pm f_2$ or $2 \cdot f_2 \pm f_1$ is displayed versus the level of the large signals at frequencies f_1 and f_2 generating the intermodulation product. An amplifier type 1989/1, now in series production, with an output signal of 100 dBμV generates an intermodulation product of about −3 dBμV, exceeding the noise level for only 2 dB.

With higher bias current, even better results can be obtained (type 1989/2 amplifier, test sample). In both cases, the amplifier is a two-stage circuit with a small-power input stage in a common-collector circuit and a high-power output stage in a common-emitter circuit with high negative broadband feedback. In comparison, an amplifier of 1970 series production with the same output level of 100 dBμV had an intermodulation product of about 57 dBμV, thus considerably exceeding the noise level and deteriorating the overall sensitivity of the receiving system. The respective circuitry was a common-emitter single-stage bipolar amplifier with small reverse feedback. In addition, limiting effects are found with the 1970 amplifier at about 106 dBμV, whereas with today's technology signals up to 125 dBμV are not compressed. Consequently, the maximum output signal found in New York is amplified without limitation.

Of course, intermodulation distortions are also produced by the receiver and contribute to the total signal. Therefore, a comparison of the linearity of antenna amplifiers and receivers is required. The performance of top radios produced for the American market is plotted in Figure 5.28 (dashed curves). An input selectivity is applied, and therefore the suppression of strong undesired stations at f_1 and f_2 increases with increasing frequency separation between the frequency of the intermodulation product to which the receiver is tuned and the interfering signals. The maximum frequency difference between an intermodulation product and the adjacent interfering signal within the FM band is therefore about 9 MHz. In this case, maximum intermodulation protection by means of selectivity is obtained as shown in Figure 5.28. This proves that, even under this assumption, the intermodulation product being generated by the active antennas is 10 dB (1989/1) and 18 dB (1989/2) below that contributed by the receiver. With a frequency difference of only 2 MHz between f_1 and f_2, the effect of selectivity is small and cannot protect the receiver from intermodulation. In this case, the contribution of the receiver to the

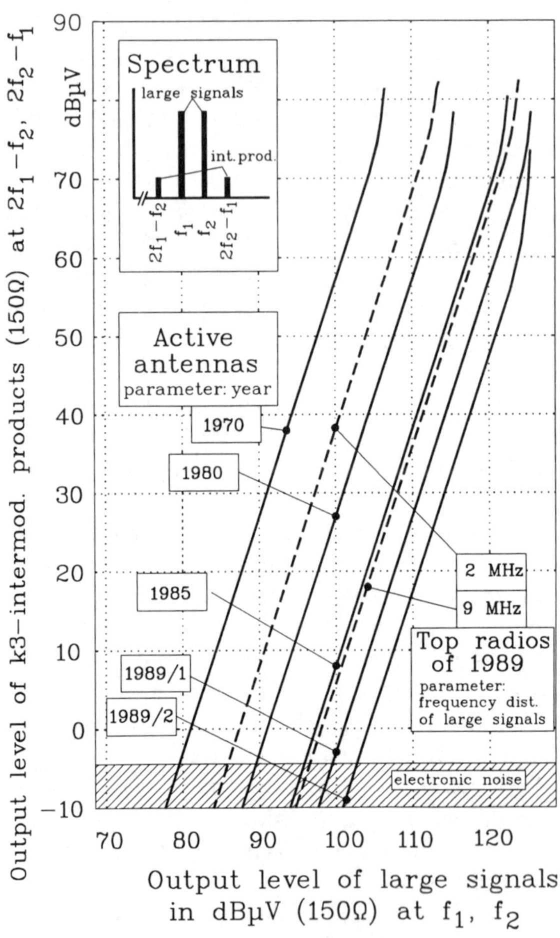

Figure 5.28 Third order intermodulation distortion in the FM range with active antennas and receivers.

third-order intermodulation product is so high that it exceeds the intermodulation level of a 1980-model antenna amplifier by 42 dB.

For this reason, in the absence of the required selectivity, most types of modern car receivers are protected from nonlinearity by an electronically controlled attenuator at their input. This has the disadvantage of desensitization of the receiver at the signal to which it is tuned if an interfering signal at a different frequency exceeds a tolerable threshold. In contrast, improvements of antenna amplifier linearity have been developed under the restriction that the amplifier sensitivity has not deteriorated.

Active Antennas With Short Monopoles

Several types of active antennas are available on the market making use of a 40-cm-long rod. The equivalent capacity of this rod is very small (4 pF) in the AM range, and its effective height is about 20 cm. Due to (5.17), any additional capacitive load (for example, resulting from the FM transformation network) deteriorates the AM performance. The circuitry of Figure 5.29 is a solution to this problem. A first FM resonance circuit is formed by the capacitive impedance of the antenna element, the inductance of coil L_1, and the input capacity C_F of the FET, which is also the input stage of the AM path. A second resonance circuit is formed by the inductance L_2 and the capacitance C in parallel. Thus, a bandpass filter is formed in the FM band by appropriate coupling between the coils L_1 and L_2. Due to the magnetic coupling of the FM signal, the sensitivity in the AM is not reduced.

Active Antenna in the Front Screen

The design of this antenna is displayed in Figure 5.29. The structure has been optimized with respect to optimum performance with horizontally polarized FM signals, as are exclusively used in most countries in Europe, and with vertically and circularly polarized FM signals as are frequently used in the United States and United Kingdom, together with AM. The antenna element is realized by a thin

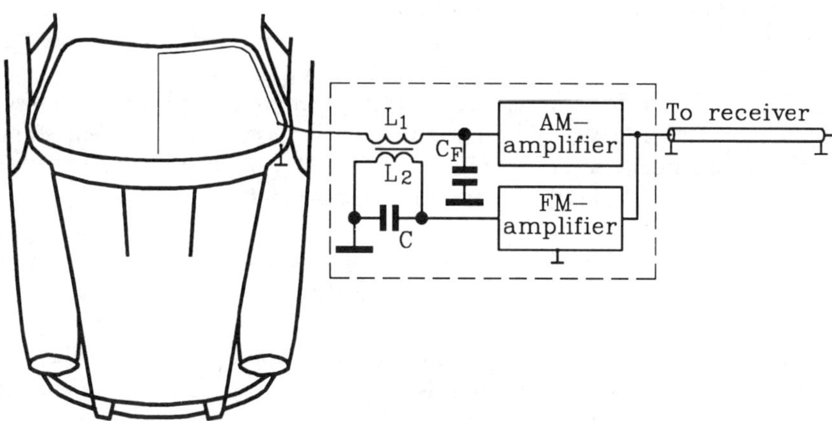

Figure 5.29 Active windscreen antenna optimized with respect to horizontal, vertical and circular FM polarization and for AM.

and hardly visible enameled copper wire embedded between the layers of the laminated windshield and may be regarded as a probe coupled to the electric field, which is considerably influenced by the shape of the aperture of the window. In the FM range, this aperture is roughly half a wavelength and, in consequence, a resonance is found forming a maximum electric field intensity in the middle of the aperture where a section of the antenna wire is located. Due to the inclined glass, this resonance is excited by both horizontally polarized and vertically polarized waves. The amplifier is mounted below the dashboard at the lower end of the A-post, near the outlet of the antenna wire at a readily accessible location.

With this antenna type, extensive measurements and subjective tests had been performed in Europe and in the United States before this antenna was introduced in the complete series production of Porsche cars in 1984. Those tests took place in the very vicinity of strong AM and FM stations, and cross- and intermodulation distortion at the fringe of the service area was investigated.

In the FM range, the average excitation of this passive antenna wire has been measured to be about -2 to -4 dB, with horizontal polarization compared to a 1m whip. With the noise-adapted FM amplifier, an improvement to typically $+1$ dB in mean signal-to-noise ratio has been achieved with a related output signal level of about $+2$ dB. Therefore, the mean performance of the active antenna is at least equivalent to the whip. In Figure 5.24(a), the output signal is displayed versus time for a short test drive with a car driven at very low speed. The distance between the fadings is about half a wavelength, which is about 1.5m in the FM range. The similar shape of the signal level curves of the two different antennas proves the equal dynamic behavior of both antennas. This statistically evaluated result can be extracted with higher accuracy from the curves of the exceeding probability function, which are displayed in Figure 5.24(b). These have been obtained from a long test drive, from which Figure 5.24(a) shows only a short section.

With a change from horizontal to vertical or circular polarization, assuming equivalent (ERP) and reception in the very vicinity of the transmitting station, an increase of the mean output signal level of 7 dB is obtained for the whip antenna and 2 dB for the windshield antenna. With increasing distance from the transmitter, the difference in mean signal level between both antennas becomes negligible. At the fringe of the service area, where the obtainable signal-to-noise ratio is most important, the active windshield antenna has been measured to be equivalent to the whip.

In order to obtain objective results for the total distortions in the receiving system, another test setup has been developed and is described later on. Results obtained in a Porsche 928 are displayed in Figure 5.30. Figure 5.30(a) shows a short section of a longer test drive along the Tegernsee in Bavaria, Germany. The distortion level is displayed versus time. Overall, there is no difference in the characteristic behavior between both antenna signals. This can be taken from Figure

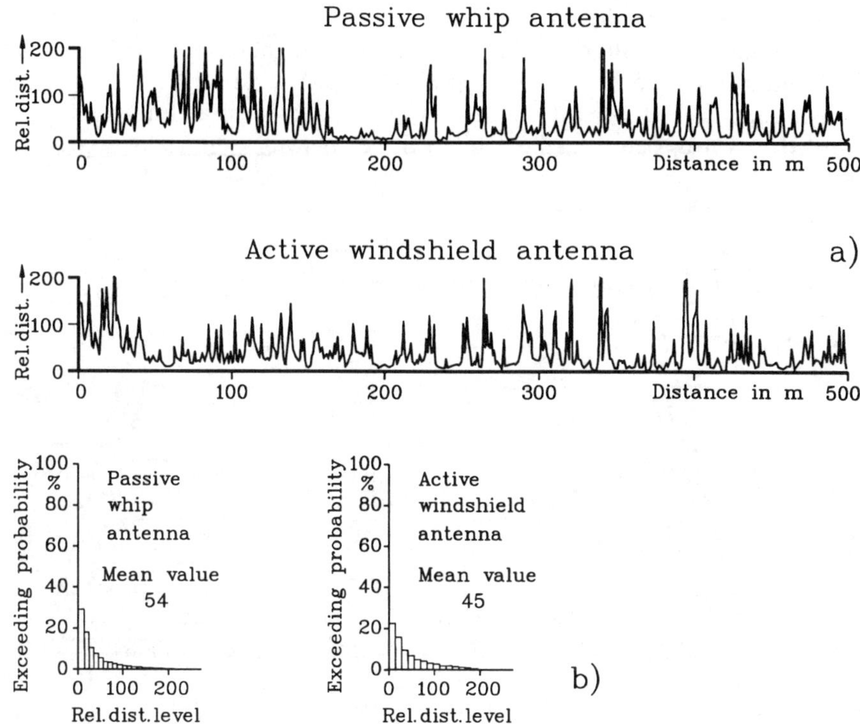

Figure 5.30 Overall distortion versus time (a) and the respective exceeding probability function (b) measured with the active windshield antenna of Figure 5.29 and with a passive whip antenna.

5.30(b) as well, where the total test drive has been evaluated by statistical means. Consequently, even with respect to the total interference in a driven car, the new active windshield antenna comes out to be slightly superior to the whip, a fact that has been proved by subjective judgment as well.

Active Antenna in the Rear Window

With modern cars, electromagnetic interference due to microprocessor-operated components applied, for example, in motor control units, in the instrument panel, or in the onboard computer, is a great problem for all kinds of antennas. Antennas mounted in the front of the car are mainly subject to this interference. In the rear, the interference is present, too, but is typically about 6 to 12 dB smaller. Consequently, an antenna with the same excellent overall performance as the above front screen antenna was required for the rear window. The new antenna design, which is displayed in Figure 5.31, has been applied to BMW cars since 1989 and meets the requirements. Extensive tests and measurements of this antenna were also performed in Europe and in the United States before its introduction to series production.

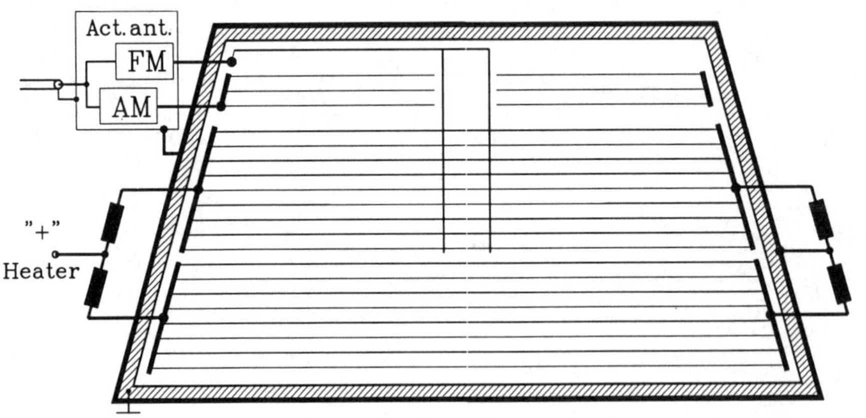

Figure 5.31 Active rear screen antenna for all FM polarizations and for AM.

Different antenna structures are used for AM and FM. Both structures are printed onto the glass with the same technology as the heaters. Thus, both the heater and the antenna conductors can be simply printed onto the glass in one stage of manufacture. The FM antenna structure containing vertical antenna conductors can be in galvanic contact with the horizontal heater conductors. Since the connecting points between the antenna structure and heaters are arranged on equipotential lines, the defrosting of the rear window is not influenced. Only in the upper part of the antenna structure, above the uppermost heater, does a shunt with a comparatively high resistance reduce the heater current in the middle, but only to an insignificant extent.

The AM structure is located in the free strip above the heaters without any galvanic contact with them. Therefore, the problems involved with antennas making use of the heaters for AM reception are avoided, since there is no need for broad band RF insulation by large coils (for up to 30A of heating current). However, with the expenditure of greater technical effort, an excellent antenna can be realized by using the heaters for AM, too. This has been explained in [7] in detail.

From research done with respect to the AM structure of Figure 5.31, the conclusion was drawn that the optimum signal-to-noise ratio is obtained if the AM structure is centered with respect to the height of the free strip, and that the optimum width of the structure depends on the capacitive load of the antenna amplifier [7]. With a load of 10 pF obtained with the antenna amplifier of Figure 5.23, the optimum width is 40% of the height of the free strip. Taking a free strip of about 14 cm high, for example, the ultimate sensitivity of the antenna is about 3 dB better than with a standard whip in the rear of the car.

5.1.5 FM Antenna Diversity Systems

From the curves in Figure 5.24, it can be learned that the minima of RF field amplitude with different antenna locations on a car do not generally occur at the same time. In the past, many attempts have been made to make use of this effect by selecting the receiving signal from that antenna which momentarily provides the best signal. In the absence of a sufficiently fast operating distortion detector, no antenna diversity system was presented in the past that could make use of the advantage of two or more antennas on a fast car.

An antenna diversity system implementing a distortion detector of extremely short distortion-detection time is described here. In the following, the efficiency of such a system using such a distortion detector is investigated as a function of the number of antennas applied [4].

Available Improvement With a Multitude of Antennas

Multipath propagation amplitude fadings create distinct amplitude minima in the received signal during mobile reception. As displayed in Figure 5.32, different kinds of distortions, such as noise and cochannel, adjacent channel, and intermodulation interference may occur during fading periods. The superposition of waves with a time-delay difference greater than 3 μs, as previously discussed, also leads to audible distortions if an amplitude condition is satisfied, which is marked by the shadowed range. Therefore, it is obvious that the distortions are strongly correlated with amplitude minima.

To obtain a realistic judgment of the distortion, the definition of a subjectively acceptable upper limit of audible distortion is necessary for a reference. The presence of a distortion is detected by comparing the instantaneous value with this reference, as shown in Figure 5.33 for two antennas. With p_1 and p_2 representing the fraction of time for a so-defined distortion in signal 1 and signal 2, respectively, the likelihood of the simultaneous occurrence of a distortion in both signals is the joint probability $p_d = p_1 \cdot p_2$. Assuming an extremely fast operating distortion

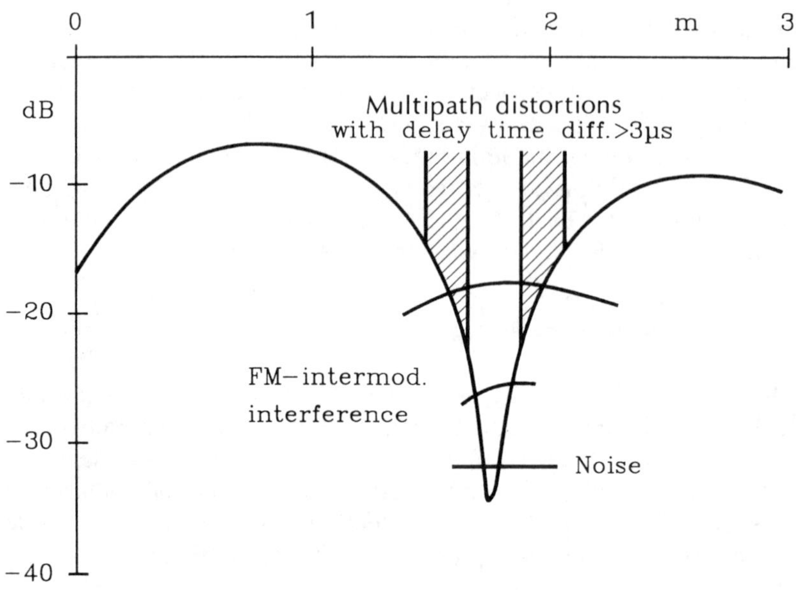

Figure 5.32 Different kinds of distortions in a level minimum.

indicator, a scanning diversity system may be applied, now making use of a multitude of receiving antennas on a car with only one receiver. The signal quality Q available with reception from only one antenna can be defined as

$$Q_S = 20 \cdot \log(1/p) \qquad (5.21)$$

with p representing the related fraction of time with distorted reception. Under the assumption of equal signal quality Q_S of each antenna and with negligible correlation between the distortions on the N antennas, the related fraction p_d of time under distorted reception is found by the joint probability

$$p_d = \prod_{i=1}^{N} p_i \qquad (5.22)$$

If the improvement factor q_{dB} is defined as p/p_d, its logarithmic value

$$q_{dB} = 20 \cdot \log(p/p_d) \qquad (5.23)$$

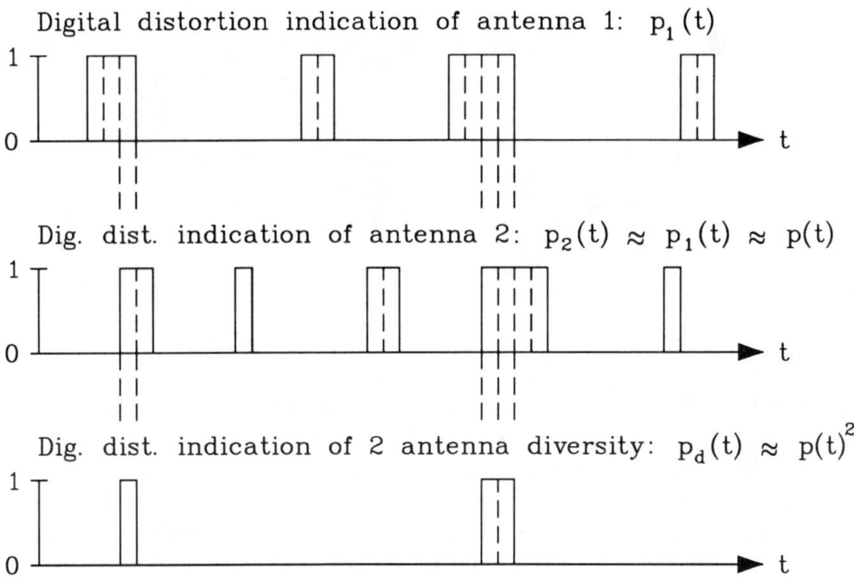

Figure 5.33 Digital distortion indication and distortion reduction with antenna diversity.

becomes

$$q_{\mathrm{dB}} = (N - 1) \cdot Q_S \tag{5.24}$$

This factor shows that the improvement factor is considerably increased by each additional antenna and that the fraction of distortion time is reduced exponentially by the number of antennas applied.

For this reason, up to 12 antennas installed on a car have been measured. The installation of these antennas is shown in Figure 5.34, where four antennas (1, 2, 5, 6) are mounted on the fenders, six antennas (7 to 12) are mounted on the shown locations on the roof of the car, and an additional pair of windshield antennas (3, 4) was fixed onto the front screen. With space diversity, it is well known that the signal correlation factor between similar antennas increases with decreasing distance between adjacent antennas. In order to find the improvement factor for distortion suppression as a function of different antenna locations on a car, special measurement equipment had to be developed.

In order to obtain objective measuring results, a radiated test signal was required, because multipath distortions are a function of the modulation. For this reason, in cooperation with the Bavarian Broadcasting Corporation, a 1-kHz mod-

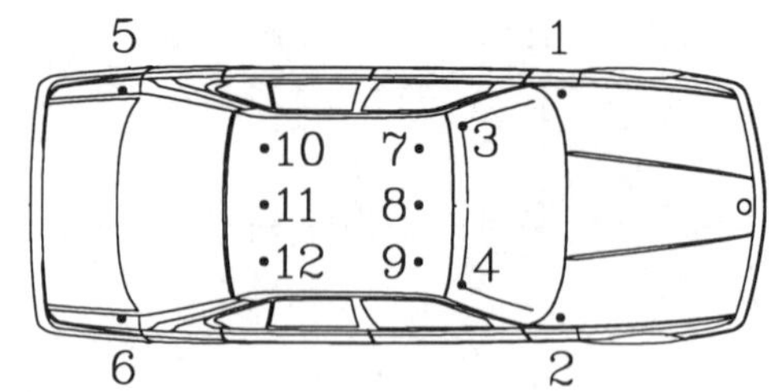

Figure 5.34 Test car with 12 antennas.

ulation signal with a maximum frequency deviation of 35 kHz has been radiated during the night with a power of 100-kW ERP from several stations along the north rim of the Bavarian Alps. With the flat area north of the Alps and with Munich being a densely built up area and with a mountainous topography near the Alps all important different receiving conditions relevant in practice were available. Therefore, studies could be made in the flat areas with low multipath distortions and in high multipath city areas with small delay-time differences between the superimposed waves, and in high multipath areas in the mountains with great delay-time differences between superimposed waves.

The measuring device in the car consists of a scanner, which sequentially switches the antennas to the receiver for one period of the audio frequency (Figure 5.35). The receiver contains a broadband, highly linear FM demodulator and an AM demodulator of a wide dynamic range for amplitude detection. The mixed-down FM signal is demodulated and filtered by a low-pass of 500 kHz and a signal trap of 1 kHz, and this AF signal is then integrated. The instantaneous amplitude is also integrated and is available by means of a sample and hold circuit at the same time at the input of the selector. This selector alternately reads the FM and AM values into the system voltmeter, which digitizes the values for the system computer. After each scanning cycle, the instantaneous error of the received signal of each antenna is stored in the computer for statistical evaluation.

Figure 5.36 shows the improvement factor q_{dB} versus the total number N of antennas applied. Antenna 1 of Figure 5.34 serves as the main antenna, while the auxiliary antennas were antennas 2, 5, and 6 for $N = 2$, 3, and 4 respectively. Curve 1 describes the q_{dB} value for different values of N and is found to be almost a straight line, as calculated by (5.24). This curve has been measured in the Munich area with a radio signal being radiated from a weak transmitting station 80 km away from Munich. The most interesting aspect is the fact that the improvement factor increases with a median value of roughly 17 dB per additional antenna; hence, more than two antennas are beneficial. Curve 2 in Figure 5.36 represents a measured result obtained in the city area at a different location and on a different

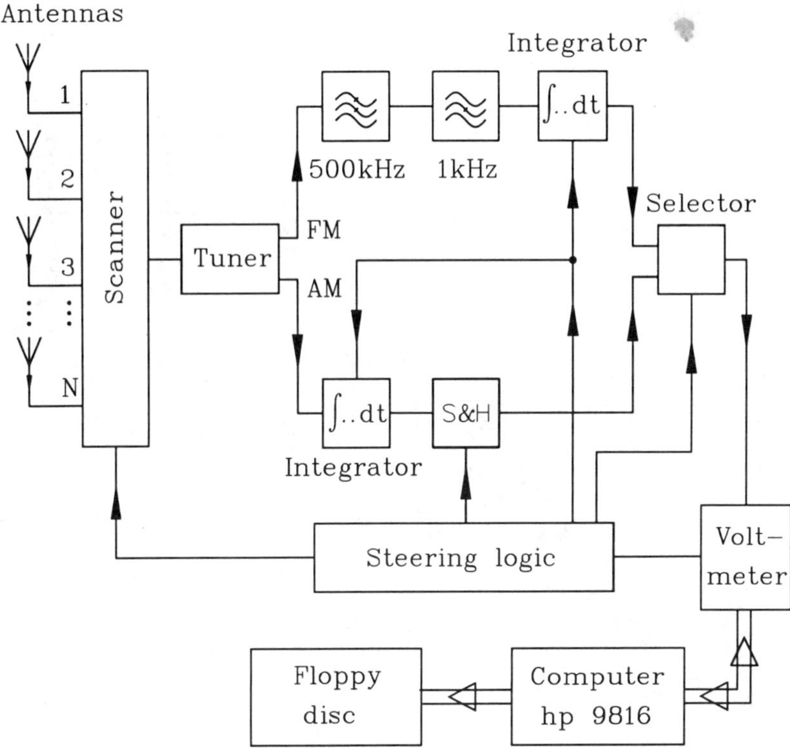

Figure 5.35 Measuring device for efficiency evaluation with a multitude of antennas.

day with the additional antenna 7 on the roof of the car. The results show a curve which shows improvements with increasing N, but the increase of the q_{dB} factor is less pronounced with N greater than 3. In this context, it should be noted that the available improvement factor is a function of the signal quality being received with the single antenna system.

In city areas, the time-delay difference between superimposed waves is comparatively small, and the number of waves of different time delays contributing to multipath distortions is usually comparatively low. This is in contrast to the situation in mountain areas, where the median value of the difference in delay time between the superimposed waves is on the order of 40 μs corresponding to a detour of 12 km in propagation. Due to this fact, the average multipath distortions are much more severe in mountain areas. The improvement factor of curve 3 has been measured in a heavily multipath-distorted region around the lake of Tegernsee in Bavaria. This curve is almost a straight line, indicating the law of equal increase of the

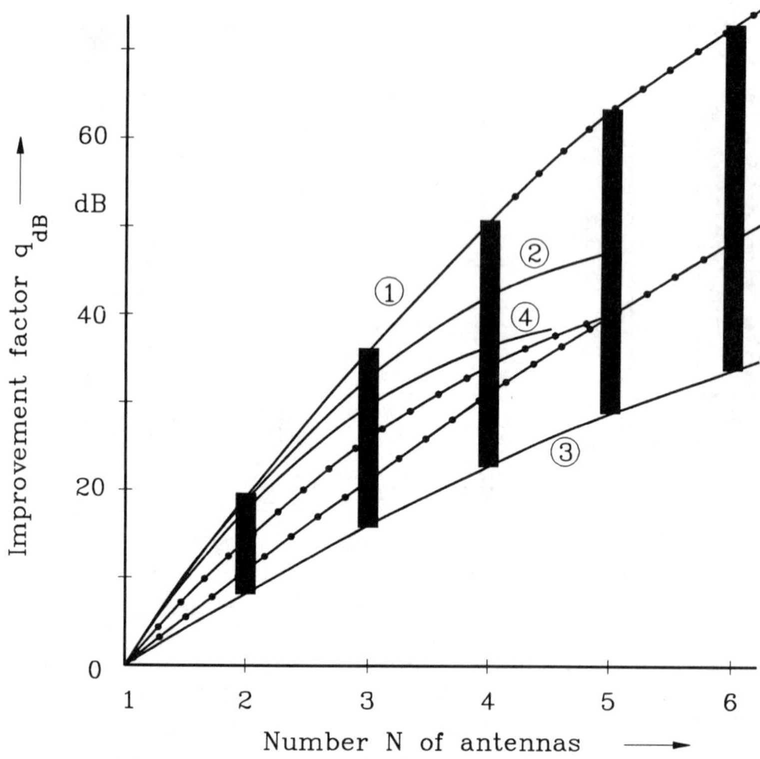

Figure 5.36 Improvement factor in city areas and mountainous areas versus the number of antennas.

logarithmic improvement factor per additional antenna (5.24). Even though there is only a little space between adjacent antennas, the cross-correlation factor between the output voltages is still low enough to get the measured increase of q_{dB} of about 8 dB per additional antenna. In this case, the signal supply is very poor, and therefore the signal quality, which is the percentage of time in which the reception is undistorted, is low. As a result of the poor signal supply, the increase of the improvement factor per additional antenna is less than that in city areas.

In a mountain valley where good signal supply was found, curve 4 has been measured. With low numbers of N, the increase of the improvement factor is roughly 20 dB per additional antenna and is found to be as large as that in city areas. This is true in spite of the large differences in delay time between superimposed waves. As a result of the good signal quality, however, the further increase of the improvement factor per additional antenna is less if N is a large number.

Among the measured results, the curves marked with dots in Figure 5.36 have been found in other receiving situations. Therefore, the improvement factor q_{dB} for a given number N of antennas is presented by the dark columns, which can be expected in a large variety of receiving areas of different characteristics.

Improvement Factor Using Linear Combinations of the Antenna Signals

In car production, the number of antennas that can be deployed is restricted, and the use of only a limited number of FM antennas has been investigated. The sum and difference of the N antenna output signals are generated in a matrix with M signal outputs, and in Figure 5.37 $N = 3$ antenna signals were used to form $M = 6$ linear combinations. These combinations were found to be especially effective in mountain areas (Figure 5.38, dashed curve), where the likelihood of multipath distortions due to great time-delay differences between superimposed waves is high. With the matrix combination of signals, different directional patterns are obtained, reducing the likelihood of distortions of this kind. Therefore, the improvement (22 dB) obtained from all signal combinations together is greater than with one additional antenna (approximately 15 dB). In urban areas (Figure 5.38, solid curve), however, the gain in signal quality by means of linear combinations is less, because in such regions noise and adjacent and cochannel interference during the deep level fadings with respect to the Rayleigh field are dominant, and the linear combinations during the deep fading of the signal does not help to improve the reception.

Equalizing Techniques for the Radio Channel

The transmission two-port describing the transfer function $\mathbf{S}(f)$ (5.1) between the transmitting antenna and the receiving car antenna gives rise to the idea of compensating for the phase and amplitude errors by means of an adaptive equalizer implemented in the FM radio. It corrects the total frequency response from transmitter aerial to receiver IF output and removes echoes and the related distortions from the signal. Normally such an equalizer requires an adaptation time of about 4 ms and can handle one echo up to a 70-μs delay-time difference [8]. Therefore, the equalizer delivers good results in receiving areas where only one echo is dominant. In receiving situations with more echoes and with echo amplitudes of the same order of magnitude, convergence problems exist, and therefore this method cannot eliminate the distortions in this case. Therefore, a time-variant antenna system has been developed in order to minimize the distortions with multipath reception.

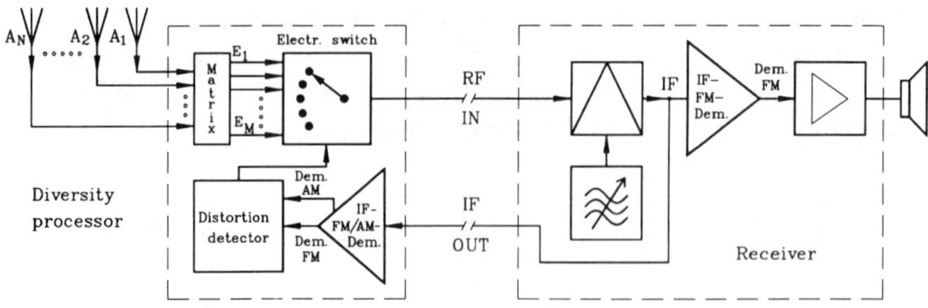

Figure 5.37 Scanning diversity system with a matrix to form linear combinations of the antenna signals.

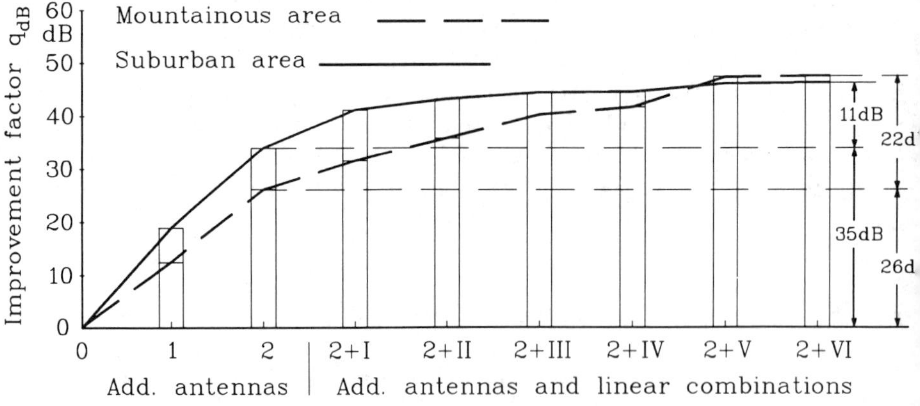

Figure 5.38 Improvement factor in a mountainous (- - -) and an urban (————) area using linear combinations.

Diversity Processor

All the kinds of interference previously discussed are associated with distinct erroneous pulsing of the instantaneous frequency deviation of the frequency-modulated RF carrier in combination with sharp erroneous pulses modulating the envelope of the carrier and forming deep minima, as displayed in Figure 5.6; this applies to a standing car if exposed to interference. The coherence between the maxima of the frequency deviation pulses and the amplitude minima serves as an excellent and extremely fast criterion for distortion recognition in the diversity processor shown in Figure 5.37.

The block diagram of the distortion detector is displayed in Figure 5.39. The IF signal, representing the RF signal behind the mixer, is demodulated in amplitude and frequency in different branches of the circuit. Then the signals are shaped by increasing their steepness by means of nonlinear circuits. In this way, the significant erroneous pulses in frequency deviation and RF envelope curve are pronounced. Both signals are led to a comparator circuit, the outputs of which are connected to the input of a binary logic. The logic AND circuit indicates the simultaneous presence of a pulse in the frequency deviation and a deep notch in the RF envelope. If a distortion pulse is recognized in the frequency deviation and the envelope of the IF carrier at the same time, the binary logic moves the signal selector immediately into a different position, searching for a less distorted antenna signal. By means of an appropriate feedback of the sequence of antenna switching to the thresholds of the two comparators, a dynamic adaptation of the distortion detector to the receiving situation can be obtained. This means that in areas with high average distortions, the detector reacts in a less sensitive way so that the movement of the antenna selector does not exceed a tolerable level of activity. This distortion detector, which includes the driver circuits for the antenna switches, is now available as an integrated circuit. The noise of the switching process itself is suppressed by the automatic interference canceler regularly available in the car radio. The distortion detector is implemented in the diversity processor, which provides N antenna inputs, one RF output, and an IF input. The diversity processor includes pin diodes in order to switch the antenna signals.

Comparison of Different Diversity Systems

In Figure 5.40, the efficiency of the realized antenna scanning diversity system using the above diversity processor is compared to a scanning combiner with low fixed threshold and an ideal selection combiner. The scanning combiner and the ideal selection combiner have been realized with a computer evaluation. In the case of the selection combiner, the less distorted signal from the three antennas is selected. For a scanning combiner, a less distorted antenna signal is selected in case the low fixed threshold indicating a distortion is exceeded. The difference between the measured distortions associated with the scanning diversity system and the idealized selection combiner is negligible. This is true in spite of the fact that the scanning diversity system only needs one receiver, in contrast to the selection diversity system where the number of receivers required equals the number of antennas deployed.

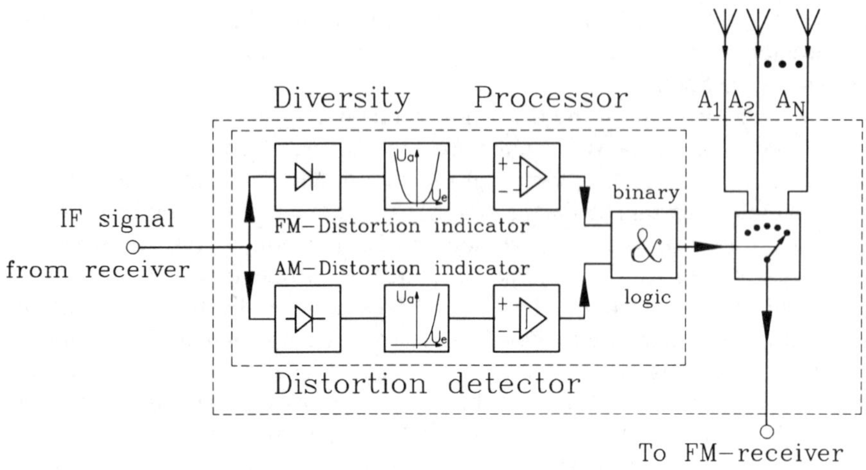

Figure 5.39 Block diagram of the distortion detector within the diversity processor.

5.1.6 FM Multiantenna Systems

The results presented in Section 5.1.5 recommend the application of more than two antennas with antenna diversity systems. It is obvious that antenna diversity on the basis of two or more standard whip antennas would not be accepted, due to esthetic, optical, and aerodynamic constraints. Therefore, investigations were made with multiantenna arrangements in windows in spite of the fact that most theories on diversity systems claim a certain distance between the antennas in order to obtain the necessary decorrelation between the antenna signals [9].

The correlation factor between the signal levels of two antennas, however, is not the only relevant criterion to estimate the obtainable diversity improvement. With the quality of reception being at least satisfactory, audible distortions occur only within 1% of the time. Consequently, the contribution of those small time periods to the correlation factor is only 1%, too. Therefore, the probability of a simultaneous occurrence of interference with both antennas has to be considered.

With weak signals and in regions with small values of delay time between superimposed waves, the AF distortions mainly result from noise if the signal level fades (Figure 5.32). For these, an analysis of the signal level versus time obtained with different antennas under test is suitable for judgment of diversity performance.

With multipath distortion due to time-delay differences, with τ_i greater than 3 μs, and with cochannel or adjacent-channel interference, a high antenna output

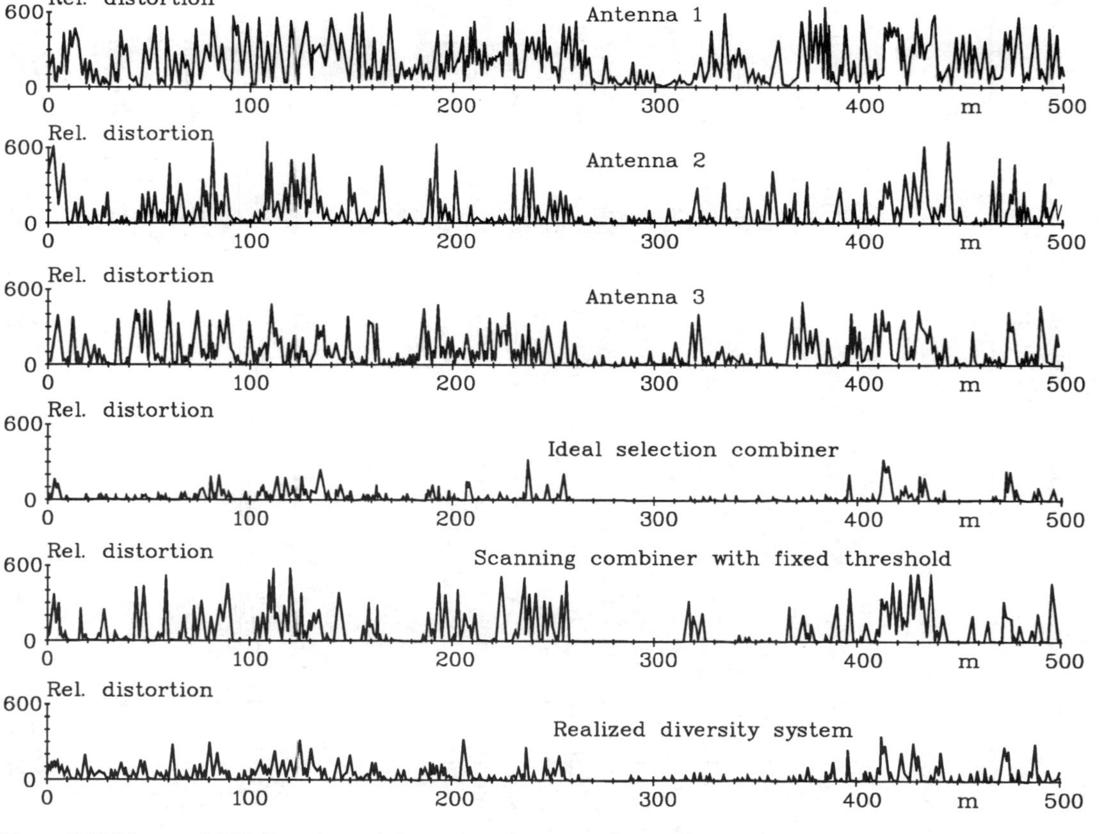

Figure 5.40 Measured AF distortions of three antennas versus time and computed reduction of distortions with an ideal selection combiner and scanning combiner with low fixed threshold compared with the measured distortions of the realized scanning diversity system.

signal may actually be more disturbed than a weaker signal, so the AF distortions versus time and not the signal level have to be measured and analyzed.

The antenna systems presented below consist of two, three, or four antennas per window. In most cases, the mutual coupling between the antenna structures mounted on a single screen cannot be neglected; therefore, active antennas are advantageously used to provide antennas with the required decoupling.

Four-Antenna System Arranged in Front and Rear Windows

Figure 5.41 shows the thoroughly optimized antenna configuration that has been installed in several types of cars for further investigations. In the rear window, two FM antennas are obtained by subdividing the heater structure horizontally into two partial heating areas, and four FM resonant decoupling networks are required. The connection points for the FM amplifiers on the bus bars are applied at the opposite limits of the bus bars on the same side. AM reception is performed by a separate structure corresponding to that previously described.

In the windshield, in addition to the structure corresponding to Figure 5.29, a second structure with an approximately 40-cm-long wire at a distance of 5 cm from the frame is used. Both windshield antenna amplifiers are located near the dashboard: one on the left side, the other on the right side.

First, an analysis of the signal level of the respective antennas will be discussed, from which the obtainable signal-to-noise ratio improvement in weak signal areas can be predicted. The results displayed in Figure 5.42 are obtained from longer

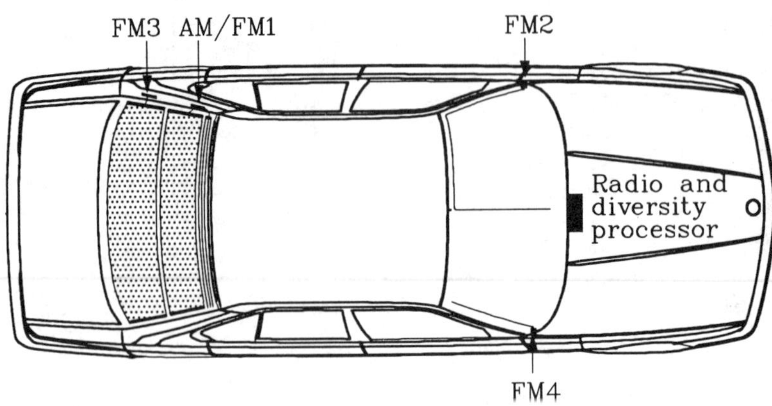

Figure 5.41 Scanning diversity system with a four-antenna arrangement.

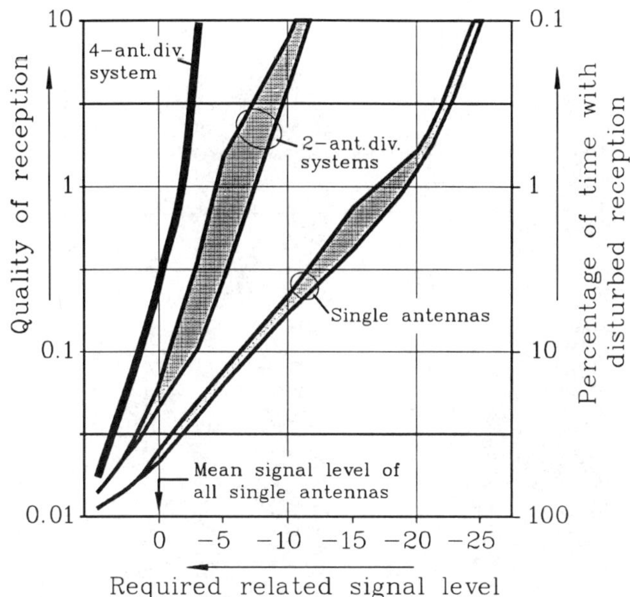

Figure 5.42 Reception quality versus required related signal level.

test drives on different frequencies. The quality of reception, which is defined as the inverse of the percentage of time with disturbed reception, is plotted versus the actual required signal level. If, for example, a signal level below the mean signal level of each of the antennas (0 dB) is assumed to be the threshold between undisturbed and disturbed reception (for example, by noise), a single-antenna system would be disturbed typically 41% to 46% of the time; the actual value within this range depends on the actual antenna. Considering all two-antenna diversity systems, which can be realized by any two of the four available antennas, it is found that the system would be disturbed 16% to 22% of the time, whereas the four-antenna diversity system would be disturbed only 3% of the time.

With a required signal level of −3 dB, the percentage of time there is disturbed reception is as follows: no distortion at all with the four-antenna system, 2.8% to 9% with any of the two-antenna systems, and 20% to 24% with any of the single antennas.

In the following, the improvement achieved by the antenna system of Figure 5.41 is discussed. At the received station, the mean output signal level was good and the recorded distortions displayed in Figures 5.43(a–d) result from multipath propagation with large τ_i and from adjacent-channel interference in a car driven in a defined narrow circle.

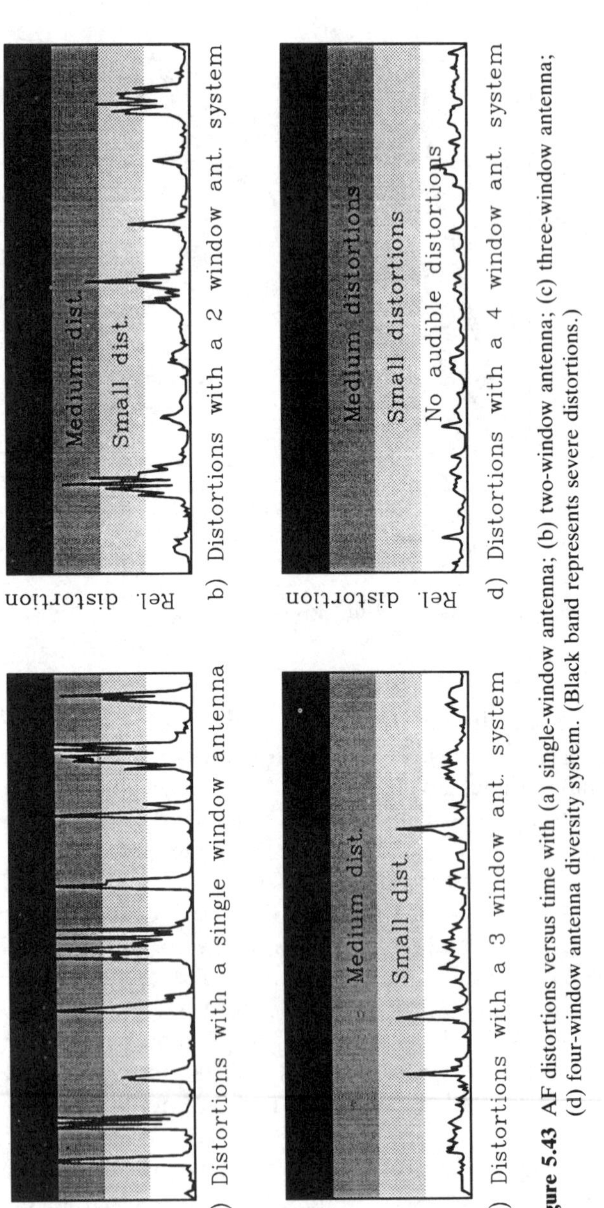

Figure 5.43 AF distortions versus time with (a) single-window antenna; (b) two-window antenna; (c) three-window antenna; (d) four-window antenna diversity system. (Black band represents severe distortions.)

In Figure 5.43(a), only one antenna was applied, and the curve displayed represents the distortions versus time. The scale on the vertical axis is related to the threshold of audibility of any distortion. Another region in this figure marks medium, and a further region indicates severe distortions. In Figure 5.43(a), several spikes exceed the threshold of audibility leading to the well-known short and hard distortions. If a second antenna is applied to the system (Figure 5.43(b)), the number of spikes is already significantly reduced. With the three-antenna diversity system (Figure 5.43(c)), only three short spikes exceed the threshold. In the recorded situation with four antennas applied to the system (Figure 5.43(d)), the distortions have been reduced to the extent that none of the distortion spikes exceeds the threshold of audibility. These measured curves are a good illustration of the contribution of any additional antenna to the reduction of distortion.

Four-Antenna System in the Rear Window

The four-antenna diversity system in Figure 5.41 consists of a large number of components distributed over the car and connected via complex wiring. For car production, the antenna assemblies would be subcontracted out to a component supplier, and more appropriate solutions better adapted to the manufacturing process of cars are required.

To that end, the antenna arrangement of Figure 5.44 was designed and optimized, because for the scanning diversity system, the technical effort on the diversity processor side is independent of the number of antennas applied. Consequently, the additional costs per antenna have to be kept small in order to allow the appli-

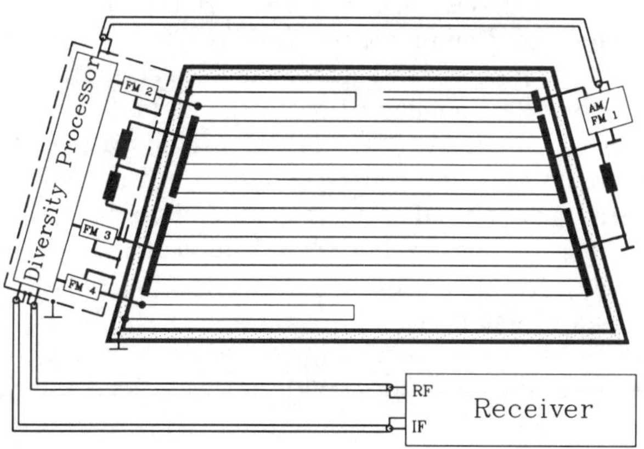

Figure 5.44 Four-antenna arrangement in the rear window.

cation of a large number of antennas. A cost-effective arrangement of antennas is to place them close together on a window, which results in minimal costs for housing and wiring. Under this assumption, Figures 5.45(a,b) give an impression of how costs and the reception quality will increase with the increasing number of antennas applied. As previously discussed in Figure 5.45(a), it is assumed that the distortion percentage for each of the antennas is 30%, corresponding to a signal quality of 10 dB. The respective values for the related time with audible distortions can be seen from Figure 5.45(a), too.

In contrast to this, the cost of additional technical effort increases considerably if any diversity system is to be used, since a distortion detector, additional wiring, and a second antenna are required. The additional technical effort for additional antennas, however, is small if an arrangement as displayed in Figure 5.44 (or similar) is used.

The antennas have different receiving signals due to different types of antenna structures and different RF loads at the heater bus bar opposite the respective amplifier. In the following, the receiving performance of a car equipped with four rod antennas, as shown in Figure 5.46 is investigated in comparison to the antenna arrangement in Figure 5.44 in the rear window. In Figure 5.47(a), the related distortion level of a single standard whip antenna is plotted versus time, and Figure 5.47(b) shows the distortion level of one of the active window antenna systems of Figure 5.44. Figure 5.47(c) presents the audio distortions with the diversity system using four whips, and Figure 5.47(d) display the curves with the four-window antenna system. The probability of a simultaneous occurrence of interference on all four closely packed screen antennas is not greater than that with the four whip antennas considerably spaced apart.

As a criterion for judging the capability of an antenna diversity system in mobile communication, the correlation factor of the time-dependent output signal levels of the various antennas is considered significant. This has caused the working group of the International Committee on Radio (CCIR) to recommend a distance of 2.8m between the antennas for FM antenna diversity [9]. The above documented results contrast with this recommendation. In [9], a time-invariant noise level is assumed and a distortion occurs if the noise threshold exceeds the instantaneous signal level. With mobile broadcasting reception, however, the interference due to time-variant intermodulation, adjacent-channel and cochannel interference, and multipath distortion are important. For the reduction of audio distortions by means of antenna diversity, the correlation factor between the actual distortions in the received signals is most significant.

In Figure 5.48, the correlation factor c_{ij} has been evaluated with respect to the signal levels and to the distortions. With the distortions, the correlation factor c_{ij} is

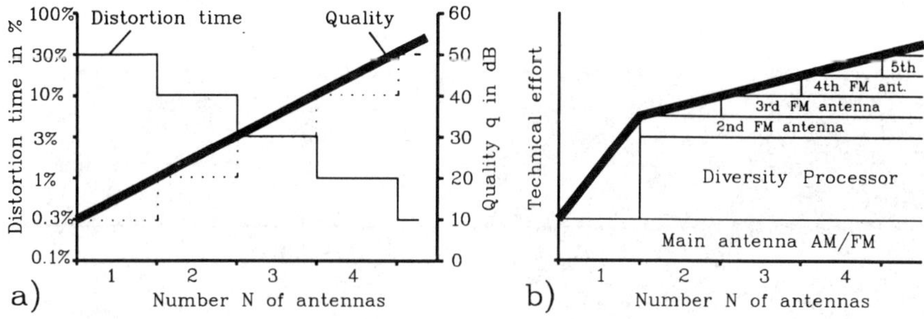

Figure 5.45 Improvement in signal quality and corresponding technical effort as function of antennas applied.

$$c_{ij} = \frac{\int_0^{t_0} [V_i(t) - \overline{V_i(t)}] \cdot [V_j(t) - \overline{V_j(t)}] \mathrm{d}t}{\sqrt{\int_0^{t_0} [V_i(t) - \overline{V_i(t)}]^2 \mathrm{d}t \cdot \int_0^{t_0} [V_j(t) - \overline{V_j(t)}]^2 \mathrm{d}t}} \tag{5.25}$$

where V_i is the distortion level of the ith and V_j of the jth antenna. $\overline{V}_i$ and $\overline{V}_j$ represent the mean values of the appertaining AF distortion levels during the drive time t_0. In Figure 5.48, c_{ij} is plotted for measured values of V_i and V_j. This correlation factor of the distortion level is represented by the dark columns. The other columns show the correlation factor of the appertaining signal level for comparison, which, in FM broadcasting, is regularly much greater than the correlation factor of the distortion levels. Therefore, in broadcast reception, the correlation factor of signal levels is not a valid criterion for judging the capability of an antenna arrangement.

Three-Antenna System in the Rear Window

Having successfully passed extensive tests comparable to those performed with the single-window antennas, antenna diversity systems were introduced in series production in the new BMW car types 5 and 7 in Europe in autumn of 1989 and in autumn 1990 in the United States. The designs of the antenna structures for Europe and the United States are not identical. Here, the United States version will be described.

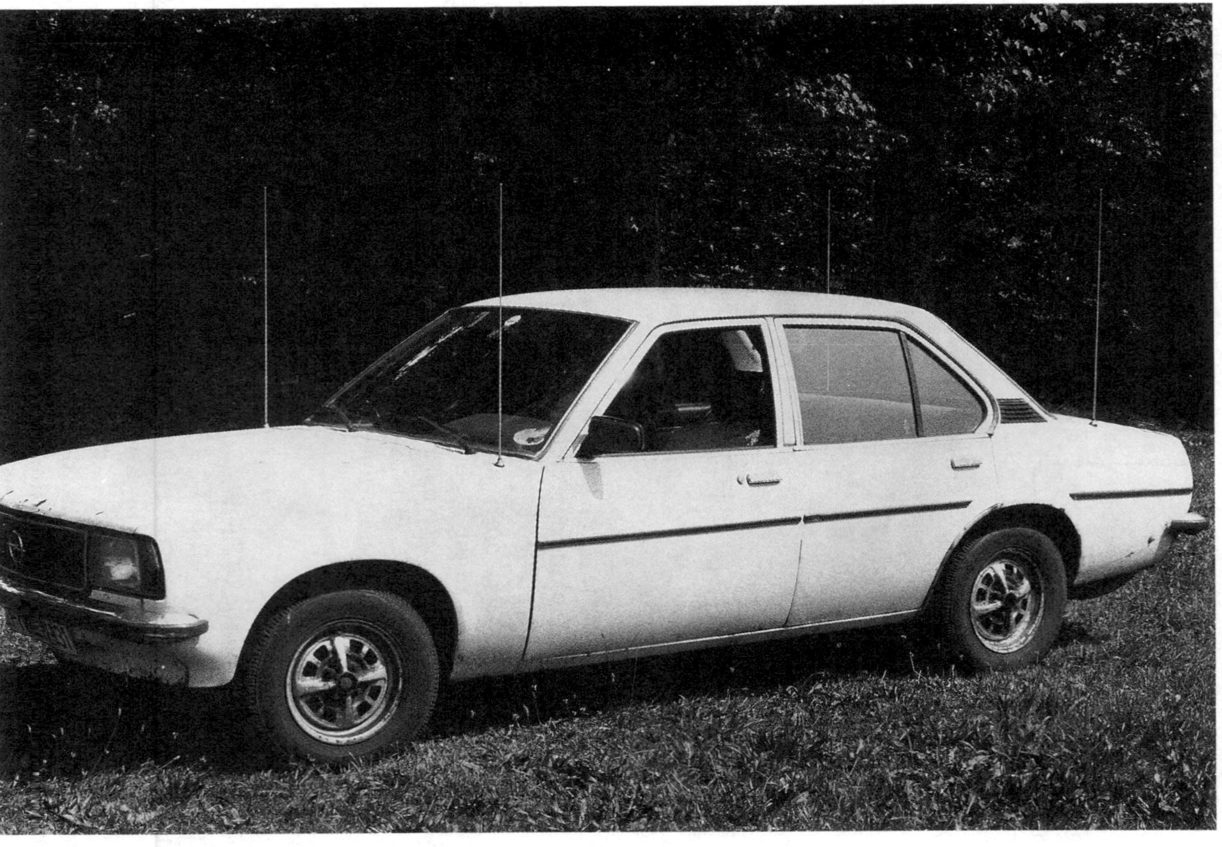

Figure 5.46 Four rod antennas on a car.

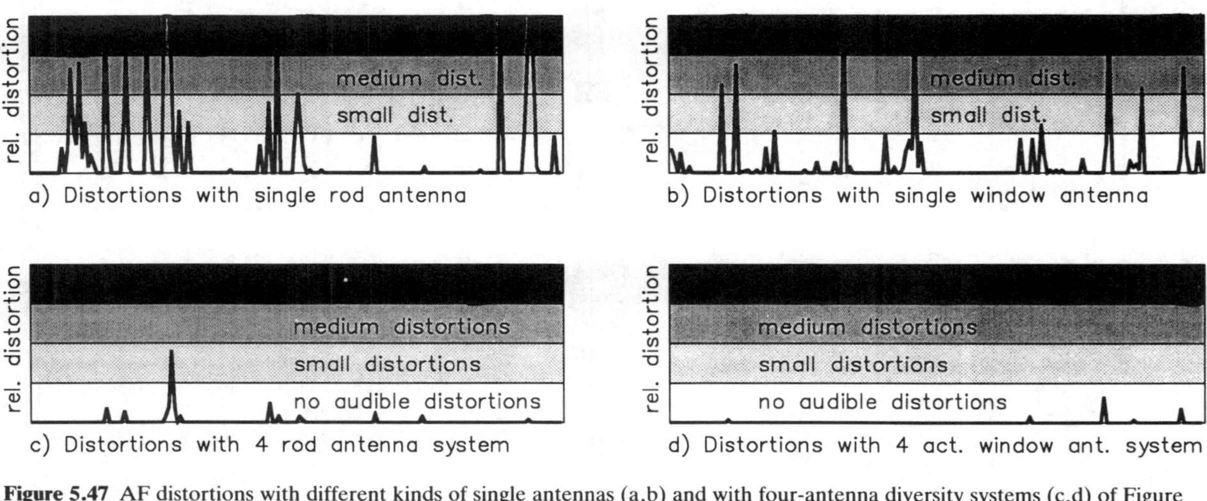

Figure 5.47 AF distortions with different kinds of single antennas (a,b) and with four-antenna diversity systems (c,d) of Figure 5.44 (d) and Figure 5.46 (c). (Black band represents severe distortions.)

The structures printed on the glass are the same as those of the single antenna of Figure 5.31, so an easy upgrading from a single-window antenna to the three-antenna diversity system is possible by an exchange of the electronic components right and left of the window and by adding two further coaxial lines as displayed in Figure 5.49. The two additional FM antennas are provided by the AM antenna structure and the lower part of the heater element. The FM1 amplifier, the AM/FM2 amplifier, the feeding and decoupling elements for the dc current-FM signal separation with respect to the heaters, the diversity processor, and the pin-diode switches are arranged in one housing (size 15 × 4.5 × 1.2 cm) mounted on the left C-post of the car. The FM3 amplifier and the respective feeding elements are arranged in a second housing mounted on the right C-post of the car. Additional

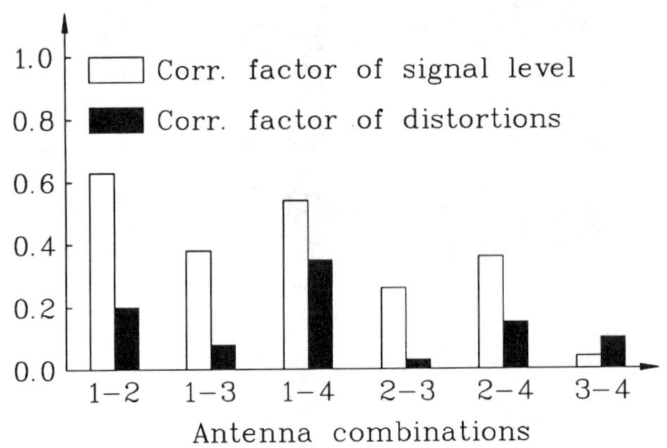

Figure 5.48 Correlation factor of signal level and corresponding AF distortions with various antenna combinations.

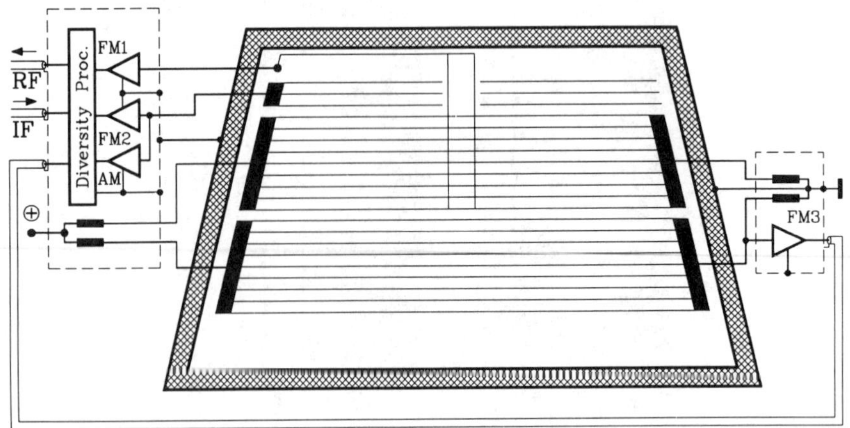

Figure 5.49 Three-antenna diversity system now in series production.

coaxial cables are required as the connection from the FM3 amplifier to the diversity processor and from the radio (IF signal) to the diversity processor. Thus, with a small technical effort, maximum performance is obtained.

Naturally, the performance of an optimized three-antenna diversity system is inferior to a four-antenna diversity system. However, in comparison to the performance of a standard system with only one antenna (Figure 5.50(a)), the reduction of distortions on the same drive is impressive (Figure 5.50(b)). The mean signal level value for Figures 5.50(a,b) is about 45 dBμV. In situations with strong multipath reception and with adjacent-channel or intermodulation interference, the improvement in signal quality is very high.

5.2 ANTENNA SYSTEMS FOR TV RECEPTION IN CARS

A printed antenna system in conjunction with a monopole mounted on the rear fender has been used for AM, FM, and TV reception where the diversity reception

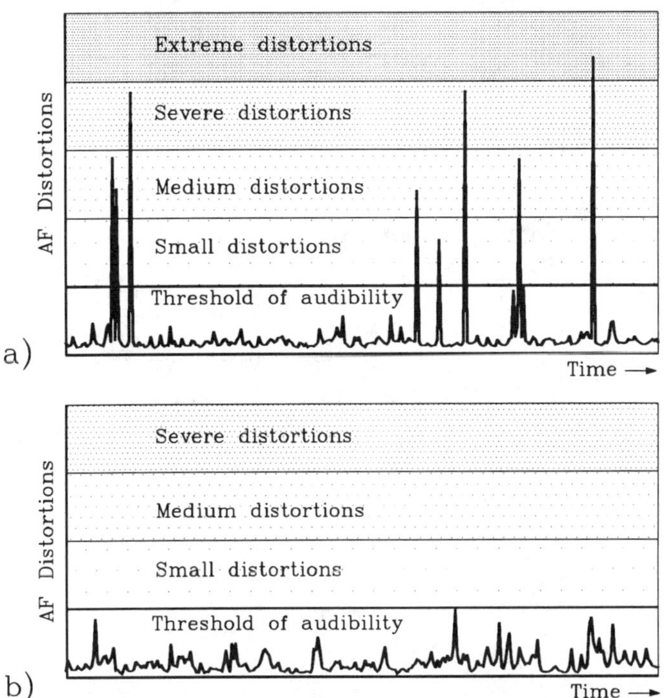

Figure 5.50 AF distortions with a single antenna (a) and with three-antenna diversity system of Figure 5.49 (b).

for FM and TV broadcasts is performed by combining the monopole and the printed elements. An example of an antenna system composed of four printed elements and a monopole is shown in Figure 5.51, where the table shows the function of each element. Space diversity is achieved by combining both a monopole element and a printed element for FM broadcast reception and by combining a monopole element and printed elements for TV broadcast reception in the VHF and UHF bands. Glass is not perfectly transparent to high-frequency radio waves. An analysis has shown [10] that the transmissivity through a glass sheet in the VHF and UHF

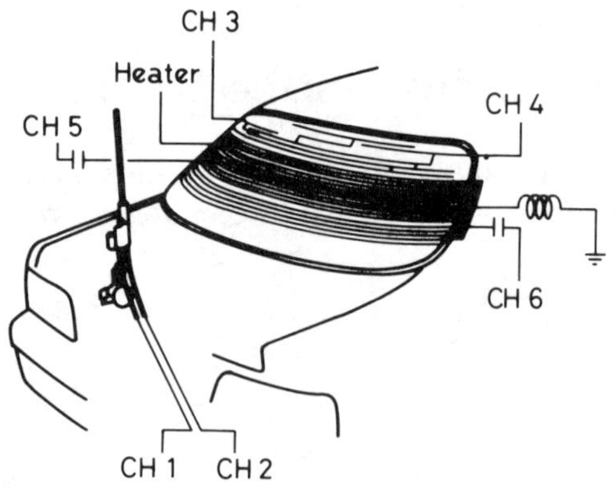

Antenna function

Ant CH	Radio		TV		
	AM	FM	VHF 1 ~ 3 ch	VHF 4 ~12ch	UHF 13~62ch
1	◎	◎			
2			◎	○	○
3			○	○	◎
4		○			
5			○	◎	○
6			○	○	○

◎ Main
○ Sub

Figure 5.51 A monopole (motor driven) and four printed antennas on the rear window.

frequency region is 95% or greater when the thickness d of the glass is less than 6 mm. Figure 5.52 shows the transmissivity versus frequency, where the thickness d of the glass is taken as the parameter and the dielectric constant ϵ_r of the glass is assumed to be 6.5. Since the thickness of glass currently used in ordinary automobiles is 3.5 mm, glass transmissivity is not a problem.

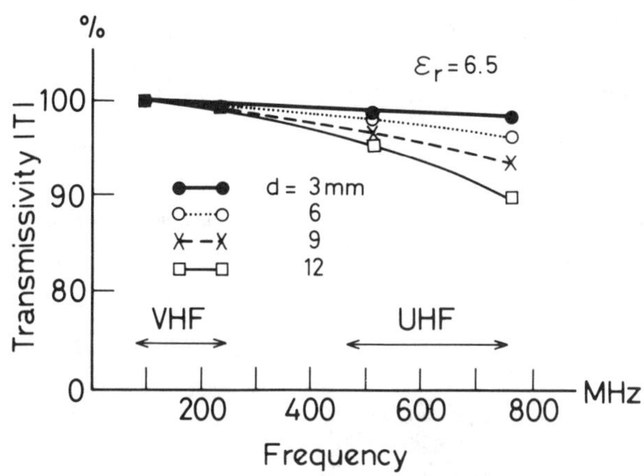

Figure 5.52 Glass transmissivity against frequency. (From [10], © IEEE.)

5.2.1 Antennas Printed on the Rear Quarter Windows

The antenna structure is shown in Figure 5.53, along with the amplifier and the TV tuner [10]. The antenna system consists of four elements. Output of these antenna elements are compared, and the one that provides the highest quality is selected by the switching circuit. Figure 5.54 is a simple block diagram of the receiver. The switching is performed during each blanking period of the TV video signals to prevent the picture from fluttering.

The antenna patterns printed on the rear quarter window are illustrated in Figure 5.55. Antenna 1 is composed of two elements: a slanted element consisting of three thin lines, to which a trap is attached to maximize sensitivity, and a horizontal element, to which a slanted element is connected. These elements, connected in parallel at the antenna output terminals, are used for the VHF band. Another, shorter horizontal element is attached to Antenna 1 for the UHF band. Antenna 2 consists of a horizontal element with a trap and a slanted element.

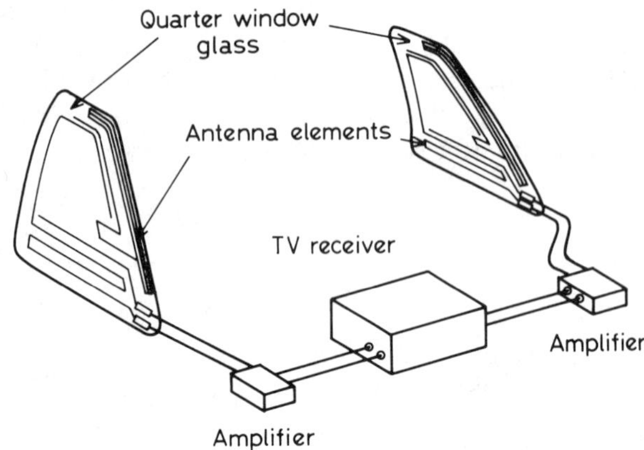

Figure 5.53 Antenna structure printed on the quarter-window glass, antenna amplifier, and TV receiver. (From [10], © IEEE.)

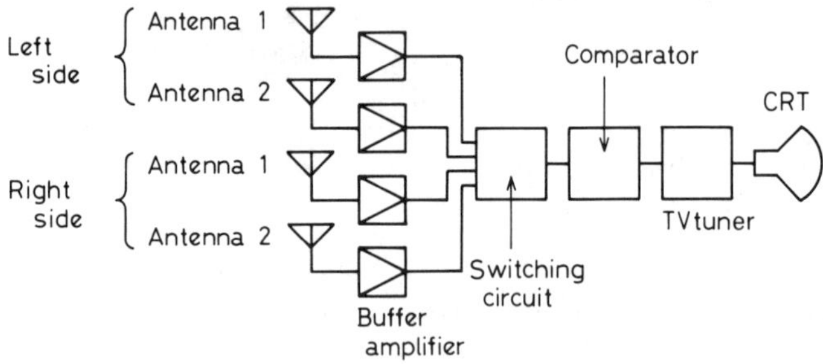

Figure 5.54 Antenna circuit and TV receiver to perform diversity reception. (From [10], © IEEE.)

Antennas 1 and 2 are capacitively coupled to make the sensitivity difference in the two antennas as small as possible. The antenna elements are printed symmetrically with the same patterns on the right and left quarter windows. These antenna patterns are based on extensive analyses performed to determine the specified sensitivity. The measured antenna sensitivity is shown in Figure 5.56, where a comparison is made with the sensitivity of a conventional monopole antenna. The antenna sensitivity is defined as the induced voltage at the antenna's open terminals under a field strength of 60 dBμ (0 dBμ = 1 μV/m).

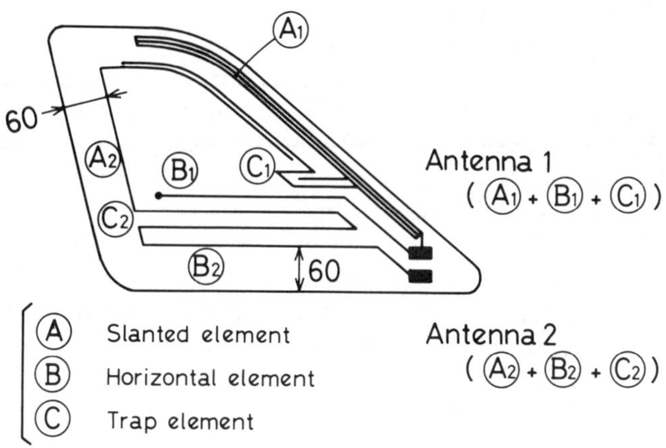

Figure 5.55 Antenna patterns printed on the quarter-window glass. (From [10], © IEEE.)

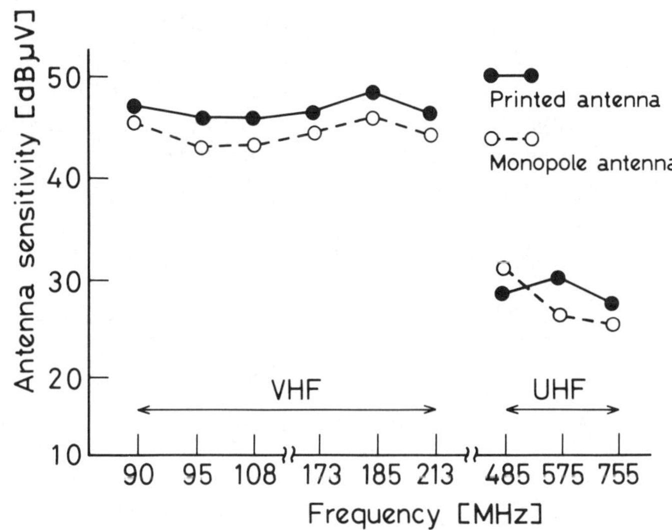

Figure 5.56 Antenna sensitivity for horizontal polarization. (From [10], © IEEE.)

The directional characteristics of the antenna are shown in Figure 5.57. To determine the antenna sensitivity, the averaged value of the receiving pattern over the entire horizontal plane is taken.

For TV reception in a car, antenna sensitivity is specified to be greater than 45 dBμ in the VHF ranges (90 to 108 MHz and 170 to 220 MHz) and greater than 30 dBμ in the UHF bands (470 to 766 MHz). A conventional monopole antenna has an antenna sensitivity of 43 to 45 dBμ in VHF bands and 25 to 30 dBμ in UHF bands. The VSWR characteristics are shown in Figure 5.58. The quality of TV reception in the field is evaluated by using a five-rank rating, the results of which are shown in Table 5.1 [10], where the evaluations of both sound and video quality are provided. The rank represents the quality with interference: rank 5 corresponds to "no interference at all," rank 4 to "slight interference but not distractive," rank 3 to "interference being distractive but not obstructive," rank 2 to "interference being heavily distractive and obstructive," and rank 1 to "unable to receive."

5.2.2 A Three-Element Antenna System for AM, FM, and TV Reception

An example of an antenna system for AM, FM, and TV reception will be described [11]. All the technical data are provided courtesy of the Nissan Motor Co., Japan. The antenna system is composed of three elements: a monopole element and a printed element on the rear window are used for AM and FM reception, and printed elements on both rear and quarter windows are used for TV reception. Diversity reception for FM broadcasts is performed by the monopole and printed elements on the rear window, and TV diversity reception is achieved by the two printed elements. The antenna pattern printed on the rear window glass is shown in Figure 5.59, where a scale shows the actual length.

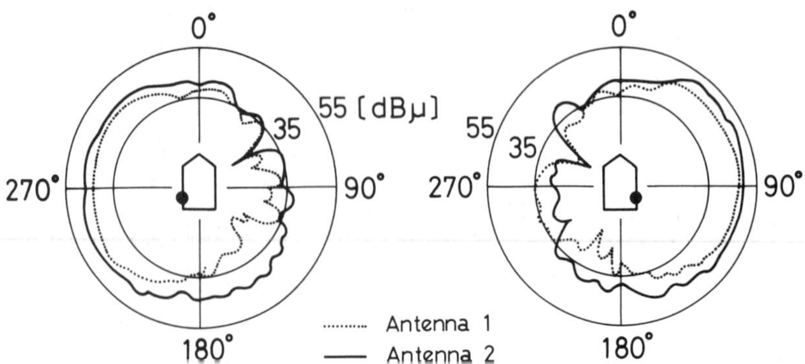

Figure 5.57 Antenna sensitivity patterns. (From [10], © IEEE.)

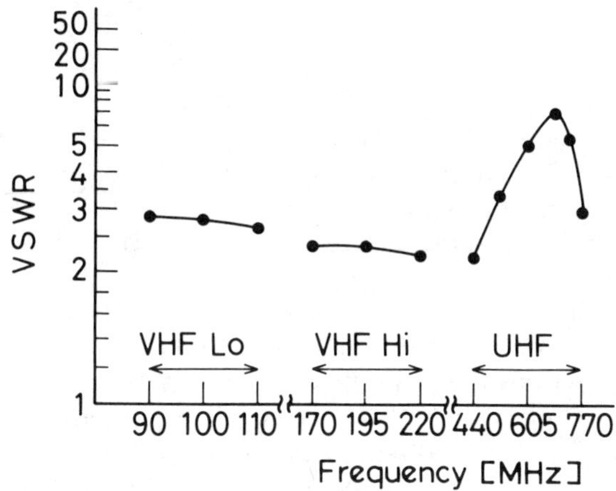

Figure 5.58 VSWR characteristics. (From [10], © IEEE.)

Table 5.1
Five-Rank Evaluation (Average) of TV Broadcast Reception by the Five-Rank Method

	Audio		*Video*	
	Printed Wire	*Monopole*	*Printed Wire*	*Monopole*
VHF	4.4	3.5	3.8	3.5
UHF	4.4	3.7	3.5	3.5

Note: Evaluation for the audio quality was performed during driving, while that for the video quality was performed when the car made a stop.
Source: After [10].

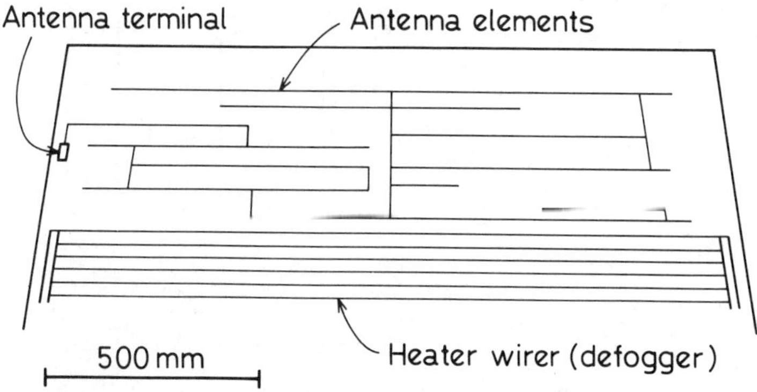

Figure 5.59 Antenna pattern printed on the rear-window glass for FM/AM and TV reception.

AM Reception

The sensitivity of the receiver is defined to be the minimum input field strength at which the specified output *S/N* is obtained. The *S/N* performance at 525 kHz is shown in Figure 5.60, where variations of the output signal power *S* and noise power *N* are given against the input signal field strength. The signal level gradually varies under the effect of automatic gain control (AGC). When the engine is operating, the output noise power increases due to the addition of ignition noise as shown by the variation of N_{ign}, the dotted line in the figure. The input to the receiver is an AM wave with a 30% tone modulation of 1 kHz, and the received power is measured at the speaker terminals of 4Ω (0 dB in this figure means the output power is 0.5W at 1 kHz). The sensitivity varies depending on the type of receiver, but typical values specified for ordinary receivers are 43 to 48 dBμ to achieve the S/N_{ign} of 12 dB. The *S/N* performance is measured in a shielded room with a nearly homogeneous electromagnetic field, and with a car placed in the middle of the room.

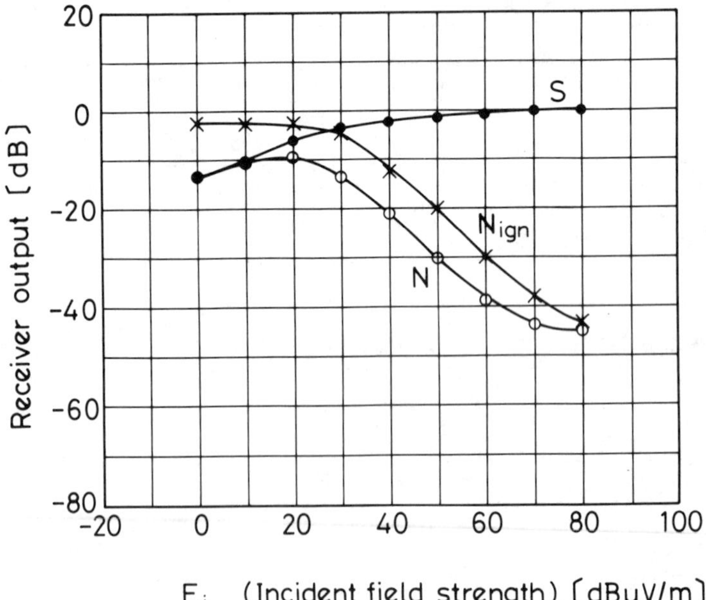

Figure 5.60 S/N characteristics in the AM reception.

FM Reception

The gain versus frequency is shown in Figure 5.61, where dark lines express mean values and dotted lines the standard deviation with respect to the mean values, which were taken from more than 70 measured data. The gain is defined as the ratio of the received power of an antenna mounted on a car to that of a standard dipole antenna in free space. Since the antenna system has directional characteristics, the receiver output varies as the direction of the car changes. Hence, in order to determine the gain, the averaged receiver output over the entire horizontal plane, taken from the receiving pattern, is compared to the received level of the standard dipole antenna (0 dB in the receiving pattern). The measured receiving patterns at four frequencies are shown in Figure 5.62. The receiving pattern is measured in an anechoic chamber containing a car, and the output (received) power of the antenna mounted on the car is measured. The impedance characteristics of the antenna are shown in Figure 5.63. The gain and impedance characteristics of the higher VHF bands (90 to 110 MHz and 170 to 220 MHz) are shown in Figures 5.64 and 5.65, respectively.

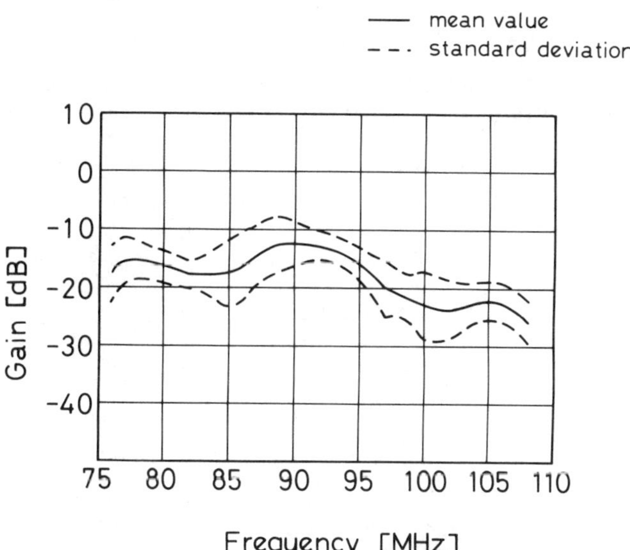

Figure 5.61 Gain characteristics in the frequency range of 76 to 108 MHz (horizontal polarization).

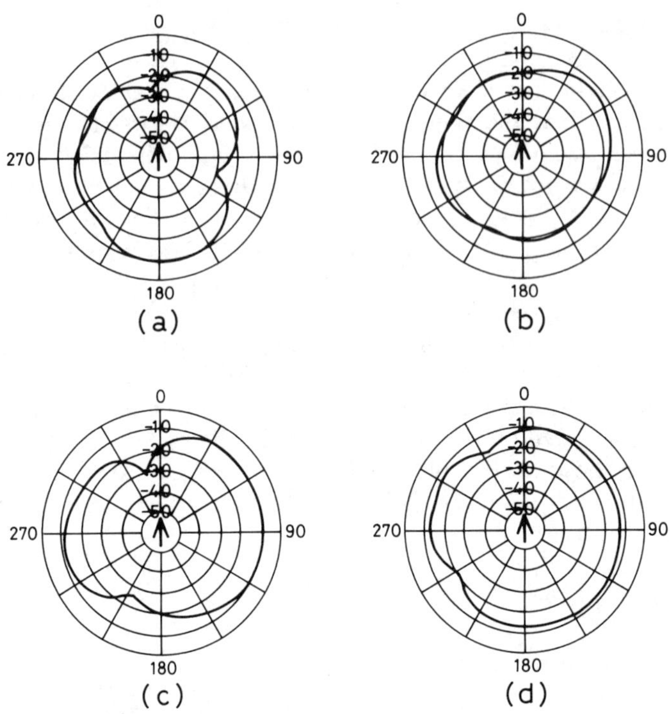

Figure 5.62 Receiving patterns in VHF (FM) bands: (a) 76 MHz (−17.7 dB); (b) 82 MHz (−18, 1 dB); (c) 87 MHz (−15 dB); (d) 93 MHz (−13.6 dB). Number in parentheses is the gain.

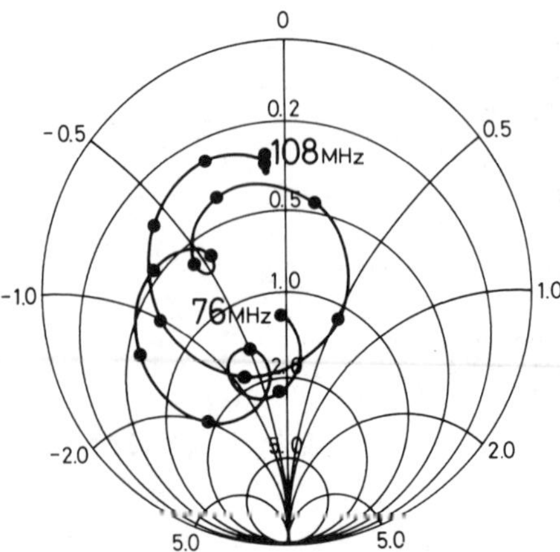

Figure 5.63 Impedance characteristics in the frequency range of 76 to 108 MHz (2-MHz intervals).

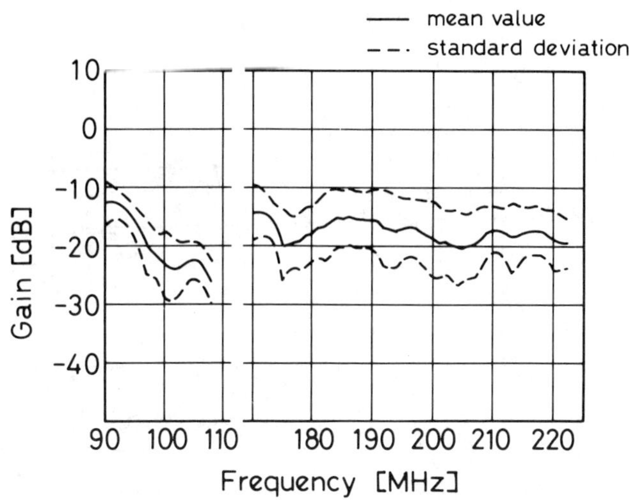

igure 5.64 Gain characteristics in VHF (TV) bands (90 to 110 MHz and 170 to 222 MHz).

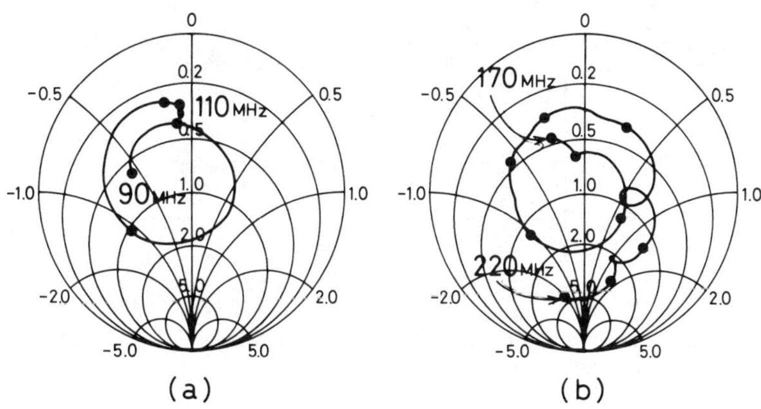

igure 5.65 Impedance characteristics of VHF (FM) bands: (a) low channel, 90 to 110 MHz; high channel, 170 to 220 MHz.

'V Reception

'he gain, receiving patterns, and impedance characteristics in the UHF bands (470) 770 MHz) are shown in Figures 5.66, 5.67, and 5.68, respectively.

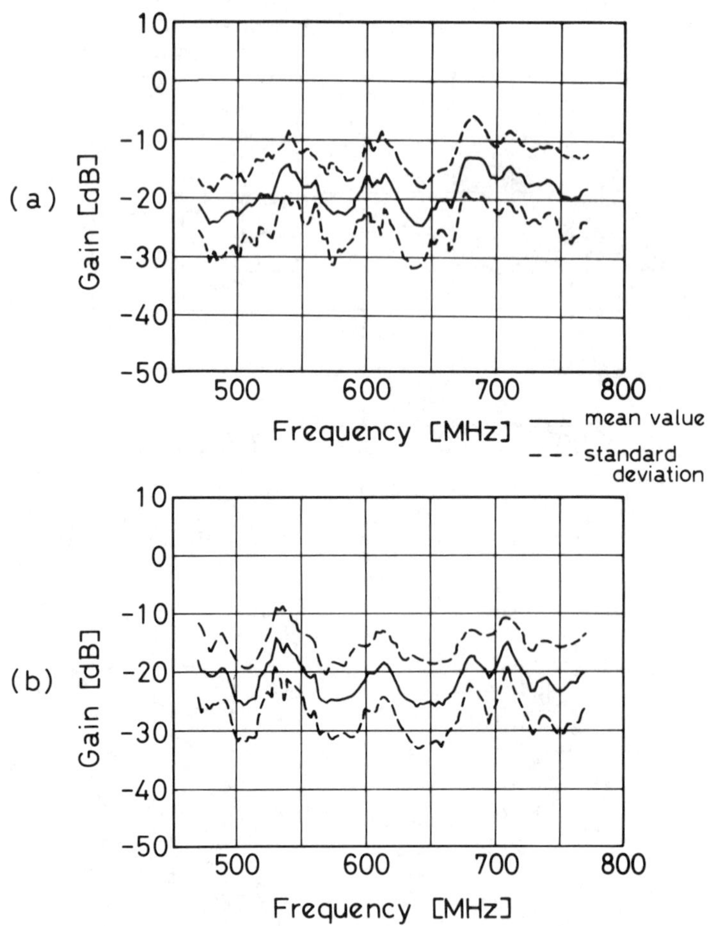

Figure 5.66 Gain characteristics in UHF (TV) bands (470 to 770 MHz): (a) horizontal polarizatio (b) vertical polarization.

Diversity Reception

FM receiver performances of a monopole element (antenna 1) and a printed ele ment (antenna 2) on the rear window are shown in Figure 5.69. They illustra receiver outputs which fluctuate heavily due to the effect of multipath propagatio The dark line shows the received level of antenna 1 and the dotted line that antenna 2. The selective diversity scheme has been used in which the antenn branch with the higher output signal is selected by means of switching. Figure 5. shows the cumulative probability distribution of the received power against th

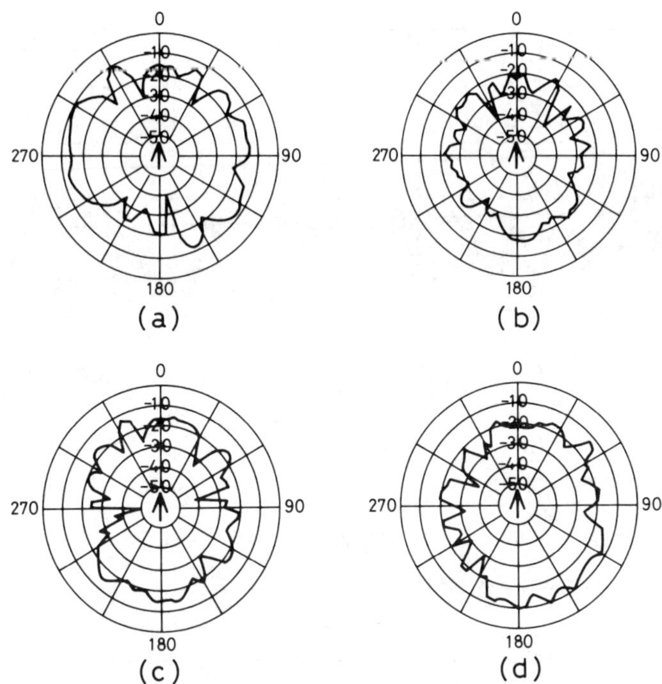

Figure 5.67 Receiving patterns in UHF (TV) bands: (a) 470 MHz (−18.2 dB); (b) 569 MHz (−25.5 dB); (c) 671 MHz (−21.7 dB); and (d) 770 MHz (−20.1 dB). Number in parentheses is the gain.

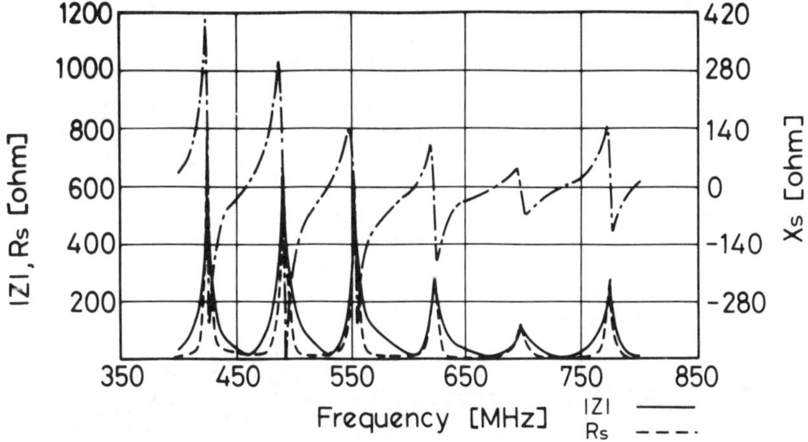

Figure 5.68 Impedance characteristics in UHF (TV) bands.

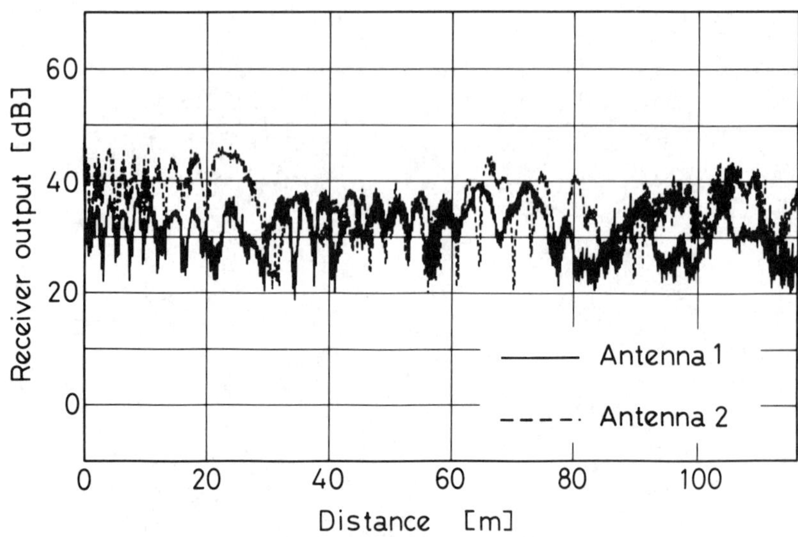

Figure 5.69 Receiver outputs at each of two elements.

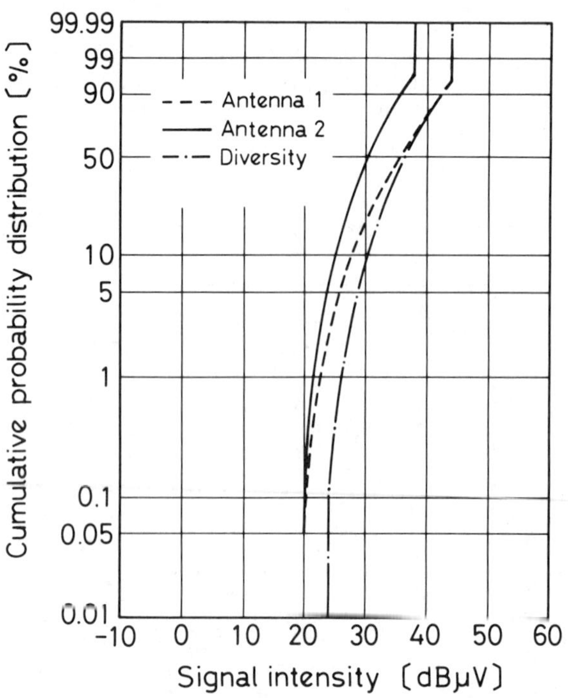

Figure 5.70 Diversity performance.

input signal strength for three cases in which a monopole (dark line), printed elements (dotted line), and diversity antenna (chained line) are employed. The effect of diversity can be clearly observed in the figure.

5.3 ANTENNA SYSTEM FOR SHINKANSEN (NEW BULLET TRAIN)

5.3.1 Introduction

Leaky coaxial cables are used for the train communication systems of Tohoku (Tokyo-Morioka (535 km)) and Joetsu (Tokyo-Niigata (334 km)) Shinkansen (Japan Railways (JR) new bullet trains) [12–16]. They are coaxial cables whose outer conductors have slots (an example is shown in Figure 5.71) radiating a part of the transmitted energy [17–20]. LCX cables were first developed for train communication systems in a tunnel for Tokaido (Tokyo-Osaka (553 km)) and Sanyo (Osaka-Fukuoka (624 km)) Shinkansen. The advantages of LCX cable systems compared to radio communication systems were then realized. They offer more stable and better train communications compared to radio communication systems because LCX cables radiate weak electromagnetic energy and the environmental effects on their radiation characteristics are small [21,22]. One LCX cable is installed their radiation characteristics are small [21,22]. One LCX cable is installed along each railway; thus, there are two LCX cables along the railway for inbound and outbound train communications, as shown in Figure 5.72. Trains have slot antennas that transmit and receive electromagnetic energy to and from the LCX cables.

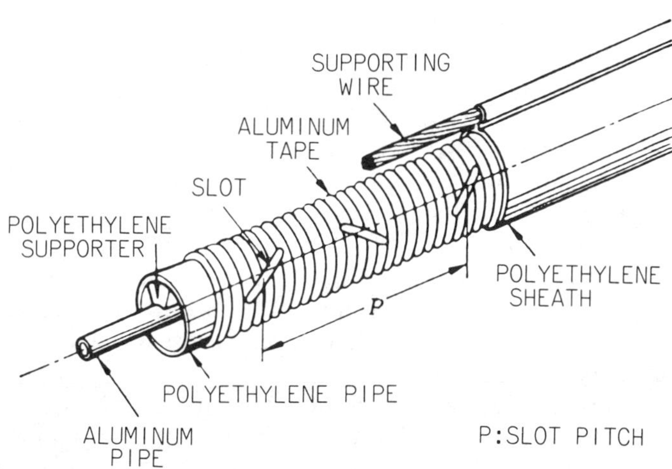

Figure 5.71 LCX cable structure. (From [13], © *IECEJ*.)

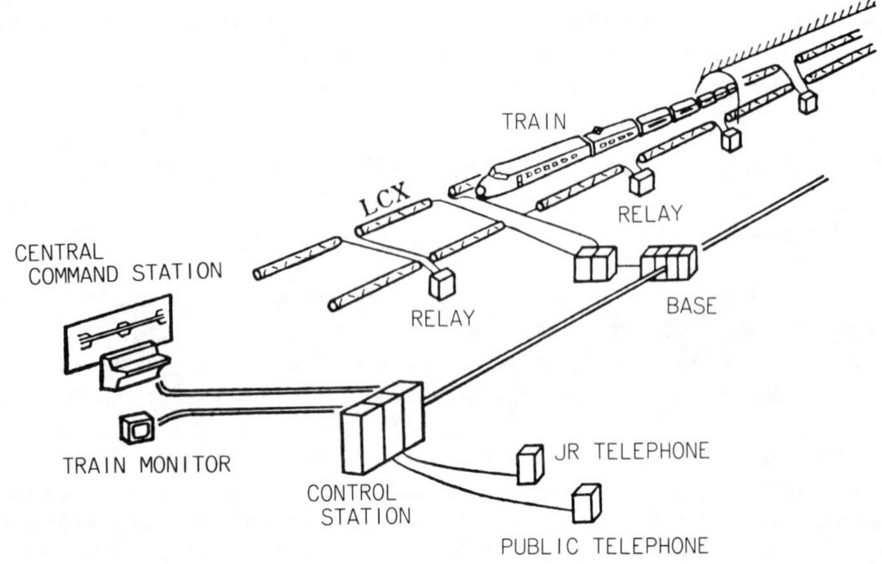

Figure 5.72 Outline of an LCX cable communication system. (From [13], © *IECEJ*.)

5.3.2 Train Radio Communication Systems

In the LCX system there are 24 radio communication channels. They are divided into four systems: train operation commands (4 channels), commands related to passengers (2 channels), business and public telephones (6 channels), and data communications (3 channels). The rest (9 channels) are for spare channels. Train operation commands are sent through direct channels between the central command station in Tokyo and motor men. Business commands are also sent through direct channels between the central command station and train conductors. Two public telephones in a train are connected to NTT public telephones all over Japan through JR's telephone exchange networks. Data channels are frequency-shift keying (FSK) 1,200 bps and are used to monitor train running conditions in order to deal with an emergency. Each train can always use six channels to cover the above four communication systems. The bandwidths are 900 and 700 kHz from base station to trains and from trains to base station, respectively. Each channel occupies 25 kHz. The advantages of the LCX system are stable channel quality, efficient channel utilization, and small environmental effects. The last two advantages are, as mentioned, due to weak radiation from the LCX slots.

Each base station is located at each train station and one base station covers one service area (average 20 km). The LCX relay system is shown in Figure 5.73. When there is a communication problem in one LCX route, a route is changed to

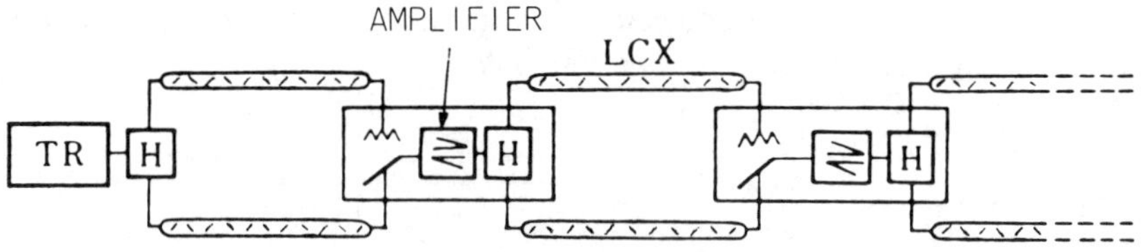

Figure 5.73 LCX relay system. (From [13], © *IECEJ*.)

the other LCX at the relay station to prevent the trouble from spreading to other places. There is a maximum of 20 relay stations between the two train stations and four kinds of LCX cables with different coupling loss; #488 (loss = 75 dB), #487 (loss = 65 dB), #486 (loss = 55 dB), and #485 (loss = 50 dB), between a train antenna and an LCX cable. They are connected properly to reduce the received signal level change (less than 10 dB with respect to the center level) between the two relay stations, as shown in Figure 5.74. The maximum distance between the

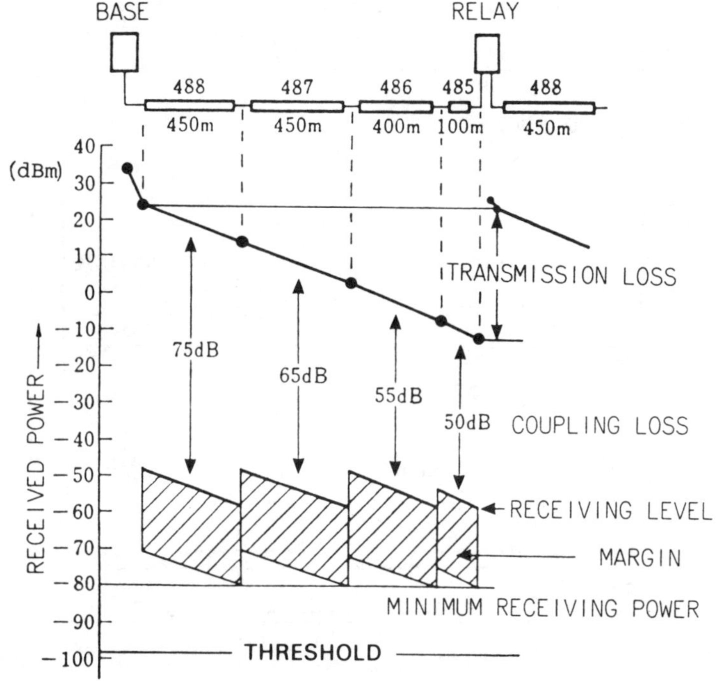

Figure 5.74 LCX cable combination for signal attenuation grading. (From [14], © *IECEJ*.)

two relay stations is 1.5 km. Each relay station has a 400-MHz bilateral amplifie[r] with a 42-dB gain, and the output is about 1W.

5.3.3 Antenna System

The slot array shown in Figure 5.75 consists of an array of three slots with differen[t] inclination angles. The separation of two sets of three slots is *P*. They are designe[d] to work in a 400-MHz band, and in future operation they will work in 800-MH[z] bands [23]. LCX cables and train antennas are located as shown in Figure 5.76[.] There are two LCX cables along the railway and an array of four folded slo[t] antennas on each side of the train. The LCX cables at the left side of the trai[n]

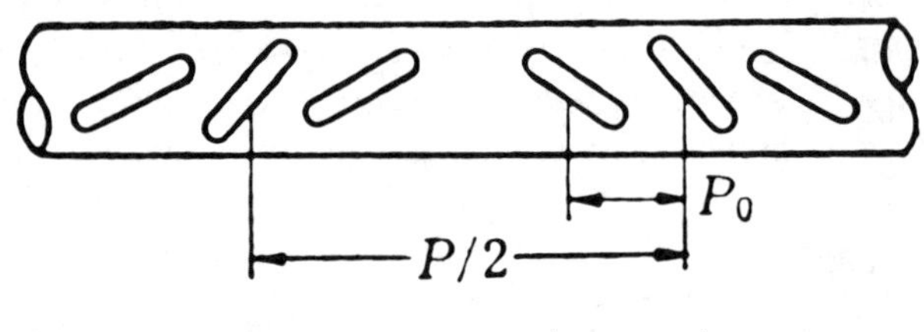

P:52~53cm P_0:P/8 or P/6

Figure 5.75 LCX slot configuration. (From [14], © *IECEJ*.)

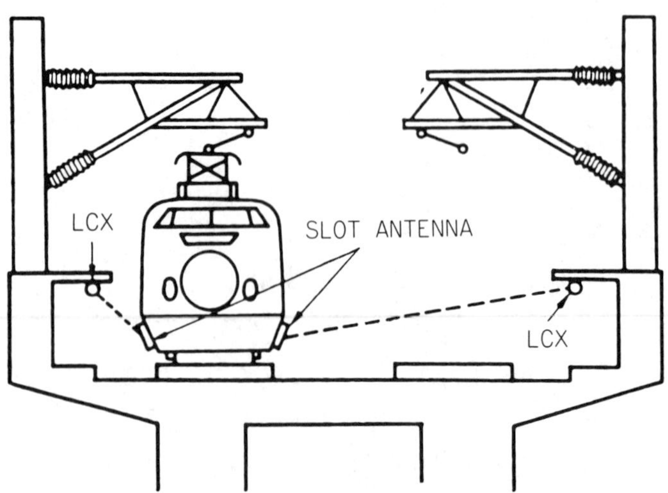

Figure 5.76 LCX cable and train antenna location. (From [14], © *IECEJ*.)

toward the direction of travel (trains keep to the left in Japan) is usually used for communication. When there is a fault, the communication channels are changed to the right-side cable.

The vertical radiation pattern of LCX cables has a broad half-power beamwidth, and a measured vertical radiation pattern is shown in Figure 5.77. The maximum radiation of LCX cables in the horizontal direction occurs slightly backward (i.e., toward a transmitter direction), and the peak radiations of the LCX cables for inbound and outbound trains are slightly different. Thus, the radiation pattern of an antenna has two peaks corresponding to those of the LCX cables, as shown in Figure 5.78(a). On the other hand, vertical radiation patterns of a train antenna must have a broad half-power beamwidth (Figure 5.78(b)) coping with LCX cable locations having different heights and the inclination of the train on a curve. The half-power beamwidth of the vertical plane of the train antenna is about 110 deg. The gain is about 5 dB with respect to a half-wave dipole antenna.

Antennas on the train are as follows. One folded slot antenna is shown in Figure 5.79 and the four equiphase folded slot array assembly shown in Figure 5.80 is placed on each side of the train, as in Figure 5.76. The four elements are equally fed by coaxial cables of the same length. The spacing between the adjacent elements is 50 cm. A matching circuit is needed for different frequencies: 412 MHz (transmitting) and 452 MHz (receiving).

There are two four-element folded slot arrays on each side of a front car, as shown in Figure 5.81. A train transmitter will be connected to arrays on the side

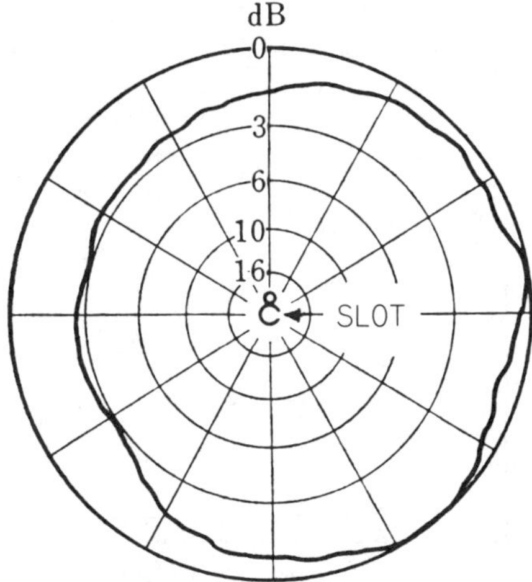

Figure 5.77 Vertical radiation pattern of an LCX cable. (From [13], © *IECEJ*.)

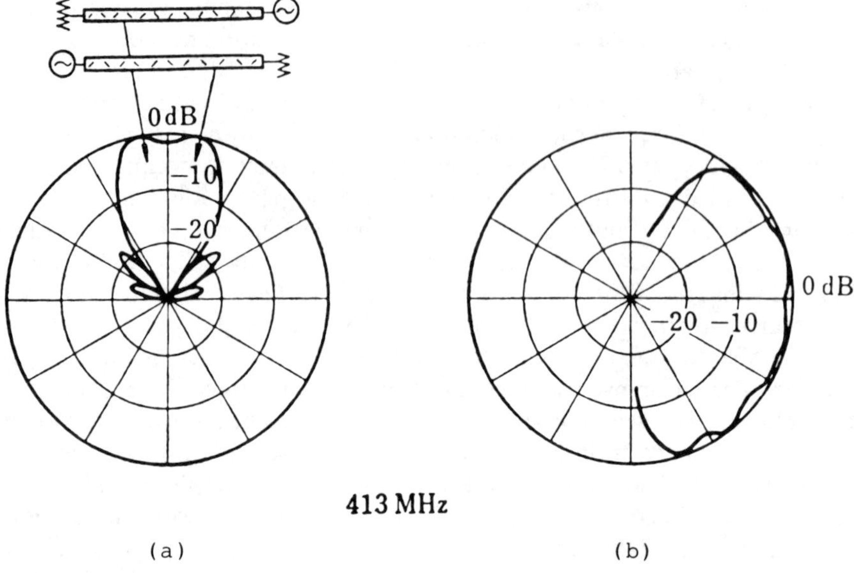

Figure 5.78 Train antenna radiation pattern: (a) horizontal pattern; (b) vertical pattern. (From [13], © *IECEJ*.)

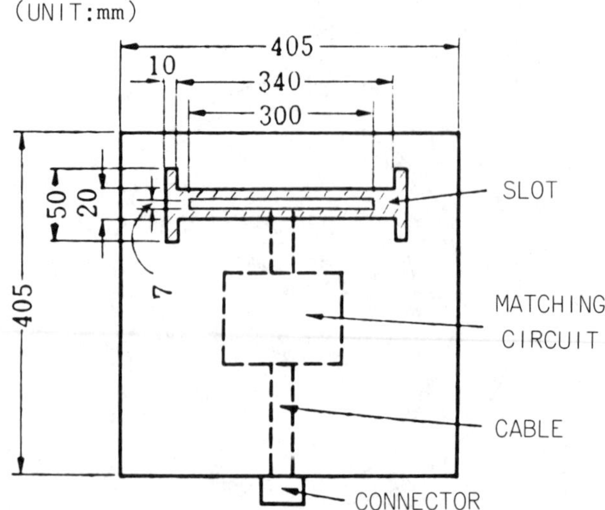

Figure 5.79 Radiating element structure of a train antenna. (From [14], © *IECEJ*.)

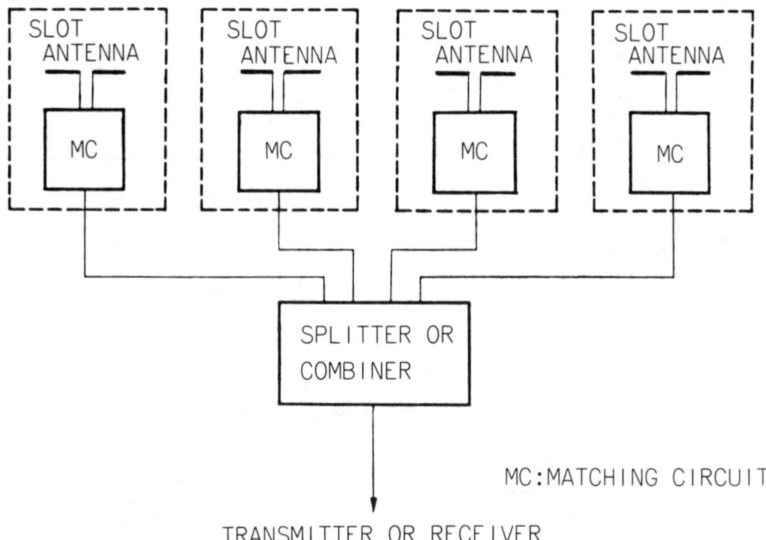

Figure 5.80 Train antenna configuration. (From [14], © *IECEJ*.)

where the received power level is higher. Thus, communication can be maintained when one of the LCX cables has a communication problem. The power level of an auxiliary array on each side is 20 dB lower than that of the main array. The auxiliary array is combined with the output of the main array to prevent the power level from abruptly dropping due to environmental effects, such as steel poles along the railways.

5.4 ANTENNA SYSTEMS FOR CITY BUS OPERATION

5.4.1 Introduction

In the past, it has been generally understood that ferrite antennas can generally be used for receiving purposes only. This section introduces an application of ferrite antennas to land mobile communication systems using an inductive communication system [24] and employing ferrite antennas for transmission as well as reception.

The principle of inductive radio communication is information transmission by means of electromagnetic near-field coupling between transmitter and receiver antennas in low-frequency (typically low- and middle-frequency) bands. The inductive radio communication system can be distinguished from the ordinary VHF/UHF radio communication system, since the communication zone is confined to

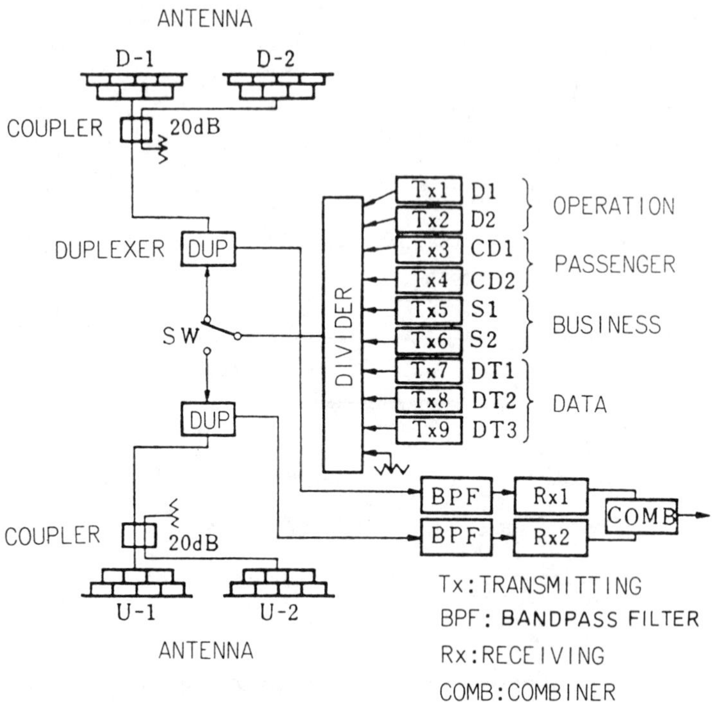

Figure 5.81 Communication system structure on a train. (From [14], © *IECEJ*.)

rather limited areas. This is due to the nature of near-field electromagnetic fields, which typically decay in proportion to the inverse second to third power of the distance. The communication zone is usually formed in an area of several tens of square meters around the transmitter antenna. This feature is advantageous for the formation of a spot communication zone, which can be used for locating a vehicle and providing information to individual vehicles. This can be applied to various vehicle communication applications, such as navigation and traffic information systems.

In practice, inductive radio communication can be used for signpost systems in which small base stations located on the roadside produce narrow communication zones and communicate with vehicle stations during their passage through the zones. The base station is called either a *roadside station* or a *signpost station*. Figure 5.82 shows an image of the system configuration. Signals received by signpost stations are sent to the central station by wire lines. Since the communication zone is very narrow, voice transmission is not possible and only digital data are exchanged

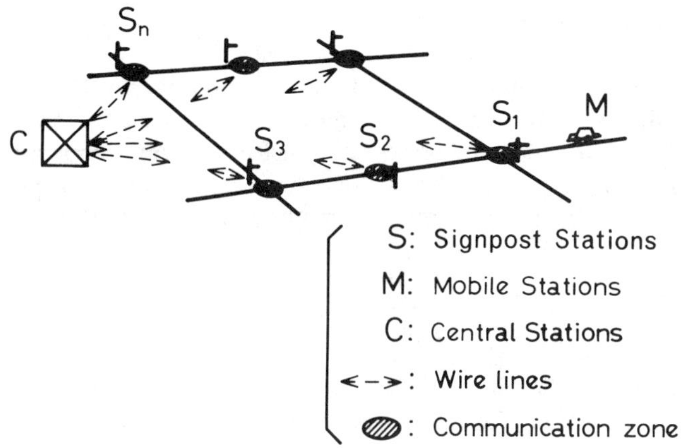

Figure 5.82 Signpost communication system.

between the signpost and vehicle stations. Data gathered at the central station are processed by a computer and used for system operations. In turn, information from the central station is transferred to vehicles via the signpost stations. Because of the narrow communication zone, information exchange is normally performed between only one vehicle station and one signpost station. It is also significant to note that the rapid decay in field strength dramatically affects the system performance; that is, both the reception of noise and interference and the causation of interference in other systems can be reduced remarkably. This has been proven in practice by experiments performed in the middle of the greater Tokyo metropolitan area, where a bit error rate as low as 10^{-8} was obtained in 200-bit data transmissions at a rate of 4.8 kbps.

To produce very narrow communication zones, small ferrite antennas are quite useful because of their magnetic coupling. In addition, ferrite antennas have the following notable features: (1) the size is small, (2) a high coupling efficiency between two antennas can easily be realized, and (3) a desired zone pattern may be easily synthesized by the arrangement of transmitter and receiver antennas. The pattern of a communication zone is principally determined by the product of the field produced by a transmitter antenna and that of a receiver antenna. The direction of the antenna axes and the location of the two antennas are essential factors for determining the zone patterns.

In the inductive radio communication system, the transmitter power is normally low, and hence ferrite antennas for transmission have become practical, even though ferrite antennas used to be employed only for receiving purposes.

5.4.2 Design of an Antenna System

The inductive radio communication link can be modeled by the circuit shown in Figure 5.83, where the transmitter and receiver are coupled magnetically by two antennas [25,26]. The efficiency of power transfer between the transmitter and the receiver is the most important factor for achieving good system performance. It depends on the coupling factor η between the transmitter and the receiver antennas, where η is defined to be

$$\eta = P_r/P_t \tag{5.26}$$

where P_r is the power output of the receiver antenna and P_t is the power input to the transmitter antenna. The communication link, equivalently expressed by a magnetically coupled circuit is represented by the two-port network shown in Figure 5.84, where antennas are represented by coils. The coupling factor η can be written in terms of antenna parameters as

$$\eta = (\omega M)^2/4r_r r_t \tag{5.27}$$

where ω is the operating angular frequency.

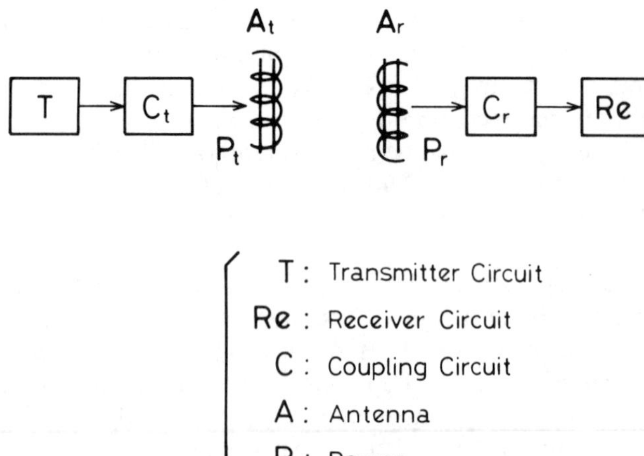

Figure 5.83 Transmitter-receiver link.

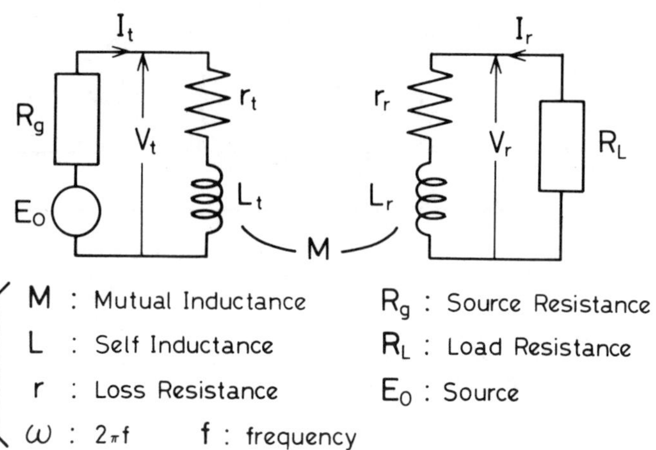

5.84 Equivalent circuit of antenna system.

An example of the antenna geometry is shown in Figure 5.85, where two magnetically coupled coil antennas are shown. The coil antenna is loaded with a ferrite core (relative permeability μ), and the number of turns is N. The mutual inductance M between the two antennas is the primary factor for determining η. M can be obtained by using Neuman's equation as

$$M = (\mu_0/4\pi)\mu_t\mu_r N_t N_r \oint_{S_t}\oint_{S_r} \mathbf{ds}_t \mathbf{ds}_r/R \tag{5.28}$$

where **ds** denotes a line element (vector), N is number of turns, R is the distance between two antennas, and subscripts t and r express transmitter and receiver, respectively. Integration is performed with respect to circumference C of a coil antenna. In the frequency range of inductive radio communication, typically below 250 kHz, ferrite antennas having fairly high permeability ($\mu \geqq 100$) and low loss resistance ($R_{\text{loss}} < 10\Omega$) can be chosen. The patterns of the communication zone are found from the output voltage V_r of the receiver as a function of the geometry:

$$V_r\,(\rho,\phi,Z) = \omega I_t M(\rho,\phi,Z) \tag{5.29}$$

It can be seen from (5.29) that the pattern of the communication zone is determined by knowing the mutual coupling M, which depends on the placement

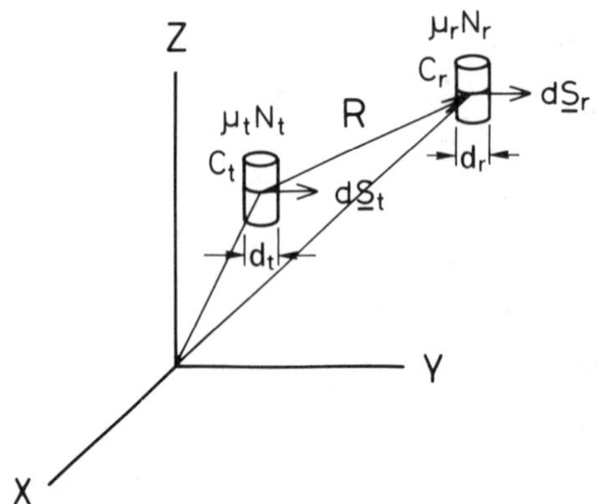

Figure 5.85 Coordinate system.

and axes of the two antennas. Several examples of calculated zone patterns are shown in Figure 5.86, where two antennas are assumed to be placed in parallel. In practice, a pattern to meet the system requirements is selected. For instance, when a zone covering only one traffic lane, as shown in Figure 5.87(a), is required (type I), a circular pattern may be the best choice. If a coverage of several lanes is desired (type II), an elliptic pattern transversally covering all lanes, but extending only a short distance in the direction of vehicle travel, as shown in Figure 5.87(b), is employed. This feature of generating zone patterns distinguishes the use of ferrite antennas from the conventional mobile antenna systems.

A practical example of the design parameter is given in Table 5.2, where both transmitter and receiver antennas are assumed to possess the same dimensions ($d_t = d_r = d$) and the same characteristics ($N_t = N_r = N$ and $\mu_t = \mu_r = \mu$). When both antennas are placed on the Z-axis with a distance of R_z, the mutual coupling M is given by

$$M = (\pi/2)\mu_0\mu^2 N^2 (d/2)^4 R_z^3 \tag{5.30}$$

The length of the ferrite core is omitted from the above factor, since it is very small compared to the wavelength. In a practical example, an antenna having a 3-cm diameter and 10-cm length has been used. Here, the coupling factor η is found to be -38.6 dB (1 − 0 dB). the usefulness of ferrite antennas when applied to mobile communication systems has been evidenced by such practical operations as bus operation control and bus location systems.

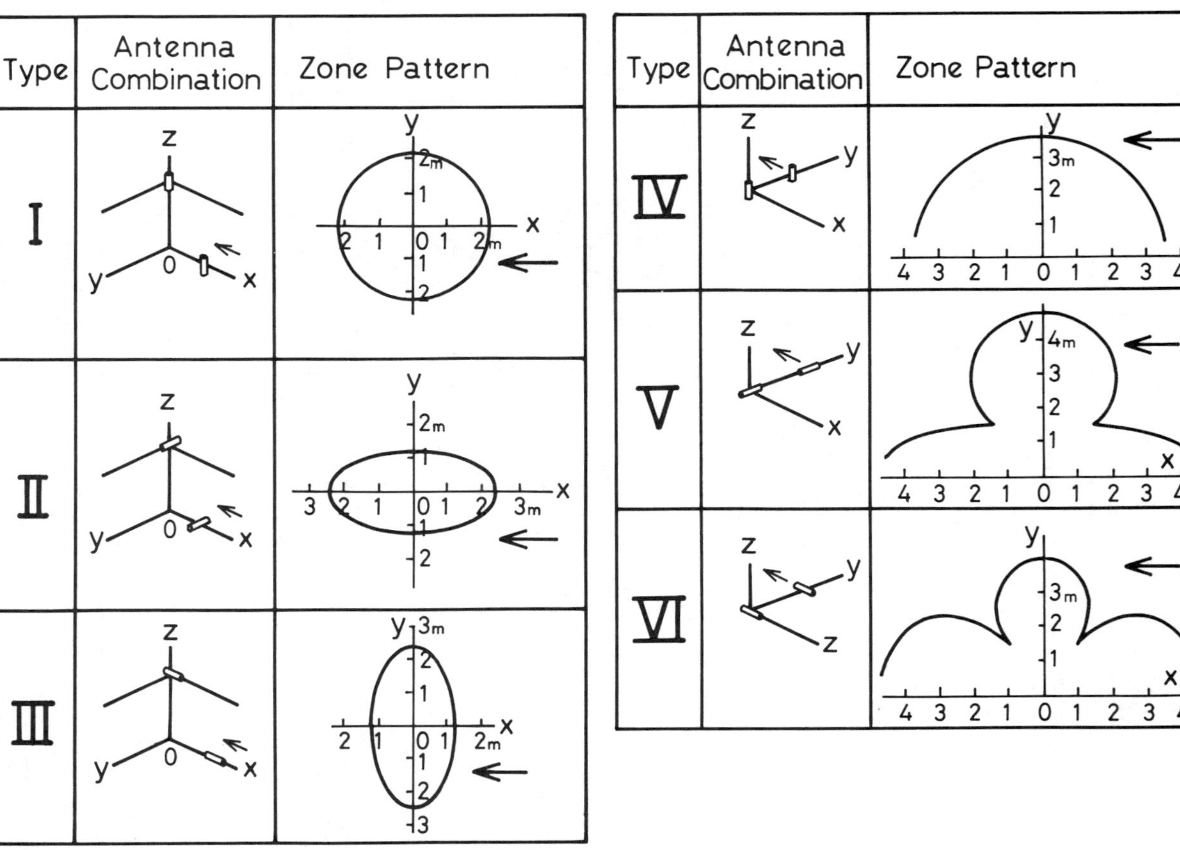

Figure 5.86 Zone patterns.

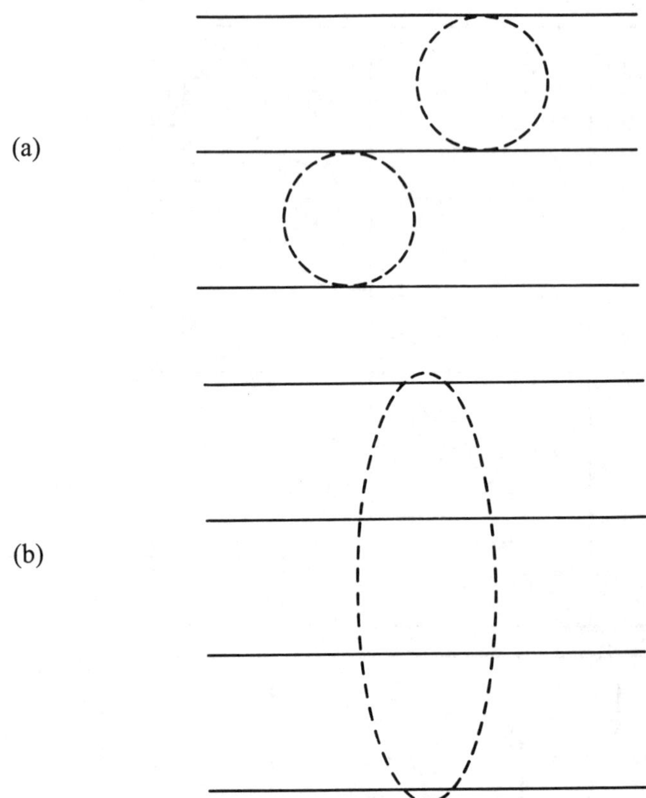

Figure 5.87 Zone patterns produced on the bus lane: (a) zone covering a single lane; (b) zone covering several lanes.

Table 5.2
A Practical Example of Design Parameters

Parameter	*Value*	*Parameter*	*Value*
f	100 kHz	d	.03m
I_t	100 mA	R_z	3m
N	100 turns	M	3.7×10^{-7}H
μ	100	η	1.37×10^{-4}
r	10Ω	V_r	12 mV

5.4.3 A Practical System

In order to improve city bus service, the Comprehensive Bus Operation Control System (CBOCS) has been developed in Japan and introduced to several major cities, including Tokyo and Yokohama [27,28]. Information on the location of buses at any given time and enroute bus operations are obtained by means of communication between buses (vehicle stations) and roadside (signpost) stations (Figure 5.88). Information (data) from buses received by roadside stations is transferred by wire lines to the district centers (bus operation centers) via depot equipment. Data gathered and processed at the district center are then sent to the central headquarters, where a main computer is installed which processes data to be used for bus operation control and management. The system also provides information

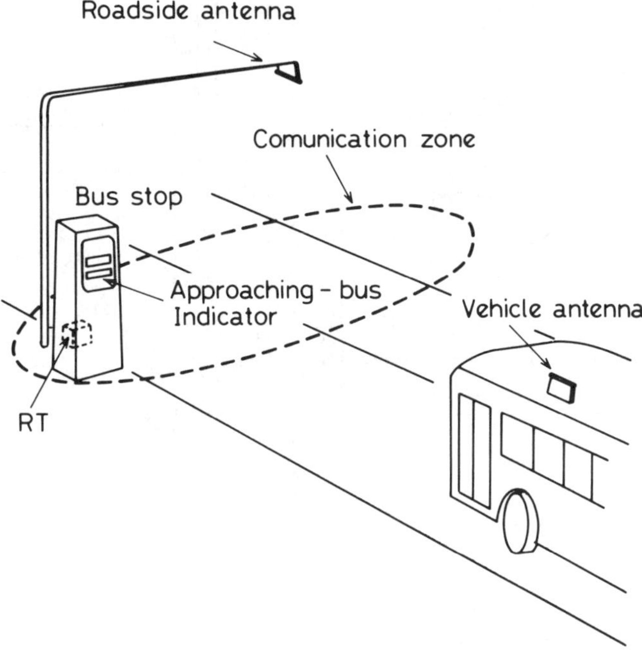

Figure 5.88 Communication system between bus and roadside stations.

about approaching buses, their destinations, approximate times of arrival, and so on at each bus stop in the form of visual display or sometimes a sound announcement (Figure 5.89). The bus stop display also provides information for the bus driver, such as the time when a previous bus left and the scheduled departure time for the bus.

The system outline of the CBOCS is shown in Figure 5.90. The system spec-

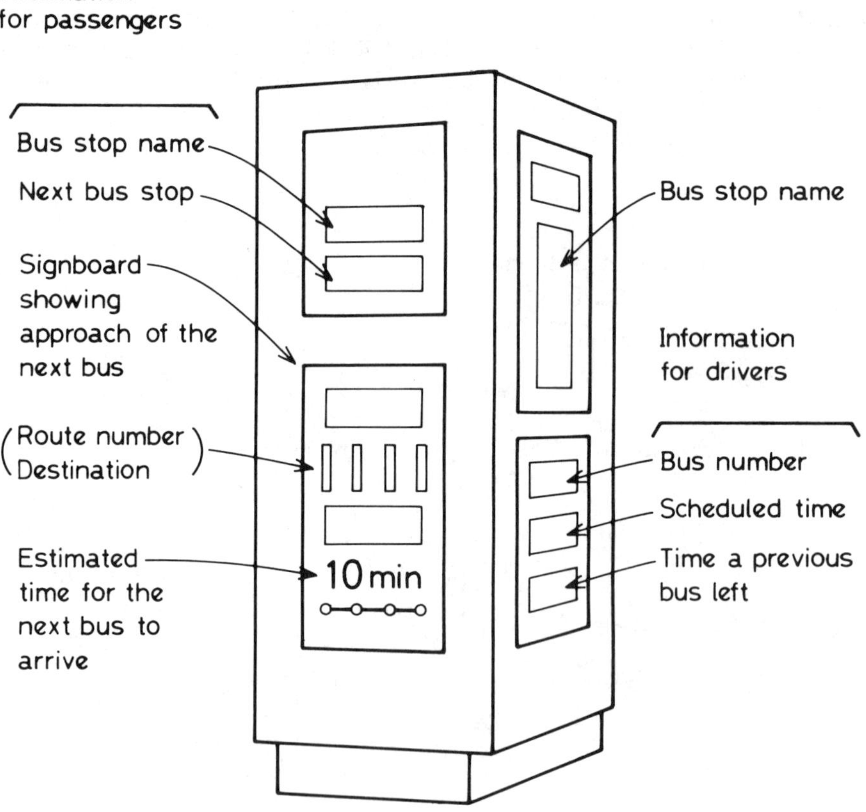

Figure 5.89 Bus stop signboard.

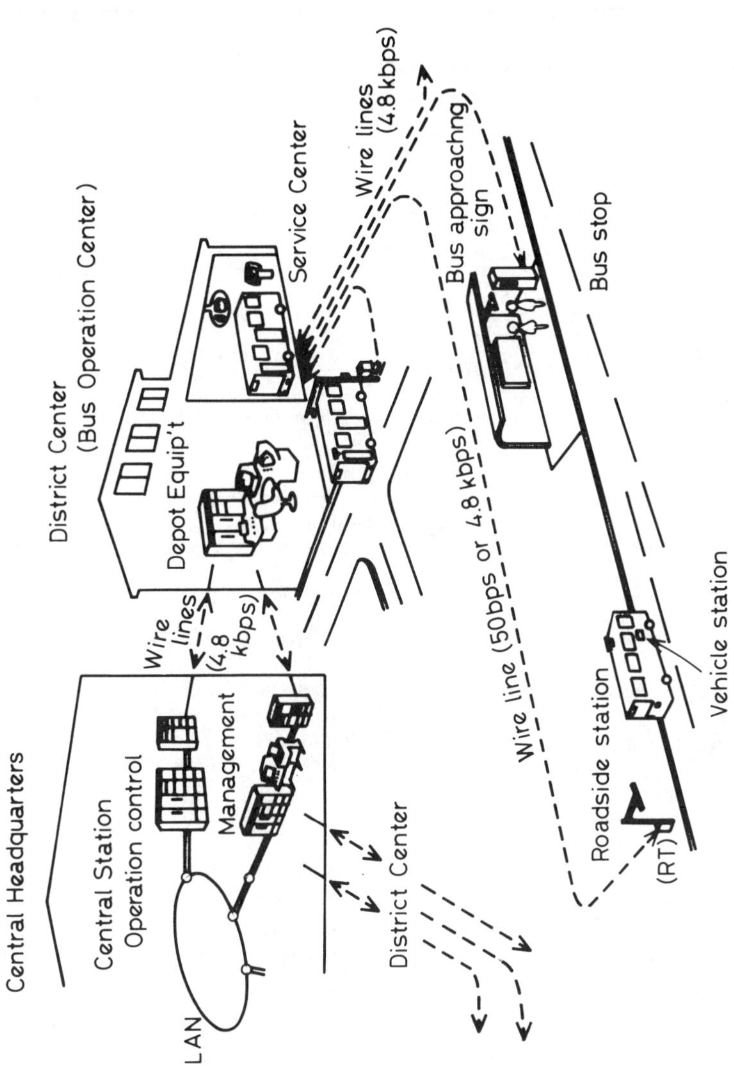

Figure 5.90 System outline of the CBOCS.

ifications are given briefly in Table 5.3. A vehicle transceiver (VT) is installed on every bus. The roadside transmitter (RT) is located at places such as depot gates, bus interchanges, and main bus stops. Data transmission rates between VT and RT are either 50 bps or 4.8 kbps, depending on the type of RT. District centers are linked with the central headquarters by dedicated wire lines with a transmission rate of 4.8 kbps.

Table 5.3

System Specifications

Category	*Specification*
Type of system	Inductive communication
Type of communication	Semidual: RT transmits trigger signals; VT responds to RT
Operating frequency	96 kHz
Data transmission rate	4.8 kbps
Transmitter power	Very small power below regulation*
Reception field strength	Road: 85 dBμ/m for RT; vehicle: 87.5 dBμ/m for VT
Modulation	FSK

Note: FSK = frequency shift keying; RT = roadside transceiver; VT = vehicle transceiver.

*Subject to the rule, defined as a power that produces field strength below 100 μV/m (15 μV/m) at a distance 100m (3m) from the transmitter in the frequency range below 322 MHz (the Radio Law of Enforcement, Item 6, For Restricted Radiation Devices, Ministry of Posts and Telecommunications, Japan).

Figure 5.91 is the exterior view of a vehicle antenna that is a ferrite antenna, protected by a plastic cover. The specifications of the antenna are also given in the figure. The vehicle antenna is placed on top of the bus as shown in Figure 5.88. Figure 5.92 presents the outside view of a roadside antenna and provides its basic specifications. The roadside antenna is also covered by plastic material and is usually installed at the end of a pole placed at the edge of the roadside, as shown in Figure 5.88.

Figure 5.91 Vehicle ferrite antenna using Q_2M material (μ = 350) at 96 ± 1.5 kHz with Q = (19.2 ± 3.5) and (100 ± 10) ohms input impedance. Weight including cable ≤ 1.5 kg, outside dimensions: 135-mm length and 36-mm diameter.

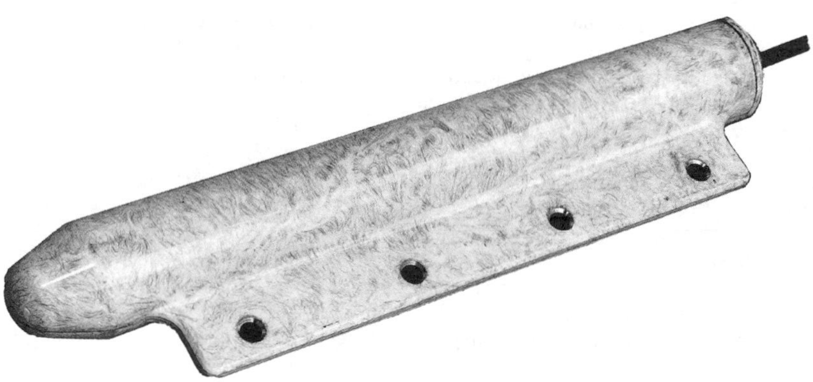

Figure 5.92 Roadside ferrite antenna using same material and frequency and input impedance as the antenna in Figure 5.91. Q = (21 ± 3.5). Weight including cable ≤ 4 kg, outside dimensions: 295-mm length and 38-mm diameter.

REFERENCES

[1] CCIR, "Planning Standards for FM-Sound Broadcasting at VHF," *Recommendations and Reports of the CCIR*, Vol. X-1, Rec. 412-3, ITU, Geneva, 1982.

[2] Lee, W.C.Y., *Mobile Communications Engineering*, McGraw-Hill, 1982.

[3] Jakes, W. C., *Microwave Mobile Communications*, John Wiley, 1974.

[4] Manner, E. J., "Distortions and Their Reduction by Antenna Diversity With Mobile FM Reception" (in German), doctoral thesis, University of the Bundeswehr Munich, 1985.

[5] Lindenmeier, H. K., J. F. Hopf, "Investigations for the Evaluation of the Minimum Required Mean Signal Level With Mobile FM Radio Reception" (in German), *Rundfunktechn. Mitteilungen*, Jahrg. 28, 1984.

[6] Lindenmeier, H. K., "Optimum Bandwidth of Signal-to-Noise Ratio of Receiving Systems With Small Antennas," *AEÜ*, Band 30, Heft 9, 1976.

[7] Lindenmeier, H. K., J. F. Hopf, and L. M. Reiter, "Active Window Antennas for Radio Reception in Cars for Single Antenna and for Antenna Diversity Application," *Journées Internationales de Nice sur les Antennes* (JINA), 1986, Nice, France.

[8] Lindenmeier, H. K., E. J. Manner, F. Sessink, "Antenna Diversity Experiments With Philips Car Radios," *Congress of the Society of Automotive Engineers* (SAE), 1985, Detroit.

[9] CCIR Working Group 10-B, "Diversity Rec. in Automobiles for Frequency Modulated Sound Broadcasts in Band 8 (VHF)," Doc. 10/160-E.

[10] Toriyama, H., et al., "Development of Printed-on-Glass TV Antenna Systems for Car," *IEEE VT-S Conf. Digest*, 1981, pp. 334–342.

[11] Technical data described in this part are provided by courtesy of Nissan Motor Co., Japan.

[12] Kishimoto, T., and S. Sasaki, "Train Telephone System," *Proc. IECEJ*, Vol. 63, No. 2, Feb. 1980, pp. 128–133.

[13] Kishimoto, T., and S. Sasaki, "LCX Communication Systems,"*IECEJ*, Tokyo, 1982.

[14] Watanabe, H., "Electronic Control and Communication System of Shinkansen," *IECEJ*, Tokyo, 1982.

[15] Hayashi, Y., "Train Radio System of Shinkansen," *Proc. IECEJ*, Vol. 65, No. 5, May 1982, pp. 541–543.

[16] Taguchi, K., et al., "Recent Train Radio Communication Systems," Mitsubishi Electric Co. Technical Report, Vol. 61, No. 2, Feb. 1987, pp. 33–37.

[17] Amemiya, Y., "Surface Wave Coaxial Cable," *Proc. IECEJ*, Vol. 48, No. 12, April 1965, pp. 131–142.

[18] Mikoshiba, K., Y. Nurita, and S. Okada, "Radiation From a Coaxial Cable and Its Application to a Leaky Coaxial Cable," *Trans. IECEJ*, Vol. 51-B, No. 10, Oct. 1968, pp. 499–505.

[19] Yoshida, K., "New Communication Systems by Leaky Coaxial Cables," *Proc. IECEJ*, Vol. 55, No. 5, May 1972, pp. 655–663.

[20] Cree, D. J., and L. J. Giles, "Practical Performance of Radiating Cables," *Radio & Electronic Engineering*, Vol. 45, No. 5, May 1975, pp. 215–223.

[21] Mikoshiba, Y., S. Okada, and S. Aoki, "Near Electromagnetic Fields Around a Leaky Coaxial Cable," *Trans. IECEJ*, Vol. 54-B, No. 12, Dec. 1971, pp. 789–796.

[22] Delogne, P., *Leaky Feeders and Subsurface Radio Communication*, London: Peter Peregrinus, Ltd., 1982.

[23] Kurauchi, N., K. Yoshida, and Y. Miyamoto, "Wideband Leaky Coaxial Cable," *Trans. IECEJ*, Vol. 54-B, No. 10, Oct. 1971, pp. 682–686.

[24] Tamura, K., and K. Fujimoto, "Analysis and Design of Two-Way Inductive Communication Systems," *National Technical Report*, Vol. 19, Feb. 1973, pp. 156–166.

[25] Tamura, K., and K. Fujimoto, "Inductive Communication Systems Applied to Mobile Systems," *National Technical Report*, Vol. 23, Oct. 1977, pp. 717–725.

[26] Fujimoto, K., "Usefulness of Ferrite Antenna Applied to Mobile Communication Systems," *Int. Symp. Ant. Propag.*, Japan, 1987, pp. 97–100.

[27] Tamura, K., "Bus Operation Control Systems," *Automobile Technology*, Vol. 39, 1985, pp. 541–546.

[28] Konda, T., "Comprehensive Bus Operating Control System," *Microcomputers in Traffic and Transport*, REAAA Workshop, Singapore, 1985, pp. 103–111.

Chapter 6
Antennas for Mobile Satellite Systems

T. Shiokawa, S. Ohmori, and T. Teshirogi

6.1 INTRODUCTION

The decade of the 1990s is expected to be an era of mobile satellite communications serviced on a commercial basis. Since 1982, the INMARSAT system has provided international maritime satellite communication services and is expanding the services to aircraft and land mobiles. On the other hand, forthcoming systems such as American Mobile Satellite Corporation (AMSC) in the United States, MSAT in Canada, and AUSSAT in Australia [1] are going to provide domestic satellite communication services mainly for land mobiles using dedicated satellites. The research and development activities on mobile satellite communications have continued since the mid-1970s in many countries and organizations. Typical research programs are MSAT in Canada [1], the ETS-V in Japan [2], MSAT-X in the United States [3], and PROSAT in Europe [4].

The mobile communication systems mentioned above are using geostationary earth orbiting (GEO) satellites, but recently new concepts have been proposed by several private sectors in the United States. The new systems are called low earth orbiting (LEO) satellite communication systems, which use a group of low-altitude orbiting satellites. The typical examples of LEO satellite systems are IRIDIUM, ODYSSEY, and GLOBALSTAR, which will use 66, 12, and 48 satellites in low earth orbits, respectively. The main advantage of LEO satellite systems from the point of view of antenna design is that mobile and handy terminals can use low-gain omnidirectional antennas, because of smaller values of free-space propagation loss compared with those of GEO satellite systems.

In implementing mobile satellite communications, a vehicle antenna is one of the most important key technologies. In this chapter, a vehicle antenna means an antenna system that is mounted on a mobile for satellite communications. This chapter describes system requirements and antenna design considerations for PROSAT, ETS-V, MSAT-X, MSAT, INMARSAT (International Maritime Satellite Organization), GPS, and broadcasting systems.

This chapter describes antenna systems for vehicles, and, to assist the reader, an outline of mobile satellite communication systems is briefly mentioned. The allocation of frequency, which is an essential factor in designing and analyzing antennas, is defined by the Radio Regulation on an international basis, depending on service types such as communication, navigation, and broadcasting.

Figure 6.1 shows the concept of a mobile satellite system (MSS), which provides communication links via a satellite between a base earth station and moving vehicles such as ships, aircraft and land vehicles. Conventional satellite communication systems are called fixed satellite systems (FSS), which establish communication channels via a satellite between fixed earth stations. A typical FSS system is INTELSAT.

A mobile satellite communication usually provides so-called bidirectional communication, such as voice, data, telex, and facsimile for mobiles, but in this chapter other types of services, such as navigation and broadcasting, are discussed. For each service the frequency bands allocated by the Radio Regulation are often designated with alphabetical symbols such as C, L, Ku, and Ka bands, as shown in Table 6.1 [5]. Some examples of typical mobile satellite services are as follows.

Communication

Typical mobile satellite communication systems are INMARSAT, AMSC, MSAT, and AUSSAT, in which the L-band (1.6/1.5 GHz) is used. The only exception is Omni-TRACS in the United States, which is serviced in the Ku-band (14/12 GHz). In Japan, use of the S-band (2.5/2.4 GHz) has been studied for a domestic system. In recent years, research on the feasibility of advanced mobile satellite communications using the Ka-band (30/20 GHz) has started, and typical research and development programs are the ACTS in the United States [6] and the COMETS in Japan [7].

Navigation

A typical system is the GPS, in which L-band (1.6 and 1.3 GHz) frequencies are used from a satellite to the earth.

Broadcasting

This service is to broadcast TV and radio programs from a satellite to the earth. Although the present systems are designed for fixed terminals, not for mobiles, an antenna system for trains and ships has been developed to receive TV programs from the broadcasting satellite (BS) satellite in Japan.

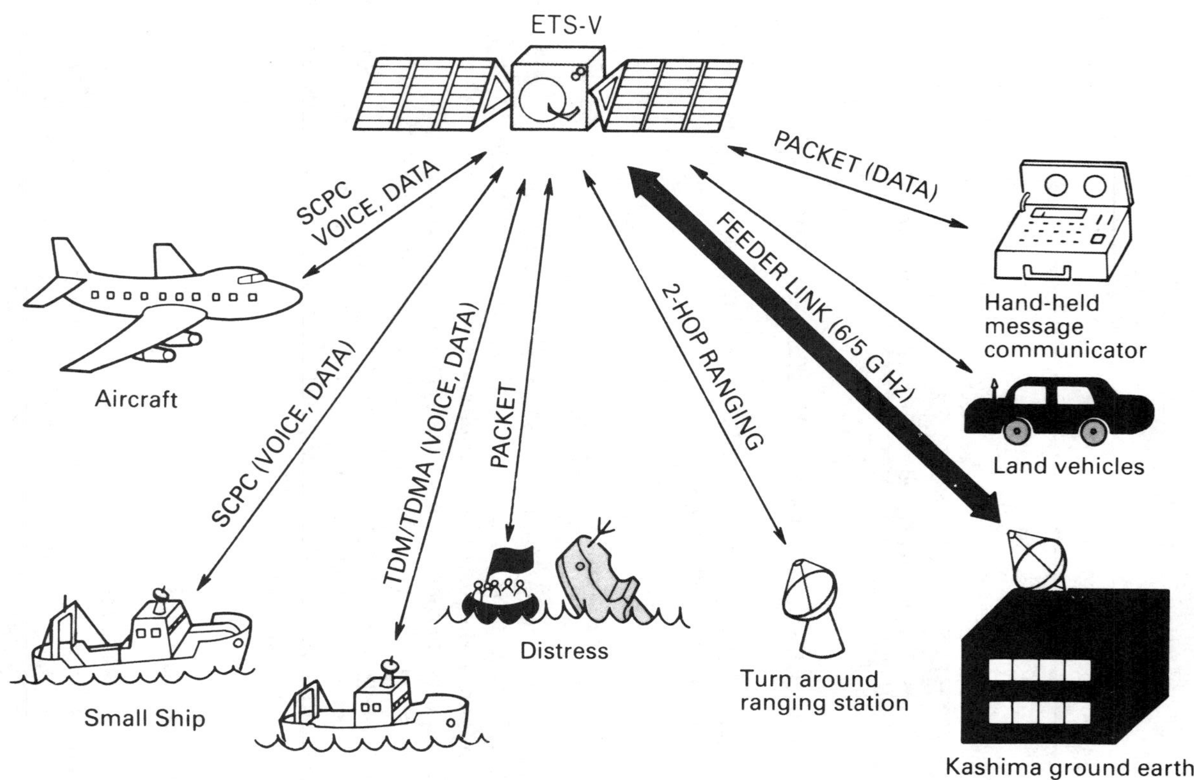

Figure 6.1 Concept of mobile satellite system.

Table 6.1
Designation of Frequency Bands

Band	*Frequency*
HF	3–30 MHz
VHF	30–300 MHz
L-band	1–2 GHz
S-band	2–4 GHz
C-band	4–8 GHz
X-band	8–12 GHz
Ku-band	12–18 GHz
K-band	18–27 GHz
Ka-band	27–40 GHz
Millimeter Waves	40–300 GHz

6.2 SYSTEM REQUIREMENTS FOR VEHICLE ANTENNAS

This section describes important general requirements for antenna systems used in mobile satellite communication systems.

6.2.1 Mechanical Characteristics

Compact and lightweight equipment is a self-evident requirement for vehicle antennas in addition to ease of installation and mechanical strength. In the case of shipborne antennas, the installation requirement is not as severe compared to that of aircraft and cars, because even in small ships there is a little space to install an antenna system. However, in the case of cars, especially for small, private cars, low-profile and lightweight equipment is an essential requirement. Although the demands are the same for aircraft, more severe conditions are required to satisfy the standards for avionics. The low air drag must be one of the most important requirements for aircraft antenna [8].

6.2.2 Electrical Characteristics

Since the gain of an antenna is theoretically decided by its physical dimensions, a compact antenna necessarily has a low gain. Accordingly, it is inevitable for vehicle antennas not to have enough performance, such as gain, radiation power, and receiving capability, because of their small physical dimensions. Furthermore, the beamwidth is wide and compact antennas have to receive undesired signals, which result in fading effects.

For the system as a whole, the relatively low performance of vehicle antennas has to be compensated for by a large antenna and a high-power amplifier onboard the satellite. So the requirements of transmitting and receiving performance of vehicle antennas mainly depend on the satellite capability.

The INMARSAT satellites of the first and second generations have global beams to provide international services, and the third-generation INMARSAT satellites are scheduled to have spot beams. On the other hand, forthcoming systems such as AMSC, MSAT, and AUSSAT have spot beams to cover domestic service areas, which give higher satellite capability than the present INMARSAT satellites. However, there are basically no big differences between requirements for vehicle antennas in international systems such as INMARSAT and in domestic systems such as AMSC, MSAT, and AUSSAT.

Frequency and Bandwidth

In almost all present and forthcoming systems using GEO satellites, the L-band (1.6/1.5 GHz) is used in communication links between the satellite and mobiles. The required frequency bandwidth in L-band communication systems is about 8% to cover transmitting and receiving channels. So in using a narrow-band antenna element such as a patch antenna, some efforts have to be made to widen the bandwidth. In proposed systems such as IRIDIUM and ODYSSEY, the S-band and L-band are allocated in WARC 92. In these frequency bands, the required frequency bandwidth is about 5%.

Gain, Beamwidth, and Beam Coverage

The required gain is decided by a link budget, which is calculated by taking into account the required channel quality (expressed as C/No, which is the carrier-to-noise power density ratio) and the satellite capability. In forthcoming systems such as AMSC, MSAT, and AUSSAT, medium gain of 12 to 15 dBi is required for voice/high-speed data (about 24 kbps) channels. On the other hand, in the case of the present INMARSAT-A terminals, the comparatively high gain of about 24 dBi is required, due to the difference in satellite capabilities.

Figure 6.2 shows the relationship between gain and aperture size, and Figure 6.3 shows the relationship between gain and half-power beamwidth (HPBW), respectively. From Figure 6.3, the HPBW of a 15-dBi antenna is about 30 deg, and that of a 4-dBi antenna is about 100 deg. In general, since an antenna beam is required to cover 0 to 90 deg. in elevation and 0 to 360 deg in azimuth directions (upper hemisphere), the former antenna needs a tracking capability. In this chapter, the former is called a *directional* (high or medium gain) antenna, and the latter is called an *omnidirectional* (low gain) antenna. Table 6.2 shows a classification of

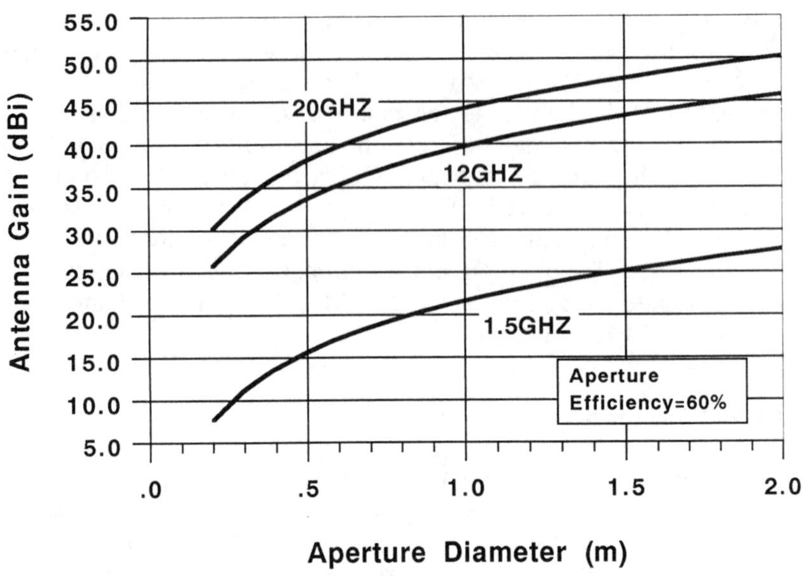

Figure 6.2 Relationship between gain and aperture size.

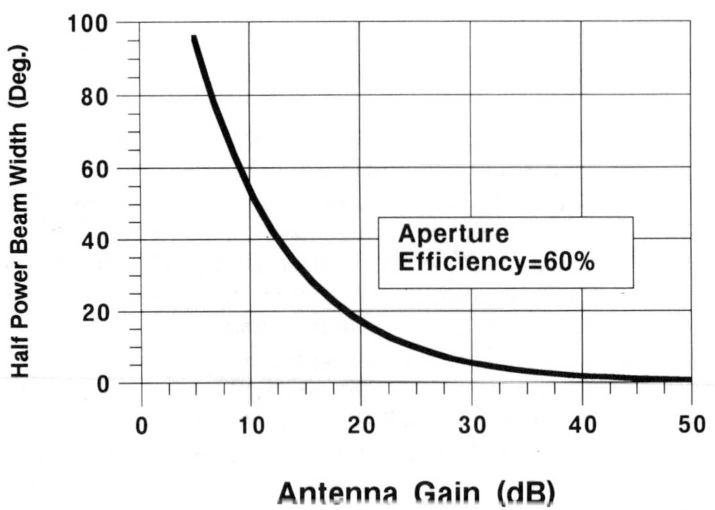

Figure 6.3 Relationship between gain and half-power beamwidth.

directional and omnidirectional antennas. In the INMARSAT system, as shown in Table 6.2, terminals have been classified by their receiving capabilities and services. The outline specification of INMARSAT terminals is given in Table 6.3.

Polarization, Axial Radio, and Sidelobes

Circular polarized waves are used in order to eliminate the need for polarization tracking. In the INMARSAT system, right-hand circular polarization has been used. In the case of aperture-type antennas such as the parabolic antenna, which

Table 6.2
Classification of L-Band Antennas in Mobile Satellite Communications

Antenna	*Gain Class*	*Typical Gain (dBi)*	*Typical G/T (dBk)*	*Typical Antenna (Dimension)*	*Typical Service*
Directional	High	20–24	−4	Dish (1 mϕ)	Voice/high-speed data Ship (INMARSAT – A,B)
Directional	Medium	12–15	−10 to −13	SBF (0.4 mϕ) Array Phased array	Voice high-speed data Ship (INMARSAT – M) LM (INMARSAT – M) Aircraft (INMARSAT – Aero)
Omnidirectional	Low	0–4	−23	Quadrifilar Drooping-dipole Patch	Low-speed data Ship (INMARSAT – C) Aircraft Land mobile

Note: LM = land-mobile; SBF = short backfire antenna.

Table 6.3
Specification of INMARSAT Terminals

Terminal	*G/T (dBk)*	*EIRP (dBw)*	*Gain (dBi)*	*Main Communication Service*	*Main Service*
A	−4	36	20–23	Voice, telex	Ship
B	−4	33	20–23	Voice,* telex, data	
C	−23	16	0–3	Message	Very small ship, land mobile
M	−10	25	15†	Voice	Small ship
	−12	25	12‡	Data	Land mobile
Aero	−13	25.5	12	Voice	Aircraft
	−26	13.5	0	Data	Aircraft

*Digital. †Maritime. ‡Land.

is commonly used as a shipborne antenna in the current INMARSAT-A terminal, an axial ratio of below 1.5 dB in a boresight direction is so easy to achieve that polarization mismatch loss is almost negligible. However, in the case of phased-array antennas, it must be taken into account because of degradation of the axial ratio caused by beam scanning.

Figure 6.4 shows the relationship between polarization mismatch loss and axial ratios of a receiving antenna, in which AR*a* and AR*b* denote axial ratios of transmitting and receiving antennas, respectively. From Figure 6.4, it is found that in order to get the polarization mismatch loss below 0.5 dB, an axial ratio of a vehicle antenna is required under 5 dB in all directions, because an axial ratio of a satellite antenna is, in general, below 2 dB. This requirement must be considered for a phased-array antenna in scanning a beam over wide angular areas.

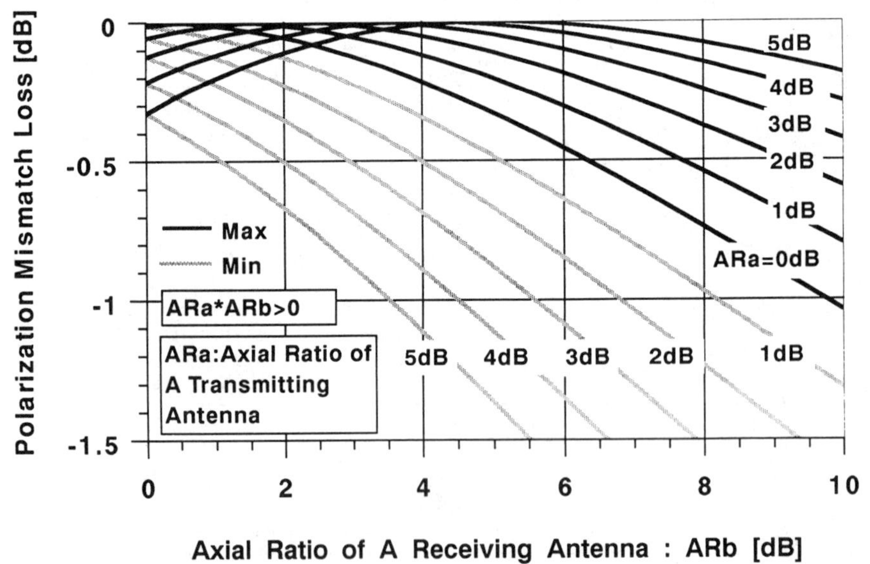

Figure 6.4 Relationship between polarization mismatch loss and axial ratio of receiving antenna.

For large antennas, with diameters over 100λ (wavelengths), a reference radiation pattern is recommended by the CCIR [9] for interference to and from other satellite and terrestrial communication systems. However, the diameters of the vehicle antennas under discussions are, in many cases, below five wavelengths in the L-band. Further CCIR action is expected to define a reference radiation pattern for vehicle antennas in mobile satellite communications.

G/T and EIRP

Although gain is an essential factor in considering antennas, the figure of merit G/T (which is the ratio of gain to system noise temperature) is more commonly specified from the standpoint of satellite communications. System noise temperature T at an input port of a low-noise amplifier (LNA) is defined by (6.1) [10], and the system gain G is defined at the same port, taking account of losses caused by tracking, feed lines, and a radome.

$$T = T_a/L_f + T_0(1.0 - 1.0/L_f) + T_r \tag{6.1}$$

T_a is an antenna noise temperature which comes from such effects as ionosphere and the earth. Although T_a depends on factors such as frequency and beamwidth, a typical value in the L-band is about 200K [11]. T_0 is the temperature of the environment, usually 300K. T_r is the noise temperature of a receiver (LNA), and its typical value is about 80K to 100K in the L-band. L_f is a total loss of feed lines and components such as diplexer, cables, and phase shifters if a phased-array antenna is used. Figure 6.5 shows a relationship between G/T and feeder loss in the case of a 15-dBi antenna, which is a typical candidate in the forthcoming systems. It is found that if G/T is required over −13 dBK, feeder loss must be under about 3 dB.

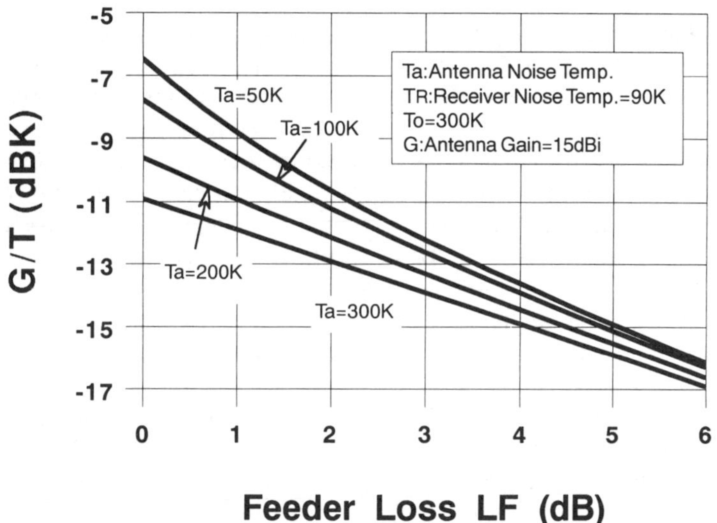

Figure 6.5 Relationship between G/T and feeder loss.

Equivalent isotropically radiated power (EIRP) is an important value in evaluating the transmitting performance of a terminal including an antenna. EIRP (dBW) is defined by a sum of the antenna gain (dB) and output power of the high-power amplifier (dBW), taking account of feed losses such as feed lines, cables, and a diplexer.

6.2.3 Propagation Problems

Telecommunications over earth-space links for maritime mobile satellite systems lead to propagation problems substantially different from those arising in the fixed satellite service [12]. For instance, the effects of reflections and scattering by the sea surface become quite severe, particularly where antennas with wide beamwidths are used. Furthermore, maritime mobile satellite systems may operate on a world-wide basis, including paths with low elevation angles. Moreover, due to the use of L-band frequencies for INMARSAT systems, the effect of ionospheric scintillation is not negligible, particularly in equatorial regions during the years of high solar activity of an 11-year cycle. On the other hand, tropospheric effects such as rain attenuation and tropospheric scintillation will be negligible for the said frequency band. Signal level attenuation due to blocking by ship superstructure is considerable problem. In the following sections, brief explanations are presented.

Multipath Fading Due to Sea Surface Reflection

Multipath fading due to sea reflection is caused by the interference between direct and reflected radio waves. The reflected waves are composed of a coherent (specular reflection) component and an incoherent (diffuse) component that fluctuate due to the motion of the sea waves. The coherent component is predominant under calm sea conditions, and the incoherent component plays an important role in rough sea conditions. If the intensity of the coherent component and the variance value of the incoherent component are both known, the cumulative time distribution of the signal intensity can be determined by statical considerations [13,14].

The amplitude of the coherent component decreases rapidly with increasing wave height, elevation angle, and frequency. Figure 6.6 shows estimates of amplitude of the coherent component for an omnidirectional antenna as a function of significant wave height for low elevation angles; the frequency is 1.5 GHz and polarization is circular. The incoherent component is random in both amplitude and phase, since it originates from a large number of reflecting facets on the sea waves. The amplitude of this component follows a Rayleigh distribution and the phase has a uniform distribution.

Since the theoretical model concerning the incoherent component is not suitable for engineering computations using a small calculator, simpler prediction

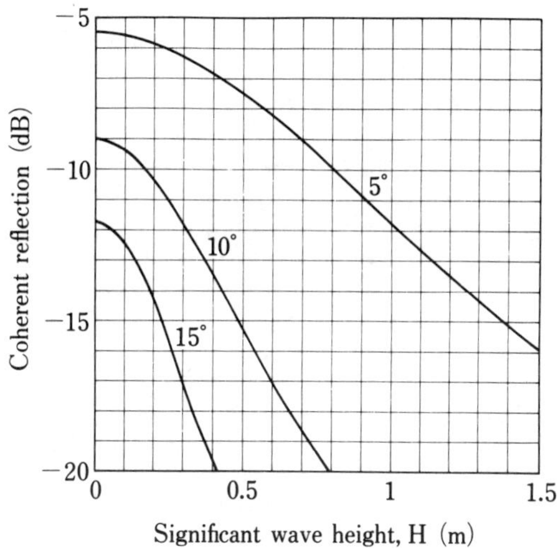

Figure 6.6 Relationship between coherent reflection and significant wave height.

models are useful for approximate calculation of fading. Such simple methods for predicting multipath power or fading depth have been developed [15,16]. Figure 6.7 shows the relationship between multipath power and elevation angle for different antenna gains based on the method in [16]. Although fading depth depends slightly on sea conditions, even if the incoherent component is dominant, as shown in [14], the simple model is useful for the rough estimate of fading depth.

Fading depth, which is a scale of intensity of fading, is usually defined by the difference in decibels between the direct wave signal level and the signal level for 99% of the time. The fading depth can be approximated by a 50% to 99% value for fading where the incoherent component is fully developed. Large fading depths usually appear in rough sea conditions, where the incoherent component is dominant [14]. Figure 6.8 shows the fading depth for antenna gains of 24, 20, 15, and 8 dBi as a function of elevation angle with fully developed incoherent component (i.e., wave heights of about 1m to 3m) [12]. The calculation is based on the theoretical method in [14]. The shaded area covers the practical range of the sea wave slope, which depends on fading depth in rough sea conditions. Values estimated from the simple method [16] give around the mean value of those in Figure 6.7.

The frequency spectral bandwidth of temporal amplitude variations increases with increasing wave height and elevation angle. Figure 6.9 shows the probable range of −10dB spectral bandwidth (which is defined by the frequency corresponding to the spectral power density of −10 dB relative to the flat portion of power spectrum) of L-band multipath fading obtained by the theoretical fading

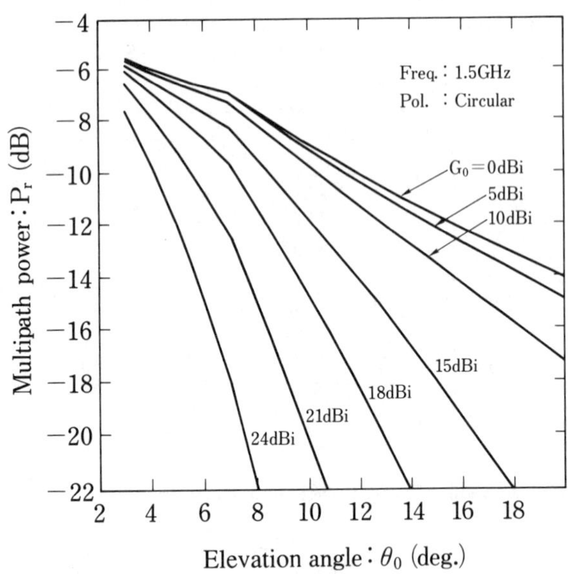

Figure 6.7 Relationship between multipath power and elevation angle.

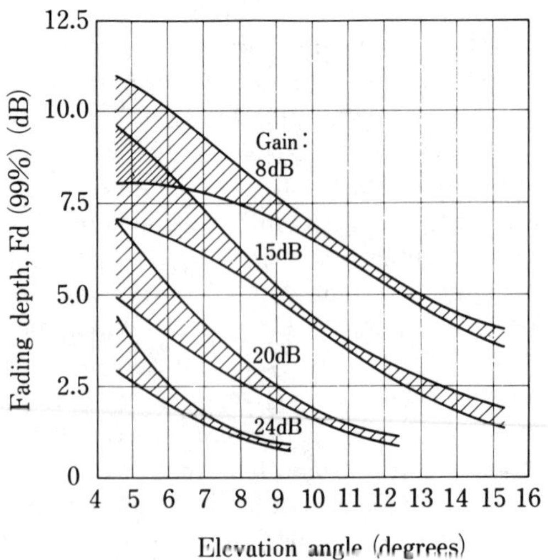

Figure 6.8 Fading depth estimated by the simple method.

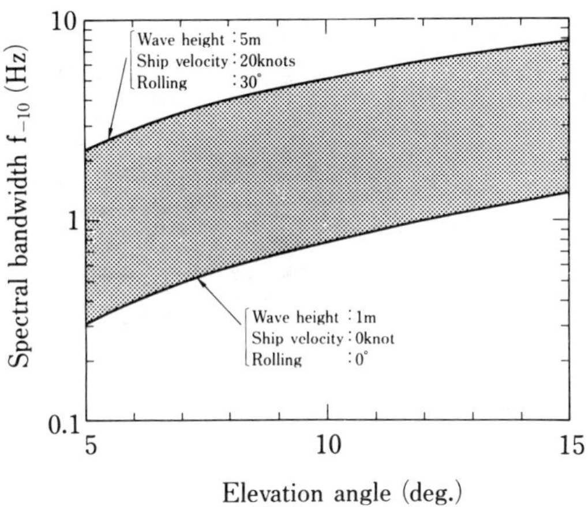

Figure 6.9 −10-dB spectral bandwidth.

model [17] as a function of the elevation angle under the usual conditions of maritime satellite communications; namely, significant wave height of 1m to 5m, ship speed of 0 to 20 knots, and rolling of 0 to 30 deg.

The error pattern in digital transmission systems affected by multipath fading is usually of the burst type. Accordingly, a firm understanding of the fade duration statistics of the burst-type fading is required. Mean values of fade duration $\langle T_D \rangle$ and fade occurrence interval $\langle T_I \rangle$ for a given threshold level as a function of time percentage, can be estimated from the fading spectrum. A simple method for predicting the mean value from the −10 dB spectral bandwidth is available in [18]. Predicted values of $\langle T_D \rangle$ and $\langle T_I \rangle$ for 99% of the time at an elevation angle from 5 to 10 deg are 0.05 to 0.4 sec for $\langle T_D \rangle$ and 5 to 40 sec for $\langle T_I \rangle$. The probability density function of T_D and T_I at any percentages ranging from 50% to 99% approximates an exponential distribution.

Blocking by Ship Superstructure

Blocking is caused by ship superstructures such as the mast and various types of other antennas. The geometry for blocking by a mast is shown in Figure 6.10. Attenuation due to blocking depends on various parameters such as diameters of the column, distance between antenna and column, and size of antenna. Based on experimental data reported so far, attenuation due to blocking caused by a column-type structure is given in Figure 6.11 for antenna gains of 20 and 14 dBi [12].

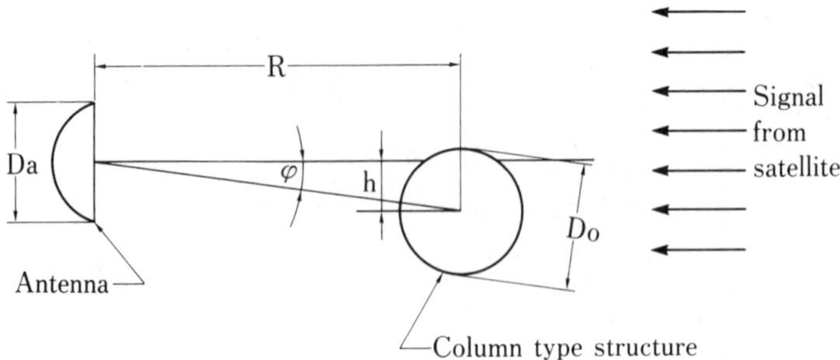

Figure 6.10 Geometry of blocking.

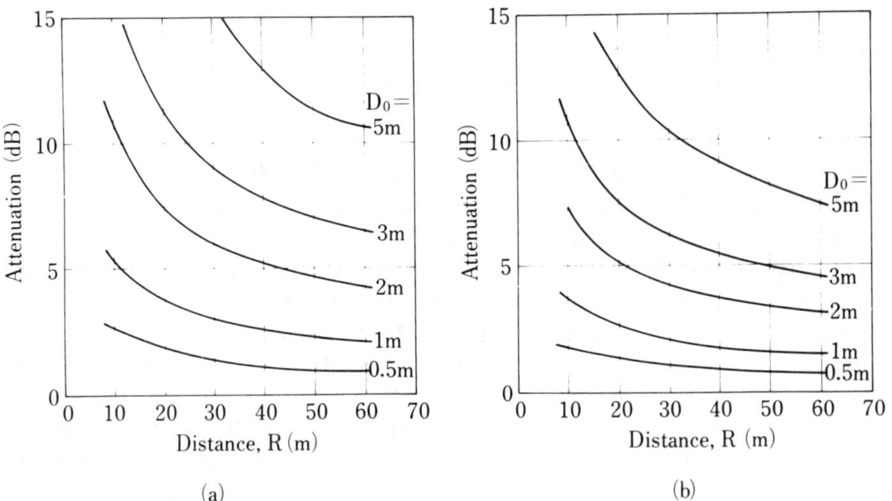

Figure 6.11 Attenuation due to blocking: (a) antenna gain: 20 dB; (b) antenna gain: 14 dB.

6.2.4 Fading Reduction Techniques

For low- and medium-gain systems, the effect of multipath fading due to sea surface reflection is a severe problem, especially at low elevation angles, as pointed out previously. In this section, possible fading reduction techniques for low- and medium-gain ship earth station (SES) antennas are surveyed, and field experimental results on the reduction effects are presented [19]. Fading reduction techniques applicable to these SES antennas are discussed in the following sections.

Diversity Method

Diversity techniques such as space, polarization, and frequency diversities have already been used practically in radio communication systems subject to severe fading. The space diversity technique needs two or more antennas, while other diversity techniques can be effected using a single antenna. In any case, the fading reduction effect largely depends on the correlation of signals with different branches concerning frequency, polarization, time difference, and so on. Figure 6.12 shows the principle of space diversity with a switch-and-stay algorithm [20]. As can be seen in this figure, the diversity output [$R(t)$] is selected by switching two signals through antenna #1 [$r_1(t)$] and antenna #2 [$r_2(t)$]. With this technique, the greatest reduction effect is expected when the correlation of the signals between the two antennas is near zero or lower.

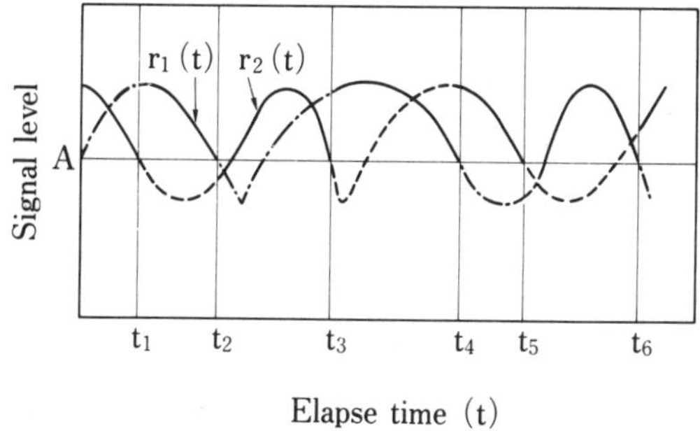

Figure 6.12 Principle of space diversity.

A space diversity experiment applying the switch-and-stay algorithm was carried out by setting up two short backfire antennas 20 cm in diameter on a small vessel under conditions with wave heights of about 1.0m to 1.5m [20]. These antennas were located with a three-wavelength separation in the vertical direction. Figure 6.13 shows the space diversity effect obtained by a computer simulation using actual data obtained by two antennas. From this figure it can be seen that the cumulative time distributions through each branch are nearly equal to each other, and these distributions almost correspond to a Rician distribution with carrier-to-interference ratio (C/I) of 6.25 dB. It can also be seen that the fading depth (= 1% value in the figure) of about 8 dB at an elevation angle of 9 deg can be reduced to about 5 dB when the threshold level is set to −5 dB.

Frequency diversity can produce a good reduction effect when the frequency difference between two signals is longer than the correlation bandwidth of the

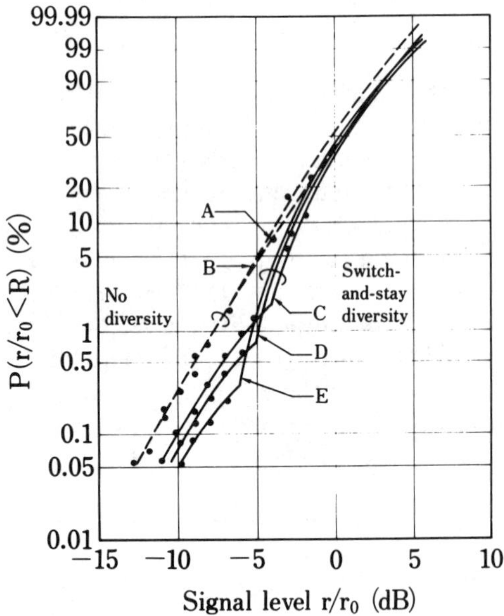

Figure 6.13 Space diversity effect.

multipath channel. Assuming that the height of the ship antenna is 15m and the elevation angle is 5 deg, the correlation bandwidth is about 20 MHz [21], the value of which is larger than the allocated bandwidth of the current maritime mobile satellite services (i.e., 15 MHz or so). Therefore, frequency diversity does not seem promising at this stage. However, if the frequency bands become widened in the future, this method could be applicable.

Polarization Shaping Method [22]

For reflections from the sea surface at 1.6/1.5 GHz, the horizontally polarized wave is almost perfectly reflected, while the vertically polarized wave is reflected with a large attenuation at grazing angles below 20 deg. Thus, the polarization of the reflected wave becomes elliptical with the opposite sense of rotation with respect to the incident circular polarization when the elevation is above 6 deg, and its major axis is nearly horizontal. (An elevation angle of around 6 deg is the Brewster angle at 1.5/1.6 GHz over sea paths.) Accordingly, if the polarization ellipse of a shipborne antenna in the direction of the reflected wave could be adjusted so that it always stays orthogonal to that of the reflected wave, the reflected wave can be suppressed. This principle can be easily applied to cross-dipole-fed antennas, such as a short backfire antenna [22]. For instance, if the cross-dipole elements are inclined by 45 deg with respect to the horizontal line and the phase is adjusted by a phase shifter inserted in one of the ports of the short backfire antenna feed, as shown in Figure 6.14(a), the axial ratio of this antenna can be arbitrarily controlled. In this case, the major axis of the polarization ellipse is always vertical, as shown in Figure 6.14(b).

Figures 6.15(a–c) show the field experimental results [23]. Figure 6.15(a) shows the effects of reduction for down-link (1.54 GHz) fading and up-link (1.64 GHz) fading with the optimum phase shift for 1.54 GHz frequency at an elevation angle of 5 deg. Although there exists about 1.8-dB loss of the mean signal level of the reception, with optimum phase (ON) relative to that of reception with the phase of 0 deg due to the polarization mismatch, the fading depth for 99% of the time, including polarization mismatch loss of 1.8 dB, is consequently reduced by about 5 dB for both up- and down-link fadings. In this way, it is confirmed that the polarization shaping method (PSM) is capable of reducing fading on the up-link as well as the down-link.

Figure 6.15(b) shows the preset value of the optimum phase for the elevation angles. Generally, these optimum phases of the phase shifter are automatically

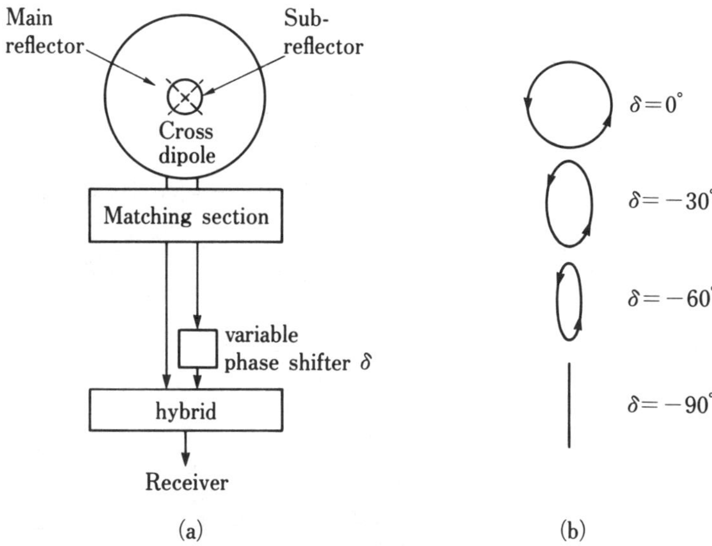

Figure 6.14 Short backfire antenna with polarization shaping method: (a) block diagram; (b) polarization ellipse of ship antenna.

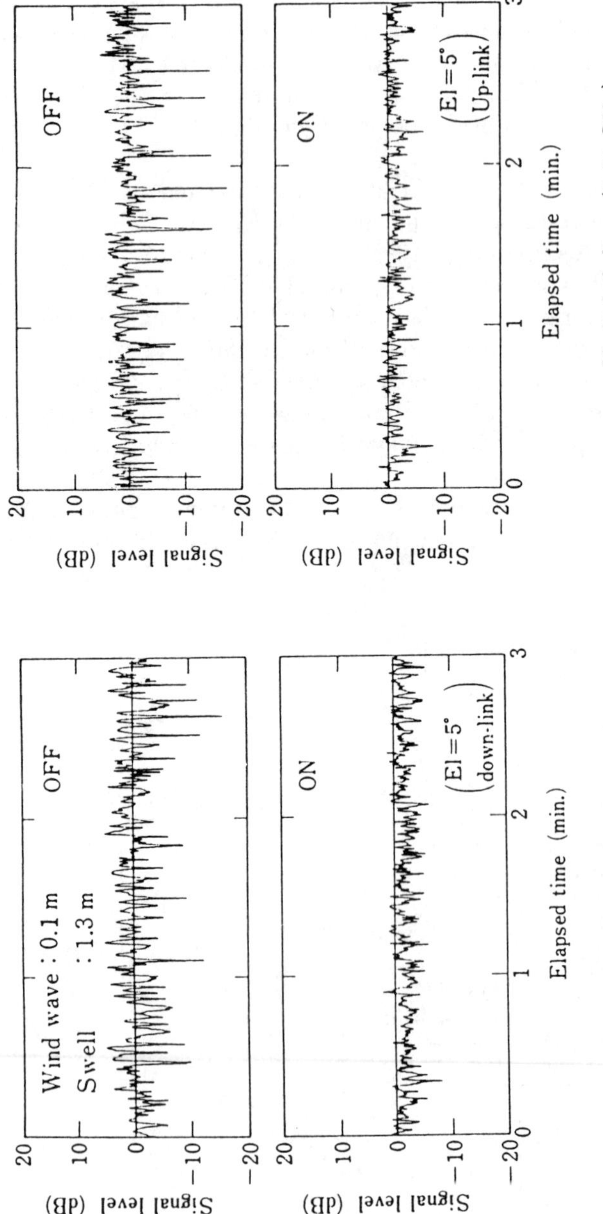

Figure 6.15a Field experimental results.

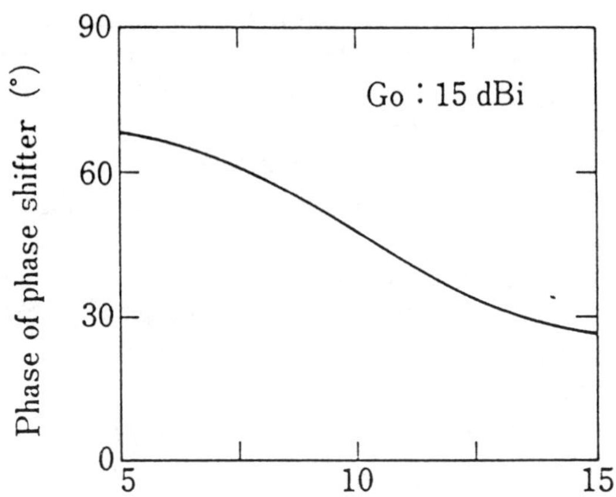

Figure 6.15b Continued.

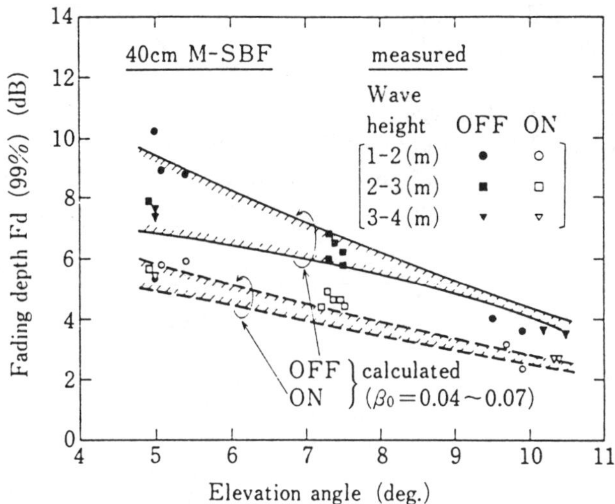

Figure 6.15c Continued.

provided with respect to the elevation angle. Figure 6.15(c) shows the fading depths as a function of the elevation angle, with (ON) and without (OFF) fading reduction. In the figure, data denoted by ON are the depths of fading received by the 40-cm modified SBF mentioned below, with optimally controlled polarization given by the additional phase as shown in Figure 6.15(b). Results indicate that multipath fading can be suppressed to the theoretically expected values for elevation angles from 5 to 10 deg and wave height up to about 3m.

This PSM is required to have a variable phase shifter to control antenna polarization according to the gradual change of elevation angle in order to realize the maximum fading reduction effect. From the viewpoint of simpler construction and easier installation, a simplified fading reduction technique combining the PSM and a beam offset method was developed (modified PSM [24]). This technique obtains almost the same fading reduction effect as the PSM for elevation angles from 5 to 10 deg by having only a fixed phase shift and making a slight beam offset.

Another reduction technique applying a polarization method has been presented in [25,26]. This technique suppresses the coherent component of sea-reflected waves, which the dominant component needed for the calm sea condition. The principle is as follows. Generally, in a circularly polarized antenna fed by two points, a hybrid combiner (3 dB-90 deg) is used as shown in Figure 6.16. The symbols T_1, T_2, T_3, and T_4 denote four terminals of the hybrid. From calculations, it is found that the right circular polarized component is on terminal T_3 and the left is on T_4, retarded by 90 deg in phase. In maritime satellite communications, right circular polarization has been adopted, so terminal T_4 is terminated with a 50Ω terminal.

On the other hand, a right circularly polarized wave is divided into two components of right and left circularly polarized waves by sea reflection. Figures 6.17(a,b) show the reflection coefficients of the waves in amplitude and phase. The electrical constants of sea water are chosen to be $\epsilon_r = 70$ for dielectric constant,

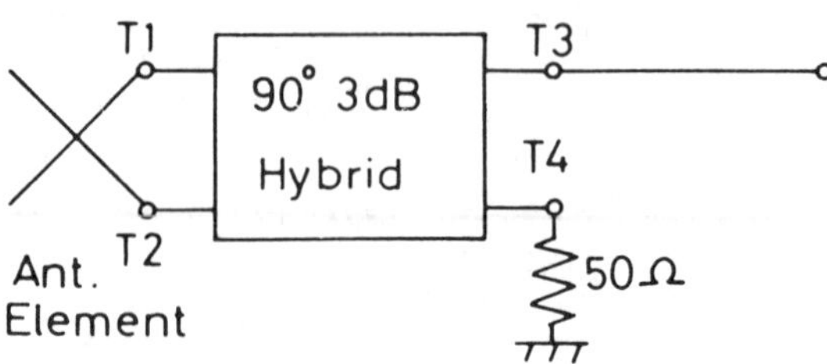

Figure 6.16 Circularly polarized antenna arrangement.

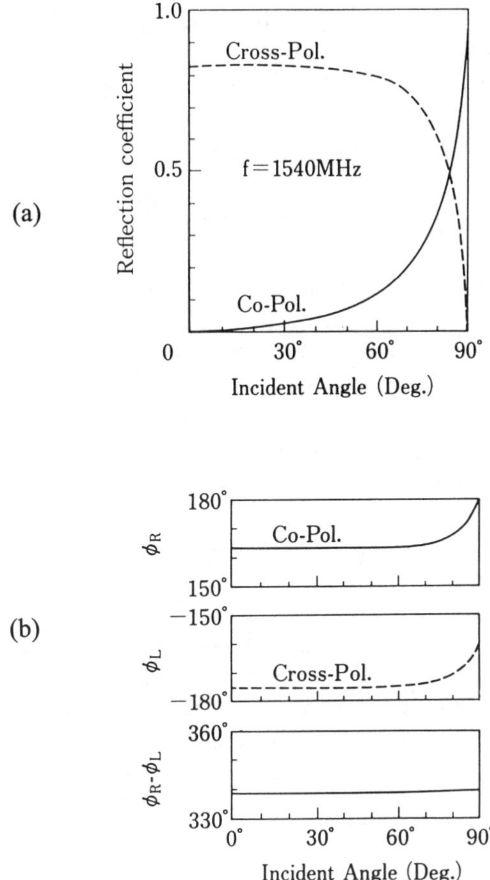

Figure 6.17 Reflection coefficient of sea surface: (a) amplitude; (b) phase.

σ = 5.5 S/m for conductivity, and f = 1.54 GHz for frequency. In the case of incident angles smaller than about 85 deg, which correspond to 5 deg in an elevation angle, the cross-polarized component is greater than the copolarized one in amplitude. The maritime satellite communication service has been provided within an ocean area where the elevation angle to a satellite is more than 5 deg. The phase difference between both components is almost constant and is independent of the elevation angles.

As mentioned above, the antenna outputs the direct copolarized component E_0 and the reflected copolarized component $E_R e^{j\Phi R}$ on terminal T_3, and the reflected cross-polarized component $E_L e^{j(\Phi_L - 90°)}$ on terminal T_4; that is,

$$T_3\text{: } E_0 + E_R e^{j\Phi R} \tag{6.2}$$

$$T_4\text{: } E_L e^{j(\Phi_L - 90°)} \tag{6.3}$$

The output power of the cross-polarized component on terminal T_4, however, has been dissipated in the conventional antenna circuit, as shown in Figure 6.18. From (6.2) and (6.3), the method is found to select only the direct copolarized component E_0 by combining two components from terminals T_3 and T_4 after making the amplitude of the reflected copolarized component E_R and that of the reflected cross-polarized component E_L to be equal, and letting the phases of the both components satisfy the equation

$$\Phi_R - (\Phi_L - 90° + \Phi) = 180° \tag{6.4}$$

where Φ is a value of phase shift.

To accomplish this fading reduction, an attenuator, a phase shifter, and a power combiner must be added to the ordinary antenna equipment, as shown in Figure 6.18. In this case, the direct component E_0 is attenuated by 3 dB because of the combiner. Figure 6.19 shows an experimental result of height-gain patterns, measured at elevation angles of 10 deg, of an ordinary SBF antenna and a proposed fading reduction SBF antenna.

Pattern Shaping Method

It is possible to suppress the reception of a reflected signal by using a shaped pattern antenna that has low-gain radiation characteristics in the direction of sea-reflected waves. This method may be realized by an array antenna or a shaped reflector

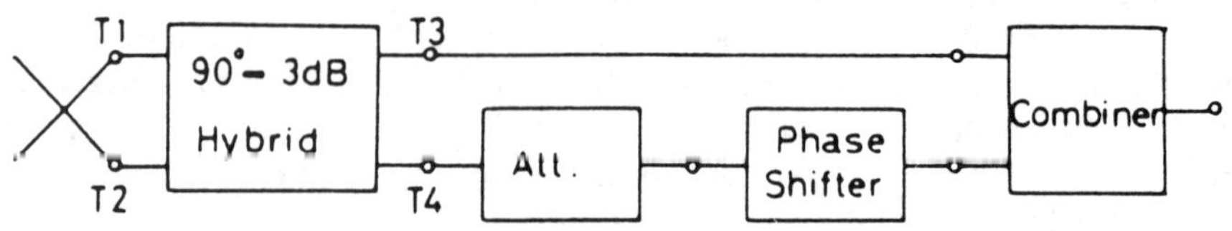

Figure 6.18 Another type of polarization shaping method.

(a)

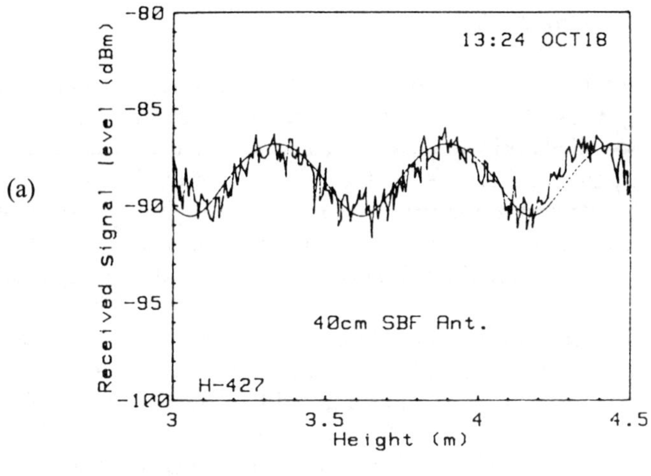

(b)

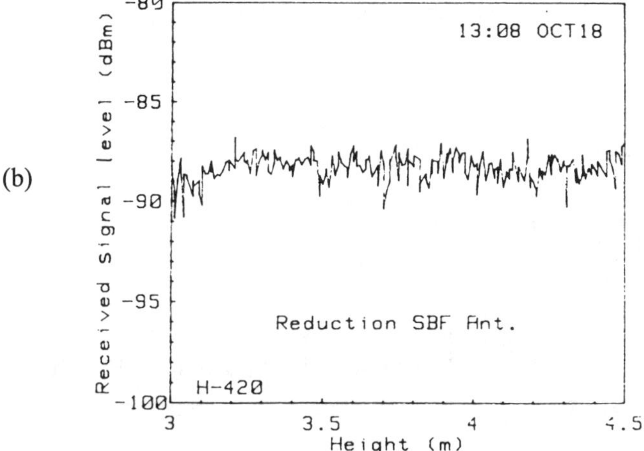

Figure 6.19 Experimental results of height pattern: (a) Ordinary SBF antenna. Smooth curve is a theoretical height pattern; (b) Reduction SBF antenna.

antenna. The radiation pattern is usually shaped so as to be flat in the main beam and to be suppressed in other directions [22]. This antenna is expected to have a good reduction effect when the reflected waves come from directions away from the main beam. The shaped pattern antenna, however, has the disadvantage that the aperture efficiency of the antenna generally becomes comparatively low.

Maximum-Level Tracking Method

If the radiation pattern of a receiving antenna is controlled so that the received signal intensity is always maintained at a higher level, the fading could be substantially suppressed. This principle can be easily applied to array antennas. The phases of variable-phase shifters inserted in the feed circuit for each antenna element are varied by a small amount in order to see whether the resultant signal level increases. The phase has to be adjusted at a rate sufficiently fast to track the speed of the fading. If the signal level increases, the control voltage is allowed to change continuously in the same direction. If not, the polarity of the control voltage has to be reversed to control the variable-phase shifter in the opposite direction. Fading can be reduced by repeating this operation. Figure 6.20 is a block diagram illustrating an example of this method as applied to a quad-helix antenna [22].

A field experiment has been carried out around a bay to receive a signal transmitted from the shore [22]. In this experiment, a quad-helix antenna (shown later on in Figure 6.42) was used for the signal reception and had a gain of about 13 dBi and an axial ratio of about 1.0 dB. The elevation angle was 7.5 deg and the wave height was 15 to 20 cm, corresponding to fairly calm sea conditions. The

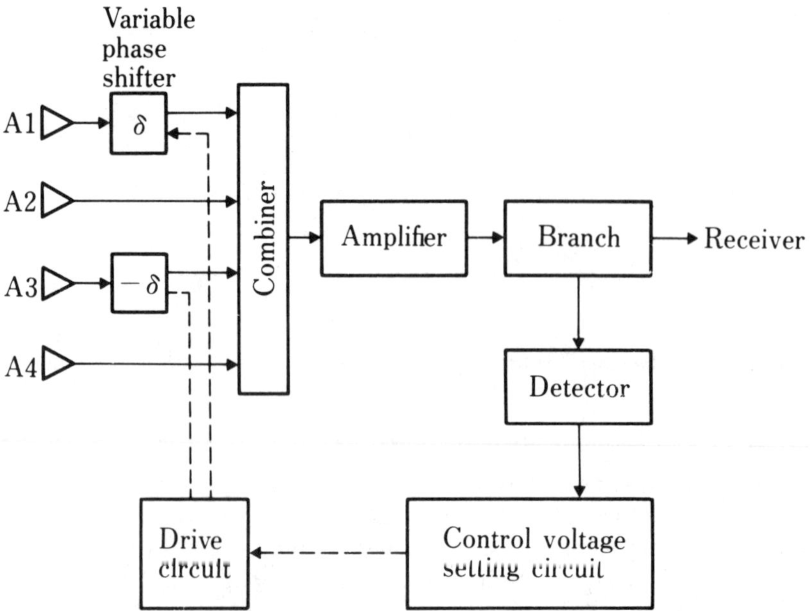

Figure 6.20 Maximum-level tracking method.

receiving antenna was fixed at the minimum-level point of the height pattern to evaluate the worst-case fading, where the received signal level is minimum. Figure 6.21 shows the cumulative time distribution of the received signal intensity. The fading depth of about 11 dB, corresponding to 99% of the time, has been reduced to about 7.5 dB.

Method Suitable for Digital Systems

It is well known that forward error correction (FEC) with bit interleaver can reduce the effect of multipath fading where fade duration is shorter than interleave period. The method is incorporated in the INMARSAT-C system. However, in the method, the fading reduction effect will be decreased when the fading with very slow fluctuations is dominant, as in very calm sea conditions. To counter such a fading, a new type of fading reduction scheme was proposed [27]. The fading reduction scheme combines an open-loop-controlled space diversity with an FEC and a block interleaver originally employed in the INMARSAT-C. The method is exceptionally useful in a calm sea when the coherent component of the reflected wave is far stronger than the incoherent component, and it has a considerable degree of fading reduction on the up-link path as well as the down-link path with a very simple configuration. This overcomes the disadvantages generally associated with diversity

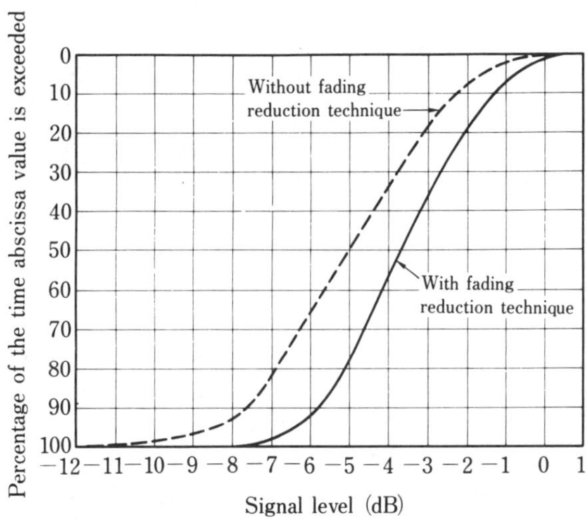

Figure 6.21 Cumulative time distribution of fading.

systems, and moreover, the open-loop-controlled diversity of this system, in principle, can shorten the interleave length.

Figure 6.22 shows a configuration of the INMARSAT-C SES with an RF circuit that implements this scheme. In the figure, the area inside the broken line is the new circuitry to be added to the current INMARSAT-C system. The two antennas (#1 and #2) are set at a vertical distance apart L_a, and the signals received by each antenna are combined in a ratio that varies with a specified period shorter than the interleave time period by means of a variable combiner/divider. Figure 6.22 also gives an example of the configuration of such a circuit. In this circuit, phase shifter #1 is used to make the direct wave components received by the two antennas in phase, and the combination ratio periodically changes by varying the phase of phase shifter #2. In this case, the circuit is composed of four very simple components (i.e., two phase shifters, one 90 deg hybrid, and one equal-ratio combiner/divider) and the loss due to its insertion is considered to be small. Effect of the fading reduction is assessed in terms of bit error rate (BER) improvement. Figure 6.23 shows the BER characteristics at an elevation angle of 5 deg (the average power ratio of carrier to multipath wave (C/M) = 7.0 dB) under calm sea conditions. The dotted line represents the ordinary FEC + interleave system in the worst case, while the solid line shows improvement by applying the method. From the figure, it can be seen that apparent BER improvement can be obtained.

6.2.5 Mount Systems

In the case of mobile communications, the antenna is always required to be pointed toward the satellite in spite of any vehicle's motions. Accordingly, the mount system is one of the key problems in designing the mobile antenna system from the technical and economic viewpoints. In this section, several types of mount system are introduced, especially those concerned with shipborne antenna systems. On the other hand, in the case of aeronautical and land mobile communications, electrical pointing systems are preferable.

Two-Axis Mount (El/Az, Y/X)

An antenna mount is a mechanically moving system that can maintain the antenna beam in a fixed direction. In maritime satellite communication systems, the mount must have a function to point in any direction on the celestial semisphere, because the ship will sail across the sea. It is well known that the mount of the two-axis configuration is the simplest mount providing such functions. There are two typical mounts of the two-axis configuration: one is the El/Az mount and the other is the Y/X mount. Simplified stick diagrams of both mounts are shown in Figure 6.24. In the El/Az mount, a full steerable function can be obtained by choosing the

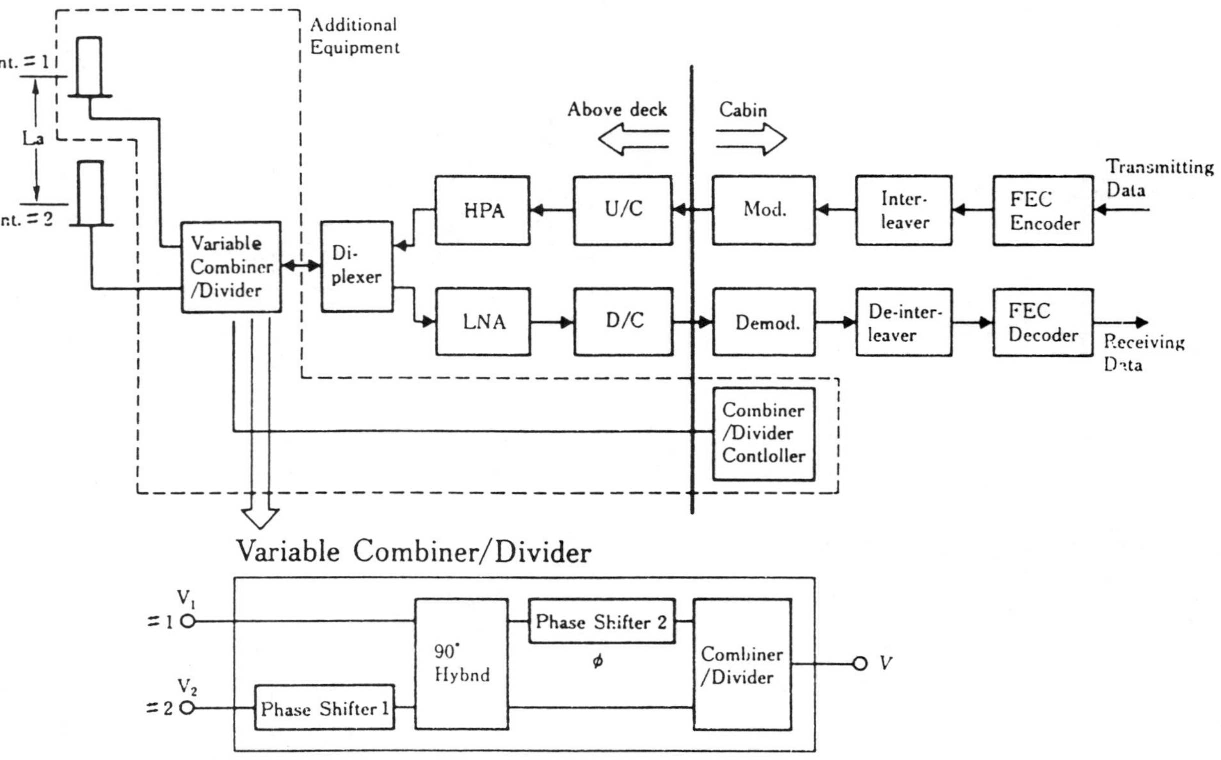

Figure 6.22 Configuration.

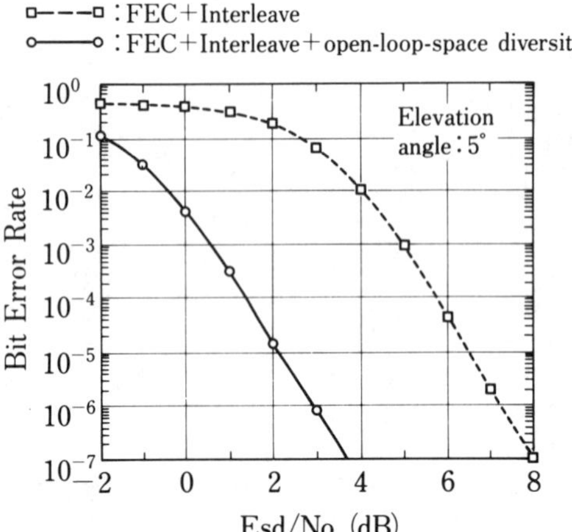

Figure 6.23 BER characteristics.

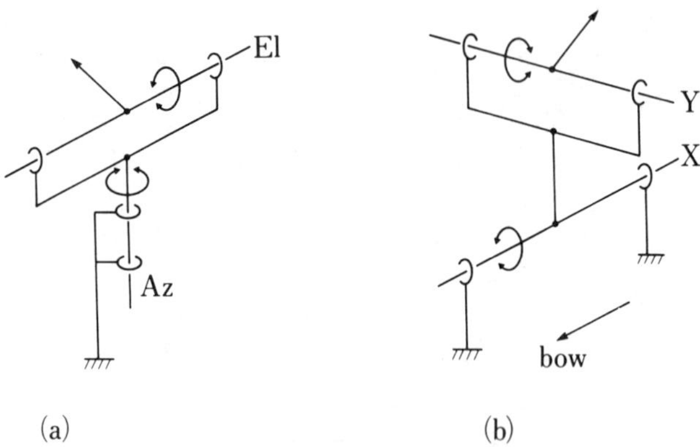

Figure 6.24 Two-axis mount systems: (a) El/Az; (b) Y/X.

rotation range of the azimuth axis (Az-axis) from 0 to 90 deg. In the Y/X mount a full steerable function is achieved by permitting the rotation angle from −90 to +90 deg to both the X-axis and Y-axis.

The basic axis configuration for the SES is also the two-axis configuration such as the El/Az mount and the Y/X mount. A special function, however, is required for the SES mount system (i.e., the compensation of ship motions due to sailing and ocean waves) to keep the antenna beam in nearly a fixed direction in space. In general, the ship motion consists of seven elements: turn, roll, pitch, yaw, heave, surge, and sway. Turn is change of headway. The components are illustrated in Figure 6.25. In these components, heave, surge, and sway are motions with the acceleration. The pointing errors due to acceleration can be made small enough for practical use by a careful adjustment of weight balance around each mount axis. The characteristics of yaw are the same as turn.

In the INMARSAT, specification of ship motions turn, roll, and pitch are the dominant components to be compensated by the SES mount system. In the case of the pointing/tracking under ship motions (see Sec. 6.2.6), the required rotation angle range of each axis is from 0 deg to more than 360 deg for the Az-axis and from -25 to $+120$ deg for the El-axis with respect to the deck, assuming that the operational elevation angle is restricted above 5 deg. Therefore, a few difficulties arise when the two-axis mount, the El/Az mount, or the Y/X mount is adopted to the mount system of SES [28]. Namely, there are operational restrictions for such two-axis mounts. The operational restrictions for the El/Az mount are as follows.

- A limitation on the Az-axis rotation is needed to avoid the cable wrapping around the Az-axis. When the rotation of the Az-axis reaches one end, the axis has to reverse the rotation at a high speed, and consequently the signal will be intermittent during the reverse rotation.
- Either slip rings or rotary joints are needed to permit the endless rotation around the Az-axis.
- In the case of the El/Az mount, very high angular rates and accelerations are required for accurate satellite tracking at a high elevation angle. This con-

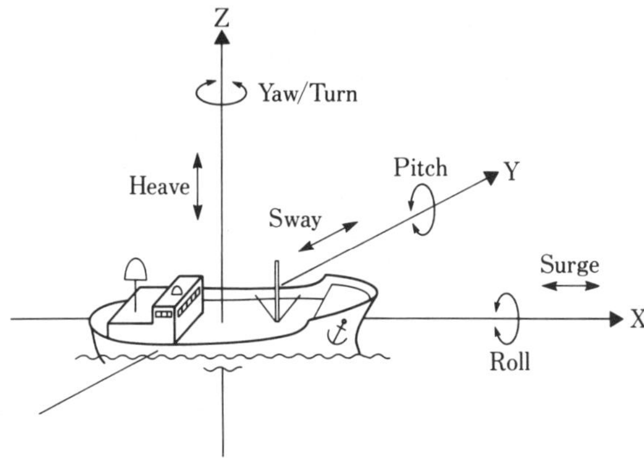

Figure 6.25 Ship's motion components.

dition is referred to as the *gimbal lock* of the El-axis. The high angular rates and accelerations require a large and powerful drive mechanism.

On the other hand, the Y/X mount has the following restrictions.

- Both the X-axis and Y-axis rotate approximately in a range of −120 to +120 deg, and the cables have to bear the bending or the twist within ±120 deg.
- At low elevation angles during ship motions such as turn and yaw, an extremely high angular rate and acceleration must be provided for accurate satellite tracking. This is the gimbal lock of the Y/X mount.

Among those constraints mentioned above, the most serious one is the gimbal lock problem, which causes large tracking errors that interrupt communications. As stated above, one method to overcome the gimbal lock is to use a mechanism of high angular rate and acceleration, but this method is not practical. Another method is to attach one or two axes to eliminate the rapid motion of a mount axis. In the case of a two-axis mount system, gimbal lock is a severe problem for mobile antenna systems to solve. As described in [29], new techniques have been proposed in order to overcome these defects.

Three-Axis Mount (El/Az/X, El'/El/Az, X'/Y/X)

The three-axis mount system is considered to be a modified two-axis mount which has one additional axis. The typical configurations of the three-axis mount are shown in Figure 6.26. The three-axis mount of an El/Az/X type shown in Figure 6.26(a) is the El/Az mount with one additional X-axis [28]. The function of the X-axis is to eliminate the rapid motion of the two-axis mount due to roll. In this system, however, the possibility of the gimbal lock for pitch is still left near the zenith, when the El-axis is parallel to the X-axis. The three-axis mount of an

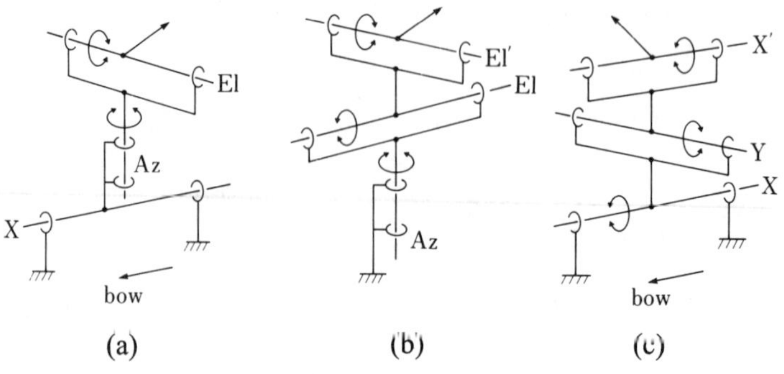

Figure 6.26 Three-axis mount systems: (a) El/Az/X; (b) El'/El/Az; (c) X'/Y/X.

El′/El/Az type shown in Figure 6.26(b) is the El/Az mount with an additional cross-elevation axis El [20]. In the mount system, the change of the azimuth angle is tracked by rotating the Az-axis, and the change of the elevation angle is tracked by a combined action of the El and El′ axes. As can be seen clearly in the figure, the El′ and El axes allow movements in two directions at a right angle. With an approximate axial control, this mount is free from the gimbal lock problem near both the zenith and the horizon.

The three-axis mount of an X′/Y/X type is the two-axis Y/X mount system with the X′-axis on it to remove the gimbal lock at the horizon. The configuration is shown in Figure 6.26(c). When the satellite is near the horizon, the X′-axis takes out the rapid motion due to yaw and turn. Note that the X′-axis can rotate within ±120 deg, so the X′-axis can only eliminate the rapid motion within the angular range. In general, the axis control of the three-axis mount is rather more complex than that of the four-axis mount, because steering and stabilization interact with each other.

Four-Axis Mount (El/Az/Y/X)

In the mount, the stabilized platform is made by the X-axis and Y-axis to take out roll and pitch, and the two-axis mount of the El/Az type is settled on the stabilized platform. The axis configuration is shown in Figure 6.27. The tracking accuracy of the four-axis mount is, of course, the best of all mount systems because the stabilization function is separated from the steering function; that is, four major components such as roll, pitch, azimuth, and elevation are controlled by its own axis individually. Therefore, the four-axis mount has been adopted in many SESs of the current INMARSAT-A system. However, the cost would increase by introducing servo mechanisms and attitude sensors for stabilization.

6.2.6 Tracking/Pointing Systems

Pointing/tracking is another important function required of the mount system. It should be noted that the primary requirements for SES tracking systems are that they be economical, simple, and reliable. Tracking performance is a secondary requirement when an antenna beamwidth is broad.

Manual Track

The manual track is the simplest method, wherein an operator controls the antenna beam to maximize the received signal level. At first, the operator acquires the signal and moves the antenna around one axis of the mount. If the signal level

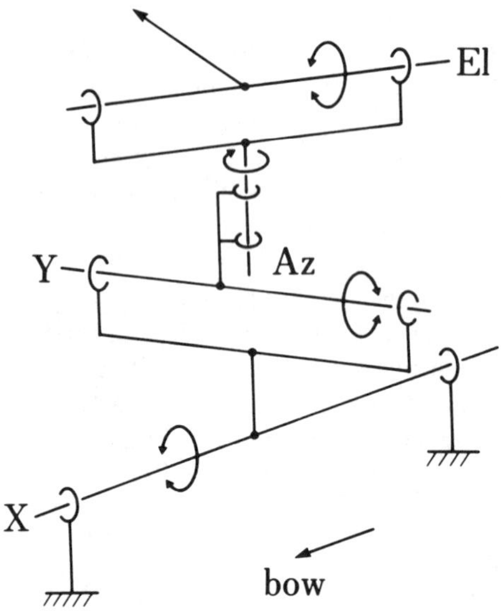

Figure 6.27 Four-axis mount system.

increases, the operator continues to move the antenna in the same direction. If it decreases, he or she reverses the direction and continues to move the antenna until the signal level is maximized. The same process is repeated around the second axis and the antenna is held after both axes are optimized. The operator will try again to find the optimum points for both axes when the received signal level decreases. This method is suitable for land mobile communication, especially portable communication and fly-away communication terminals.

Step Track

Among various autotrack systems, the step track system has recently been recognized as a suitable tracking system for SES because of its simplicity for moderate tracking accuracy. The recent development of integrated circuits and microprocessors has brought a remarkable cost reduction to the step track system. The principle of the step track system is the same as that of the manual track. The only difference is that an electric controller plays the role of an operator in the manual track. The schematic block diagram of the step track system is shown in Figure 6.28. Sample/hold circuits are used to hold the signal levels. The signal levels are compared before and after the antenna has been moved by a preset angular step

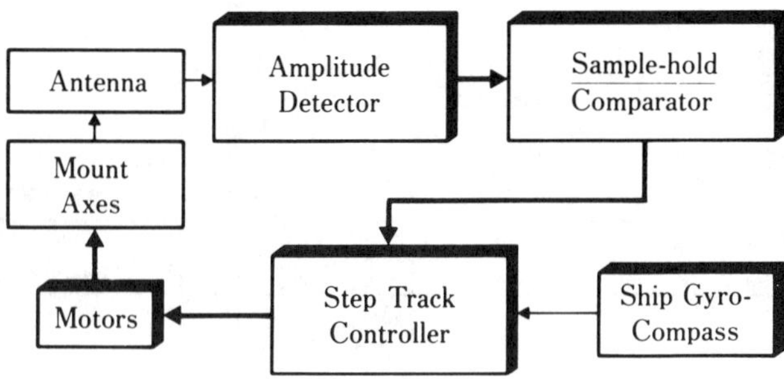

Figure 6.28 Functional block diagram of step track.

If the level is increasing, the antenna is moved in the same direction. If the level is decreasing, the direction will be reversed. This process will be carried out alternately between two axes. As a result, the beam center is maintained in the vicinity of the satellite direction. As can be clearly seen from the principle of the step track system, the accuracy depends on the comparator's sensitivity. Wrong decisions on the comparison of levels generally arise from the signal-to-noise ratio and the level changes due to the multipath fading and the stabilization error.

Program Track

The concept of the program track system is based on the open-loop control slaving to the automatic navigation equipment, such as a gyrocompass, GPS, the Omega system, and the Loran-C system. The navigation equipment, especially the Navy Navigation Satellite System (NNSS), is widely used in many ships. In the program track, the antenna is steered to the point of the calculated direction based on the position data of the navigation equipment. Since the satellite direction changes because of roll, pitch, and turn, a function to remove the rapid motions is required in the program tracking. The schematic block diagram of the program track is shown in Figure 6.29.

Recent navigation equipment has excellent accuracy. For example, the error of NNSS is some hundred meters at cruising condition and less than a hundred meters at anchor. This error is negligibly small for the program track system. Therefore, the error of the program track system mainly depends on the accuracy of sensors for roll, pitch, and turn (i.e., the stabilization error). An adequate sensor for the program track system is a vertical gyro, because it is hardly affected by the lateral acceleration. When the stabilization requirement is lenient, the conventional level sensor, such as an inclinometer, a pendulum, and a level, may be used with

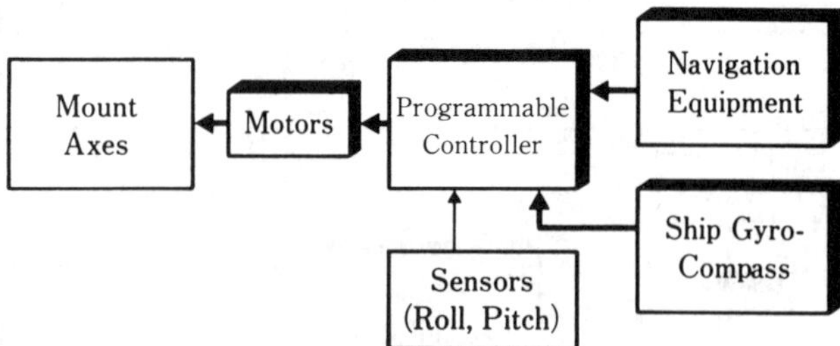

Figure 6.29 Functional block diagram of program track.

careful choice of the sensor's location. The controller calculates the direction of the satellite to compensate for the ship motions. Generally speaking, the simpler the axis configuration of the mount, the more complex the calculation procedure becomes. Since the controller has to execute calculations of many trigonometric functions, a microprocessor is a candidate for the controller. The program track system is also applicable to the four-axis mount. A combination with the step track system is more desirable, because the error of the program track system can be compensated for by the step track system, and the error of the step track due to rapid ship motions can be compensated for by the program track system.

6.3 OMNIDIRECTIONAL ANTENNAS FOR MOBILE SATELLITE COMMUNICATIONS

6.3.1 Overview

In this section, three types of basic omnidirectional antennas are introduced. As described in the previous section, antennas for mobile satellite communications are classified into omnidirectional and directional antennas. The gain of omnidirectional antennas is generally from 0 to 4 dBi in the L-band, which does not require the capability of satellite tracking. In a family of omnidirectional antennas for mobile satellite communications, there are three basic antenna elements. The first is a quadrifilar helix, the second is a crossed-drooping dipole, and the third is a microstrip patch antenna. In this section, these omnidirectional antennas, which were mainly studied and breadboarded by the Jet Propulsion Laboratory (JPL) are introduced by quoting from [30]. These omnidirectional antennas are very attractive, owing to the possible small size, light weight, and circular polarization properties. These basic antennas are also used as elements of directional antennas, which will be mentioned in later sections.

6.3.2 Quadrifilar Helix Antenna

The quadrifilar helix antenna [31] is composed of four identical helixes wound, equally spaced, on a cylindrical surface. The helixes are fed with signals equal in amplitude and 0, 90, 180, and 270 deg in relative phase. Figure 6.30 shows the elevation patterns that can be achieved from three different quadrifilar antennas for various angular coverage requirements. General characteristics are: gain, −4 dBi minimum; axial ratio, 3 dB maximum; and height, 40 cm.

6.3.3 Crossed-Drooping Dipole Antenna

The crossed-drooping dipole antenna is the most interesting for land mobile satellite communications, where required angular coverage is narrow in elevation and is almost constant in azimuth. By varying the separation between the dipole elements and the ground plane, the elevation pattern can be adjusted for optimum coverage

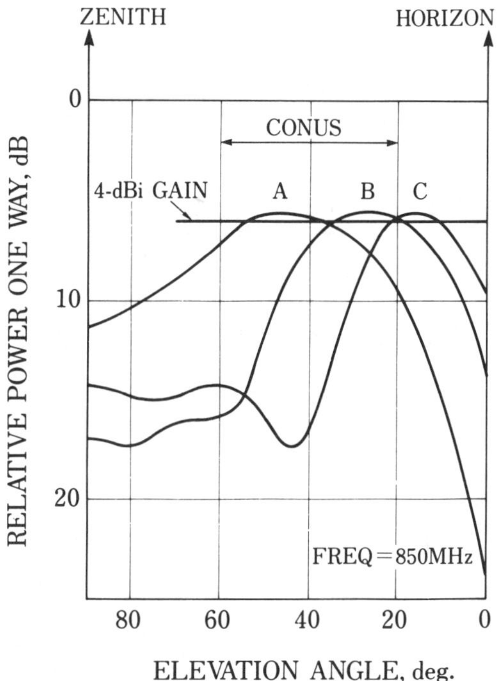

Figure 6.30 Quadrifilar helix antenna patterns. (Courtesy of JPL.)

for the coverage region of interest. Figure 6.31 [30] shows patterns of a crossed-drooping dipole antenna. By adjusting the height of the antenna from the ground plane, a 4-dBi gain can be achieved over the entire continental United States (CONUS). General characteristics are: gain, 4 dBi minimum; axial ratio, 6 dB maximum; and height, 15 cm.

6.3.4 Patch Antenna

The circular microstrip disk antenna [32] has a circular metallic disk supported by a dielectric substrate material over a ground plane. In order to produce circularly polarized waves, a patch antenna is, in general, excited at two points orthogonal to each other and fed with signals equal in amplitude and 0 and 90 deg in relative phase. A higher mode patch antenna [33] can also be designed to have a similar

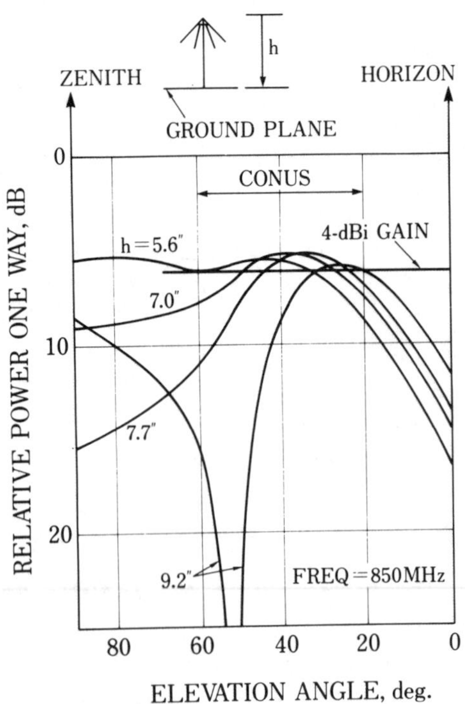

Figure 6.31 Crossed-drooping dipole antenna patterns. (Courtesy of JPL.)

radiation pattern to the drooping dipole. To produce conical radiation patterns (null on axis) suitable for land mobile satellite applications, the antenna is excited at higher order modes. Figure 6.32 shows the radiation patterns for different modes. General characteristics are: gain, 3.5 dBi minimum; axial ratio, 4 dB maximum; and height, 1 cm.

RT/Duroid with dielectric constant of 2.3 (thickness = 3.2 mm) is used as a dielectric substrate because of its good temperature characteristics. Since the available frequency bandwidth of a patch antenna is very narrow, the two-layer patch antenna [34] is adopted in which the upper and lower parts play a role for transmission and reception, respectively. Each layer antenna is individually fed at two points with a phase difference of 90 deg in order to obtain the circular polarization. The upper layer antenna is the well-known conventional circular microstrip antenna (MSA) [35], while the lower one is a circular MSA with an electric shielding ring which supplies enough space for the upper antenna to be easily fed [36]. Another useful printed antenna element is the cavity-backed cross-slot antenna (XSA). Each

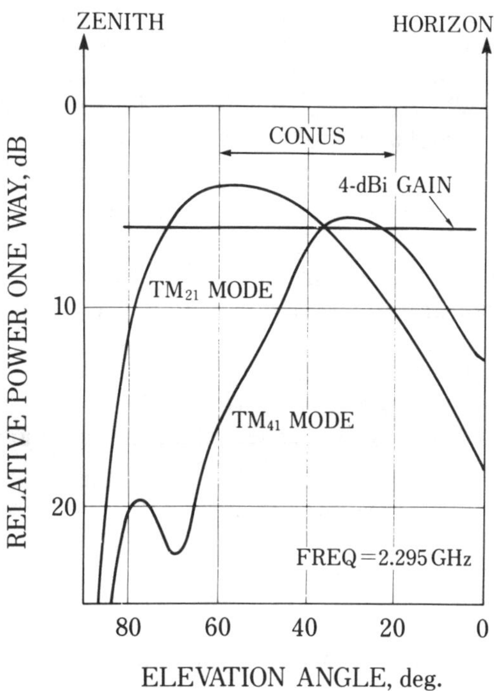

Figure 6.32 S-band circular microstrip disk antenna patterns. (Courtesy of JPL.)

slot antenna is fed with an equal amplitude and in-phase condition at two points located equally from the center [37]. As a result, the input impedance can be matched for a wider frequency band than in the case of the slot antenna. In the case of the MSA, the antenna gain at boresight is about 15.2 dBi and a gain of about 13.5 dBi is obtained at a scanning angle θ of 45 deg, while for the XSA, the boresight gain is about 15.7 dBi and the gain at $\theta = 45$ deg is about 14 dBi. It can be shown from these figures that the degradation of antenna gain for both antennas is not so large, even if the scanning angle is 45 deg.

6.4 DIRECTIONAL ANTENNAS FOR MOBILE SATELLITE COMMUNICATIONS

6.4.1 Antennas for INMARSAT

INMARSAT initiated the operation of its worldwide maritime satellite communication in February of 1982. Now INMARSAT has three standards: INMARSAT-A, INMARSAT-C, and INMARSAT-Aero. INMARSAT-B and INMARSAT-M will begin operation in the near future. Table 6.3 summarizes the specification of current and future INMARSAT standards. As can be seen, the communication qualities, such as G/T and EIRP, are different for each standard. Many types of antennas can be selected as candidates of these standards:

- Aperture antennas: paraboloidal antenna, parabolic cylinder antenna with fan beam, four-element short backfire antenna, short backfire antenna, and others;
- Wire antennas: helical antenna, cross Yagi/Uda antenna, log-periodic antenna, quadrifilar helix antenna, monopole antenna, crossed-drooping dipole antenna, and others;
- Microstrip/slot antennas; and
- Array antennas: cross-Yagi/Uda array, helical array, spiral array, microstrip/slot array, switched-element spherical array, and others.

In this section, the candidate antennas for each of the INMARSAT standards are introduced and the typical mobile antenna systems are presented.

6.4.1.1 Antennas for INMARSAT-A and -B

INMARSAT-A was the first operating standard for maritime satellite communications. It was started in 1982 using the first-generation INMARSAT satellites and provides voice (analog. compounded FM (CFM)), facsimile, and telex services.

INMARSAT-B will be started at the end of 1993 using the second-generation INMARSAT satellites. This standard is compatible with INMARSAT-A, and digtal techniques will be introduced in order to respond to the expected growth of communication demands for maritime satellite communications. The same antenna specification as that of INMARSAT-A is used; that is, G/T is required to be −4 dBK for INMARSAT-B SES, as can be seen from Table 6.3. However, the EIRP of INMARSAT-B is less than that of INMARSAT-A by 3 dB due to the fact that the slightly shaped global beam antenna is adopted in the second-generation INMARSAT satellites. Considering the desired EIRP and the noise performance of current low-noise amplifiers, the G/T of −4 dBK and the EIRP of 33 dBW can be obtained by an antenna with a gain of about 20 dBi.

For INMARSAT-A and -B systems with antenna gains of about 20 dBi, an aperture antenna such as a paraboloidal antenna is suitable because of the high aperture efficiency. The paraboloidal antenna (see Figure 6.33) is especially desirable for INMARSAT-A and -B systems, because it is simple in its structure and can be designed flexibly with respect to its antenna gain when the gain is more than around 20 dBi.

6.4.1.2 Antennas for INMARSAT-M

INMARSAT-M has been operated with a small antenna to meet the requirements of a small-size and low-cost SES, because INMARSAT-M is mainly for small ships, such as fishing boats, and land vehicles. Assuming the same satellite EIRP as that of the current CFM channel, the G/T is −10 dBK for the maritime mobile version and −12 dBK for the land mobile version in order to get a sufficient quality of planned service. Under these conditions, the antenna gains ranging from 13 to 16 dBi are necessary for getting the required G/T. This is compatible with one of the baseline assumptions on INMARSAT-M concerning the efficient utilization of satellite power.

The primary requirement for ship antennas of INMARSAT-M is simplicity (i.e., small in size, light in weight, and simple in configuration). Therefore, the antenna applicable to this system will be a single antenna of small size or an array antenna composed of element antennas with high efficiencies. For the paraboloidal antenna with an antenna gain that ranges from 13 to 16 dBi, it will become a difficult problem to illuminate the paraboloid efficiently with the primary radiator because the diameter of the paraboloid is smaller than that of a high-gain antenna. Therefore, this type of antenna is not adequate for the INMARSAT-M system compared to other types of antennas with a high aperture efficiency, such as a SBF antenna. Traveling-wave antennas such as a cross-Yagi/Uda array, a single helical antenna, and a log-periodic antenna have high efficiency for the moderate gains from 13 to 16 dBi. However, since those antennas are comparatively long in their axial direction, the volume of a radome becomes fairly large and conflicts with the requirement of INMARSAT-M. We have the options described below.

Figure 6.33 Example of antenna system for INMARSAT-A and -B paraboloidal antenna with four axis mount (El/Az/Y/X).

Short Backfire Antenna

The SBF antenna (see Figure 6.34) was developed experimentally by H. W. Ehrenspeck in the 1960s [38,39]. This antenna is well known as a highly efficient antenna of simple and compact construction. Its high-directivity and low-sidelobe characteristics make this antenna a single antenna with high-gain characteristics, which is applicable to satellite communications, tracking, and telemetry. Moreover, due to its excellent radiation characteristics, this antenna and a modified SBF antenna (mentioned below) have been recently proposed for shipborne antennas of INMARSAT-M. A theoretical study of the antenna has been recently performed

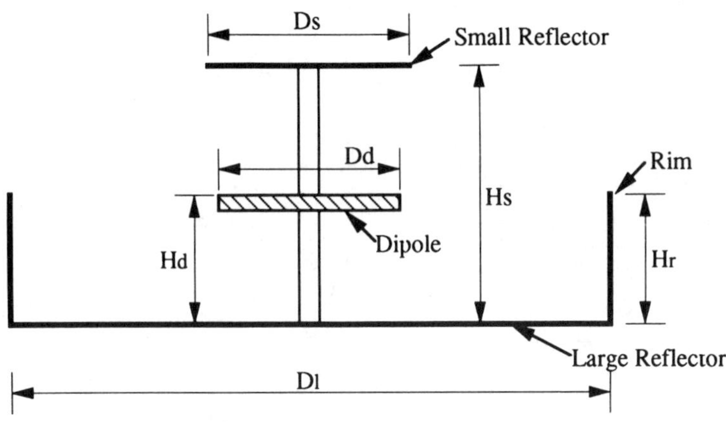

Figure 6.34 Short backfire antenna.

[40]. The SBF antenna consists of two circular planar reflectors of different diameters, separated generally by about one-half wavelength, forming a shallow leaky cavity resonator with the radiation beam normal to the small reflector. This antenna is fed by a dipole at around the midpoint between two reflectors. The antenna has almost a quarter-wavelength rim on the larger reflector. In Figure 6.34, D_l, D_s, H_r, H_s, H_d, and D_d are the diameter of large reflector, the diameter of small reflector, the rim height, the distance between small and large reflectors, the distance between large reflector and feed dipole, and the antenna length of feed dipole, respectively.

Half-power Beamwidth. Figure 6.35 shows the HPBW of a SBF antenna with D_s as parameters. Calculated results of the E-plane are presented with a solid line and of the H-plane with a broken line. Measured results are also shown. The other antenna parameters are $D_l = 2.0\lambda$, $H_r = 0.25\lambda$, $H_s = 0.5\lambda$, $H_d = 0.25\lambda$, and $D_d = 0.48\lambda$. From this figure, we can see that the HPBW is remarkably affected by the diameter size of the small reflector, and as the diameter becomes larger, the HPBW becomes smaller. There is good agreement between the calculated and measured values.

Sidelobe Level. Figure 6.36 shows the example of sidelobe level versus rim height (H_r). The numerical results of the E-plane are shown by a solid line, and that of the H-plane by a broken line. Measured results are also shown. The other antenna parameters are $D_l = 2.0\lambda$, $D_s = 0.5\lambda$, $H_s = 0.5\lambda$, $H_d = 0.25\lambda$, and $D_d = 0.48\lambda$, respectively. Note that the rim plays an effective role in suppressing the sidelobe

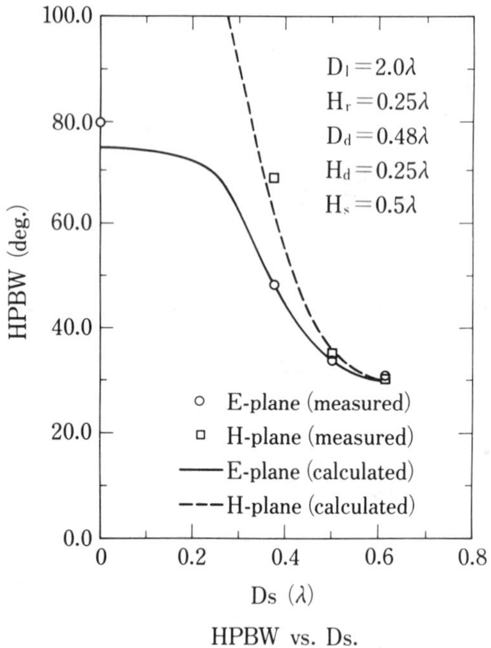

Figure 6.35 HPBW of SBF antenna.

level, and in the case of H_r more than about 0.2λ, the sidelobe level becomes less than 20 dB.

Antenna Gain. Figure 6.37 shows the antenna gain versus the large reflector's diameter with parameters of the small reflector's diameter. In this figure, the indication of aperture efficiency is also denoted by broken lines. We can see from this figure that for a size of D_l around 2.0λ, the aperture efficiency is about 80% while in the case of D_l less than about 1.5λ, more than 90% seems to be expected. The maximum gain of the SBF antenna is about 15 dBi as the size of small reflectors varies from 0.5λ to 0.7λ, and optimum D_l is around 2.24λ almost regardless of the size of the small reflector. According to our studies, the dependence of the size of H_r, H_s, and H_d on the antenna gain characteristics is relatively small.

Modified SBF Antenna

As mentioned before, the basic configuration of the SBF antenna consists of a cross dipole element, which is required to generate a circularly polarized wave, large and small reflectors, and a circular metallic rim [41]. This SBF antenna has strong

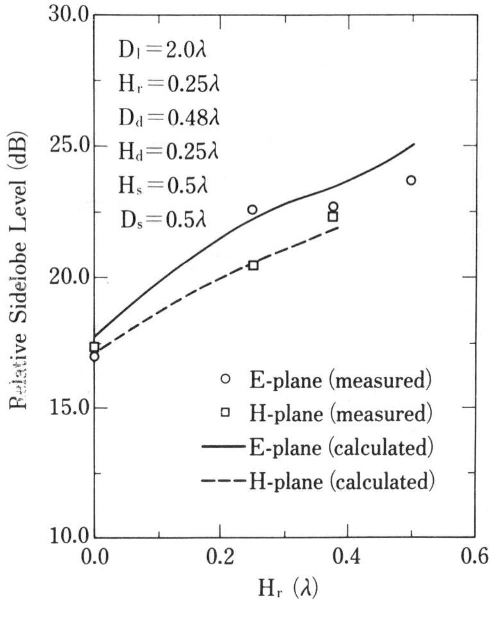

Figure 6.36 Sidelobe level of SBF antenna.

directivity normal to the reflectors, and its performance is superior to that of other types of antennas with the same diameter; however, it has the problem of narrow frequency band characteristics. A modified SBF antenna differs from the conventional SBF antenna in that there is either an additional step on the large reflector, as shown in Figure 6.38(a), or a change in the shape of the large reflector from a circular to conical plate, as shown in Figure 6.38(b), in order to improve the gain characteristics and the frequency bandwidth of the VSWR. The antenna parameters are presented in Figures 6.38(a,b). In Figure 6.38(a), D_1, D_2, h_s, d_1, d_2 and h_r are the diameter of large reflector, the diameter of the stepped inner reflector, the step height, the diameter of larger small reflectors, the diameter of smaller small reflectors, and the cylindrical rim height, respectively. Figure 6.39 shows the experimental values for the gain and axial ratio for varying step heights h_s (measured at 1.5 GHz) associated with the parameter D_2 as in Figure 6.38(a). In Figure 6.39, the value at $h_s = 0$ is that of the conventional SBF antenna. This figure makes it clear that there is a specified step height h_s corresponding to each D_2, and that a step height of 15 to 20 mm when $D_2 = 270$ mm provides improvements of about 1 dB in gain and about 0.3 dB in axial ratio over that of the conventional SBF.

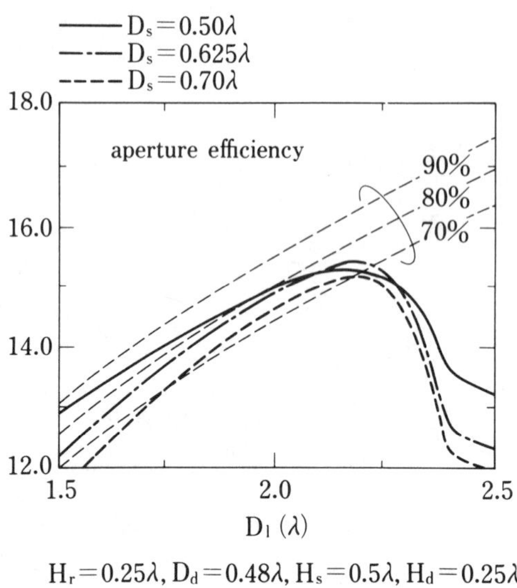

Figure 6.37 Antenna gain (dB) of SBF antenna.

On the other hand, the dual reflectors improve the input impedance characteristics covering the frequency range between transmitting and receiving. (The conventional SBF antenna is a resonant-type antenna, making the input impedance characteristics narrow in bandwidth, so wider bandwidth is required to cover the 1.6/1.5-GHz range.) The improvement in the input impedance is greatly dependent on the sizes and the separation of the small reflectors. The VSWR can be reduced from 1.7 and 1.5 (at 1.54 and 1.64 GHz) to below 1.2 for each frequency.

The optimum dimensions for the antenna are obtained experimentally and are: D_1 = 400 mm, D_2 = 270 mm, h_s = 20 mm; d_1 = 95 mm, d_2 = 85 mm, h_r = 50 mm. At the frequency of 1.54 GHz, these dimensions yield the gain and the axial ratio of 15.5 dBi and 0.7 dB, respectively. Further improvements to the SBF antenna have been described [42]. All symbols in Figure 6.38(b) are defined in Table 6.6.

Quad-Helix Antenna

Since an axial mode helical antenna has good circular polarization characteristics over a wide frequency range, it has been put into practical use as a single antenna or as array elements [43]. With respect to the structure, this antenna can be considered a compromise between the dipole and the loop antennas, and the radiation mode

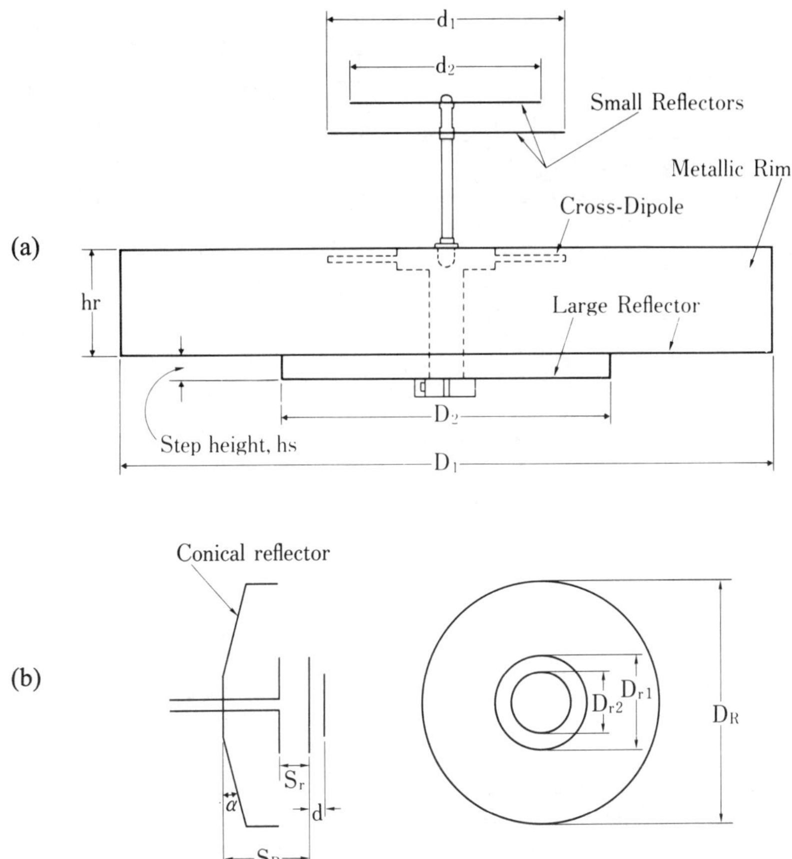

Figure 6.38 (a) Modified SBF antenna; (b) improved SBF antenna.

varies with the pitch angle and the circumference of the helix. In particular, the helix with the pitch angle α of 12 to 15 deg and the circumference C_λ of about 1λ has a sharp directivity toward the axial direction of the antenna. This radiation mode is called the axial mode, which is the most important mode in a helical antenna. Several studies were reported on the properties of the axial mode helical antenna with a finite reflector [31,33,38–40].

The current induced on the helix is composed of four major waves, which are two rapidly attenuating waves and two uniform waves along the helical wire. Such waves include the traveling wave and the reflected wave. In a conventional helical antenna, the uniform traveling wave will be dominant when the antenna

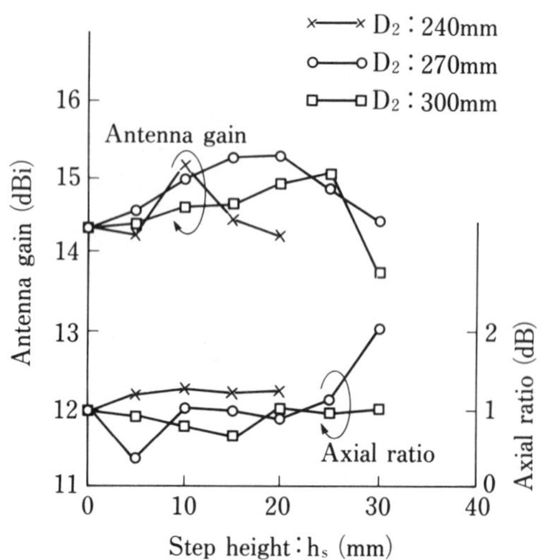

Figure 6.39 Gain and axial ratio of modified SBF antenna.

length is fairly large. On the other hand, in a short-cut helical antenna, the rapidly attenuating traveling wave will be dominant, especially in a two-turn ($N = 2$) helical antenna.

The two-turn helical antenna has a relatively high antenna gain and excellent polarization characteristics for its size. Radiation patterns with respect to E_θ and E_ϕ planes of the two-turn helical antenna are shown in Figure 6.40. In this figure, D is a diameter of the finite circular reflector of the axial mode helical antenna.

As can be seen in Figure 6.41, the antenna gain is affected remarkably by the size of the reflector. In Figure 6.41, the antenna gain is about 9 dBi and the axial ratio is about 1 dB, with the reflector diameter D around 1.0λ. Such a two-turn helical antenna has a comparatively high performance in spite of its small size. From the above mentioned consideration, a highly efficient antenna for the INMARSAT-M SES can be realized by applying the two-turn helical antenna to elements of an array antenna.

The quad-helix array antenna is composed of four two-turn helical antennas in a square arrangement whose elements are oriented in the manner shown in Figure 6.42. According to related studies [30], the effect of the mutual coupling between each element of this antenna is not negligible, and this mutual coupling mainly degrades the axial ratio. The axial ratio of a single helical antenna is about 1 dB, as shown in Figure 6.43, but this value is degraded to about 4.5 dB in the case of the array antenna with an array spacing of 0.7λ.

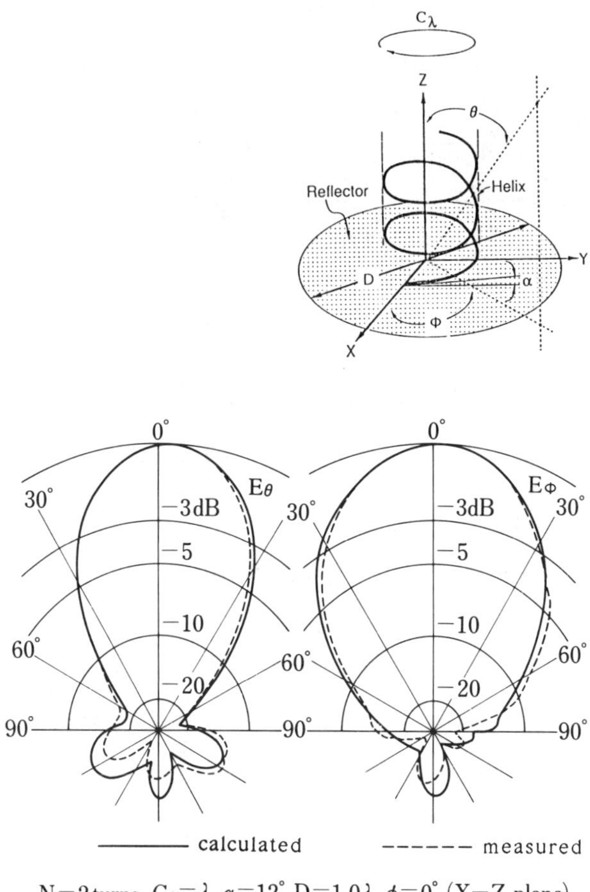

Figure 6.40 Radiation characteristics of two-turn helical antenna.

Figure 6.43 shows the experimental values of the antenna gain and axial ratio versus the rim height with an element spacing of 0.7λ and rim diameter of 0.7λ. As can be seen, the best properties of the antenna gain and the axial ratio can be obtained at a rim height of about 0.25λ. The antenna gain is improved by 0.4 dB and the axial ratio is also improved by 3.5 dB compared to that of the quad-helix array antenna without rims.

The performance characteristics of the quad-helix array antenna shown in Figure 6.42 has an antenna gain of about 13 dBi (HPBW:38 deg), an axial ratio of about 1.0 dBi, and an aperture efficiency of about 100%. It appears that the antenna is also suitable for INMARSAT-M.

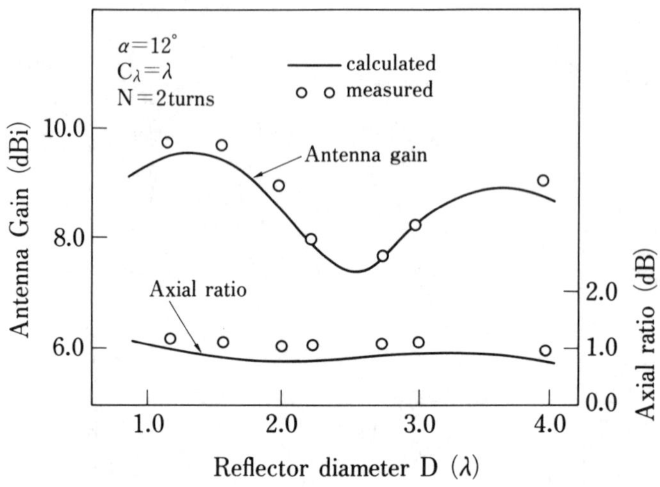

Figure 6.41 Gain and axial ratio of two-turn helical antenna.

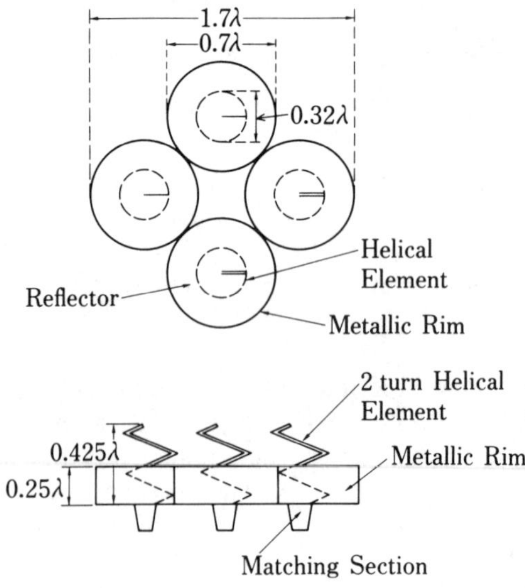

Figure 6.42 Quad-helix array antenna.

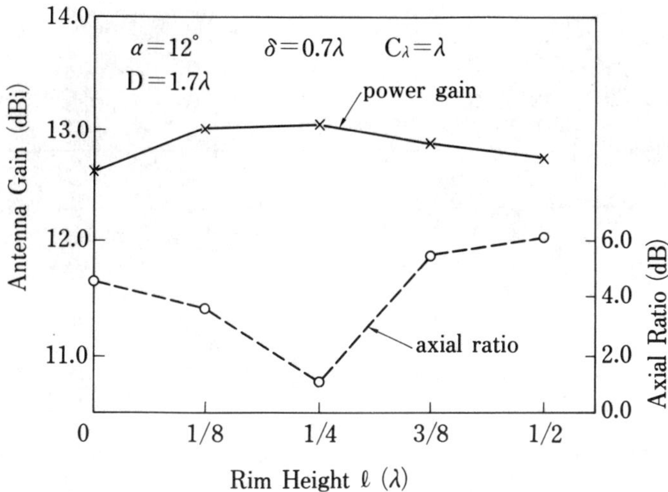

Figure 6.43 Gain and axial ratio of quad-helix array antenna.

Microstrip/Slot Array Antennas

Figure 6.44 shows photographs and element arrangements of an MSA and a XSA. The MSA is a nine-element microstrip array antenna whose element spacing is 94 mm (about a half wavelength at 1.6/1.5 GHz) and whose antenna volume is about 300 × 300 × 10 mm. The XSA is a 16-element cross-slot array antenna with 97-mm spacing and its volume is about 560 × 560 × 20 mm. As shown in these figures, the element arrangement of the MSA is a 3 × 3 square array, while that of the XSA is a modified 4 × 4 square array in order to obtain similar radiation patterns in different cut planes. In both the MSA and XSA, beam scanning is performed by controlling four-bit variable phase shifters attached to each antenna element. These antennas are also applicable to the mobile earth station (MES) antennas for INMARSAT-M as well as INMARSAT-Aero. Figure 6.45 shows the example of INMARSAT-M SES. In this antenna system, a modified SBF antenna, two-axis El/Az mount system, and program tracking are adopted for antenna, mount system, and pointing/tracking method, respectively.

6.4.1.3 Antennas for INMARSAT-C

In the case of the INMARSAT-C antenna system, the simplest and most compact configuration is required; that is, without mount systems and tracking/pointing

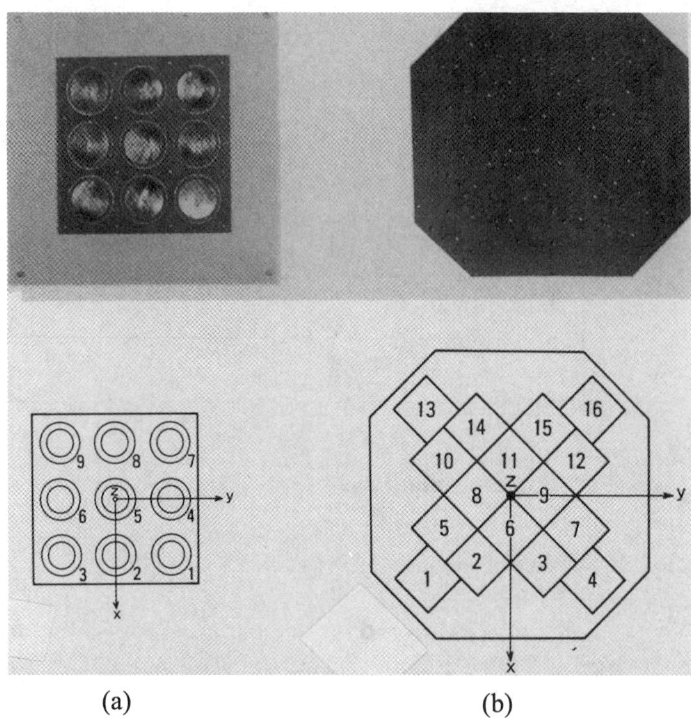

(a) (b)

Figure 6.44 (a) Microstrip array; (b) cross-slot array.

systems. Therefore, the omnidirectional antennas are the most suitable ones for this standard. These are the quadrifilar helix [44], cross-drooping dipole [45], and the microstrip patch antenna [46]. As a shipborne antenna, a quadrifilar antenna is well known because of its good performance of axial ratios in wide angular coverage. Figure 6.46 shows an example of INMARSAT-C terminals for maritime satellite communications. Its general characteristics are a minimum of about 4 dBi in gain, a maximum of about 3 dB in axial ratio, and about 40 cm in height.

6.4.1.4 Portable Terminal's Antennas for INMARSAT-A and -C

INMARSAT-A and -C are the original standards for maritime satellite communications, and their purpose was to improve the qualities of ship-to-ship and ship-to-land communications. Recently, however, INMARSAT satellites have also started new transportable and portable communications services. Using relatively small earth stations, it is possible to access INMARSAT satellites for high-quality

Figure 6.45 Example of antenna system for INMARSAT-M: Antenna: modified SBF; mount: two-axis (El/Az); fading reduction: polarization shaping method.

telephone and facsimile communications. And, in fact, INMARSAT satellite communications played an important role following the earthquakes in Iran and the Philippines, which caused heavy damage to the ground-based communications networks in those countries.

For these services, several types of mobile terminals have been developed and proposed by many companies, such as MTI Corporation, EB Corporation, Magnavox Corporation, and JRC. For example, MTI's portable earth station (see Figure 6.47) is compact enough to fit inside a suitcase, weighs only about 30 kg and can be assembled in about 5 min. By simply pointing the antenna in the direction of an INMARSAT satellite, the user can put into operation the high-quality telephone and facsimile communications. The antenna used in this service is an umbrella-type paraboloidal antenna, and its diameter is about 1m.

Furthermore, Figure 6.48 shows an example of the INMARSAT-C portable terminal (developed by TOSHIBA Corporation) with four patch antennas [47], which can be connected to a personal computer to exchange messages for about 60 min of receiving and about 30 minutes of transmitting using dry battery operation. The diameters are 34 (w) $\times$ 21 (d) $\times$ 6 (h) cm in size and 3.8 kg in weight.

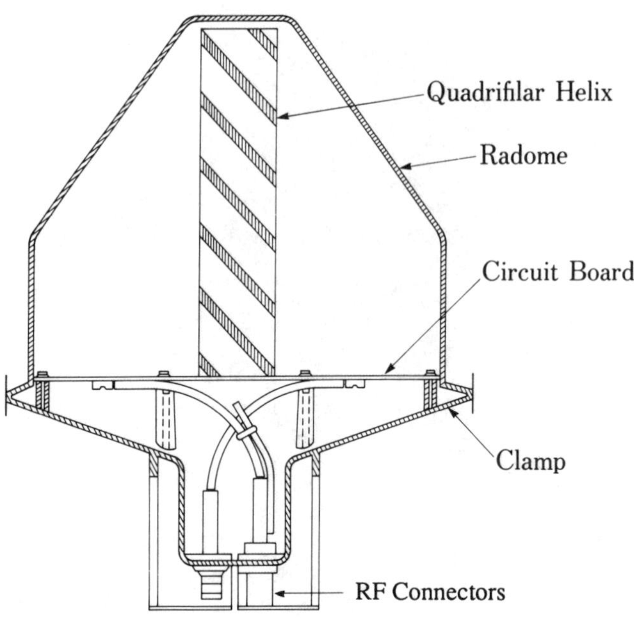

Figure 6.46 Example of antenna system for INMARSAT-C. Antenna: quadrifilar helix.

6.4.1.5 Antennas for INMARSAT-Aero

Airborne equipment for aeronautical satellite communications in the INMARSAT system is called the *INMARSAT-Aero terminal.* Two types of high-gain and low-gain antennas are defined in the INMARSAT-Aero terminal. The G/T of high-gain and low-gain antennas is required to be over −13 dBK (gain is about 12 dBi) and −26 dBK (gain is about 0 dBi), respectively. The high-gain antenna is used for public voice/high-speed data (21 kbps) communications, which can be connected to terrestrial public telephone networks. On the other hand, the low-gain antenna is used for low-speed data (600, 1,200, 2,400 bps) communications.

One of the most important issues in the INMARSAT-Aero high-gain antennas is the *key hole,* which is the direction in which the airborne antenna cannot satisfy the required G/T. The phased-array antennas are the best candidates for airborne antennas because of such advantageous characteristics as low profile and mechanical strength; however, they have a fatal disadvantage in the key hole problem. At the present time, two types of phased-array antennas have been used: one is a conformal type, which has two sets of phased arrays on both sides of a fuselage, and the other is a top-mount type, which has a set of phased arrays on the top of a fuselage. The conformal type has the advantage of low air drag because of its very low profile;

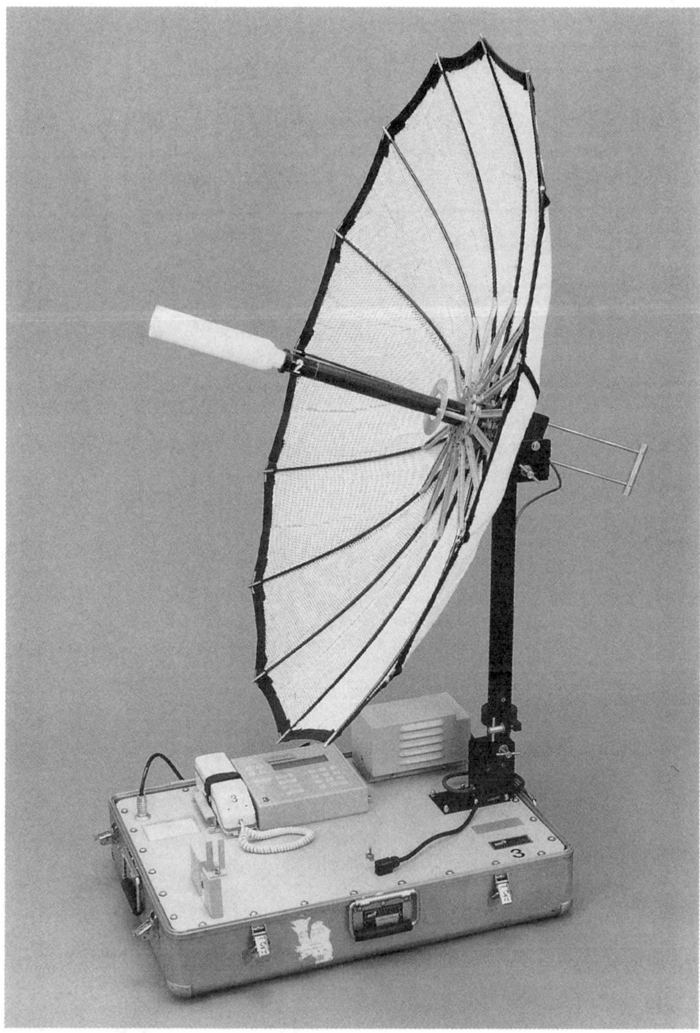

Figure 6.47 Portable terminal for INMARSAT-A. (Courtesy of MTI Corporation.)

however, it has a fatal key hole. At present, only a top-mount-type phased-array antenna has satisfied the INMARSAT specifications, which are defined by the INMARSAT Aeronautical System Definition Manual (SDM). Figure 6.49 shows a top-mount phased array for the INMARSAT-Aero terminal, which was improved for commercial use based on the research and development of the airborne phased-array antenna in the ETS-V program. Figure 6.50 shows an airline aircraft, and

Figure 6.48 Portable terminal for INMARSAT-C. (Courtesy of Toshiba.)

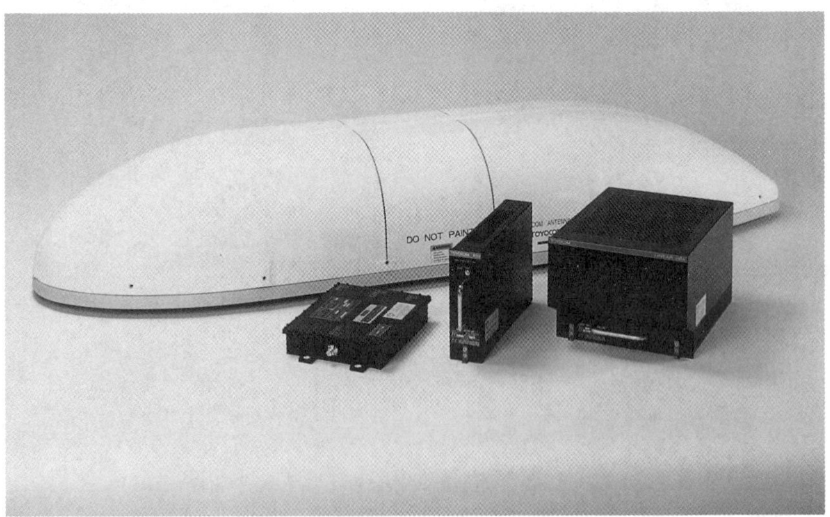

Figure 6.49 Top-mount phased array for INMARSAT-Aero. (Courtesy of Toyocom.)

on the top of the fuselage, the top-mount phased-array antenna is installed. The low-gain antennas of the INMARSAT-Aero conventional omnidirectional antennas mentioned in Section 6.3 have been used, taking into account the critical specifications such as air drag and mechanical strength.

Figure 6.50 View of top-mount phased array on aircraft. (Courtesy of Toyocom.)

6.4.2 Directional Antennas in the PROSAT Program

In this section, directional shipboard antennas developed for experiments of the PROSAT program are described. Seven European Space Agency (ESA) member states (Belgium, Federal Republic of Germany, France, Italy, Spain, United Kingdom, and Norway) decided to undertake a PROSAT program to promote mobile satellite communications. In the first stage (December 1983 to December 1984) of the PROSAT program, several kinds of antenna systems for maritime satellite communications were developed and evaluated by the European Space Research and Technology Center (ESTEC) using the MARECS satellite. In this section, three types of maritime antennas developed by ESTEC are introduced. Table 6.4 shows the main characteristics of three antennas. The description in this section is derived from the PROSAT Phase I Report, which was published by ESA in May 1986 [4].

Five-Turn Helix Antenna

Figure 6.51(a) shows a photograph of a five-turn helix antenna system. The main electrical characteristics are:

Table 6.4
Antennas Developed in PROSAT Program

Antenna	*Peak, Transmit Gain (dB)*	*Peak G/T (dBk)*	*Max EIRP (dBW)*	*Axial Ratio (dB)*	*Radome (H, D) (m)*	*Stabilization*
Helix five-turns	12.5	−16	29	3	1.2, 0.84	Gravity elevation on double-gimbaled suspension
Short backfire	15	−12	28	1	1.4, 1.45	Two gyroscope wheels rotating in opposite directions on a platform fixed through a U-point
Array of 16 crossed dipoles	17	−9.5	31.7	0.7	1.25, 1.13	Double-gimbaled mount, gravity stabilized and stiffened by one momentum wheel

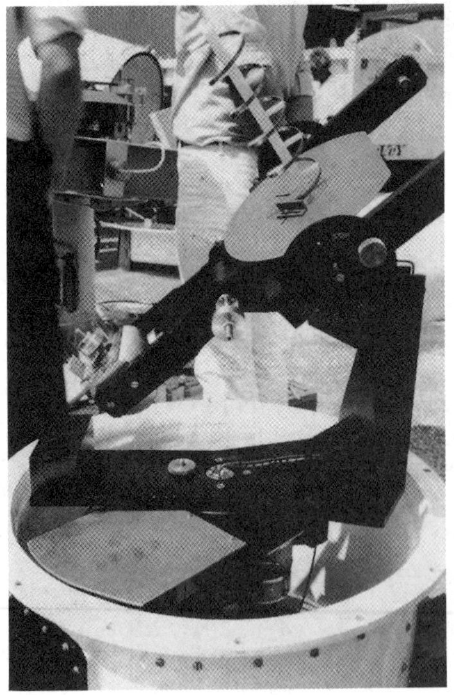

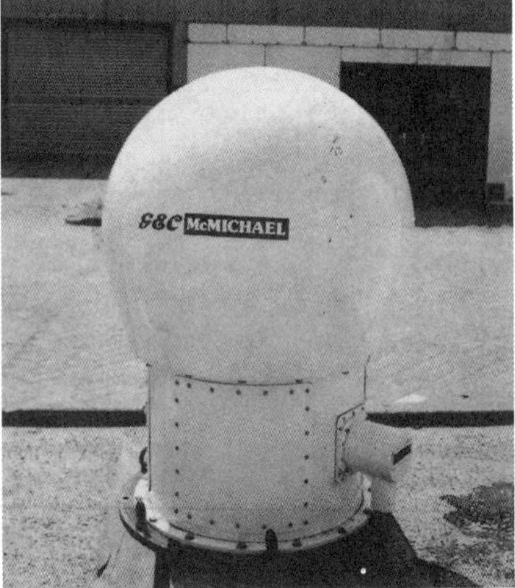

(a)

Figure 6.51 (a) Five-turn helix antenna developed by ESTEC; (b) short backfire antenna developed by ESTEC. (Courtesy of ESTEC.)

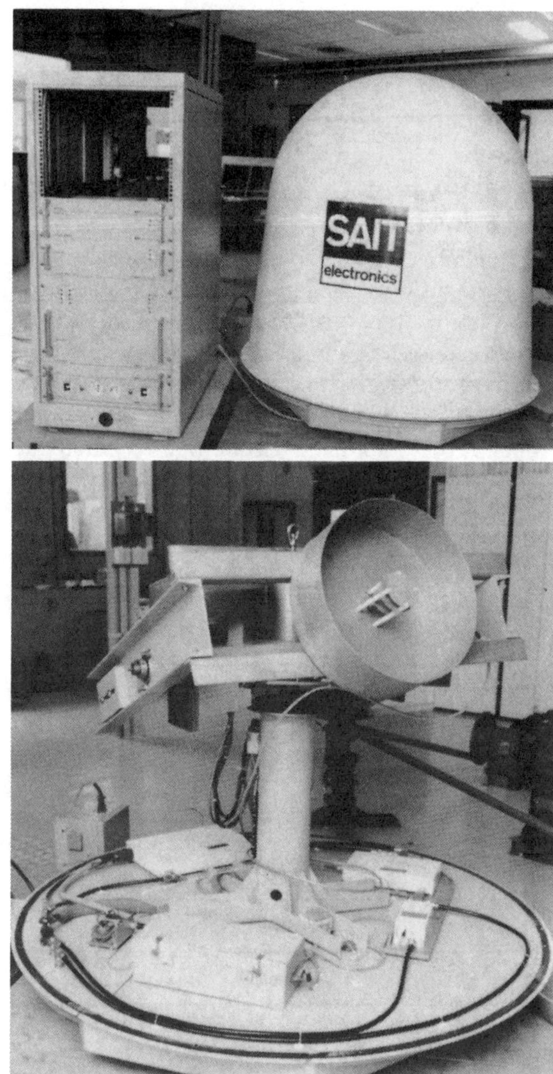

(b)

Figure 6.51 Continued.

- Gain: 12.5 dBi peak right-hand circular polarization (RHCP) transmit and 11.5 dBi receive;
- Sidelobe level: −13 dB;
- Axial ratio: 3 dB;
- Beamwidth (3 dB): −47 deg;
- Terminal G/T: −16 dBK; and
- Terminal EIRP: 29 dBW.

Stabilization is obtained by gravity elevation on a double-gimbaled suspension. The pendulum aligns itself with the vertical when not subject to other acceleration. However, because the center of rotation of the pendulum is distant from that of the ship, pitch and roll movements induce horizontal acceleration to which the pendulum is sensitive. In order to limit perturbations, the resonant frequency of the pendulum must be low with respect to excitation frequencies in pitch and roll, and the damping (friction) must be minimal. Low resonance frequency is achieved by minimizing the distance between the center of gravity of the rotating part and its center of rotation. This also reduces torques due to horizontal acceleration, but at the same time reduces the stabilizing torque due to gravity.

Short Backfire Antenna

Figure 6.51(b) shows a photograph of a SBF antenna system. The main electrical characteristics are:

- Gain: 15 dBi peak RHCP transmit;
- Axial ratio: 1 dB;
- Beamwidth (3 dB): −34 deg;
- Terminal G/T: −12 dBK; and
- Terminal EIRP: 28 dBW.

The antenna system consists of the stabilized platform (two gyroscopes), SBF antenna, diplexer, high-power amplifier, and low-noise amplifier, which are enclosed under a radome. In order to stabilize the antenna, two gyro wheels rotate in opposite directions on a platform.

Array Antenna of 16 Crossed Dipoles

Figure 6.52 shows a photograph of an array system of 16 crossed dipoles fed in phase with a peak gain of 17 dBi and with the feeding circuit behind the radiating aperture. The main electrical characteristics are:

- Gain: 17 dBi peak RHCP transmit;
- Axial ratio: 0.7 dB;

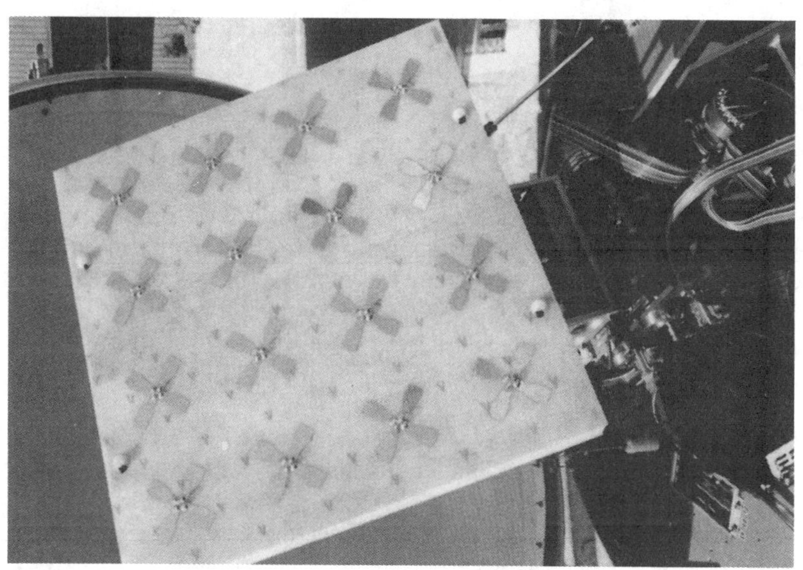

Figure 6.52 Sixteen-cross-dipole antenna developed by ESTEC. (Courtesy of ESTEC.)

- Beamwidth (3 dB): −34 deg;
- Terminal G/T: −9.5 dBK; and
- Terminal EIRP: 32 dBW.

The antenna system consists of a stabilization mechanism, array antenna, diplexer, high-power amplifier, and low-noise amplifier, which are protected by a radome. Stabilization is obtained by a single-wheel gyroscope. Azimuth pointing is controlled by the output signal from the ship's gyrocompass.

6.4.3 Directional Antennas in the ETS-V Program

In this section, directional antennas developed and evaluated in the ETS-V program are introduced. The ETS-V satellite was launched to a geostationary orbit of 150 deg E in August 1987 for the purpose of research and development of mobile satellite communications. In the ETS-V program, the Communications Research Laboratory (CRL) of the Ministry of Posts and Telecommunications, the Electronic and Navigation Research Institute (ENRI) of the Ministry of Transport, (NTT), and Kokusai Denshin Denwa Co. (KDD) carried out the experiments using ships, aircraft, and land mobiles with newly developed directional antennas.

6.4.3.1 Shipborne Antenna

Improved Short Backfire Antenna

The main research activities of the ETS-V program in maritime satellite communications have been focused on studying fading reduction using compact and high-efficiency antennas with gains of around 15 dBi. An SBF antenna [38] is a very attractive antenna for gains on the order of 13 to 15 dBi and can be mounted on small vessels in maritime satellite communications [48]. Although this antenna has many beneficial characteristics, including efficiency and the simplicity of construction, it is also considered a favorite option for a compact and high-efficiency shipboard antenna, it had the problem of narrow bandwidth of about 3% because of its leaky cavity operation. A natural impedance match bandwidth is only 3% to 5% for VSWR's under 1.5 [42].

In the ETS-V program, the electrical characteristics of a conventional SBF antenna have been improved by changing its main reflector from a flat disk to a conical [42] or a step plate [41], and by adding a second small reflector as previously described. Table 6.5 shows the comparison of electrical characteristics between a conventional SBF antenna with a plane main reflector and an improved SBF antenna with a conical main reflector. The main electrical characteristics of the improved SBF antenna at 1,540 MHz are 15 dBi in gain, −22.5 dB in first sidelobe

Table 6.5
Comparison of Electrical Characteristics of Conventional and Improved SBF Antennas

	Plane Reflector SBF Antenna	*Conical Reflector SBF Antenna*
Effective gain	14.5 dB	15.0 dB
Half-power beamwidth	34 deg	34 deg
Directive gain	14.8 dB	15.5 dB
First sidelobe level	−21.0 dB	−22.5 dB
Axial ratio	−1.3 dB	−1.1 dB
Aperture efficiency:		
Effective gain	65%	76%
Directive gain	75%	85%
Frequency bandwidth for VSWR under 1.5	3%	9% 20%*

*With a small reflector.

level, and 85% with a directive gain in the aperture efficiency. Frequency bandwidths of 20% for VSWRs under 1.5 are obtained as shown in Figure 6.53, and gain is improved by about 1 dB without changing sidelobe levels [41,42].

Physical dimensions of conventional and improved SBF antennas are shown in Table 6.6. Stabilization is obtained by a two-axis stabilized method, and satellite

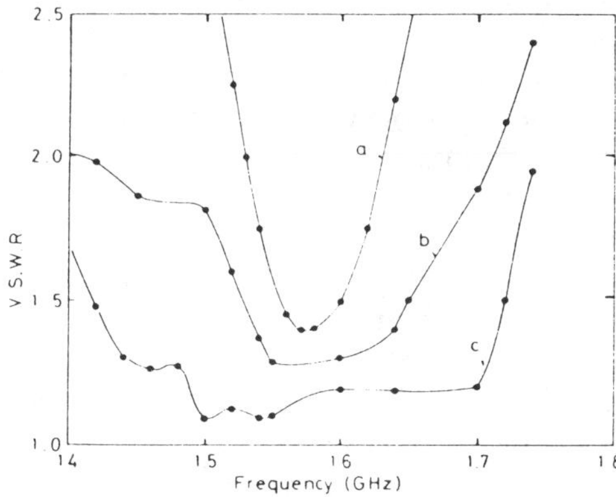

Figure 6.53 VSWR Characteristics of three types of SBF antenna: (a) conventional SBF; (b) improved SBF; (c) improved SBF with two small subreflectors.

Table 6.6
Dimensions of Conventional and Improved SBF Antennas

	Plane Reflector SBF Antenna	*Conical Reflector SBF Antenna*
Diameter of a large reflector, D_R	40 cm (2.05λ)	40 cm (2.05λ)
Diameter of a small reflector:		
D_{r1}	9 cm (0.46λ)	9 cm (0.46λ)
D_{r2}	—	8 cm (0.41λ)
Width of a rim, W_B	4.9 cm (0.25λ)	4.9 cm (0.25λ)
Distance between large and small reflectors, S_R	9.7 cm (0.5λ)	12.9 cm (0.66λ)
Distance between an exciter and a small reflector, S_r	4.9 cm (0.25λ)	5.7 cm (0.29λ)
Distance between small and second small reflectors, d	—	1.8 cm (0.09λ)
Slanting angle of a large reflector, α	0 deg	15 deg

Note: λ = wavelength at 1,540 MHz.

pointing is carried out by program tracking using output signals from the ship's gyrocompass.

6.4.3.2 Airborne Antenna

Phased-Array Antenna

In aeronautical satellite communications, a directional medium gain antenna is considered a key technology, and a major research aim is to realize not only electrical requirements, but also strict mechanical requirements, such as installation, low profile, light weight, and strength [49]. In the ETS-V program, the phased-array antenna [32], developed by CRL, was installed on a commercial jet aircraft for the first time in the world to carry out communication experiments mainly between Tokyo and Anchorage.

Taking account of the electrical and mechanical requirements of aeronautical communications, a phased array with low-profile antenna elements was chosen for a directional antenna. An MSA [50] is chosen as an antenna element because of its very low profile, light weight, and mechanical strength, which satisfy the requirements for airborne antennas [8]. However, one disadvantage is a very narrow frequency bandwidth, usually 2% to 3%. The antenna adopted is a two-frequency resonant element, because it provides a compact array and a simple feed line configuration [47]. However, this type of element has very poor axial ratios. The problem was overcome by using the sequential-array technique [51], where a thin substrate with high dielectric constant is used over a wide frequency bandwidth with excellent axial ratios. Figure 6.54 shows the antenna (see Figure 6.49) installed on the top of a Boeing 747. The volume is 76 (L) × 32 (W) × 18 (H) cm, and weight is 18 kg.

System Details

The system block diagram is shown in Figure 6.55, and the main characteristics are shown in Table 6.7. Two array antennas were adopted to provide the required wide coverage with high-gain beams, and either one is selected to operate, depending on the flight direction. An array consists of 16 elements, with 2 in elevation and 8 in azimuth. In the experiments, the required coverage angles are as narrow as +20 deg, so the beams are not steered in elevation directions, and the array plane is set at 65 deg with respect to the horizon in order to optimize the beam coverage on the flight routes. By controlling eight 4-bit digital phase shifters, the antenna beam scans in a 4-deg step within +60 deg with respect to a line perpen-

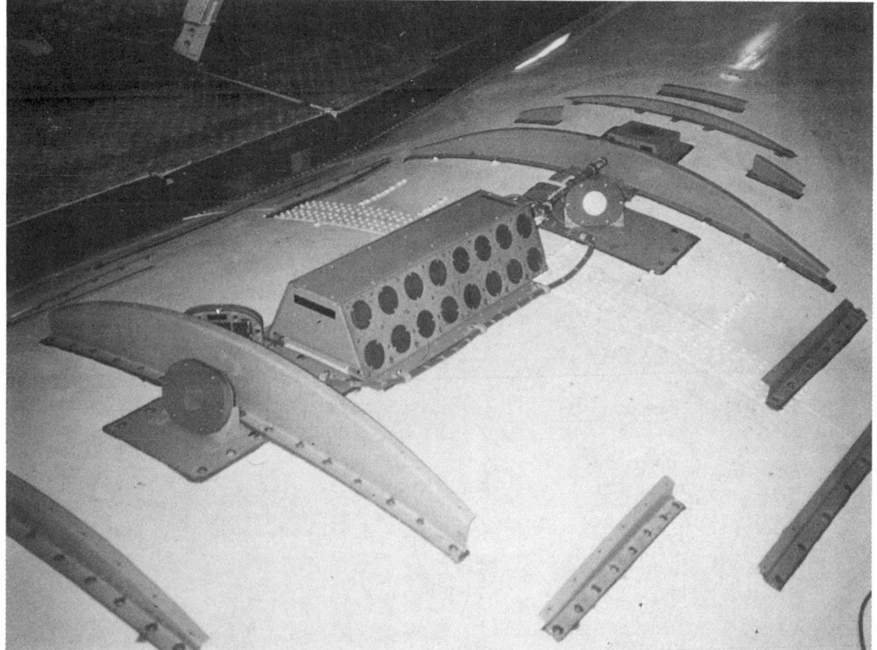

Figure 6.54 View of antenna installed on the top of Boeing 747 (radome removed).

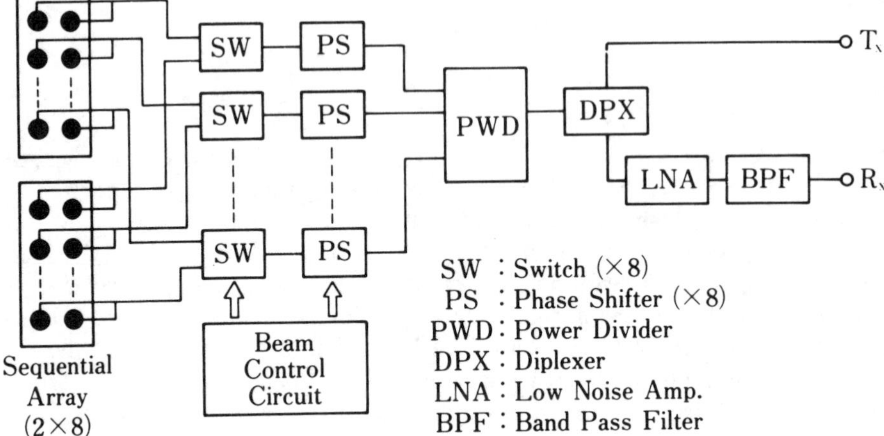

Figure 6.55 Configuration of the phased-array antenna.

Table 6.7
Characteristics of Airborne Phased Array Antenna

Characteristic	*Measurement*
Frequency:	
Transmitter	1,545–1,548 MHz
Receiver	1,647–1,650 MHz
Polarization	Left-handed circular
Gain:	
Transmitter	14.7 dBic (non-scan)
Receiver	13.5 dBic (non-scan)
G/T	−10.8 dBK (non-scan)
Element	Circular patch (one-point feed)
Array	2 × 8 sequential phased array
Substrate	Glass Teflon (ϵ_r = 2.6)
Axial ratio	>2.0 dB (non-scan)
VSWR	>1.4 (non-scan)
Tracking	Step track
Beam step	4 deg
Phase shifter	4-bit digital × 8
Weight	18 kg
Volume	760 (l) × 320 (w) × 180 (h) mm

dicular to the axis of the aircraft. A step track method was adopted to track the satellite. The characteristics are as follows.

Radiation Patterns. Radiation patterns of the antenna are given in Figures 6.56(a,b), which shows a nonscanned pattern and a scanned pattern over 60 deg, respectively. Coordinates of the measured system are also shown in Figure 6.56. The dots denote measured values and the solid line denotes a theoretical curve, which is calculated by multiplying the array factor by the theoretical radiation pattern of the circular patch element [10]. Spacing of the elements is a half wavelength to avoid grating lobes, which were not observed in the measurement.

Gain and Axial Ratio. Figure 6.57 shows the variation of the gain and axial ratio with respect to the beam-scanning angles. The gain was measured at an interim point between a power divider and a diplexer, so losses from phase shifter, RF switches, and feed lines are included in the measured gain. Gain degradation caused by a beam scanning of 60 deg was found to be about 5 dB, which agreed with theoretical prediction values. Axial ratios of the element in a boresight direction, fed from one point, were measured as 9.5 and 5.2 dB in transmitting and receiving frequencies, respectively. By using a sequential technique, the axial ratios were markedly improved to a level of 3 dB, as shown in Figure 6.57. The frequency dependence of the VSWR of the element and array are shown in Figure 6.58 with dotted and solid lines, respectively. Although the VSWR of the element shows two resonant frequencies, the adoption of a sequential-array technique was shown to improve the VSWR and make frequency bandwidth wider.

Noise Temperature and G/T. The G/T is the most important factor for the aircraft earth station. In order to evaluate G/T, the noise temperature of the antenna was measured in two elevation angles using a conventional Y-factor method. Figure 6.59 shows noise-measured temperature versus beam-scanning angles in two directions; in one the beam is not scanned, and in the other it is scanned by ± 45 deg. The noise temperature increases when the beam is scanned over large angles, more noise is received as the beamwidth increases in beam scanning. The system noise temperature was found to be about 200K, depending on the beam-scanning angle, The Y-factor [52] method is commonly used to measure the noise temperature of an antenna by comparison with a standard noise source.

Environmental Characteristics. It is very difficult to measure basic characteristics of gain, radiation pattern, and so on in environmental tests (e.g., thermal cycle test, humidity test, etc.), which are defined by the avionics standards. The evaluation of the characteristics in environmental conditions was performed by measuring the VSWRs of the array antenna, which was set in a thermal test chamber. The measured VSWRs are shown in Table 6.8 with test items and environmental conditions. The maximum variation of VSWR was found to be about 0.3, and this value can be disregarded when considering the performance of the array antenna.

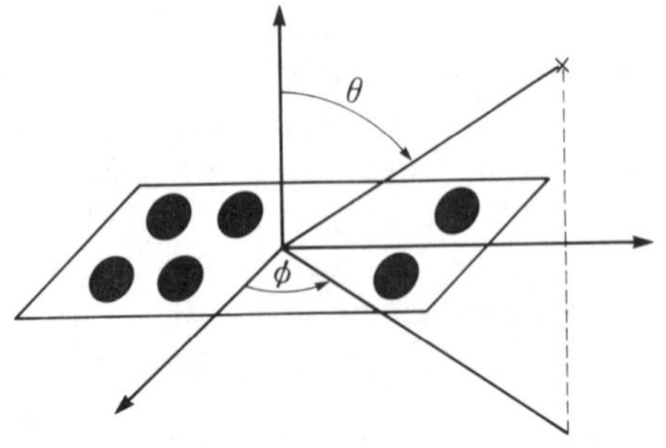

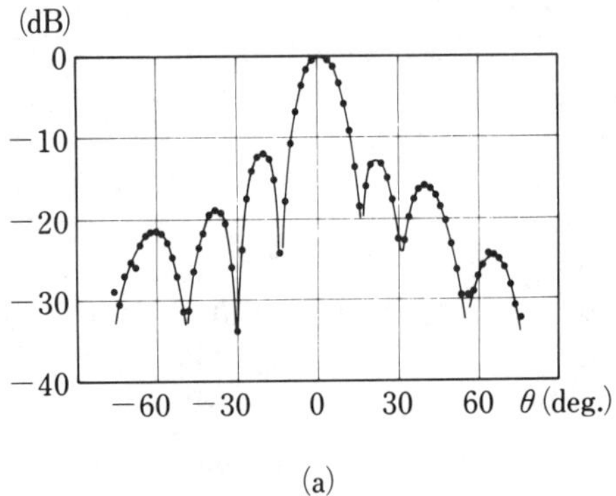

(a)

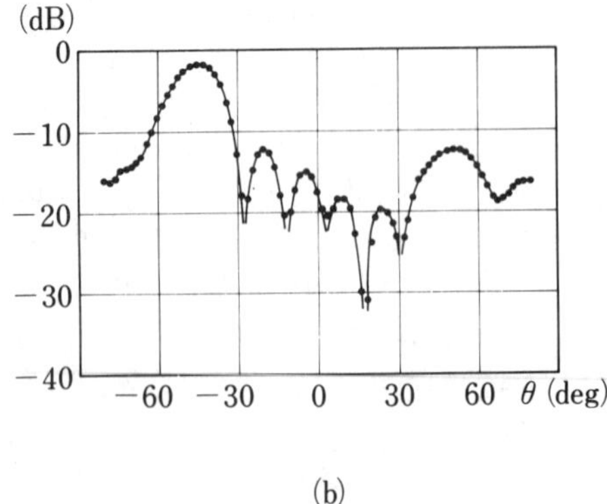

(b)

Figure 6.56 Radiation patterns: (a) scan angle 0 deg (ϕ = 90 deg); (b) scan angle 45 deg (0 = 90 deg).

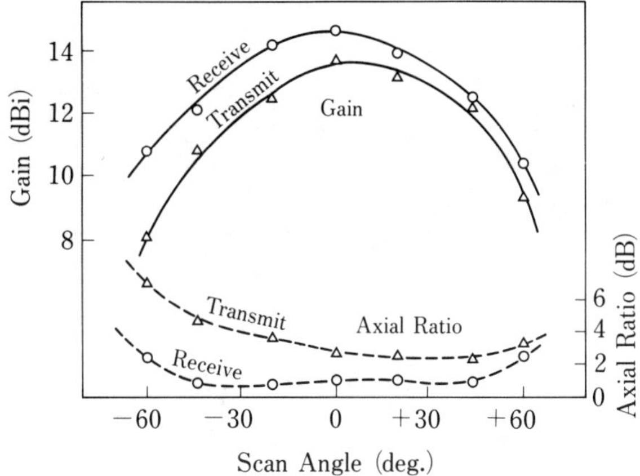

Figure 6.57 Variation of gain and axial ratio versus scanning angle.

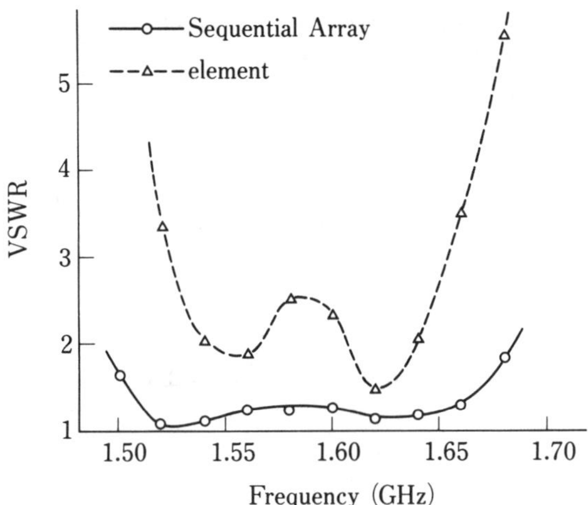

Figure 6.58 Frequency dependence of VSWR of antenna element and sequential array.

Thermal variation is considered the biggest factor affecting electrical characteristics. The dielectric constant for a temperature change of $-55°C$ through $+70°C$ varies in the range of 2.57 through 2.61. This change in the dielectric constant gives an

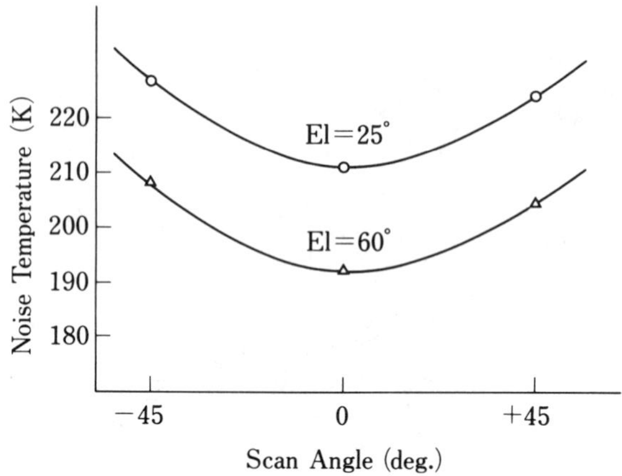

Figure 6.59 Measured noise temperature of the array.

Table 6.8
Results of Environmental Tests on Airborne Array

Test item	*Max Δ VSWR*	*Condition*
Low temperature	0.23	−55°C
High temperature	0.33	+70°C
Altitude	0.10	90 mmHg (15,000m)
Humidity	0.21	95% at 50°C
Vibration	0.05	3g

impedance variation of about 1%; this fact shows that thermal effects on the array can be ignored.

Tracking Error

Figure 6.60 shows the radiation patterns of the airborne antenna measured by the inflight experiment. The dots are the data measured at the transmitting frequency from an aircraft to a base station, and the circles are the data measured at the receiving frequency from a base station to an aircraft. It is found that there is an angular difference of about 8 deg between f_R and f_T frequencies. This means that even if an aircraft transmits the signal to the satellite at the maximum level, it cannot receive the signal from the satellite at the maximum level. This effect causes

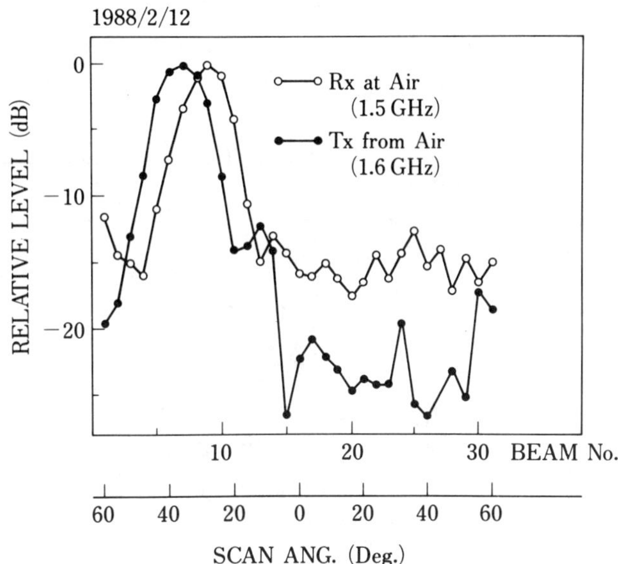

Figure 6.60 Measured radiation patterns of the phased array.

not only channel quality imbalances between transmitting and receiving channels, but also interferences between and from other satellite systems.

The relationship between the beam-scanning angle and scanning error is given by (6.5) [53], where Θ and $\Delta\Theta$ are angles of beam scanning in f_T and scanning error, and f is the frequency difference between f_T and f_R.

$$\Delta\Theta = -\Theta + \sin^{-1}[\sin\Theta/(1.0 + f/f_T)] \tag{6.5}$$

Figure 6.61 shows a comparison between the experimental and theoretical results, which are calculated by (6.5). The solid lines show theoretical values in the case of frequency differences of 1%, 5%, 6.8%, and 10%. The value investigated in the experiment is 6.8%. The circles are the measured data of the antenna. The tracking error was found to depend on the beam-scanning angle and the ratio of f_T and f_R.

The reason the beam-scanning errors are observed between transmitting and receiving frequencies is because the same phase shifters are used at both frequencies. To eliminate the scanning errors, the following equation must be satisfied, where Φ_R and Φ_T denote the value of phase shifts at f_R and f_T, respectively.

$$(\Phi_R/f_R) = (\Phi_T/f_T) \tag{6.6}$$

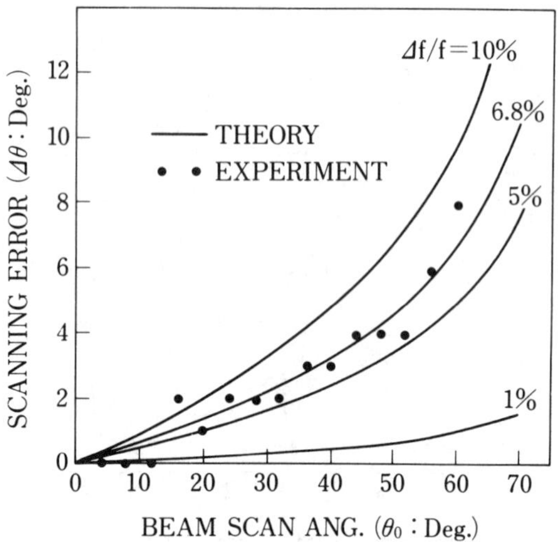

Figure 6.61 Theoretical and experimental results of tracking errors caused by frequency differences.

From (6.6), it is found that the adoption of frequency-dependent phase shifters can potentially reduce the tracking errors between the transmitting and receiving frequencies.

6.4.3.3 Land Mobile Antenna

Directional antennas have been expected to provide voice/high-speed data links not only for long-haul tracks but also for small, private cars. From that point of view, cost is a very important factor to be taken into consideration in designing antennas. In the early stage of land mobile satellite communications, a mechanical steering antenna system is considered the best candidate for vehicles; however, it will be replaced by a phased-array antenna in the near future, because a phased-array antenna has many attractive advantages, such as low profile, high-speed tracking, and potential low cost. In recent years, the main research activities have been focused on phased-array antennas and the tracking control method, but first we will note the characteristics of mechanically steered antenna systems.

Antenna

The photograph shown in Figure 6.62 is of a mechanical steering antenna with eight spiral elements, which gives about 15 dBi in system gain. A closed-loop tracking method is adopted. The volume is 30 cm in radius, 35 cm in height, and

1,500g in weight. The array consists of 2 × 4 elements, and it forms a fan beam having a half-power beamwidth of 21 deg in the azimuth plane and 39 deg in the elevation plane at 1,545 MHz. Its peak gain is about 15 dBi, including the feeder loss, and is suitable to track the satellite for mobile satellite communications, because elevation angles to the satellite are not as varied as those of azimuth angles; the antenna beam direction can be shifted in two azimuth directions by switching the pin-diode phase shifters. The difference between the received signals in both directions is used to drive the antenna toward the satellite. The beam-shifting angle is set to about 4 deg.

Tracking System

Figure 6.63 shows a block diagram of the tracking system. The newly developed single-channel tracking system is applied to this mobile antenna. This system is

Figure 6.62 Mechanical steering antenna with eight-spiral antennas.

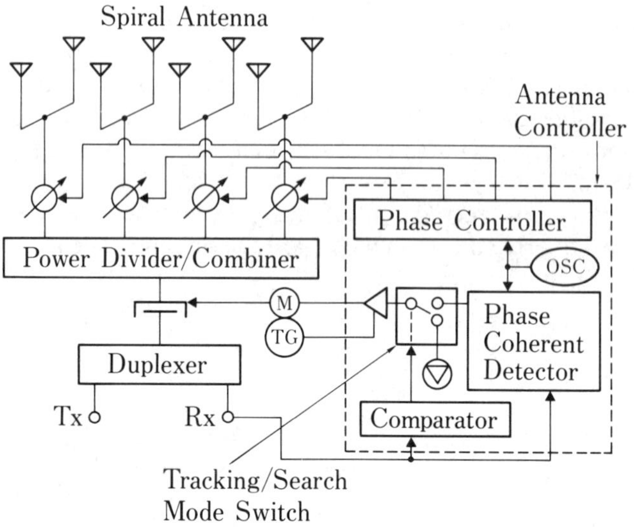

Figure 6.63 Mechanical tracking antenna.

similar to the conical scan tracking system, which shifts the antenna beam direction in order to get the angle difference between the antenna boresight and the satellite direction. In this antenna system, the received signal is modulated by shifting the beam directions, and its modulation period coincides with the beam-shifting period (400 Hz). So, the information on beam-tracking error can be obtained by demodulating the received signals with reference to the beam-shifting signal. We will now note the characteristics of electronically steered antenna systems.

Main Features

Although mechanical tracking antennas have the advantage of wide beam coverage, they have severe disadvantages for small cars. The problems are large size, slow tracking speed, and high cost. Based on the research experiences of the airborne phascd-array antenna [32], a new phased-array car antenna has been proposed by CRL [54]. Figure 6.64 is a photograph of the phased-array antenna without a radome, and Figure 6.65 shows a block diagram of the phased-array antenna system. The main features of the phased-array antenna are as:

- Antenna elements are excited by electromagnetic coupling with microstrip feed lines. Because of its easy configuration, production cost will be reduced greatly for land vehicles, including small, private cars.

Figure 6.64 Vehicle phased-array antenna developed for the ETS-V experiments.

- The total number of digital phase shifters was reduced to one-half by adopting newly developed 3-bit pin-diode phase shifters. The phase shifter is printed on a substrate, which makes the present complex feed lines simple.
- Tracking error between transmitting and receiving frequencies, which is inevitable for a phased-array antenna, is potentially eliminated using a frequency-dependent phase shifter.
- Excellent performance of satellite tracking has been provided by the proposed open-loop method, which uses an optical fiber gyro and a geomagnetic sensor.

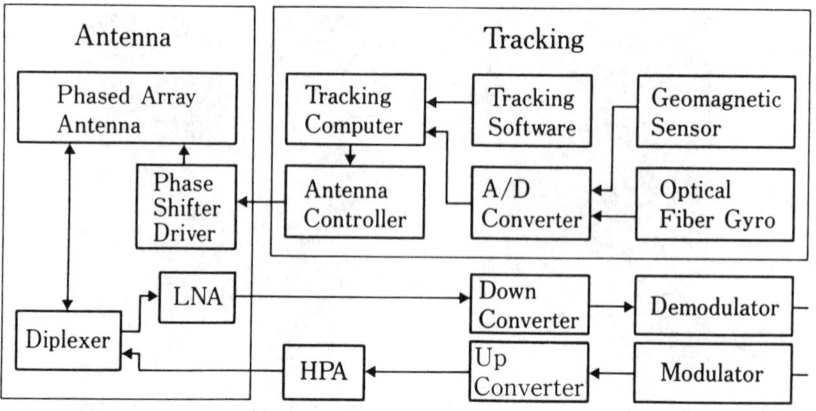

Figure 6.65 Phased array.

Antenna Element

Figure 6.66 shows a configuration of the antenna element. An electromagnetic coupled antenna is adopted as an antenna element because it will reduce procedures of assembling the antenna system and it will reduce the cost of the phased-array antenna and thus compete with a conventional mechanical tracking antenna. A feed line is a microstrip line printed on a substrate, which excites the radiating elements by electromagnetic coupling through a coupling aperture. A radiating element, which radiates circularly polarized waves, is also printed on a thin-film substrate.

Phased Array

The array consists of 19 elements, which are printed on a very thin film. Figure 6.67 is a photograph of feed lines and phase shifters printed on a substrate installed on the backside of the plate of exciting apertures. The gain of the phased-array antenna in a nonscanned direction (elevation angle is 90 deg) is about 18 dBi, and 10 dBi in a scanned angle of 60 deg (elevation angle is 30 deg). The main characteristics are shown in Table 6.9. Figure 6.68 shows the elevation angular dependence of G/Ts, system noise temperature, and temperature measured at the input port of an LNA. A G/T is about −13 dBK at the elevation angles of 40 to 50 deg.

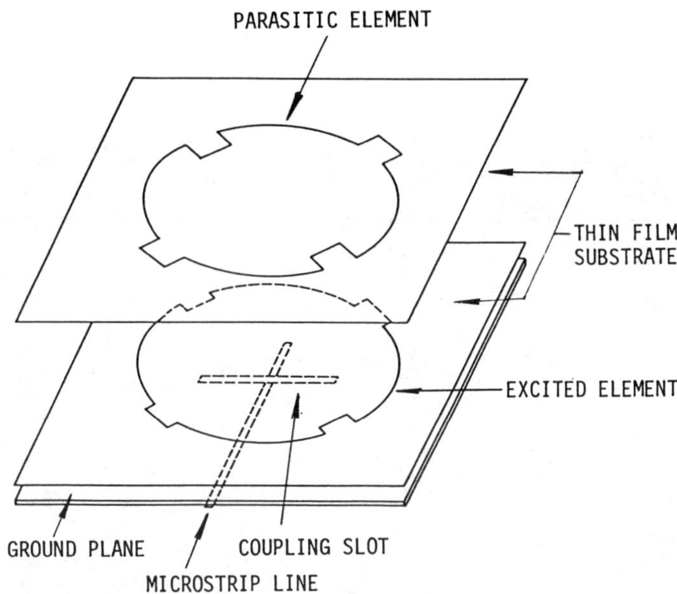

Figure 6.66 Antenna element of the phased array.

Tracking Performance

The phased-array antenna system has adopted an open-loop tracking method with the hybrid use of a geomagnetic sensor and an optical-fiber gyro [42]. A geomagnetic sensor is very attractive because of its capability of determining an absolute direction and because of its low cost. However, its performance is affected by environmental conditions such as buildings, bridges, and structures made of metal. On the other hand, an optical-fiber gyro has excellent performance in determining relative directional variations independent of environmental conditions; however, it cannot determine the absolute direction.

In the present open-loop method, an optical-fiber gyro is mainly used to give the information of vehicle motions, and a geomagnetic sensor gives an absolute direction to calibrate the accumulative angular error of the optical-fiber gyro at an appropriate time interval. The optical-fiber gyro has been designed to have the directional error within 1 deg for a time period of 1 hour.

Figure 6.69 shows an experimental result of tracking performance. The antenna installed on a test van tracks the ETS-V satellite at an elevation angle of 47 deg. The test van changes directions, but, the received signal from the satellite is found to be almost constant except for shadowing and blocking effects. An output

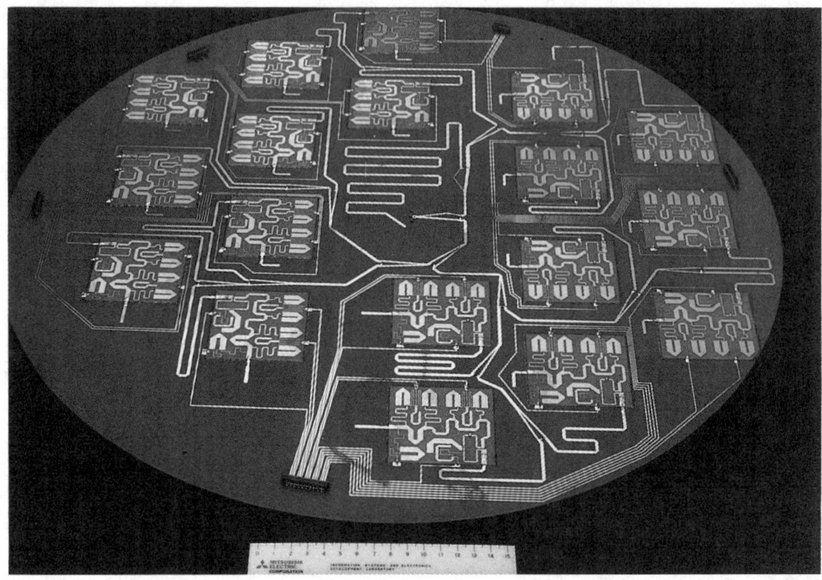

Figure 6.67 Feed lines and phase shifter printed on a substrate.

Table 6.9
Characteristics of Phased-Array Antenna

Characteristic	*Measurement*
Frequency:	
Receiver	1,530.0–1,559.0 MHz
Transmitter	1,626.5–1,660.5 MHz
Polarization	Left-hand circular
Scanned angle	El 30–90 deg, Az 0–360 deg
Gain	18 dB: (El = 90 deg)
	10 dB: (El = 30 deg)
System temperature	200°K
Axial ratio	4 dB (El = 30 deg)
Volume	60 cm ϕ × 4 cm (H)
Weight	5 kg

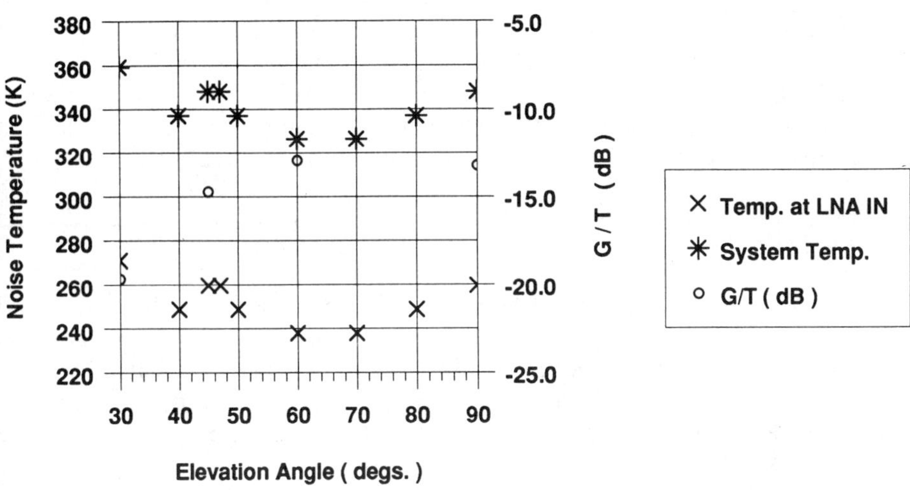

Figure 6.68 Measured G/T and noise temperatures of the array.

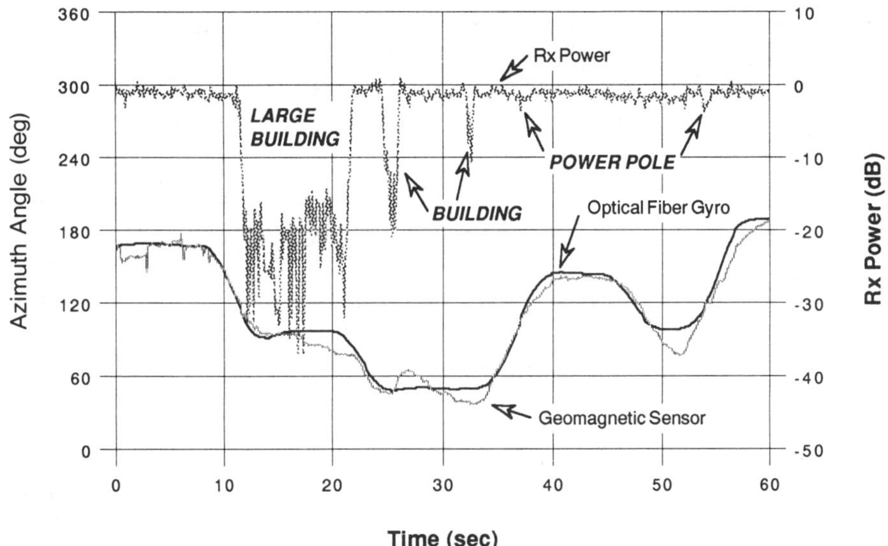

Figure 6.69 Experimental results of received signal levels by an open-loop tracking method using an optical-fiber gyro.

of a geomagnetic sensor is also recorded in Figure 6.69, showing an error of about 20 deg compared to that of the optical-fiber gyro. Experiments confirm that open-loop tracking is better than a closed-loop method [55]. A phased-array antenna with 19 elements [56] in a different feed system and an electronic steering antenna with a spherical shape [57] have also been proposed.

Briefcase Antenna

The main feature of future mobile satellite communications will be *portability,* which means that a person can directly access a satellite to establish communication channels using a very small handheld terminal. Even in the present L-band system, great efforts have been made to develop transportable and portable terminals. Figure 6.70 is a photograph of a briefcase portable terminal [58] developed in the ETS-V program. The main characteristics are:

- Antenna: two microstrip patch antennas;
- Gain: 6 dBic;
- G/T: −21 dBK;
- EIRP: 6 dBW;
- Tracking: hand-pointed; and
- Weight: 13 kg.

The terminal can transmit and receive messages via a satellite with two separate conventional microstrip patch antennas on a lid of a case for transmitting and receiving. The reason for adopting separate antennas is to eliminate a diplexer, which is too large and too heavy to realize a compact and lightweight terminal. The beamwidth of the antenna on the lid is wide enough to point to a satellite by manual pointing.

6.4.4 Directional Antennas in the MSAT-X Program

The National Aeronautics and Space Administration (NASA) and JPL have studied and developed several L-band directional vehicle antennas, mainly for land mobile satellite communications [59]. These include a mechanically steered, tilted 1 × 4 patch array antenna and two electrically steered, planar-phased-array antennas. All three of these antennas feature beams that are narrow in azimuth; hence, they require azimuth steering to keep the beam pointed toward the desired satellite as the vehicle changes its azimuth orientation. These medium-gain antennas provide 9 to 12 dBi gain, reject multipath signals outside their beam pattern, and allow two satellites separated by 30 deg in a geostationary orbital arc to reuse the same frequency to cover CONUS. The following description is mainly derived from [59] and [60]. We commence with the mechanically steered antennas.

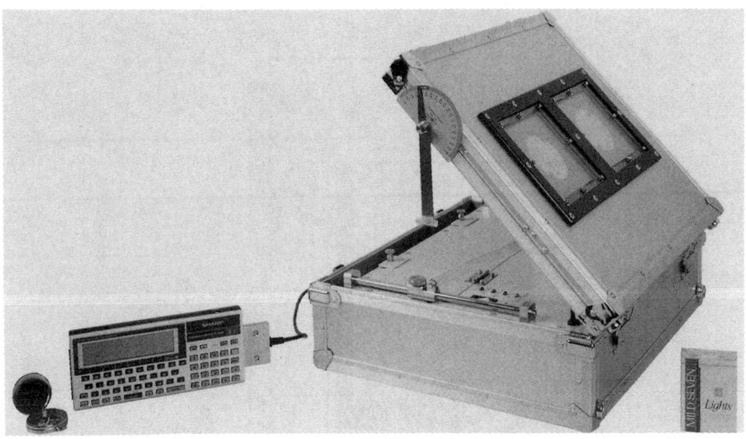

Figure 6.70 Briefcase terminal message communicator with 2 patch antennas.

Antenna

Two mechanically steered vehicle antennas have been developed at JPL [60,61]. The radiating part is a linear array of four square microstrip patches tilted with respect to the ground plane to provide elevation coverage from 20 to 60 deg with a minimum of 10 dBi gain. The rotating antenna platform is mounted on a fixed platform that includes the motor drive and pointing system hardware. The success of the breadboard version led to additional antenna development aimed at the reduction of the antenna height from 9 in down to 5 to 6 in and the integration of some of the discrete RF components into the stripline-fed 1 $\times$ 4 microstrip patch array. Figure 6.71 shows a side view of the reduced-height mechanically steered antenna.

Tracking

Both antenna systems employ the same monopulse technique, a kind of a close-loop tracking, for tracking the satellite in azimuth. The full antenna system with either antenna installed is effectively described by the pointing system block diagram shown in Figure 6.72. The antenna is divided into two identical subarrays. Since the phase centers of the two halves are physically separate, the phases of their received signals will be offset by an amount proportional to the off-boresight angle. The signals from the two subarrays pass through a sum/difference hybrid

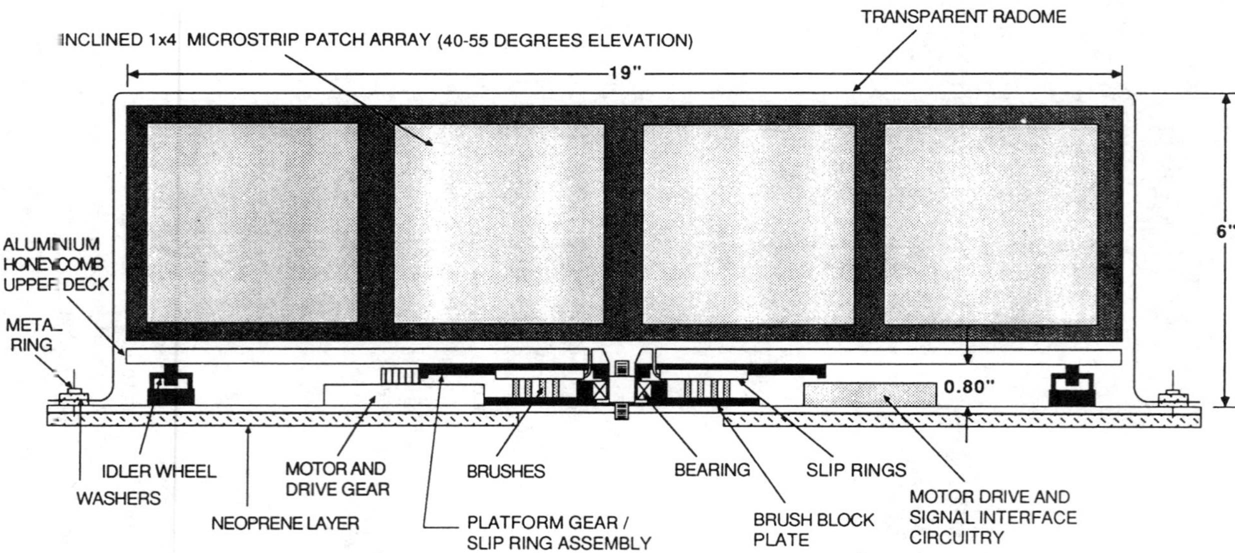

Figure 6.71 Side view of the reduced-height mechanically steered antenna. (Courtesy of JPL.)

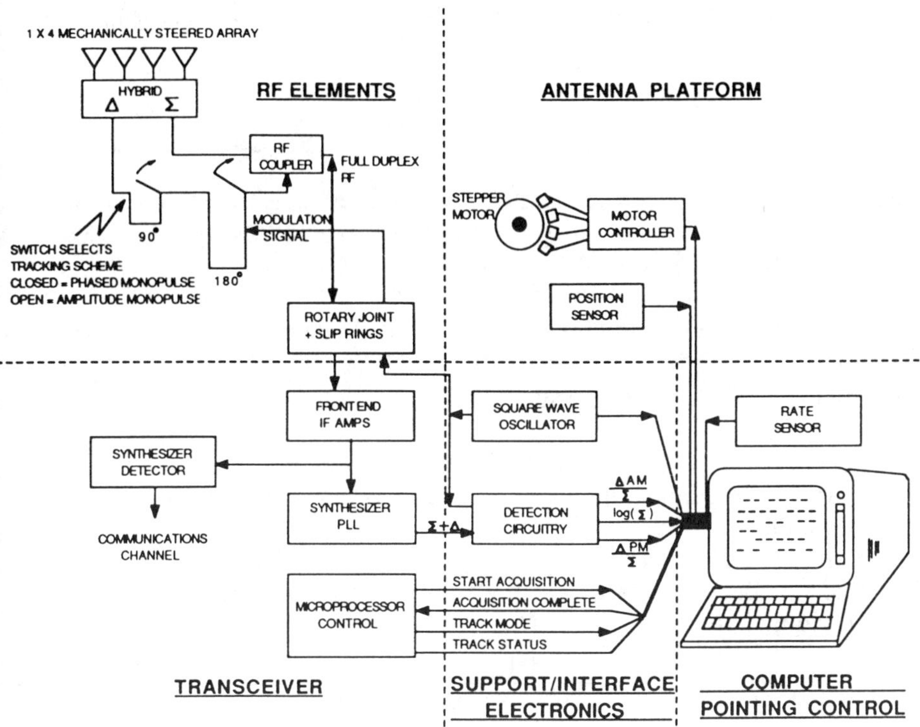

Figure 6.72 Antenna system. (Courtesy of JPL.)

whereby their sum and difference are obtained. The difference of *error* signal is then canceled through a 1-bit (0 to 180 deg) phase shifter, which will be modulated at a rate of between 200 to 4,800 Hz. It is then passed through a 30-dB ferrite isolator. The difference signal is then combined with the sum signal via a 10-dB coupler in which the difference signal strength is reduced by approximately 10 dB before being combined with the sum signal. The combined output of the coupler is then passed to the single-channel rotary joint. Since the difference signal is modulated before addition to the sum signal, it can be extracted from the composite signal at the receiver by the standard filtering technique. This normalized error signal is then supplied to the pointing control circuitry, which commands a stepper motor to rotate the antenna array in order to maintain pointing toward the source of the incoming signal. The azimuth rate sensor is also used to maintain pointing during the tracking phase when the received signal fades below a prescribed threshold. This constitutes open-loop tracking. We now consider the electrically steered antennas.

Array Antenna

Phased-array antennas were developed principally to provide a thin antenna that can be installed conformal to the top of the vehicle. These antennas are well known for their complexity and high cost. As a result, emphasis was placed on the selection of manufacturing techniques, materials, and component types, in addition to meeting the RF and pointing requirements, to keep the cost down. The breadboard phased-array antennas [62] developed are to meet the following performance goals:

- Frequency: 1,545.0 to 1,559.0 MHz receive and 1,646.5 to 1,660.5 MHz transmit;
- Spatial coverage: 20 to 60 deg elevation and full 360 deg azimuth;
- Gain: 10-dBic minimum above 30 deg in elevation and 8-dBic at 20 deg in elevation;
- Half-power beamwidth: 25 deg in azimuth and 35 deg in elevation;
- Intersatellite signal isolation: 20 dB between two geostationary satellites separated by approximately 35 deg;
- Beam pointing accuracy: +5 deg.

Two phased-array antennas were developed, as shown in Figure 6.73, and they exhibit several common features. The overall antenna system is shown schematically in Figure 6.74. The components enclosed by the dashed line are for rooftop mounting. The components outside the dashed line are mounted inside the vehicle to keep the array height at a minimum. Each antenna consists of 19 low-profile radiating elements, with 18 3-bit diode phase shifters. The left model of Figure 6.73 uses a dual resonant stacked circular microstrip element [62] to cover both the transmit and receive bands, while the right model employs a stripline crossed-slot radiator.

Tracking

For the antenna's beam-pointing system, the initial acquisition of the satellite is accomplished by a full azimuth search for the strongest received signal. An angular rate sensor is used to establish an inertial reference point when the acquisition is performed while the vehicle is turning. Tests show that the antenna can acquire a reference pilot signal in 2 sec from a random spatial position. After the desired satellite signal has been acquired, the antenna tracks the satellite by a closed-loop sequential lobing technique. In the event of a severe signal fade due to shadowing, the sequential lobing can no longer function properly. In this case, the open-loop

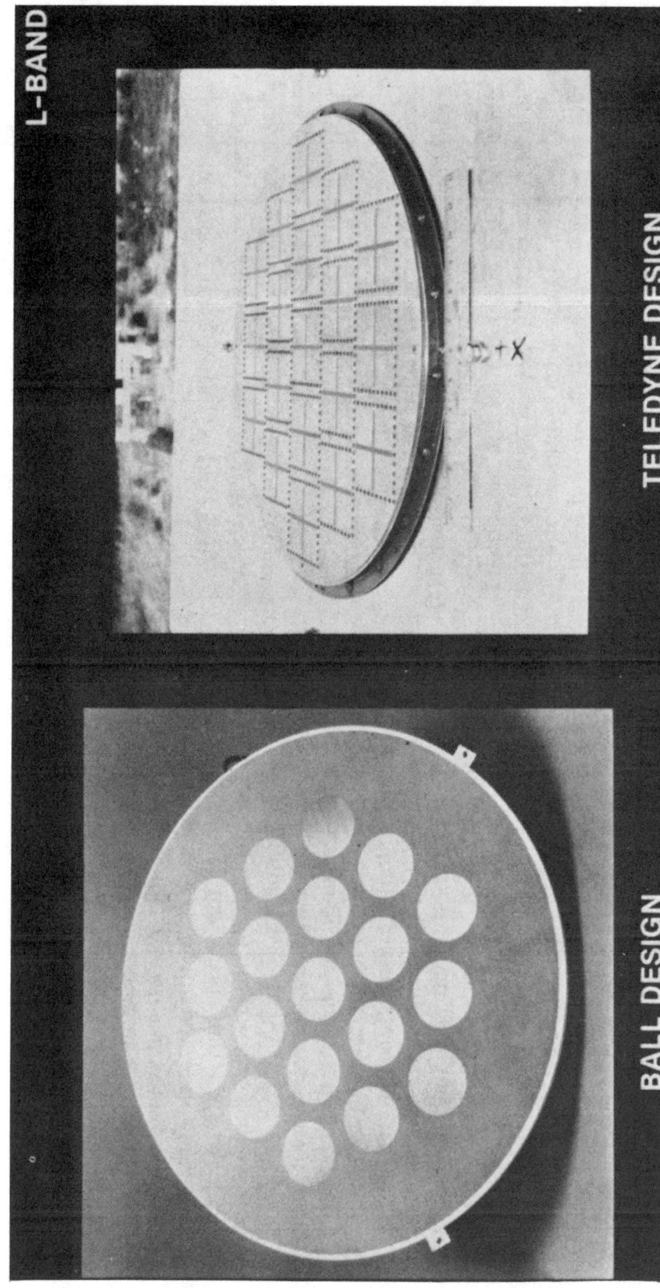

Figure 6.73 Electrically steered phased arrays. (Courtesy of JPL.)

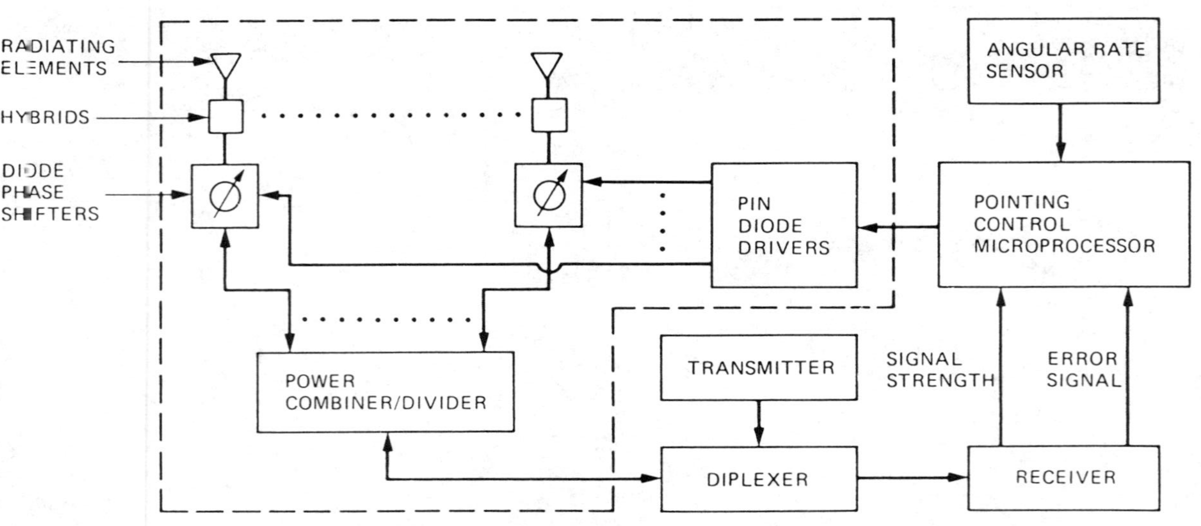

Figure 6.74 Electrically steered phased-array antenna system.

angular rate sensor takes over the pointing for approximately a 10-sec period until the sensor drifts away.

6.4.5 Directional Antennas in the MSAT Program

The Communications Research Center (CRC) has developed an antenna for land vehicle operation. Linearly and circularly polarized array antennas have been developed and evaluated in extensive field trials using the MARECS-B satellite. The description in this section is derived from [63].

Adaptive Array

The linearly polarized adaptive antenna array consists essentially of a driven quarter-wave monopole surrounded by concentric rings of parasitic elements all mounted on a ground plane of finite size. The parasitic elements are connected to ground via pin diodes. With the application of suitable biasing voltages, the desired parasitic elements can be activated and made highly reflective. The directivity and pointing of the antenna beam can be controlled both in the elevation and azimuth planes using high-speed digital switching techniques. The use of a circular polarizer in the linearly polarized design can realize an increase in gain at the expense of an increase in antenna height.

The polarizer has an elliptical cross section, a diameter of 40 cm, and a height of 20 cm. It consists of a number of conformal scattering matrices to achieve the 90 deg differential phase shift between two orthogonal polarizations.

A five-ring linearly polarized antenna is shown in Figure 6.75. The antenna incorporates sufficient electronics to control the radiation patterns and pointing on command. It is designed to be mounted on the metallic roof of a vehicle, where the effective ground plane can significantly enhance antenna gain at low elevation angles.

The circularly polarized array, shown in Figure 6.75, is obtained by adding the linearly polarized array. The main characteristics are:

- Frequency: 1,530 to 1,660 MHz;
- Spatial coverage: 15- to 50-deg elevation and 360-deg azimuth;
- Gain: linearly polarized model—9 to 11 dBi and circularly polarized model—10 to 13 dBic;
- Size: linearly polarized model—24 in (d), 2.5 in (h), 11 lb (w) and circularly polarized model—24 in (d), 8 in (h), 16 lb (w).

Tracking. The satellite is initially acquired with a closed-loop method of stepping through 16 azimuth beam positions and selecting the beam with the strongest signal. In the event that the signal falls below a given threshold, the acquisition sequence

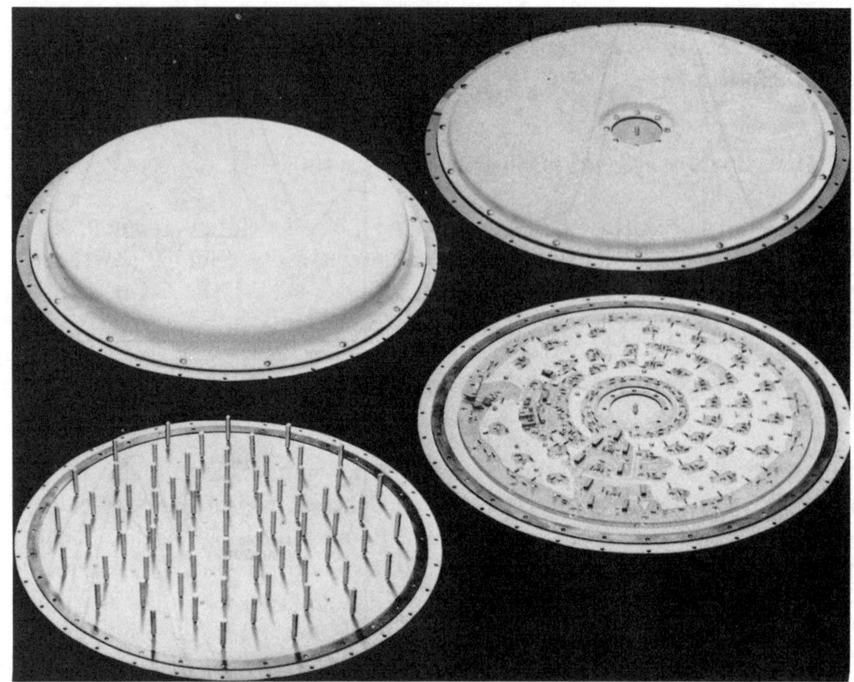

Linearly Polarized Antenna

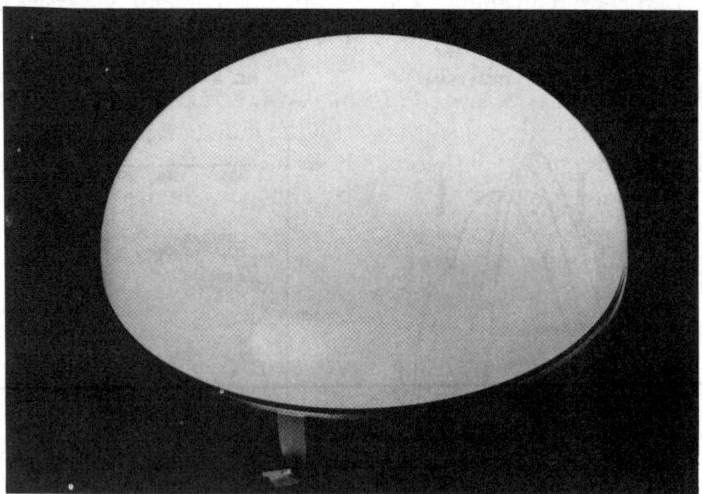

Circularly Polarized Antenna

Figure 6.75 Linearly and circularly polarized antenna developed by CRC. (From IMSC'90 with permission.)

is again initiated until the signal is required. The speed of operation is determined by the terminal C/N_0 ratio, and the signal-to-noise requirements in the control loop bandwidth. Currently, it takes less than 0.1 sec to acquire the satellite after initial phase lock. A dc signal proportional to the RF signal is derived at the terminal receiver. The satellite is tracked by periodically switching on either side of the current beam position and selecting the beam with the strongest signal. A number of algorithms have been devised to minimize any perturbation of the communications signal to less than 1% of the time. The maximum phase transients in azimuth can be kept to less than ±10 deg over the required angular coverage and operating frequency bands.

6.5 ANTENNA SYSTEMS FOR GPS

6.5.1 General Requirements for GPS Antennas

Several navigation systems using satellites have been used very widely around the world. The oldest system is NNSS, which was originally developed as a U.S. Navy military satellite, but has been open to the civil sector since 1967. The position of a mobile terminal is determined by measuring the Doppler shifts of 400-MHz signals transmitted from the NNSS satellite. The dual frequencies are used in order to compensate for a refraction error due to ionospheric propagation. The Navigation System with Time and Ranging/Global Positioning System (NAVSTAR/GPS) is the most widely used navigation system at present. A GPS satellite transmits two frequencies, L_1 (1,575.42 MHz) and L_2 (1,227.6 MHz). There are two kinds of positioning applications: an absolute or single positioning and a relative or differential positioning.

The absolute positioning, which provides the absolute three-dimensional position of the GPS receiver with an accuracy of several tens to a hundred meters, is carried out by receiving the signals from four GPSs and decoding the navigation codes, called the *coarse acquisition* (C/A) code. On the other hand, the relative positioning can precisely measure the distance between two GPS receivers by detecting the time delay (usually the phase difference) between the received signals. This system provides accurate land surveying over a long distance; for example, an accuracy of less than 1 cm is possible for a base line of several hundred kilometers.

The Global Orbiting Navigation Satellite System (GLONAS) is a system of the former USSR, and now it is in operation using five satellites. The system is very similar to NAVSTAR/GPS, but the frequencies are slightly different: L_1 is 1,597 to 1,617 MHz and L_2 is 1,240 to 1,260 MHz. Requirements for receiver antennas for these navigation systems seem to be similar. On GPS receiver antennas, the following requirements are imposed [64].

Amplitude Radiation Pattern and Gain

An antenna should provide uniform response over approximately the entire upper hemisphere over which the satellites may be visible. This limitation of the coverage is imposed on the antenna because the reception is excluded when GPS satellites are below a specified elevation angle (e.g., 10 deg) in order to avoid severe multipath and tropospheric effects. The hemispherical beam can remove the need for a tracking mechanism for the antenna; therefore, simple antenna systems become available.

The uniform amplitude response over the coverage region is required because the signal level must be sufficient for all angles of the desired view so that the receiver electronics can maintain signal lock with a required signal-to-noise ratio. Beyond this, it is necessary that the antenna be effectively "blind" to all signals arriving from outside the coverage range; that is, the pattern cutoff must be sharp with no backlobes.

For maritime GPS applications, however, antennas are required to have the coverage extending to negative elevation angles to compensate for the vehicle motion due to pitching and rolling. Typically, the gain of −5 dBi or more should be maintained up to −20-deg elevation.

The power flux density of radio waves emitted by a GPS satellite is about 4×10^{-14} (W/m^2) on the surface of the earth. Since most receiving antennas are omnidirectional with gains around 0 dBi or less, LNAs are necessary to amplify the weak received signal. Typical GPS receivers in commercial use seem to have noise figures less than 3 dB and gains between 20 and 40 dB. When the antenna and receiver are located a distance apart and are connected with a long cable, a preamplifier should be implemented at the antenna output in order to compensate for the transmission loss of the cable.

Phase Pattern

In the case of relative positioning systems using direct phase measurements, the phase difference of the antenna outputs corresponding to different directions of the satellites causes considerable position errors, which are unacceptable for accurate land surveying. Within the region of coverage, the antenna should ideally provide uniform response not only in amplitude, but also in phase. The requirement for uniform phase response over the coverage area of the antenna is critical, particularly in phase tracking receivers. The phase measured at the antenna output changes as a signal source moves around the phase center at a constant range. These phase changes produce an apparent displacement of the phase center from its on-axis position, which is dependent on the direction of the signal source as seen from the antenna. In effect, the antenna seems to move, depending on where the signal source is. A true phase center is defined as the center of curvature of an equiphase contour for a defined observation plane, polarization, and frequency [65]. If the phase changes associated with the signal source direction are not considered in the data analysis process, the antenna position determined from the measurements will be incorrect.

A uniform phase response is generally much more difficult to realize than is a uniform amplitude response, particularly in the case of the added constraint of a sharp horizon pattern cutoff. The pattern cutoff (or null) results from interferences of radio waves; consequently, it generates a nonuniform phase response for the antenna. In this respect, small, compact antennas produce a more uniform phase response than longer or more spatially distributed antennas, although this is achieved at the cost of poorer amplitude pattern cutoff characteristics.

Frequencies and Polarization

The GPS satellite transmits two radio waves, L_1 and L_2, with right-hand circular polarizations. In the simple case, a GPS receiver is used either at L_1 or at L_2. For more accurate positioning, however, the dual frequencies are used to compensate the excess delay of radio waves due to ionospheric propagation. The operation in this case requires an antenna that can operate equally well at both frequencies. Omnidirectional antennas introduced in Section 6.3 are potential candidates for GPS reception. Actually, quadrifilar helix antennas (QHA) and MSAs have been most widely used owing to their simplicity, small size, low cost, and satisfactory electrical characteristics. It should be noted, however, that the design parameters of GPS receiver antennas must be determined so that their radiation patterns may be as uniform as possible over the upper hemisphere, which is somewhat different from the omnidirectional antennas for satellite communication described in Section 6.3.

6.5.2 Quadrifilar Helix Antennas

The QHA was invented by Gerst, and some of its characteristics were reported [66]. By changing the parameters, these antennas may show widely different radiation patterns, so many manufacturers of the GPS receiver use volutes of different size and construction. The literature [67] is suitable for selecting or designing QHAs for many applications. Data from Kilgus [67] and Adams [31] are particularly useful in obtaining desired amplitude pattern responses; however, owing to the lack of phase data for these antennas, these references are not directly applicable in evaluating the various antenna design choices or in identifying the particular design features that are most critical in determining antenna performance for GPS applications.

Recently, Tranquilla [64] presented a detailed analysis of a QHA using the method of moments technique and various design considerations for GPS applications. In the paper, the phase performance of the antenna as well as the amplitude was taken into account, and the use of the antenna for dual-frequency operation was discussed. Resonant fractional turn volutes are particularly attractive for GPS applications due primarily to their small size and wide beam elevation pattern coverage.

Figure 6.76 shows a model of the QHA with two independent bifilar structures, each having a feed generator at the center of the radial conductor. The parameters of three models examined are summarized in Table 6.10. The design frequency is L_1. Figure 6.77 shows the computed polar amplitude pattern (circular polarization) for these three designs. Note that all three provide adequate upper hemispherical coverage; however, only the shortest volute ($L_{ax} = 0.20\lambda$) has a suppressed backlobe. The figure also shows the phase patterns of both linear polarized components (θ and ϕ) in a fixed observation plane. Figure 6.78 shows the computed phase center location for ϕ linear polarization for the half-wavelength QHA. The longer, slimmer volutes are advantageous in their reduced phase center movement at extremely low elevation angles. This arises due to the small separation distances between helical arms as viewed from near broadside.

For dual-frequency operation, a single QHA is not available, because it is a resonant-type antenna and is inherently too narrow-band to accommodate both L_1 and L_2. Therefore, the dual-band operation can only be achieved through the incorporation of two antennas into one structure by coaxially mounting them in either an enclosed (L_1 inside L_2) or a *piggyback* (L_1 atop L_2) fashion, as shown in Figure 6.79.

Figures 6.80 and 6.81 show the amplitude and phase patterns of the enclosed and the piggyback types, respectively. For the enclosed volute design, neither the L_1 nor the L_2 performance is as good as with the single volute, the gain reduction being approximately 3.5 and 1.5 dB at L_1 and L_2, respectively, over the upper hemisphere, and a slight increase of the backlobe level at both L_1 and L_2. Phase center performance is also somewhat poorer at off-boresight angles for the compound antenna. From Figure 6.81 it can be seen that the coaxial piggyback arrangement has the greatest difference between L_1 and L_2 performance in the compound arrangement. In particular, the L_1 amplitude and phase response deteriorates significantly (with respect to L_2) at elevation angles below 45 deg. In this case, the radiation from the L_1 antenna is considered to excite the L_2 element and to generate a considerably large backlobe.

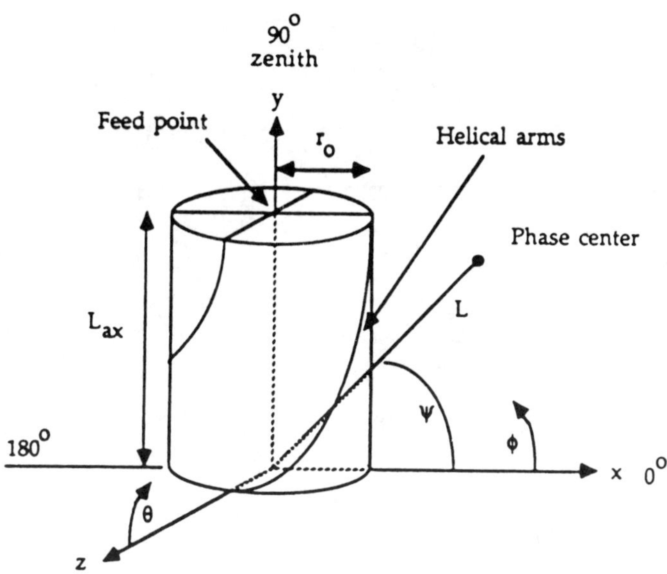

Figure 6.76 Geometry of the quadrifilar helix antenna. (From [64], © 1990 IEEE.)

Table 6.10
Design Parameters of Three Quadrifilar Helix Antennas

	Axial Length (L_{ax}) *(λ)*	*Volute Radius* (r_0) *(cm)*
Antenna 1	0.20	1.60
Antenna 2	0.27	1.39
Antenna 3	0.35	1.04

6.5.3 Microstrip Antennas

Owing to their low-profile and compact structure, MSAs currently account for almost half of the production of GPS antennas. A number of papers and several monographs [50,68] on MSAs have been published so far. One of the effective design methods for obtaining a broadbeam for GPS reception using a typical microstrip antenna is a reduction of the size of its ground plane [69,70].

A circular patch (operating in the TM_{11} mode) with a finite ground plane is shown in Figure 6.82, and the computed radiation patterns using the method of moments are shown in Figure 6.83 [69]. In the upper graph, showing the H-plane

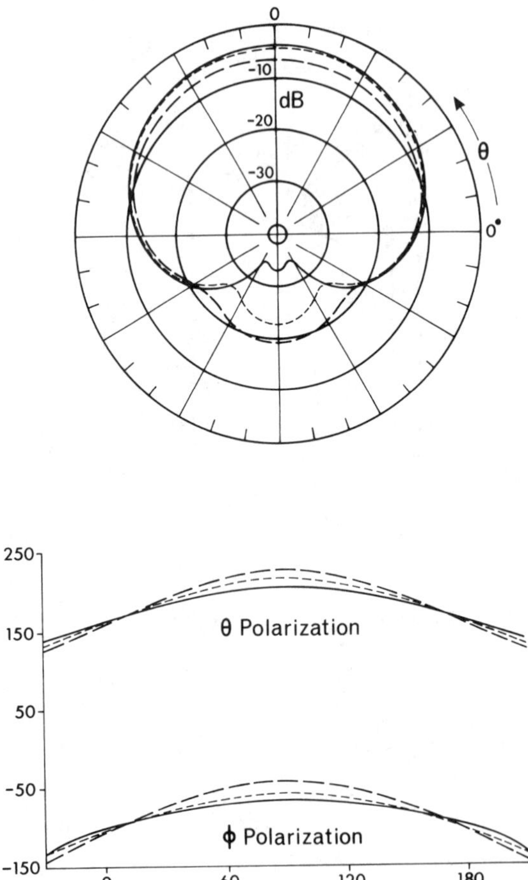

Figure 6.77 Calculated amplitude and phase patterns of the three QHAs of half-wavelength helixes a L_1, with $L_{ax} = 0.2\lambda$ (—), 0.27λ (----), and 0.35λ (– –). (From [64], © 1990 IEEE.)

pattern, the cross-polarization pattern in the $\phi = 45$-deg plane is also shown. The beamwidth of the H-plane pattern decreases by increasing the ground-plane size Consequently, the infinite ground plane has the sharpest cutoff pattern. In the E plane, on the other hand, the beamwidth decreases initially by increasing the ground-plane radius g, but increases for $g > 0.7\lambda$. The infinite ground plane give the broadest beam, which approaches -6 dB at the horizontal plane. These result show that the assumption of an infinite ground plane in approximate analysis o

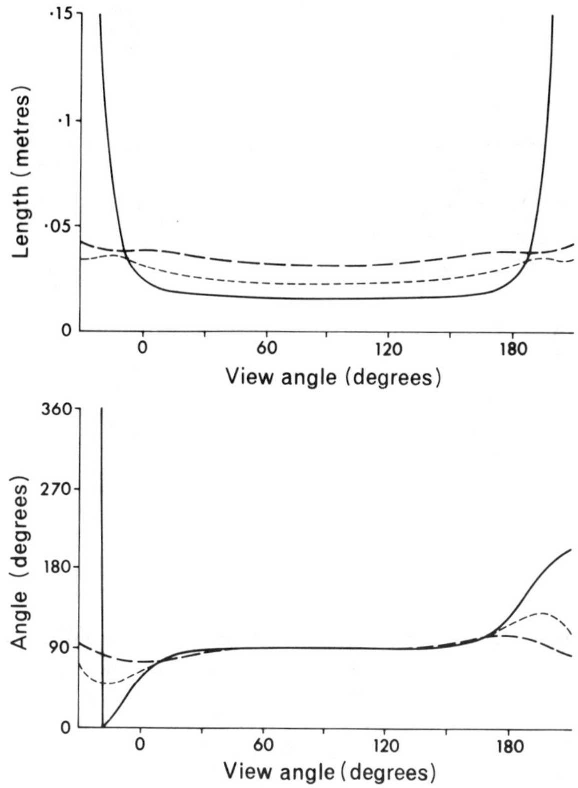

Figure 6.78 The phase center locations of the three QHAs for ϕ polarization component with L_{ax} = 0.2λ (—), 0.27λ (----), and 0.35λ (– –). (From [64], © 1990 IEEE.)

MSAs will have a serious effect on the correct prediction of the radiation patterns, particularly for angular ranges beyond 45 deg off the main beam.

For a rectangular microstrip patch with a finite ground plane, Huang [70] presented radiation patterns calculated on the basis of slot theory and the uniform GTD. Both the E- and H-plane radiation patterns of a rectangular microstrip patch, whose geometry is illustrated in Figure 6.84, are shown in Figure 6.85 with the measured results. The antenna dimensions in inches are given in the figure, and the operating frequency is 2.295 GHz (wavelength = 5.146 in). The agreement between the measurement and the calculation is quite good. From the figure, it can be seen that the gain reduction for the low elevation angles is relatively large, particularly in the H-plane. These characteristics seems to be similar to those of

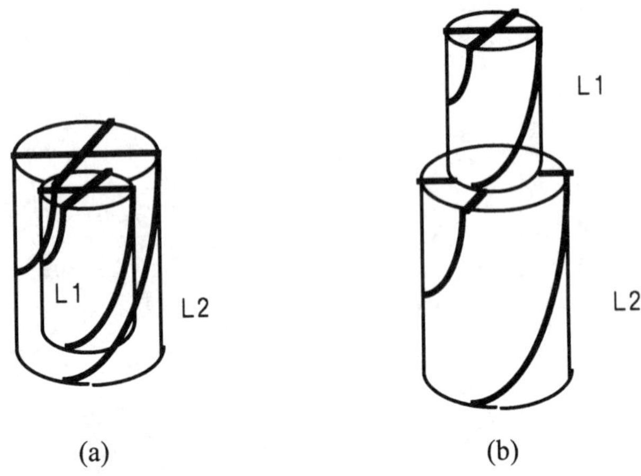

Figure 6.79 Dual-frequency quadrifilar helix antennas: (a) enclosed type; (b) piggyback type.

the circular microstrip patches; therefore, for GPS application, especially for its operation in the low elevation angle, a QHA may be more preferable for an MSA.

6.5.4 Array Antenna for GPS Reception

The deployment of a directional-array antenna for GPS reception obtains a highly precise pseudorange through the high signal-to-noise ratio of the received signal. The CRL developed a precise relative positioning system using GPS, called PRESTAR (Precise Satellite Ranging) [71]. By measuring the phases of the modulated clock signal and the carriers of L_1 and L_2, PRESTAR provides a three-dimensional relative position (i.e., baseline vectors) of up to several hundred kilometers with an accuracy of 10^{-7}, and also the clock offset between two receivers. Other features are an observation time requirement of less than five minutes and short processing time, and an ability to determine the precise orbit of the GPS satellite from accurate carrier range data. Because the antenna of PRESTAR is steered mechanically to direct its beam to one of GPS satellites for a moment and then switch to another, PRESTAR is a single-channel sequential system.

The antenna is a dual-frequency microstrip array antenna as shown in Figure 6.86. Eight elements are used for L_1, while six elements are used for L_2. Each element is a single-feed circular microstrip patch with two notches to generate circular polarization. In order to maintain good circular polarization characteristics,

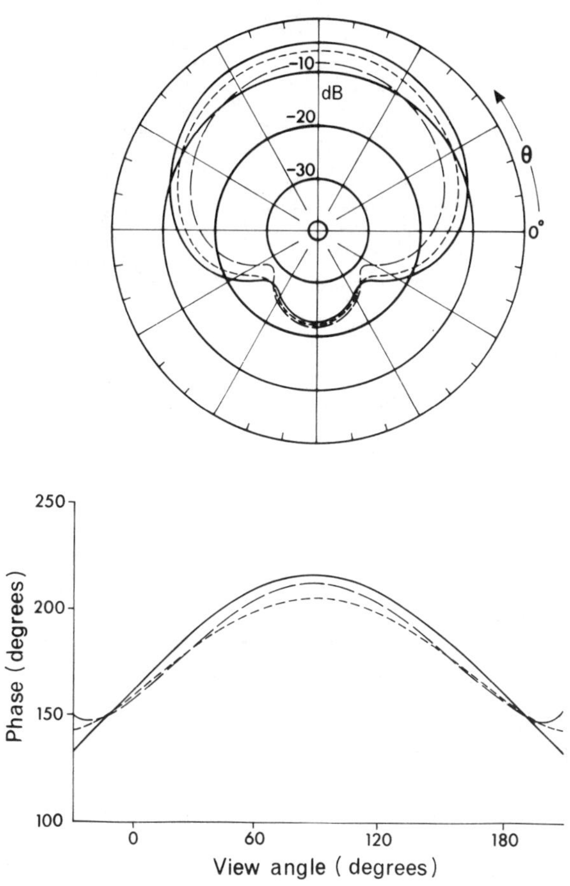

Figure 6.80 Amplitude and phase patterns of the enclosed-type QHA (the L_1 volute is inside the L_2 volute) at L_1 (– –), at L_2 (----), and of a single QHA (—). (From [64]), © 1990 IEEE.)

the sequential-array technique [51] was adopted, whereby the proper rotation and differential phase shift are given to each element. The numbers written on the elements show the relative rotations and phase shifts to be assigned.

Figures 6.87 and 6.88 show the measured radiation patterns at L_1 and L_2, respectively. From these figures, it can be seen that the excellent circular polarized patterns are obtained by the sequential technique. The measured gains are 18.7 dB at L_1 and 13.1 dB at L_2. The PRESTAR array is shown in Figure 6.89.

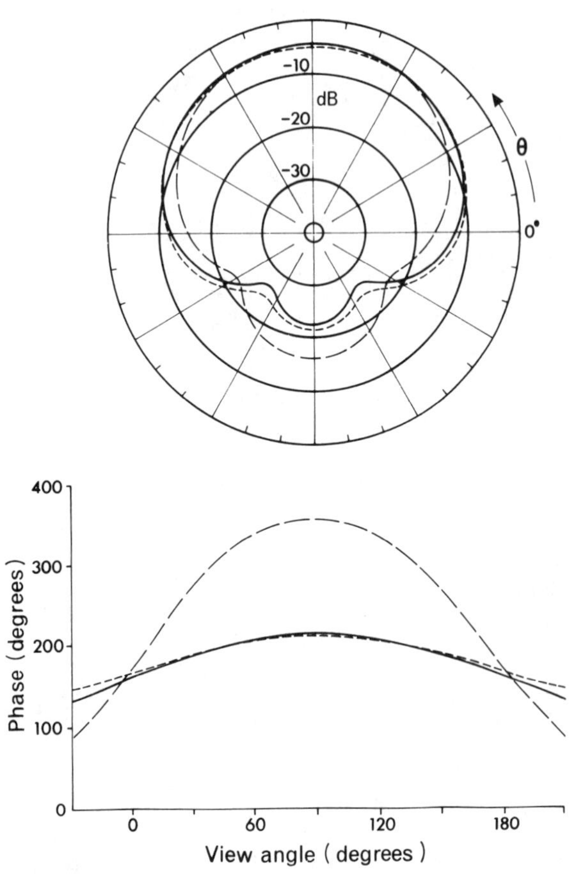

Figure 6.81 Amplitude and phase patterns of the piggyback-type QHA (the L_1 volute is above the top of the L_2 volute) at L_1 (– –), at L_2 (----), and of a single QHA (—). (From [64], © 1990, IEEE.)

6.6 ANTENNA SYSTEMS FOR SATELLITE BROADCASTING

6.6.1 Requirements for Mobile Antennas for DBS Reception

As satellite broadcasting services at the 12-GHz band become more widespread, the need for TV reception by relatively large mobiles, such as ships, trains, buses, and trucks, is growing. Since a satellite broadcasting system covers a wide area, it is inherently suitable for mobile reception. Direct Broadcasting Satellite (DBS) reception by mobiles has advantages in propagation conditions compared to its

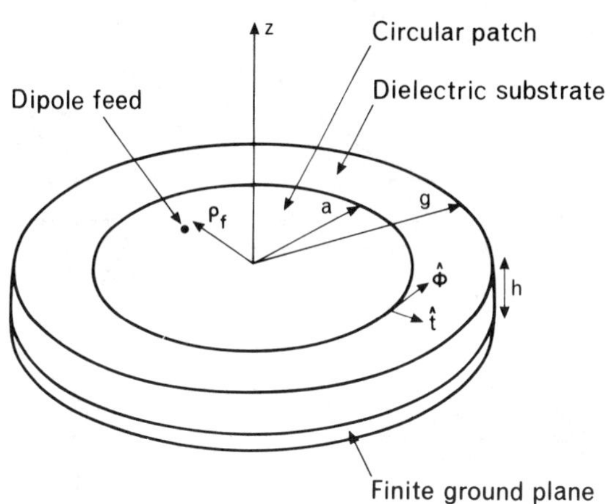

Figure 6.82 Circular microstrip antenna with a finite ground plane. (From [69], © 1986 IEEE.)

terrestrial TV broadcast counterparts, because the former suffers less interference due to multipath propagation as commonly experienced in the latter. In order to receive satellite broadcasts with good quality, the following electrical, mechanical, and structural requirements are imposed on mobile antennas, particularly antennas for land mobiles.

Accurate Beam Pointing

In the existing DBS system, the gain of a receiving antenna is required to be more than 32 dBi; thus, the beam must be considerably narrow (i.e., about 2 deg). Therefore, an accurate beam-pointing capability is required.

High-Speed Tracking

Since a mobile moves fast and changes its direction very frequently, the mobile receiver must track the satellite quickly, not only for beam steering but also for compensation of the vehicle vibration. In this sense, electronic scanning may be promising.

Countermeasure for Satellite Signal Interruption

In the monopulse tracking system, the receiving antenna can point its beam to the satellite as long as the signal from the satellite is received. If the signal, however,

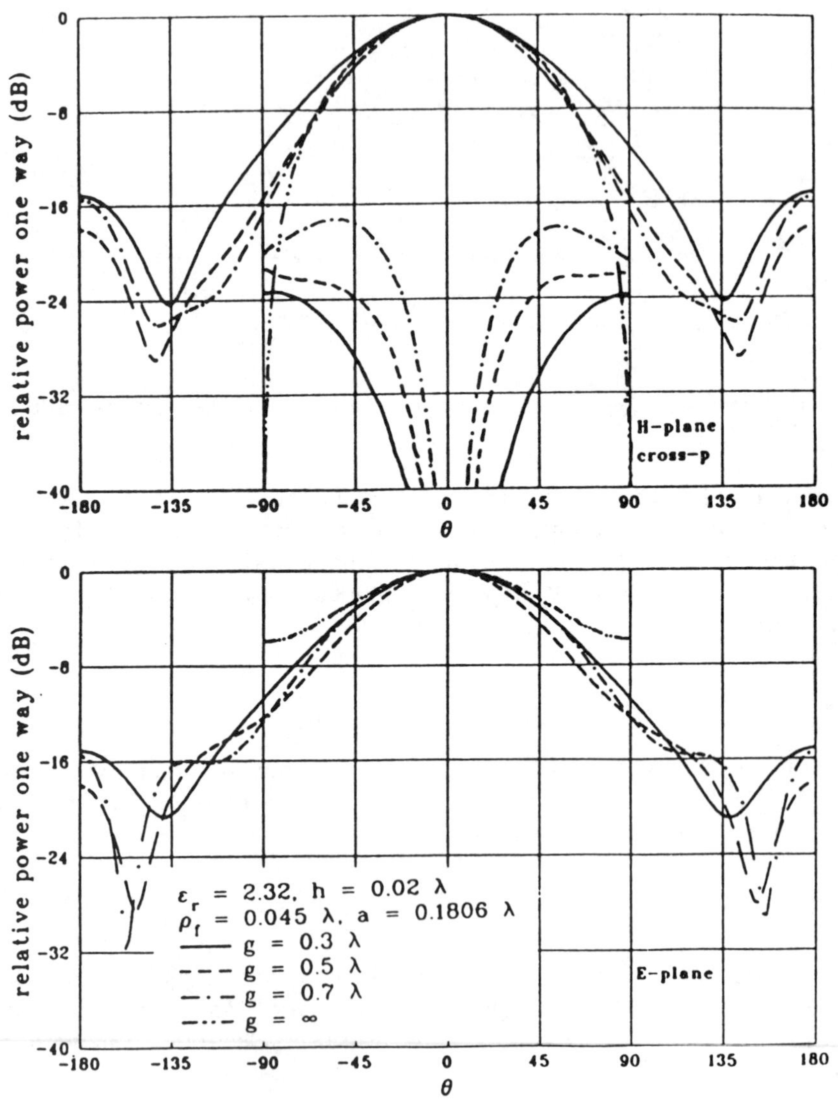

Figure 6.83 The radiation patterns of a circular patch with different ground plane radius *g*. (From [69], © 1986 IEEE.)

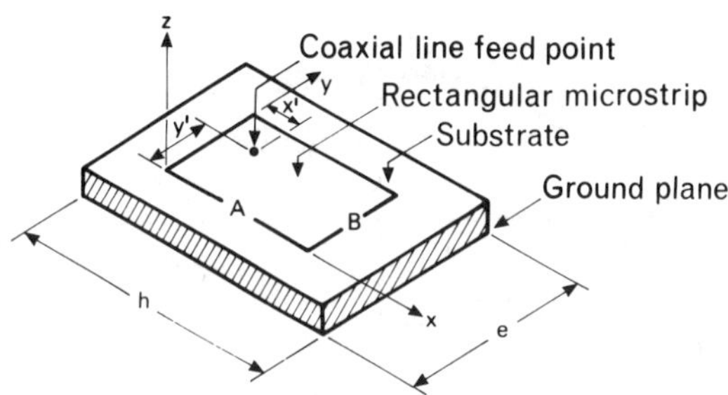

Figure 6.84 Rectangular MSA with a finite ground plane. (From [70], © 1983, IEEE.)

might be interrupted by surrounding obstacles, tracking becomes impossible. In order to cope with satellite signal blocking, several techniques are required: space diversity; holding the antenna direction using a gyroscope or a magnetic sensor; and high-speed mechanical searching of the satellite signal for recapture.

Height Limitation

Receiving systems as a whole must be compact, light-weight, and especially, not too high to prevent accommodation on the vehicle. Usually, the antenna equipment on the vehicle should be less than a foot high. In order to satisfy these requirements, several kinds of mobile antennas for DBS reception were developed, and some of them are now put into practical use for trains.

6.6.2 Overview of Antenna Tracking Methods

There are two kinds of antenna tracking, namely, automatic tracking and program tracking, but the latter may not be suitable for DBS reception by mobiles because they frequently move around and change their direction. Automatic tracking can be considered to be practical for these applications. The tracking methods are classified as two elementary techniques, mechanical and electrical trackings, as shown in Figure 6.90; but from a practical point of view, mechanical/electrical combined methods may be effective.

A mechanical tracking system drives an antenna mechanically to direct its beam to the satellite using the direction of the incident wave, which is detected by some means. Although the mechanical system cannot track the target with high speed, it provides omnidirectional tracking easily; therefore, it is suitable for tracking in the azimuthal direction.

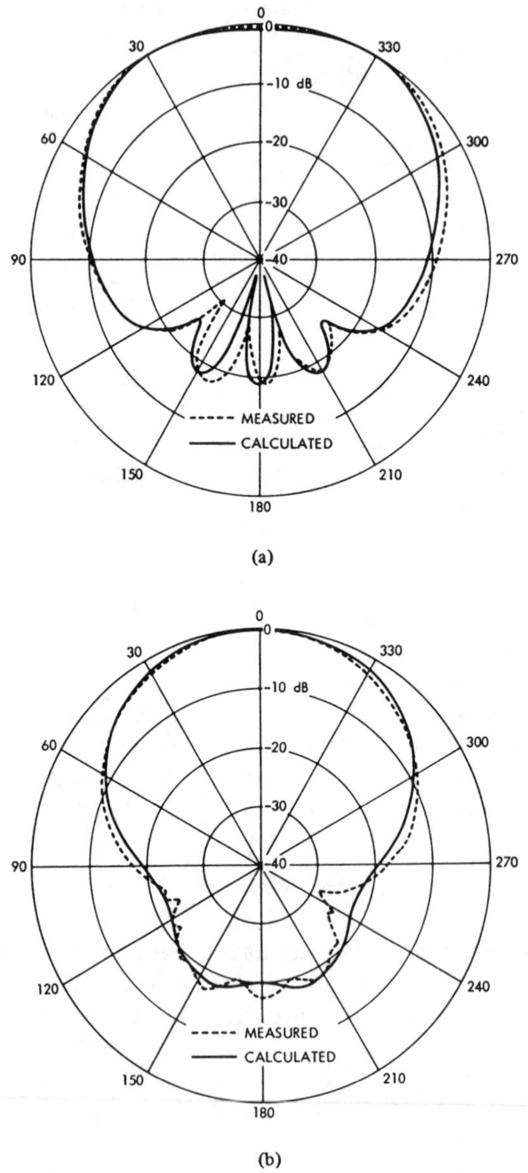

Figure 6.85 E-plane (a) and H-plane (b) radiation patterns of a rectangular MSA. (A = 2.126 in, B = 1.488 in, e = 10.5 in, h = 14.0 in, substrate thickness = 0.125 in, ϵ_r = 2.55, frequency = 2.295 GHz). (From [70], © 1983 IEEE.)

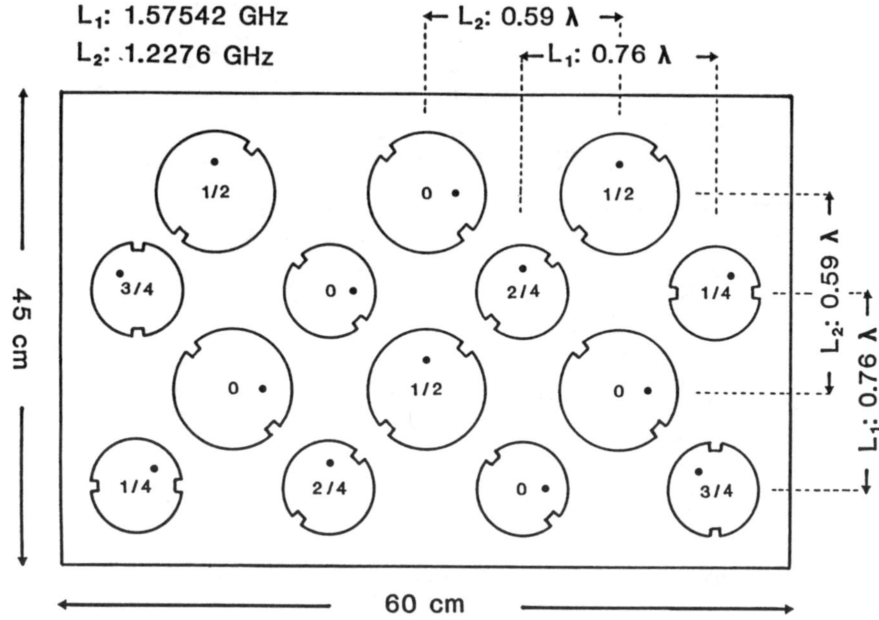

Figure 6.86 Configuration of the microstrip array antenna of GPS receiver PRESTAR.

Electrical tracking methods control the excitation phase of each element of an array to be in-phase in the direction of the satellite. The self-phasing antenna method, one of electrical trackings, makes all the outputs from antenna elements in-phase automatically by using phase lock loop (PLL) circuits. Usually this operation is carried out at the IF stage. On the other hand, in the phased-array antenna method, a conventional monopulse system can be used, in which digital phase shifters are controlled so that the difference signal of the monopulse may become null. A fully electrical tracking system however, has not yet been developed, because the required antenna may be a nonplanar array with a large number of elements to steer a high-gain beam over the hemisphere. Consequently, it must be electrically large, and would therefore be too expensive for civil use. In particular, the cost of the phase shifters could be very high. Electrical tracking provides high-speed operation, but it would be impractical for wide scanning; accordingly, it is suitable for limited scan or for the compensation of vehicle vibration, which needs a quick response.

The combination of mechanical and electrical tracking methods, which utilizes the advantages of both, can realize a rather simple and fine tracking system; some examples are shown in a later section. Usually, a mechanical system is employed

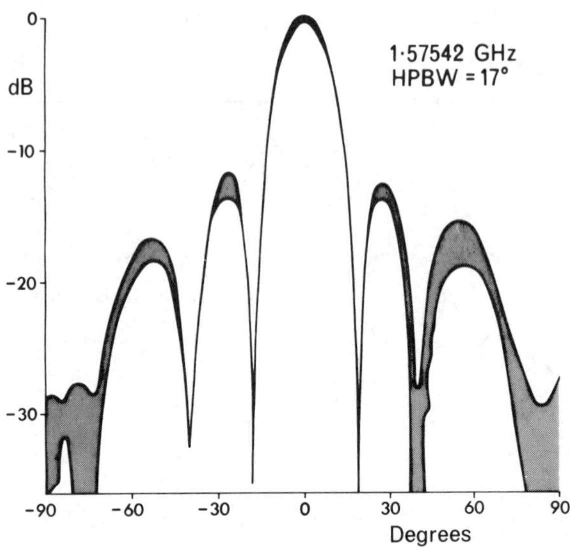

Figure 6.87 The radiation pattern of the PRESTAR at L_1.

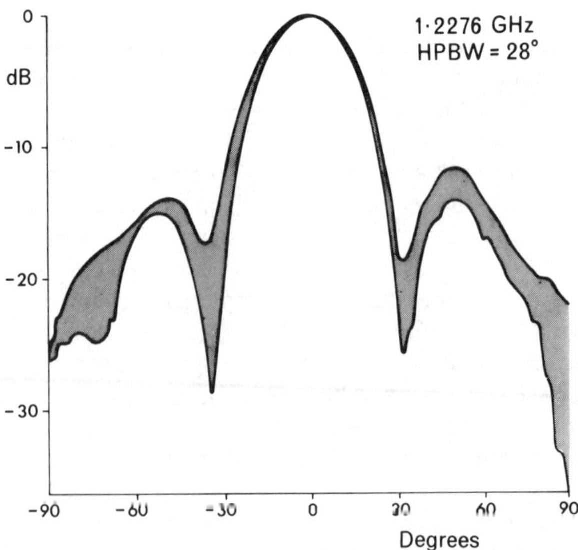

Figure 6.88 The radiation pattern of the PRESTAR at L_2.

Figure 6.89 The view of the PRESTAR. (Courtesy of CRL.)

for azimuthal tracking owing to its wide-scan capability, while an electrical system is used for elevation tracking due to its high-speed operation.

In automatic tracking systems, once an acquisition of the satellite is performed, it is preferable to hold its azimuthal direction. If the system has such a function, then the antenna points to the satellite all the time, and when the satellite transmission is interrupted by obstacles, the TV reception can be restored instantly after the interruption is removed. A gyroscope widely used for ships and airplanes for the purpose of memorizing the azimuthal data has high accuracy, but it is too expensive for civil use. On the other hand, a magnetic azimuth sensor is inexpensive but less accurate, so it is suitable for supplementary use in the search operation.

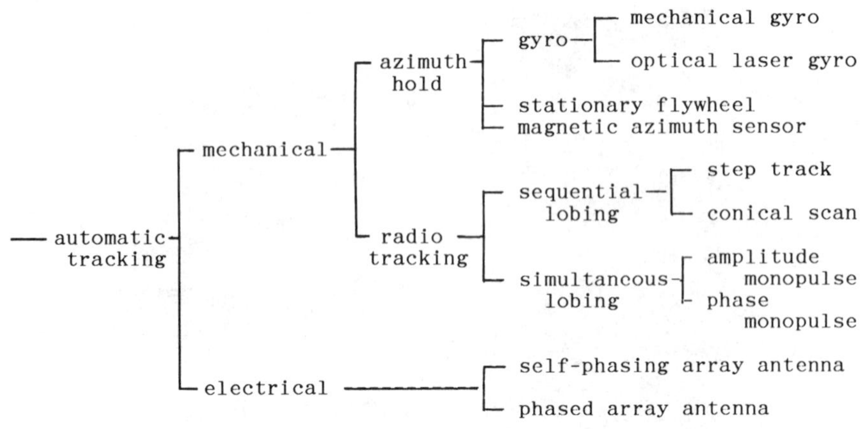

Figure 6.90 Antenna tracking techniques.

6.6.3 Mechanically Tracking Antennas

An example of mechanical tracking is the train antenna for receiving DBS, which was developed in Japan [72]. The beam of a planar-array antenna is mechanically controlled both in the azimuth and elevation directions. The configuration of the antenna system is shown in Figure 6.91. It applies a combination of open- and closed-loop controls for satellite tracking. For the open-loop control, program tracking is employed which directs the antenna beam to the satellite based on the information of the position and the direction of the train and the cant of the rail, known by the distance from the start point. The data are stored in the random access memory (RAM) of the antenna driver. Meanwhile, for the closed-loop control, a kind of lobing method is used. In this mode, the beam is swung in four directions (two in the azimuth and the other two in the elevation) by 2-bit digital phase shifters driven with a high frequency. When the antenna boresight is offset from the direction of the satellite, the received output of the swung beam is modulated. By demodulating the output, the tracking error to steer the antenna is obtained. In order to simplify the configuration, an AGC signal of the TV receiver is used for monitoring the modulated receiving level.

The antenna consists of four planar antennas and is mounted on 88×88 cm^2 aluminum honeycomb plate. At first, a crank-type microstrip line antenna element (or a rampart line microstrip antenna) printed on a glass-fiber reinforced polytetrafluoroethylene (PTFE) was employed. In order to make the antenna height lower to reduce wind pressure and keep the beam to the elevation of the geostationary

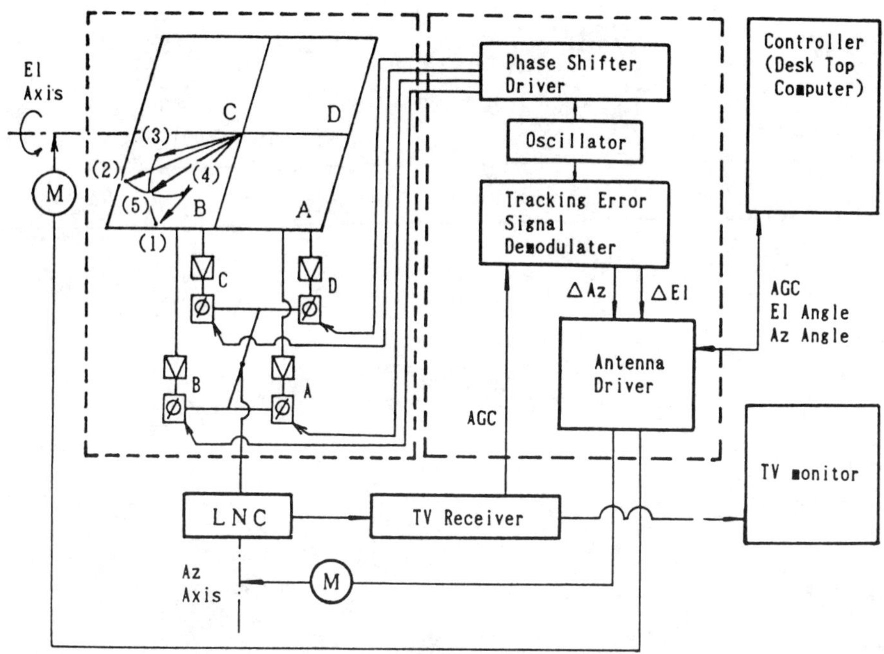

Figure 6.91 Configuration of the mechanical tracking train antenna for DBS reception. (From [72].)

satellite, the antenna was designed to provide a beam tilt of 30 deg. More recently, these antenna elements were replaced by a waveguide slot array [73] because the waveguide slot array can realize a large beam tilt, and consequently a low-profile antenna can be obtained. Figure 6.92 shows the slot-array antenna developed. The antenna consists of four subarrays, each of which is a 16 $\times$ 24-element cross-slot array. The overall dimension of the antenna is 662 $\times$ 646 mm^2 and the tilt angle of the beam is 52 deg.

Each antenna has an LNA to reduce C/N degradation due to feeder loss. The phase shifters connected behind the LNAs are switched by the phase shifter driver from 100 to 400 Hz to swing the antenna beam by up to 0.2 deg sequentially in the four directions. The antenna gain is 35 dBi and the gain and the noise figure of the LNA are 22 dB and 1.2 dB, respectively. The LNAs and phase shifters are mounted on a PTFE board and attached at the rear side of the planar antenna.

In the case of the train application, especially the bullet train, the highest wind pressure to be born exceeds 800 kg/m^2 in a tunnel; therefore, a radome is used to protect the antenna. The radome is fabricated with three layers (two FRP layers, and one polyurethane layer), and its transmission loss is less than 0.5 dB.

Figure 6.92 Slot array train antenna. (Courtesy of NEC.)

Figure 6.93 shows the antenna and the radome mounted on the Japanese bullet train Shinkansen. The dimensions of the radome are 143 cm in diameter and 63 cm high.

The major antenna performance parameters are summarized in Table 6.11 and Figure 6.94 shows the tracking errors during rolling tests. The errors are maintained within 0.17 deg at the speed of 3 deg/sec in azimuth and within ±0.2 deg at 2 deg/sec in elevation plane, respectively.

6.6.4 Antennas Using a Combination of Mechanical and Electrical Tracking

Since land mobiles such as cars and trains move fast and change their direction frequently, electrical tracking is preferable, but in order to fully cover 360 deg in azimuth, some facility for mechanical tracking is used. Two kinds of antennas of this type were studied and developed by NHK (Japanese Broadcasting Corporation). The first one uses a gyroscope which is always calibrated with a phase

Figure 6.93 Antenna and radome mounted on the Japan's bullet train Shinkansen. (Courtesy of NEC.)

Table 6.11
Performance of Mechanical Tracking DBS Receiving Antenna

Characteristic	*Measurement*
Gain	35 dB/11.95 GHz
Polarization	Right-hand circular polarization
Beam tilt	30 deg
Dimension	Φ 1,430 × 630 mm
Operation	25–58 deg (El)
Range	±120 deg (Az)

Note: From [72].

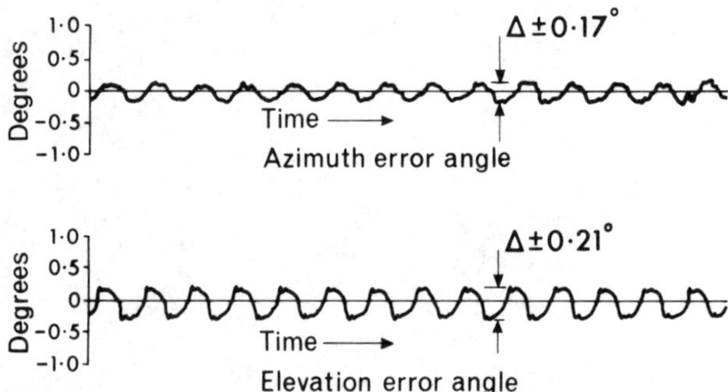

Figure 6.94 Tracking error in rolling tests.

comparison monopulse method for mechanical tracking, and with an adaptive beam-forming method for electrical tracking [74]. Although automatic tracking using the phase comparison monopulse method [75,76] provides highly accurate detection of the direction of the satellite, the DBS signal is interrupted often due to environmental obstacles, resulting in the need for the gyroscope in conjunction with the phase comparison monopulse.

Figure 6.95 illustrates the mobile DBS television receiving system. Four flat-panel antenna units are divided into two pairs to reduce the system height. The system is housed in a cylindrical sandwich-structure radome of 350 mm in height and 862 mm in diameter. The system weighs 50 kg. The flat-panel antenna unit consists of an 8 × 16-element circularly polarized printed array. Each element is a rectangular microstrip patch with dual feeds formed on a two-layer structure of the substrate, as shown in Figure 6.96 [77,78]. The two-layer structure makes an optimum design possible for both the feed line and the radiation element separately, and it provides a low-loss and wide-band microstrip antenna.

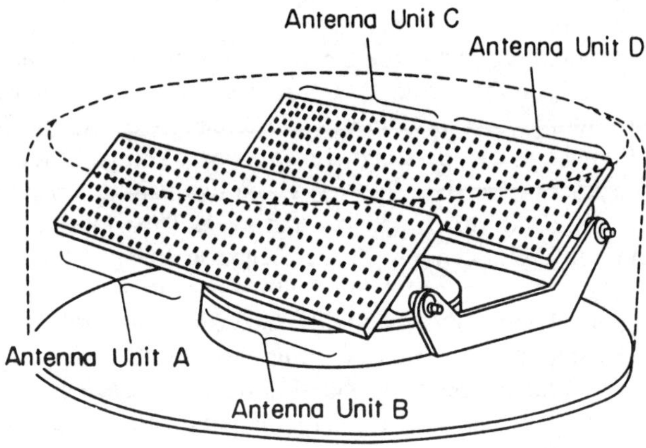

Figure 6.95 Mobile DBS receiving system using mechanical and electrical combined tracking. (From [74], © 1989 IEEE.)

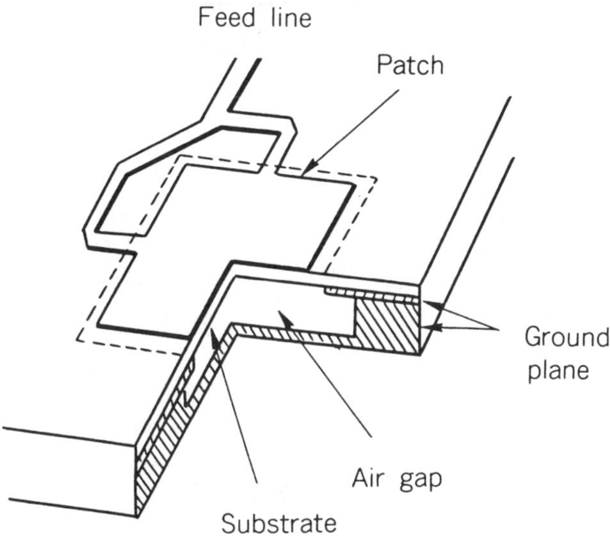

Figure 6.96 Circularly polarized microstrip antenna with two-layer structure. (From [78].)

The receiving antenna system shown in Figure 6.95 employs mechanical steering in both azimuth and elevation for coarse tracking and the electrical beam steering by the adaptive technique for fine tracking. The antenna units A and B, which are mounted on the turntable and rotate around the azimuthal axis, produce the phase difference corresponding to the direction of the incident wave. Using the phase difference as the error signal of the phase comparison monopulse, we can form a feedback loop. On the other hand, the two pairs of antenna units, A/B and C/D, rotate around the different elevation axes. Therefore, the phase difference between the pairs is independent of the elevation angle of them. This means that the elevational mechanism, controlled by the phase comparison monopulse, would not form a feedback loop. The phase difference, however, relates to the elevational direction of the satellite. We can determine the elevation angle of the two antenna panels using the information of the phase difference between them; consequently, we can direct the beam peak of the antenna panels to the satellite.

The mobile DBS television receiving system employs, for simplicity, the phase comparison monopulse method in the second IF stage. Signal processing in the second IF stage is far easier than in the first IF stage or RF stage. One of the features of this system is that all the outputs from four antenna units are automatically combined to be in phase. In-phase combination, instead of simple combination, of the four antenna outputs may not only increase the antenna gain, but also make the antenna units work as a so-called adaptive antenna with a wide beamwidth equivalent to a single antenna unit. Figure 6.97 shows the azimuthal pattern of one antenna unit accompanied by the received C/N obtained by the in-phase combination of four antenna units.

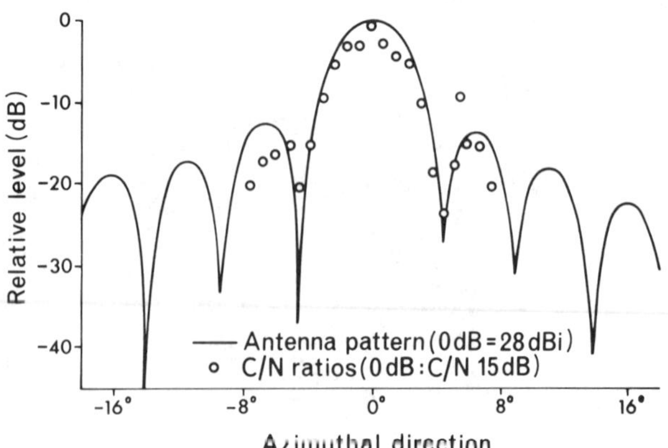

Figure 6.97 Measured azimuth radiation pattern of the antenna unit and the received C/N in-phase combined by four units. (From [74], © 1989 IEEE.)

The adaptive beam-forming antenna units electronically support the tracking mechanism when the latter produces pointing errors, which are caused by feeble, yet very fast vibrations as well as transient effects during vehicle startup, acceleration, and stop. In order to perform adaptive beam forming, the outputs from antenna units A,B,C, and D must always be combined in phase, irrespective of the phase difference among these outputs. The in-phase combination is also carried out at the second IF stage (center frequency, 402.78 MHz) using the circuitry as shown in Figure 6.98. As can be seen from the figure, the phase of signal B fed to terminal B is locked with the reference phase of signal A coming from terminal A. Then signal B is simply combined with signal A for output.

The overall system configuration is shown in Figure 6.99. In order to minimize signal handling between the inside and outside of the vehicle, all RF processing equipment is mounted outside the vehicle, and the combined second IF signal is transmitted inside the vehicle through a rotary coupling transformer (or rotary joint) for demodulation.

To maintain coherency, the output signals from four flat-panel antenna units A,B,C, and D are separately converted into the first IF band (1.0 to 1.3 GHz) by BS converters excited by a common local oscillator. The signals are further converted into the second IF with BS tuners, tuned by a common voltage-controlled oscillator (VCO). The four second IF signals are divided into two pairs for antenna tracking and signal reception. The received signals from antenna units A,B,C, and D are sent in pairs to two azimuthal in-phase combining circuits. The output signals A + B and C + D are given to the elevational in-phase combining circuit.

Meanwhile, the control signals from antenna units A and B are sent to the azimuthal monopulse circuit. At the same time, outputs A + B and C + D from the two azimuthal in-phase combining circuits are fed to the elevational monopulse

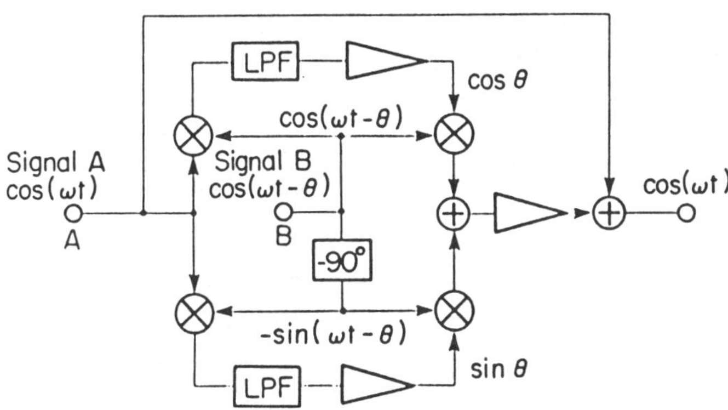

Figure 6.98 In-phase combining circuitry. (From [74], © 1989 IEEE.)

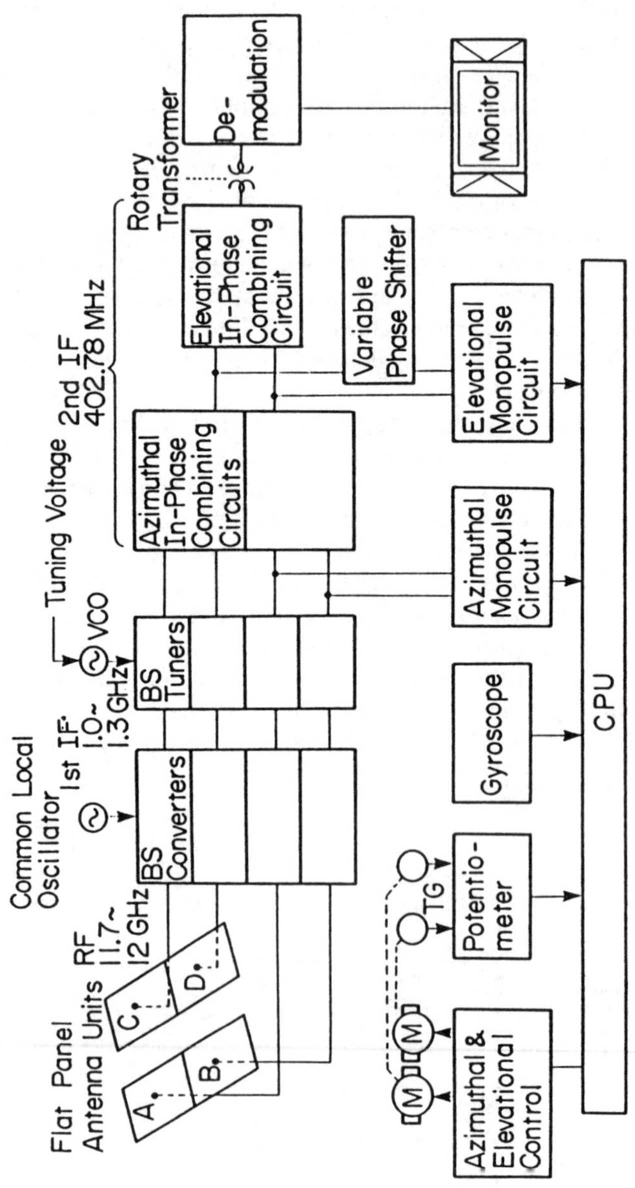

Figure 6.99 The overall configuration of the mobile DBS receiving system. (From [74], © 1989 IEEE.)

circuit, where the output from one is phase-shifted by a variable phase shifter corresponding to the antenna units' elevational rotary drive, while the output from the other is not shifted. Then the azimuthal and elevational monopulse signals are provided to the Central Processing Unit (CPU). Outputs from the gyroscope and the potentiometer are also provided to the CPU, which generates the signals to control the antenna units' azimuthal and elevational angles. The CPU completes one session of processing within 5 ms. The drive system rotates at a maximum speed of 15 rpm in azimuth and 12 rpm in elevation, in ranges of 0 to 360 deg in azimuth and 25 to 90 deg in elevation.

Another example of a combined tracking system was also developed and presented by NHK [79]. The antenna employs a mechanical control for the azimuthal coarse tracking, but for the elevational tracking and the azimuthal fine tracking, it uses an electrical adaptive beam-forming technique. Figure 6.100 shows the system configuration of the mobile receiver with an adaptive array. The receiver consists of eight receiver units, which include a flat-panel antenna unit and a frequency converter unit. Most of the adaptive array is mounted on the turntable, which is placed on the roof of the vehicle.

Figure 6.101 shows the mobile receiver when the top cover of the radome is removed. Each of the two large panel antennas in the photograph includes two antenna units. The left-side antenna unit is divided into two subunits, which are used for the phase comparison monopulse to detect the azimuthal direction as well

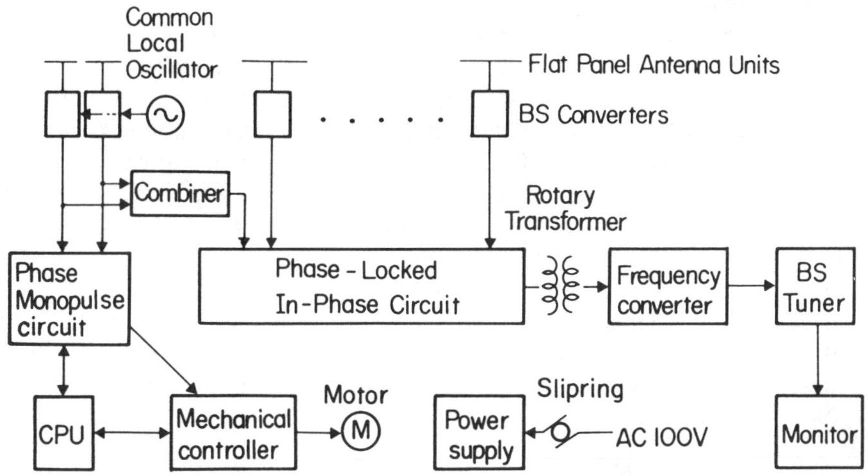

Figure 6.100 System configuration of the mobile DBS receiver with eight-element adaptive array. (From [80].)

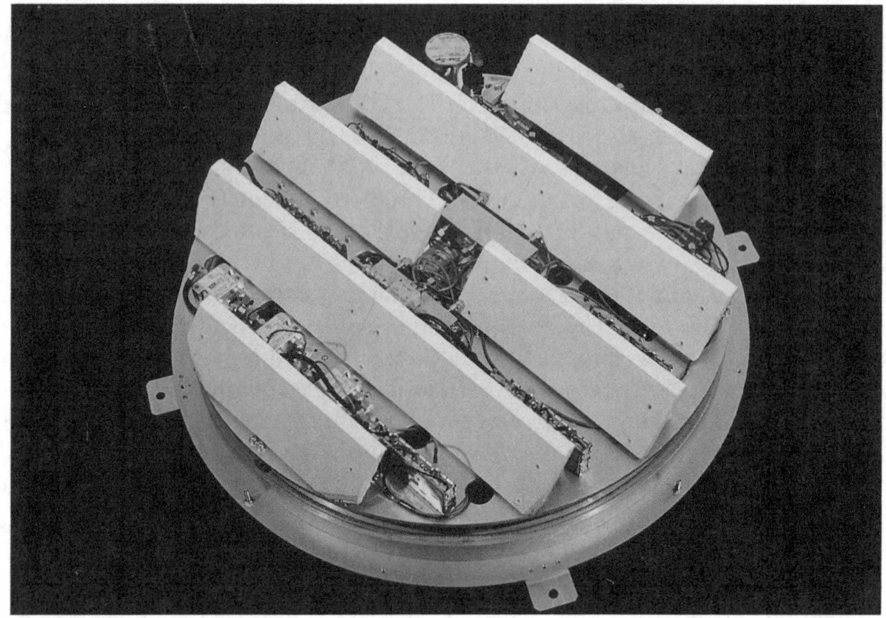

Figure 6.101 Eight-element moble antenna. (Courtesy of NHK.)

as the DBS signal reception. The flat-panel antenna, whose design is basically the same as the previous example, consists of a circularly polarized 4 × 16-element printed patch array with dual feeds. A foamed polyethylene substrate is used to reduce dielectric loss. The beamwidth of the antenna unit is 20 deg in elevation and 5 deg in azimuth. The monopulse signal received by each subunit is converted into the 1.3-GHz band by two frequency converters that have a common local oscillator. The converted signals are then divided into the phase detection and phase synthesizing circuits. The monopulse antenna has a detection range of 16 deg in azimuth. A CPU examines the output of the phase monopulse circuit and increases the mechanical lock-in range.

The adaptive system in this DBS receiver employs a phase lock loop circuit to perform the in-phase combination of the signals received by element antennas. The specifications and major performances of the DBS mobile receiver are shown in Table 6.12.

Table 6.12
Performance and Data on Mobile DBS Receiver Antenna With Eight Adaptive-Array Elements

Characteristic	*Measurement*
Size	200 mm in height, 700 mm in diameter
Weight	30 kg
Power requirement	100W
Tracking range:	Omnidirectional in azimuth
Mechanical	±2.5 deg in azimuth
Electric	±10 deg in elevation
Antenna type	Adapative-array antenna using eight flat-panel antennas
Unit antenna	64-element microstrip antenna array, 23.5 dBi, 76 x 272 mm
Gain	32.5 dBi

Note: from [80].

REFERENCES

[1] *2nd Mobile Satellite Conf.,* Ottawa, Canada, June 1990.

[2] Hamamoto, N., S. Ohmori, and K Kondo, "Results on CRL's Mobile Satellite Communication Experiments Using ETS-V Satellite," *Space Communications,* 1990, pp. 483–493.

[3] A series of *MSAT-X Quarterly,* Jet Propulsion Laboratory, No. 1 (1984), to No. 25 (1990).

[4] "PROSAT Phase I Report," European Space Agency, ESA STR-216, May 1986.

[5] "IEEE Standard Radar Definitions," IEEE, NY, 1982.

[6] Gedney, R. T., and R. J. Scherfler, "Advanced Communications Technology Satellite (ACTS)," *Int. Conf. Communications '89* (ICC '89), 1989, pp. 1566–1577.

[7] Takeuchi, M., S. Isobe, N. Hamamoto, S. Ohmori, and M. Yamamoto, "Experimental Advanced Mobile Satellite Communications System in MM-wave and Ka-Band using Japan's COMETS," *IEEE GLOBECOM '92,* Orlando, Dec. 1992, pp. 443–446.

[8] "Environmental Conditions and Test Procedures for Airborne Electronics, Electrical Equipment and Instruments," Radio Technical Commission for Aeronautics, RTCA Do-160A.

[9] CCIR Recommendation 465–1.

[10] Blake, L. V., Chs. 8 and 9 of *Antennas*, Artech House, 1984.

[11] Bell, D., K. Dessouky, P. Estabrook, M. K. Sue, and M. Bobb, "MSAT-X Antennas: Noise Temperature and Mobile Receiver G/T," *MSAT-X Quarterly,* No. 16, 1988, pp. 12–17.

[12] CCIR, "Propagation Data for Maritime Mobile-Satellite Systems for Frequencies Above 100 MHz," Rep. 884-2, ITU, 1990.

[13] Beckmann, R., and A. Spizzichino, *The Scattering of Electromagnetic Waves From Rough Surfaces,* New York/Oxford: Pergamon Press, 1963.

[14] Karasawa, Y., and T. Shiokawa, "Characteristics of L-Band Multipath Fading Due to Sea Surface Reflection," *IEEE Trans. Ant. Propag.,* Vol. AP-32, No. 6, 1984, pp. 618–623.

[15] Sandrin, W., and D. J. Fang, "Multipath Fading Characterization of L-Band Maritime Mobile Satellite Links," *COMSAT Tech. Rev.,* Vol. 16, No. 2, 1986, pp. 319–338.

[16] Karasawa, Y., and T. Shiokawa, "A Simple Prediction Method for L-Band Multipath Fading in Rough Sea Conditions," *IEEE Trans. Communications,* Vol. Com-36, No. 10, 1988, pp. 1098–1104.

[17] Karasawa, Y., and T. Shiokawa, "Spectrum of L-Band Multipath Fading Due to Sea Surface Reflection," *Trans. Inst. Electron. Comm. Engrs. Japan,* Vol. J67-B, No. 2, 1984, pp. 171–178.
[18] Karasawa, Y., and T. Shiokawa, "Fade Duration Statistics of L-Band Multipath Fading Due to Sea Surface Reflection," *IEEE Trans. Ant. Propag.,* Vol. AP-35, No. 8, 1987, pp. 956–966.
[19] CCIR, "Fading Reduction Techniques Applicable to Ship Earth-Station Antennas," SG 8, Rep. 1048-1, ITU, 1990.
[20] Kozono, S., and M. Yoshikawa, "Switch and Stay Diversity Effect on Maritime Mobile Satellite Communications," *Trans. Inst. Electron. Comm. Engrs. Japan,* Vol. J64-B, No. 5, 1981.
[21] Karasawa, Y., and T. Shiokawa, "Space and Frequency Correlation Characteristics of L-Band Multipath Fading Due to Sea Surface Reflection," *Trans. Inst. Electron. Comm. Engrs. Japan,* Vol. 67-B, No. 12, 1984, pp. 1347–1354.
[22] Shiokawa, T., and Y. Karasawa, "Ship-borne Antenna Suppressing Multipath Fading in Maritime Satellite Communication," *IEEE Ant. Propag. Society Symp. (AP-S),* New Mexico, 1982.
[23] INMARSAT 7th ACTOM, 1981, Tokyo.
[24] Yasunaga, M., Y. Karasawa, and T. Shiokawa, "A Simplified Fading Reduction Technique in Maritime Satellite Communications," *Trans. Inst. Electron. Comm. Engrs. Japan,* Vol. E69, No. 2, 1986, pp. 83–85.
[25] Ohmori, S., and S. Miura, "A Fading Reduction Method for Maritime Satellite Communications," *IEEE Trans. Ant. Propag.,* Vol. AP-31, No. 1, 1983, pp. 184–187.
[26] Ohmori, S., et al., "Characteristics of Sea Reflection Fading in Maritime Satellite Communications," *IEEE Trans. Ant. Propag.,* Vol. AP-33, No. 8, Aug. 1985, pp. 838–845.
[27] Iwai, H., M. Yasunaga, and Y. Karasawa, "A Fading Reduction Technique Using Interleave-Aided Open Loop Space Diversity for Digital Maritime Mobile-Satellite Communications," *Trans. Inst. Electron. Inform. Comm. Engrs.,* Vol. E74, No. 10, 1991.
[28] Giorgio, F., I. Knight, and R. Matthews, "A Maritime Mobile Terminal for Communications Satellite Application," *ICC'74,* 29C1–29C6, 1974.
[29] Shiokawa, T., Y. Karasawa, and H. Yuki, "Reduction Method of Pointing Error of El/Az Mount for Maritime Satellite Communications," *IEICE,* Vol. J69-B, No. 8, pp. 833–841.
[30] MSAT-X Technical Brochure, "Low-Cost Omnidirectional Vehicle Antennas for Mobile Satellite Communications," Jet Propulsion Laboratory.
[31] Adams, A. T., et al, "The Quadrifilar Helix Antenna," *IEEE Trans. Ant. Propag.,* Vol. AP-22, March 1974, pp. 173–178.
[32] Taira, S., M. Tanaka, and S. Ohmori, "High Gain Airborne Antennas for Satellite Communications," *IEEE Trans. Aerospace and Electronic Systems,* Vol. 27, No. 2, March 1991, pp. 354–360.
[33] Nakano, H., K. Vichien, T. Sugiura, and J. Yamauchi, "Singly-Fed Patch Antenna Radiating a circularly Polarized Conical beam," *Electronic Letters*, Vol. 26, No. 10, May 1990, pp. 638–640.
[34] Yasunaga, M., F. Watanabe, T. Shiokawa, and M. Yamada, "Phased Array Antennas for Aeronautical Satellite Communications," *5th ICAP,* March 1987, p. 1.47.
[35] Howell, J. Q., "Microstrip Antennas," *IEEE Trans. Ant. Propag.,* Vol. 23, January 1975, pp. 90–93.
[36] Kuribayashi, M., and N. Goto, paper at *IECE of Japan,* March 1982, p. 643.
[37] Itoh, K., H. Baba, Y. Ogawa, F. Watanabe, and M. Yasunaga, "L-Band Airborne Antenna Using Crossed Slots," IECE of Japan Technical Report, AP85-101, Jan. 1986.
[38] Ehrenspeck, H. W., "The Short Backfire Antenna, A New Type of Directional Line Source," *Proc. IRE,* Vol. 48, Jan. 1960, pp. 109–110.
[39] Dod, L. R., "Experimental Measurements of the Short Backfire Antenna," Rep. S-525-66-480, Goddard Space Flight Center, Greenbelt, MD., Oct. 1966.
[40] Takeuchi, K., M. Yasunaga, and T. Shiokawa, "Radiation Characteristics of Short Backfire Antenna Applicable to Mobile Communications," *IEICE Trans.,* Vol. E74, No. 10, Oct. 1991.

[41] Shiokawa, T., and Y. Karasawa, "Compact Antenna Systems for INMARSAT Standard-B Ship Earth Stations," *IEE 3rd Int. Conf. on Satellite Systems for Mobile Communication and Navigation,* London, June 1983.

[42] Ohmori, S., S. Miura, K. Kameyama, and H. Yoshimura, "An Improvement in Electrical Characteristics of a Short Backfire Antenna," *IEEE Trans. Ant. Propag.,* Vol. AP-31, No. 4, July 1983, 644–646.

[43] Shiokawa, T., and Y. Karasawa, "Array Antenna Composed of 4 Short Axial-Mode Helical Antennas," *IECE of Japan,* Vol. J65-B, No. 10, 1982, p. 1267.

[44] Kumar, A., *Fixed and Mobile Terminal Antennas,* Norwood, MA: Artech House, 1991, Chapter 5.

[45] MSAT-X Technical Brochure, "Low-Cost Omni-Directional Vehicle Antennas for Mobile Satellite Communications," Jet Propulsion Laboratory.

[46] Derneryd, A. G., "Analysis of the Microstrip Disk Antenna Element," *IEEE Trans. Ant. Propag.,* Vol. AP-27, No. 5, Sept. 1979, pp., 660–664.

[47] Long, S. A., "A Dual-frequency Stacked Circular-disk Antenna," *IEEE Trans. Ant. Propag.,* Vol. AP-27, March 1979, pp. 270–273.

[48] Heckert, G. P., "Investigation of L-Band Shipboard Antennas for Maritime Satellite Applications," Automated Marine Int. Rep. NASW-2165, Feb. 1972.

[49] Stricland, P. C., "Low Cost, Electrically Steered Phased Array for General Aviation," *Proc. 2nd Int. Mobile Satellite Conf.,* Ottawa, 1990, pp. 169–171.

[50] James, J. R., P. S. Hall, and C. Wood, *Microstrip Antennas—Theory and Design*, IEEE Electromagnetic Wave Series 12, Peter Peregrinus, 1981.

[51] Teshirogi, T., et al. "Wideband Circularly Polarized Array With Sequential Rotations and Phase Shift of Elements," *Int. Symp. Ant. Propag.,* Tokyo, Aug. 1985, pp. 117–120.

[52] Blake, L. V., *Antennas,* Artech House, 1986, pp. 389.

[53] Ohmori, S., S. Taira, and A. Austin, "Tracking Error of Phased Array Antenna," *IEEE Trans. Ant. Propag.,* Vol. AP-39, No. 1, Jan. 1991, pp. 80–82.

[54] Ohmori, S., K. Tanaka, S. Yamamoto, M. Matsunaga, and M. Tsuchiya, "A Phased Array Tracking Antenna for Vehicles," *Proc. 2nd Int. Mobile Satellite Conf.,* Ottawa, June 1990, pp. 519–522.

[55] Yamamoto, S., T. Tanaka, H. Wakana, and S. Ohmori, "Antenna Tracking System for Land Mobile Satellite Communications," IEICE Technical Report, SANE90-51, Dec. 1990.

[56] Nishikawa, N., K. Sato, and M. Fujino, "Phased Array Antenna for Land Mobile Satellite Communications," *Trans. IEICE,* Vol. J72-B-II, No. 7, July 1989, pp. 323–329.

[57] Hori, T., N. Terada, and K. Kagoshima, "Electrically Steerable Spherical Array Antenna for Mobile Earth Station," *5th Int. Conf. Ant. Propag.,* (ICAP'87), March 1987, pp. 55–58.

[58] Maruyama, S., K. Kadowaki, and Y. Hase, "Experiments on Message Communications With Hand-Held Terminal," *IEICE Trans. B-II,* Vol. J72-B-II, No. 7, July 1989, pp. 269–275.

[59] Woo, K., J. Huang, V. Jamnejad, D. Bell, J. Berner, P. Estabrook, and A. Densmore, "Performance of a Family of Omni and Steered Antennas for Mobile Satellite Applications," *Proc. 2nd Int. Mobile Satellite Conf.,* Ottawa, June 1990, pp. 540–546.

[60] Bell, D., V. Jamnejad, M. Bobb, and J. Vidican, "Reduced Height, Mechanically Steered Antenna Development," *MSAT-X Quarterly,* No. 18, JPL410-13-18, Jan. 1989.

[61] Jamnejad, V., "A Mechanically Steered Monopulse Tracking Antenna for PiFEx: Overview," *MSAT-X Quarterly,* No. 13, JPL 410-13-13, Jan. 1988.

[62] Huang, J., "L-Band Phased Array Antennas for Mobile Satellite Communications," *IEEE Vehicular Technology Conf.,* Tampa, Florida, May 1987.

[63] Milne, R., "An Adaptive Array Antenna for Mobile Satellite Communications," *Proc. 2nd Int. Mobile Satellite Conf.,* Ottawa 1990, pp. 529–532.

[64] Tranquilla, J. M., and S. R. Best, "A Study of the Quadrifilar Helix Antenna for Global Positioning System (GPS) Applications," *IEEE Trans. Ant. Propag.,* Vol. AP-38, Oct. 1990, pp. 1545–1550.

[65] Carter, D., "Phase Center of Microwave Antennas," *IRE Trans. Ant. Propag.,* Vol. AP-4, 1956, pp. 597-600.

[66] Gerst, C., and R. A. Worden, "Helix Antenna Take Turn for Better," *Electronics,* Aug. 1966, pp. 100–110.

[67] Kilgus, C. C., "Shaped Conical Radiation Pattern Performance of the Backfire Quadrifilar Helix," *IEEE Trans. Ant. Propag.,* Vol. AP-23, May 1975, pp. 392–397.

[68] James, J. R., and P. S. Hall, eds., *Handbook of Microstrip Antennas,* IEE Electromagnetic Wave Series 28, Peter Peregrinus, 1989.

[69] Kishk, A. A., and L. Shafai, "The Effect of Various Parameters of Circular Microstrip Antennas on Their Radiation Efficiency and the Mode Excitation," *IEEE Trans. Ant. Propag.,* Vol. AP-34, No. 8, Aug. 1986, pp. 969–976.

[70] Huang, J., "Finite Ground Plane Effect on the Microstrip Antenna Radiation Patterns," *IEEE Trans.,* Vol. AP-31, No. 4, July 1983, pp. 649–653.

[71] Sugimoto, Y., N. Kurihara, H. Kiuchi, A. Kaneko, F. Sawada, T. Shirado, and Y. Saburi, "Development of GPS Positioning System PRESTAR," *IEEE Trans. Instrument. Measurement.* Vol. IM-38, No. 2, April 1989, pp. 644–647.

[72] Imafuku, H., S. Ogawa, K. Seki, N. Endo, R. Shimizu, Y. Kaneko, and A. Kuramoto, "Mobile TV Receiving Antenna System of Direct Broadcasting Satellite for Train Application," *Proc. 1989 Int. Symp. Ant. Propag.,* (ISAP'89, Tokyo), Aug. 1989, pp. 545–548.

[73] Kuramoto, A., N. Endo, A. Kawaguchi, R. Shimizu, Y. Furukawa, S. Oyaizu, K. Maehara, and Y. Suzuki, "Mobile DBS Receiving Antenna System," *1991 IEICE National Convention Rec.,* Vol. B-51, March 1991, p. 59.

[74] Itoh, Y., and S. Yamazaki, "A Mobile 12 GHz DBS Television Receiving System," *IEEE Trans. Broadcast,* Vol. 35, No. 1, March 1989, pp. 56–62.

[75] Rhodes, D. R., "Introduction to Monopulse," New York: McGraw-Hill, 1959. Reprint, Dedham, MA: Artech House, 1982.

[76] Sherman, S. M., "Monopulse Principles and Techniques," Dedham, MA: Artech House, 1984.

[77] Murata, T., and K. Ohmaru, "A Flat Panel Antenna With Two-Layer Structure for Satellite Broadcasting Reception," NHK Laboratory Note, No. 374, Dec. 1989, pp. 1–12.

[78] Murata, T., and K. Ohmaru, "A Printed Antenna With Two-Layer Structure For Broadcasting Reception," *Trans. IEICE of Japan,* Vol. J72-B-II, No. 6, June 1988, pp. 236–244.

[79] Takano, K., and Y. Itoh, "Mobile Receiver for Direct Satellite Broadcasting," NHK Laboratory Note, No. 393, July 1991, pp. 1–8.

[80] Itoh, Y., S. Yamaguchi, and K. Ohmaru, "Mobile Receiving Test for Direct Broadcasting Satellites," *J. ITE of Japan,* Vol. 43, No. 12, Dec. 1989, pp. 1375–1380.

Chapter 7
Antenna Systems for Aeronautical Mobile Communications

Y. Suzuki

Many different types of antennas are mounted on aircraft today to carry out numerous individual functions. Table 7.1 shows a list of typical antennas and the associated avionic systems, which include navigation, identification, and radar. In this chapter, propagation problems peculiar to aeronautical systems are noted prior to itemizing common requirements for the airborne antennas and the information necessary to design them. After that, examples of current airborne antennas for navigational avionics and identification are introduced, categorized by frequency. Several circularly polarized antenna elements for satellite communications [1] are summarized, along with concise design procedures, in the final section.

7.1 PROPAGATION PROBLEMS

The propagation problems in aeronautical mobile radio links can be analyzed by the same treatment as those in the maritime mobile case, except that the mobile station is flying at high altitude. The problems are, in general, divided into two categories. One is an interferential propagation due to the direct and reflected signals, and the other is diffraction and scattering, but the distinction between these two sources of problems is not clearly defined.

In the interference region, the propagation path is defined within the line of sight, and the field strength on the propagation path is estimated well enough by the calculation based on a two-ray interference model. When the aircraft flies over land, however, significant fading may not be caused and the propagation can be approximated by that in free space when estimating the local mean value of the field strength.

The region out of sight may be divided into the diffraction and scattering regions. The spherical diffraction waves mainly propagate in the diffraction region

Table 7.1
Antennas for Aircraft

Frequency Band	*Equipment or System*	*Representative Antenna*
VLF/LF/MF	Comunications, ADF, Loran	Wire, loop, plate
HF	Communications	Wire, tail cap, probe, notch
VHF/UHF	Communications, glide path (glide slope), Locarizer, VOR, marker beacon, homing, telemetry and command, DME/TACAN, ATC transponder	Sleeve monopole, whip, folded dipole, ramshorn, tail cap, annular slot, notch, blade, loop
UHF/SHF	radio altimeter, radar	Horn, reflector, slot (array)

and create a field strength that decreases rapidly as the distance increases. After the diffraction waves become very weak, the troposcatter appears and the feeble waves, scattered by the small, random irregularities or fluctuations in the refractive index of the atmosphere, dominate. There was considerable interest in troposcatter propagation during the decade of the 1950s. Many theories have been developed, and the useful performance data have been gathered. With the development of satellite communication systems, however, there is now less need for troposcatter systems.

The field strength on the propagation path for aeronautical mobiles may suffer considerable fluctuation due to the seasonal change of the refractive index of the atmosphere, and the theoretical prediction is difficult. For example, the monthly mean value of refractivity at an altitude of 300m is quoted at about 300 N-units in February, but it changes from 340 to 370 N-units in August. The fluctuation of refractivity causes the change in equivalent earth radius. The refractivity also varies according to altitude and decreases as the altitude increases. So, when the aircraft flies at high altitude, it should be noted that the equivalent earth radius becomes smaller than that assumed on the earth.

7.2 GENERAL REQUIREMENTS AND REMARKS

In this section, the general requirements for airborne antennas are briefly surveyed, along with some useful information necessary to design them. Typical avionic systems that require antennas are illustrated for two aircraft in Figure 7.1.

Airborne antennas must satisfy more severe environmental conditions than the other mobile antennas and the following measurements are typical:

- Temperature: −85°C to 71°C,
- Vibration: 5 to 2,000 Hz with amplitude from 0.01 to 0.4 in, and
- Acceleration: 1G to 20G.

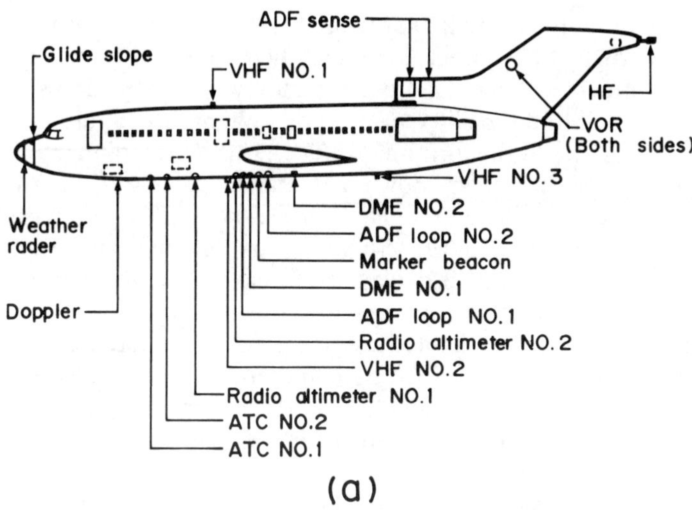

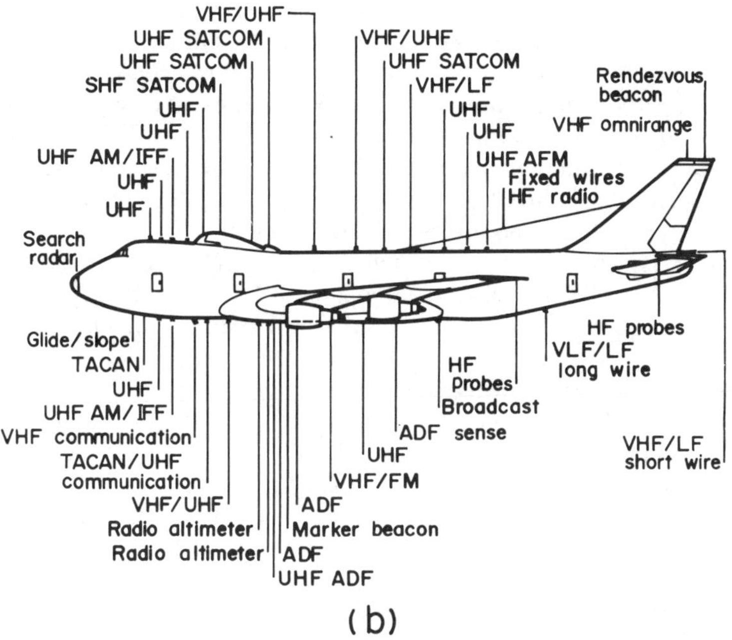

Figure 7.1 Layouts of antennas mounted on aircraft: (a) for B-727 (after [21]); (b) for E-4B.

Also, in order to ensure higher reliability, the antennas must be designed and manufactured ruggedly, without degrading their electrical characteristics, by considering the influence of sand and dust, humidity, salty fog, drops in atmospheric pressure at high altitude, static electricity, and lightning. Static electricity means triboelectric charging of the airframe surface. This is generated by dust or precipitation particles and has been known as P-static. The P-static causes corona discharges, which may produce extreme electrical noise. The corona discharge may also limit the power-handling capacity of the antenna in transmitting operations.

Airborne antennas are also required to have structures that do not increase aerodynamic drag, and this demands light weight, small size, low profile, and conformity. Conformity is defined as constraining an antenna to conform to a surface whose shape is determined by considerations other than electromagnetic (e.g., aerodynamic or hydrodynamic). A conformal antenna should be distinguished from a flush-mounted antenna constructed into the surface on the fuselage, without affecting the shape of that surface.

From the discussions above, it can be said that the main task of airborne antenna designers is realizing rugged antennas that satisfy individual electrical requirements constrained by environmental and aerodynamic conditions. However, the individual electrical requirements are ordinarily different according to each purpose. For example, the antennas for communication or navigation, which are mainly used at frequencies lower than the VHF band, are required to have an azimuthally omnidirectional radiation pattern. In this case, the radiation pattern and other electrical parameters are strongly influenced by the size and shape of an airframe. The current induced on the airframe surface interferes with the original field of the antenna itself, and the overall effect depends on the aircraft's electrical dimensions. A typical variation of a radiation pattern as the frequency is varied, is shown in Figure 7.2 for a tail-cap antenna on a DC-4 aircraft [2]. This example illustrates the frequency dependence of the influence of the airframe on the radiation pattern.

In the low- and medium-frequency bands, below 3 MHz, the wavelengths are more than 100m and are considerably larger than the maximum dimensions of most aircraft; for example, the Boeing 747LR/SR is 70.5m in length. Since the airframe size is generally small compared to the wavelength, the radiation efficiency remains very low. A quality factor Q of the antenna itself will be high, so some tunable matching systems may be necessary for impedance matching over the frequency band.

In the high-frequency band, from 3 to 30 MHz, the wavelength is comparable to the major dimensions for most aircraft. In this case, the airframe becomes a good radiator, so the HF antenna can be designed to maximize the electromagnetic coupling to the airframe. If the antenna succeeds in getting maximum coupling,

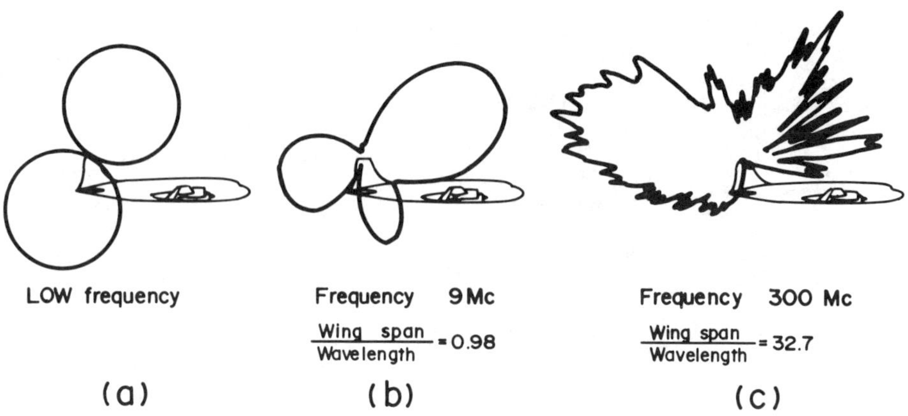

Figure 7.2 Typical radiation patterns of a tail-cap antenna on a DC-4 aircraft (after [2]):(a) a case of low-frequency range; (b) a case of longitudinal resonance range; (c) a case of transverse resonance range.

then the currents on the airframe surface exhibit strong resonant phenomena. The resonant currents can be divided into two modes, as shown in Figure 7.3 [2]; one is the symmetric mode and the other is the antisymmetric mode. The antennas should be designed to couple to either of them, so that the system may benefit from this additional degree of design freedom at HF.

In the very high and ultrahigh frequency bands, from 30 to 3,000 MHz, the wavelength is shorter than the airframe size, so the antenna itself or a part of the airframe becomes resonant and the degrees of freedom in both antenna design and installation increase. However, it should be noted that omnidirectional patterns, as required for short-range communication, distance measuring equipment (DME), and air traffic control (ATC), are perturbed by the airframe. The airframes cause shadowing and reflection and can result in major distortions of the primary pattern of the antenna, as shown in Figure 7.2(c). These effects can be analyzed by using the method of moments [3–7], based on a solution of the electromagnetic wave-scattering integral equations, when the airframe is not large compared with the wavelength. At higher frequency, the diffraction due to the airframe edge can be approximately calculated by using the geometrical optics techniques [8–14].

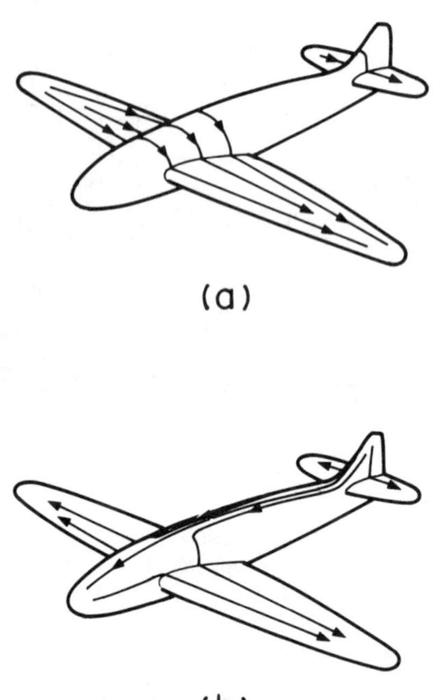

Figure 7.3 Resonant current modes (after [2]): (a) symmetric mode; (b) antisymmetric mode.

7.3 CURRENT AIRBORNE ANTENNAS

Although airborne antennas can be classified from various points of view, the classification according to the operational frequency seems to be the most rational, as listed in Table 7.1.

7.3.1 VLF/LF/MF Antennas

The typical avionics systems operating in this frequency band are VLF/LF communications, automatic direction finder (ADF), and loran (long range navigation). Aeronautical mobile communications operating in the VLF/LF band are limited to long-range communications for military purposes and are seldom used for civil communications. The ADF is a receiving system similar to the loran system and it is used to find the incident direction of the radio waves radiated from the nondirectional beacon (NDB) ground station. Although the NDB frequency is from 200 to 415 KHz, it has been necessary for the ADF antenna to be able to receive the radio waves in the 190- to 1,750-KHz frequency range. Loran has been developed from loran-A to a loran-C, although loran-B has not yet reached practical use. Four frequencies of 1,750, 1,850, 1,900, and 1,950 KHz are assigned to loran-A in the world. Loran-A is called a standard loran, or simply a loran, and it has only been used for international aircraft flying over the ocean, while loran-C has not been used yet.

VLF/LF Communication Antennas (Long-Wire Antennas)

In this frequency band, the so-called *trailing-wire* antennas are mainly used and Figure 7.4 shows two kinds of VLF/LF transmitting antennas. Figure 7.4(a) is a dual-wire (or counterpoise) antenna and Figure 7.4(b) is a trailing-wire antenna [15]. Both antennas are tuned by changing the wire lengths. The trailing-wire antenna is, in general, driven against the aircraft with the antenna gap at the point where the wire leaves the aircraft. The dual-wire antenna is effectively fed at the aft end of the fuselage where the short wire exits from the aircraft. The long wire is grounded to the aircraft skin and serves as a counterpoise wire to increase the antenna admittance at the operating frequency. In the 17- to 300-KHz frequency band, the wires are of the order of 0.4 to 1.4 km and 2.4 to 8.4 km in length, respectively, and severe technical difficulties are involved in extending such long wires. In particular, the speed and direction of aircraft affect the antenna radiation

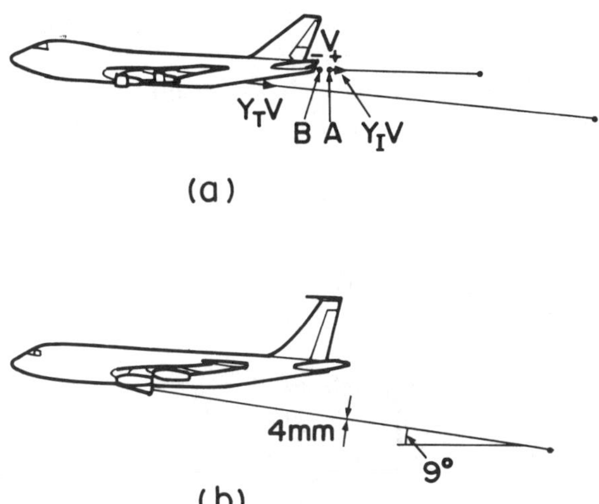

Figure 7.4 VLF/LF aircraft wire antennas (after [15]): (a) dual-wire antenna; (b) trailing-wire antenna.

characteristics [16]. That is, if flight conditions such as speed and direction allow this long wire to drop vertically, then the vertically polarized component of the radiated power is greatly increased.

ADF Antenna (Small Monopole and Loop Antennas)

The operational principle to find the direction is the same as that for a shipborne direction finder, but the ADF system for airborne adopts the switched cardioid method to automatically determine the direction using a pair of sense and loop antennas. A large-sized aircraft is obliged to install a dual-ADF system, and Figure 7.5 shows the dual sense antennas mounted on a B-727 aircraft [17]. The loop antenna consists of two elements, one being the north-south (NS) loop and the other the east-west (EW) loop. An example of the main structure is shown in Figure 7.6 [17,18], where a foam styrene block has four ferrite slabs on it, upon which the NS and EW loops are wound. Figure 7.7 shows the schematic radiation patterns for the loop and sense antennas. Each cardioid pattern, drawn as a solid or broken line, can be obtained by combining the radiation pattern of the sense antenna with that from one of the NS or EW loop antennas. The pair of cardioid patterns is then used to find the incident direction of the NDB radio waves.

The ADF antenna must receive signals with a frequency below about 2 MHz, so that the operating wavelength becomes more than 150m, which is considerably greater than the maximum dimensions of most aircraft; hence, the sensitivity of such an antenna strongly depends on the airframe. In the case of a small monopole antenna like a sense antenna, it is necessary to consider the distortion caused by the airframe in the electric field component of the incident waves. Figure 7.8 shows the electric field fringing produced by the airframe for incident fields polarized in the three principal directions: vertical (a), longitudinal (b), and transverse (c) [2]. A small antenna placed on the airframe is found to respond to all three of these

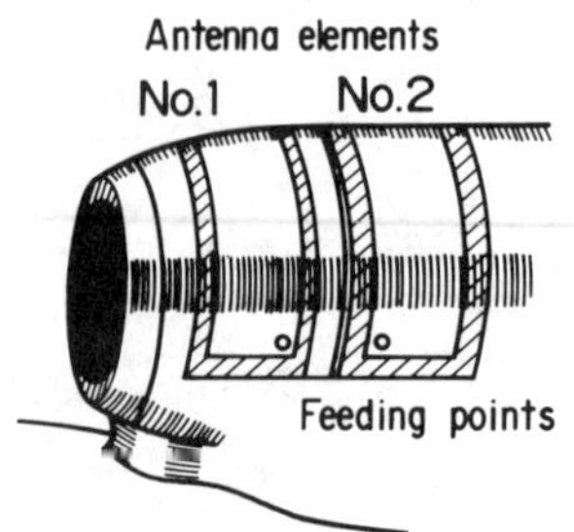

Figure 7.5 ADF sense antenna on aircraft. (After [17].)

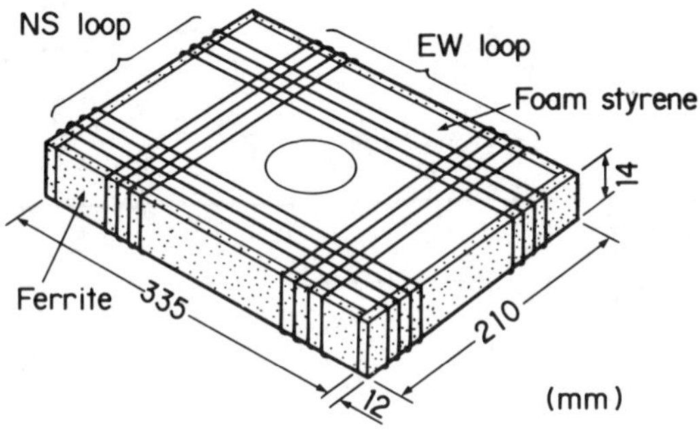

Figure 7.6 One example of structure of ADF loop antenna. (After [17].)

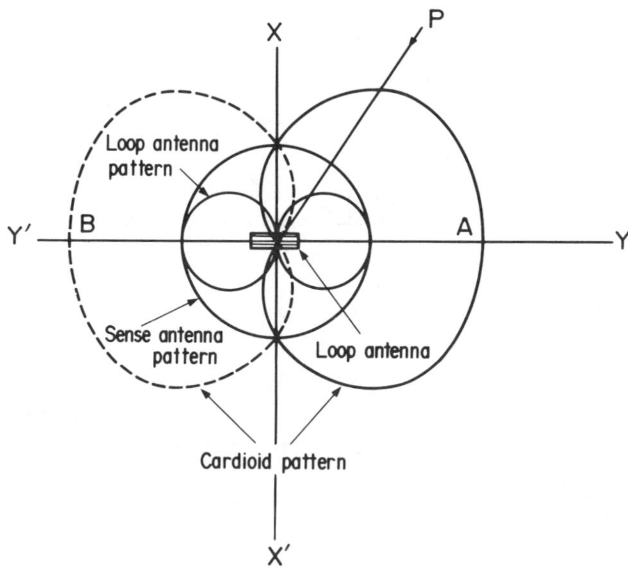

Figure 7.7 Radiation patterns of ADF sense and loop antennas.

field components. The airframe effect on antenna sensitivity is, in general, estimated by a curvature factor defined as an index expressing the increase in effective height. The curvature factor is approximately equal to the ratio of the effective height of an antenna on the airframe to that on a flat ground plane. The shape of the airframe, in the case of the small monopole antennas, also influences the polarization of the radiation. The remaining important parameter for the small

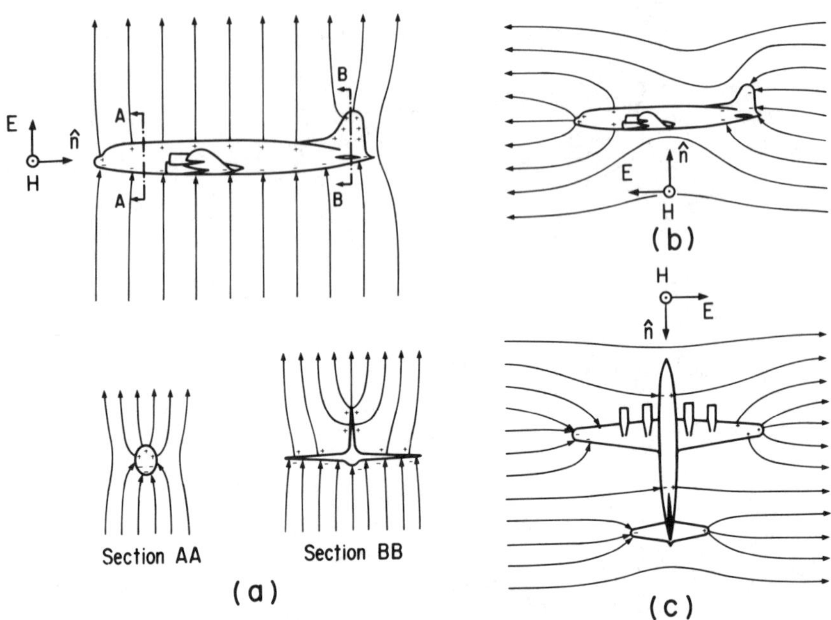

Figure 7.8 Field fringing and airframe charging produced by incident electric fields (after [2]): (a) for vertical electric field; (b) for longitudinal electric field; (c) for transverse electric field.

antennas is the quality factor Q, and it depends on the effective height and capacitance. The effective height on the airframe can usually be estimated by multiplying the effective height on the flat ground plane by the appropriate curvature factor. The capacitance, however, may be assumed to be the same as that of the antennas on a flat ground plane.

It is also necessary to consider the distortion caused by the airframe in the magnetic field component of the incident waves. Figure 7.9 shows the schematic magnetic field distortion produced by the conducting airframe for two polarization cases [2], where the longitudinal and transverse polarizations correspond to the cases in which the signal arrives from the side of the aircraft and from the front or rear of the aircraft, respectively. From this figure, it may be observed that the local field intensity is greater than the original incident field intensity in most locations near the top and bottom center line. This field enhancement contributes to increasing the signal level induced in a loop antenna. The ratio of the local magnetic field intensity on the airframe surface to the incident magnetic field intensity can be defined as a_{xx} for transverse polarization and as a_{yy} for longitudinal polarization, respectively. In the case of the ADF systems, the discrepancy between

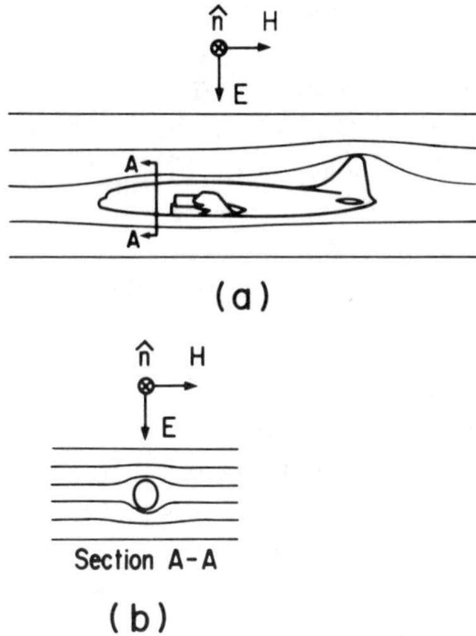

Figure 7.9 Magnetic field distortion caused by a conducting airframe (after [2]): (a) for longitudinal magnetic field; (b) for transverse magnetic field.

a_{xx} and a_{yy} causes bearing errors, because the relationship between true and apparent directions to the signal source is given by

$$\tan \phi_t = \frac{a_{xx}}{a_{yy}} \tan \phi_a \tag{7.1}$$

where ϕ_t and ϕ_a are the true and the apparent bearings for the signal source, respectively. This relationship implies that the bearing error, defined as $(\phi_t - \phi_a)$, is a function of ϕ_t and the ratio of a_{xx}/a_{yy}. From (7.1), it is found that the maximum bearing error is obtained by

$$(\phi_t - \phi_a)\big|_{\max} = \tan^{-1}\left(\frac{\tan^2 \phi_t' - 1}{2 \tan \phi_t'}\right) \tag{7.2}$$

where

$$\phi_t' = \tan^{-1}\sqrt{(a_{xx}/a_{yy})} \tag{7.3}$$

If the maximum bearing error is not over 20 deg, it can be compensated for by modifying the immediate environment of the actual loop element to equalize the coefficients a_{xx} and a_{yy}.

When the polarization is vertical, the fringing field is found to be particularly intense near the tip of the vertical stabilizer as can be seen in Figure 7.8(a). This region is one of the ideal locations for LF/MF band antennas, and this includes electrically small communication antennas.

7.3.2 HF Communication Antennas

Today, the HF band is mainly used to enable an aircraft in midocean to communicate with land-based facilities, and the frequencies from 2 to 30 MHz are used in conventional HF aeronautical mobile communications. These HF-band airborne antennas are especially required to have acceptable high-power transfer performance. This requirement can be solved by invoking resonant conditions and selecting a feed point such that the best excitation is obtained for the desired airframe mode. The feed point must also be selected to minimize the variation in input impedance with respect to the frequency in order to keep the losses in the tuning unit to a minimum. The induced currents flowing on the airframe surface usually dominate the impedance and radiation pattern behavior in this frequency band, and the antenna must be designed allowing for the electromagnetic coupling to the airframe.

The influence of the ionized layers existing between 90 and 1,000 km above the earth's surface cannot be ignored in this frequency band. The effective antenna gain must be treated in terms of the total power density, because the polarization is rotated by the reflection characteristics of the ionosphere. For frequencies below 6 MHz the ionosphere and ground reflection effects enable full use to be made of the radiated power for communications. Above 6 MHz, the propagation path is restricted to cones 30 deg above and below the horizon.

In order to overcome losses in the aircraft antenna systems, substantial transmitted power may have to be handled. Since an ordinary antenna system has relatively large losses including the mismatching loss, the normal transmitted power may be on the order of 100W to 1,000W watts; hence, the antenna must be designed with due consideration of the effect of reduced atmospheric pressure on voltage breakdown.

Wire Antennas

On a lower speed aircraft, HF wire antennas are often supported between the vertical fin and an insulated mast (Figure 7.10) or are trailed out into the airstream from an insulated reel. A reasonable angle between a fixed wire and the airstream is said to be less than 15 deg in order to decrease the aerodynamic drag. The wire

Figure 7.10 HF wire antenna on aircraft.

antennas are also regarded as a transmission line antenna, and some lumped reactances are conventionally inserted between the wire and the fin in order to decrease the physical length using the reactance loading effect. The decrement is, in this case, proportional to the amount of loading, but the frequency bandwidth is inversely proportional to the decrement ratio. In the 6- to 24-MHz band, as the physical wire length decreases, the shape of the radiation pattern in the horizontal plane approaches an omnidirectional format, and nearly 60% of the total radiated energy remains in the sector bounded by the cones 30 deg above and below the horizontal plane. However, it should be noted that the wire antenna efficiency is, in general, not high, due to the resistance loss in the wire itself in addition to the resistance loss in the reactance loading and the dielectric loss in the supporting insulators and masts. As an example, the resistance of wires commonly used is on the order of 0.05 Ω/ft at 4 MHz. Although the wire antennas are satisfactory enough for slow, piston-engine aircraft, they may be unfit for high-performance aircraft, where the aerodynamic drag cannot be ignored. Also, when the antennas are installed on high-altitude aircraft, it is necessary to pay special attention to RF corona breakdown, the onset of which depends on the wire diameter and nature of installation. The breakdown can be improved somewhat by coating a wire with a relatively large-diameter sheath of polyethylene.

Tail-Cap Antennas

Some HF band antennas have been produced by exciting the airframe at the fin tip or a wing tip. This kind of antenna has the advantage that it meets the aerodynamic requirements by itself. Four types of tail-cap antennas are illustrated, along with their input resistance characteristics, in Figure 7.11 [19], where the heavy arrows indicate the feed point. A large aircraft generally has several different resonance modes within the HF band, while a smaller aircraft or a helicopter has no resonance modes. Figure 7.11 shows that two kinds of resonance appear in the 2- to 20-MHz frequency range. The first peaks at near 3 MHz are due to the half-wavelength resonance for the induced currents on the entire airframe. This resonance mode is dominant in all cap-type antennas. The second peaks are associated with the half-wavelength resonance for the induced currents on the wings and fuselage.

Notch Antennas

Other types of antenna meeting the aerodynamic requirements are notch-type antennas buried in the airframe. In the HF band, the notch antennas become increasingly popular on modern high-speed aircraft because of the drag reduction

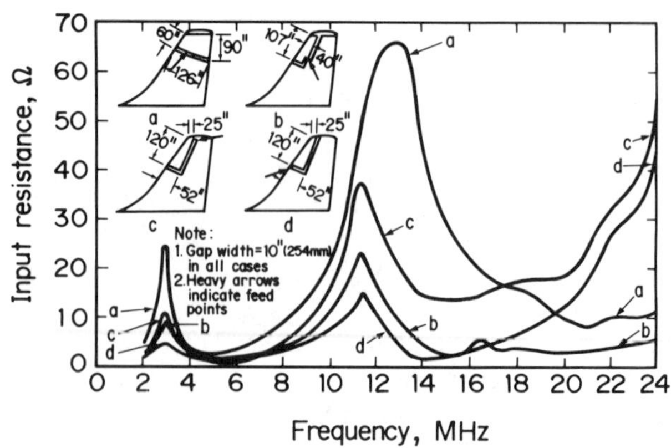

Figure 7.11 Effect of changing feed configurations on the input resistance of tail-cap antennas. (After [19].)

and higher reliability. In order to make the notch antennas operate efficiently, the notch must be coupled to the currents flowing on the airframe surface as tightly as possible. However, the electrically ideal notch positions occasionally coincide with mechanically highly stressed positions, such as a wing root or the base of the tail fin, where it is mechanically undesirable to cut out the large-sized notch. For these reasons, the notch cannot be made as large as the antenna designer would like in the HF band, so its impedance is normally inductive. Although this inductive impedance can be easily tuned to resonance by a capacitive tuner, the resultant impedance is, in general, very different from the 50Ω impedance of the feeder cable. In practice, even when the notch is optimally sited at a high current point, its radiation resistance is only on the order of 0.1Ω. The HF notch antennas must therefore have some matching transformer in addition to the capacitive tuner, and a simple transformer is not enough; an adjustable reactive matching network is necessary for the wide-band matching. The tuning and matching units may be combined in a single unit as an automatic tuning unit (ATU). Figure 7.12 shows an example of an HF notch antenna with the ATU, which is mounted in front of the normal tail fin of a Comet XV814 [20]. The antennas indicated by HF1 to HF3 in Figure 7.12 are the examples of HF wire antennas of various lengths. Figure 7.13 shows a comparison of radiation patterns at 23.265 MHz [20]. In these results, the notch antenna shows a better performance than the wire antennas. The bandwidth of usual tuned notch antennas is about 10 kHz. If higher efficiency is desired, lower loss reactances must be employed for the tuning and matching units.

Probe Antennas

The probe antennas can be regarded as a variation of the tail-cap antenna structurally and as a variation of the wire antenna electrically. The radiation patterns and impedance characteristics are also influenced by the geometry and size of the tail, fuselage, and wing assemblies. An advantage of the probe antennas over the tail-cap antennas lies in the relative ease of incorporation into the aircraft structure. An example of a probe antenna, mounted on the tail tip of a B-727, is shown in Figure 7.14 [21]. In this example, the metallic bulkhead acts as an electrostatic shield between the antenna and other metallic objects. The RF cable is connected to the antenna through a lightning arrester, which protrudes through the bulkhead into the tail cone. The lowest resonance is expected to occur at the frequency producing a half-wavelength aircraft resonance, and is dependent on the distance from wing tip to tail tip. The resistive component at this resonant frequency is on the order of 10 to 20Ω, and hence a matching unit is also necessary to obtain the maximum coupling to the airframe. In the case of the probe antenna shown in Figure 7.14, the matching unit can be advantageously mounted on the bulkhead.

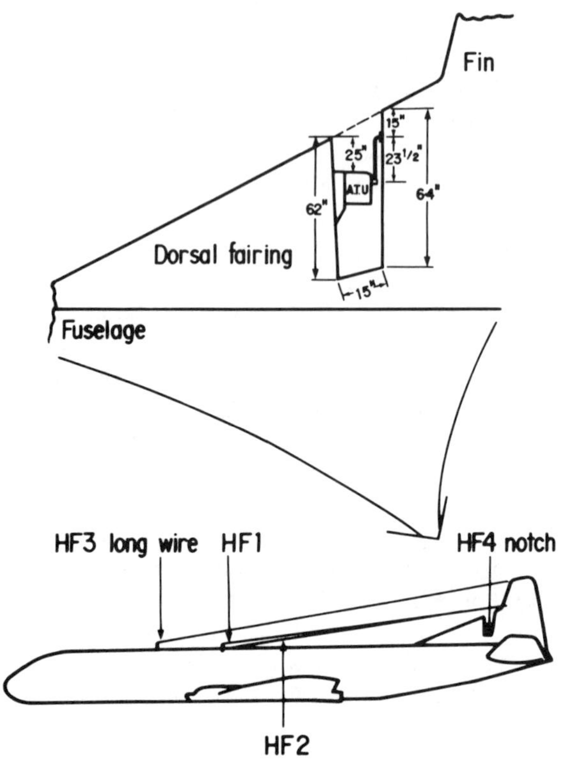

Figure 7.12 HF wire and notch antenna on COMET XV814. (After [20].)

7.3.3 VHF/UHF Antennas

In the VHF/UHF band from 30 to 3,000 MHz, various kinds of antennas are used for purposes such as communications, identification, and ATC, telemetry, navigation, and homing. The coverage requirement for the communication purpose is omnidirectional in the azimuth plane, while the elevation coverage requirement depends on the individual flight performance; for example, most nonaerobatic aircraft require an angular region from 5 to −30 deg with respect to the aircraft horizontal axes, while more maneuverable aircraft require considerably wider angular coverage. In comparison with the above communication antennas, relatively narrow-band antennas with a simple pattern, directed down or forwards from the aircraft, are required for DME and ATC, telemetry and command, navigation, and homing. Although these antennas for navigational avionics, identification, and so on do not necessarily need an azimuthally omnidirectional pattern, their design

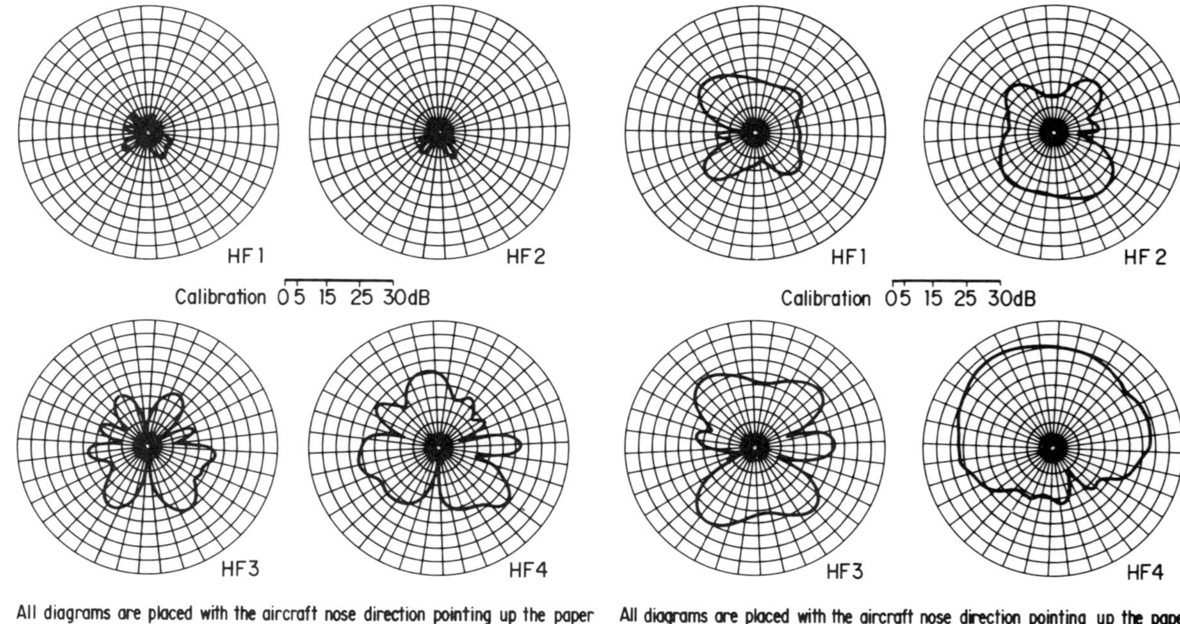

Figure 7.13 Radiation patterns of HF wire and notch antennas on COMET XV814 (after [20]): (a) for horizontal polarization at 23.265 MHz; (b) for vertical polarization at 23.265 MHz.

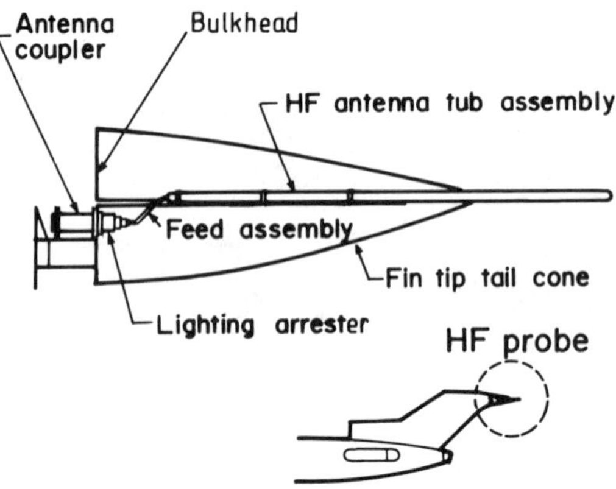

Figure 7.14 Structure of probe antenna on B-727 aircraft. (After [21].)

is often complicated by the need for conformal flush assemblies placed in locations where little choice for the antenna site exists.

7.3.3.1 *VHF/UHF Communication Antennas*

VHF/UHF communication systems for ground-to-air and air-to-air ordinarily use a vertical polarization whose frequency is in the 30- to 400-MHz band. The antenna gain requirement is modest, and a minimum level of -10 dBi within the desired coverage will normally be adequate, so that the monopole antenna and its variants are most commonly used. In this frequency band, if the transmit powers do no exceed 50W with continuous wave (CW), then the corona and flashover problems are not significant.

The 30- to 100-MHz frequency band is mainly used for military communi cations with mobile ground forces, though they tend to be required only for heli copters and small aircraft. In this frequency band, the fuselage dimension is on the order of a wavelength, and the induced currents on the fuselage flow unequally forward and behind the antenna. The resultant radiation pattern is distorted and strongly depends on the frequency. This pattern distortion and frequency-depen dence can be overcome by using two or more antennas with some pertinent selection method. A simple combination of the input/output for two or more antennas i not desirable when they are widely separated, because this causes multiple inter ference lobes, called *grating lobes*. In order to avoid the grating lobes, a diversit scheme is necessary and a space diversity technique is especially applicable in thi case.

In the 100- to 400-MHz frequency band, wide-band antennas, which need no matching circuits, are desired because it is difficult to realize the wide-band tunable unit in such a miniature size when considering wavelengths of at least 0.75m. However, if the antenna gain is sufficient, a simple switched tunable system may be feasible.

Monopole Antennas

A variety of monopole antennas are commonly used in airborne systems. The azimuth radiation pattern for a monopole antenna on a long horizontal cylinder is substantially circular if the antenna is more than about a half-wavelength away from the cylinder edge. For frequencies above 100 MHz, therefore, most monopole antennas mounted on the top or bottom surface of the fuselage are acceptable as long as they are on or near the center line to avoid tilting of the roll-plane pattern. The pitch-plane pattern can be approximated to that for an ideal monopole antenna. A position in the middle third of the fuselage length is generally preferred for a balanced pattern. In the practical application, the wings are regarded as an extended ground plane, and their effects on the radiation patterns may be calculated by a combination of reflection and diffraction mechanisms. These effects are usually minimized by arranging the antennas forward or aft of the wings. When installing on the upper fuselage positions, however, the effects of shadowing due to the tail fin must be considered, because this shadowing may cause unacceptable degradation on the radiation patterns when the aircraft is large relative to the wavelength.

A typical bent-sleeve antenna for the VHF band, which belongs to the monopole antenna family, is shown in Figure 7.15 [22]. This is an example of a ruggedly constructed antenna with respect to aerodynamic and hydrodynamic constraints. The mast consists of a cast skeleton, incorporating the mounting base, two main spars, and two subsidiary spars, to which light-alloy side plates are argon-arc welded. The top element consists of a light-alloy casting, to which a tapered tube is welded. The coaxial cable is molded into a Bakelite insulator and also sealed into the base of the mast to give a waterproof assembly.

As mentioned previously, the above type of antenna has been recommended to be mounted on or near the center line to avoid tilting of the roll-plane pattern. If it is difficult to do so, however, an alternative position, much used on high-performance aircraft, is at the top of the tail fin. In this case, the antennas are required to have a tapered airfoil cross section to minimize drag. This type of antenna is called a *blade antenna*. Although most blade antennas have very simple and convenient structures, it is necessary to pay attention to lightning and P-static, because their main frame is a metal blade. Figure 7.16 shows the blade-type monopole antenna mounted on the YS-11 aircraft [17], where it consists of a printed bent monopole antenna with stripline parallel resonant circuits and loading coil.

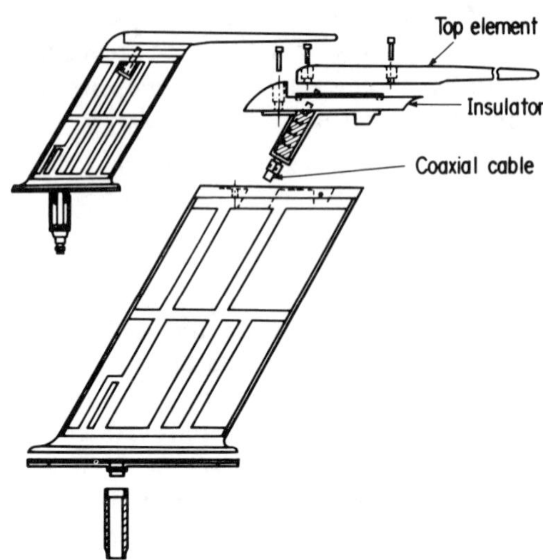

Figure 7.15 Basic configuration of bent-sleeve antenna for VHF band. (After [22].)

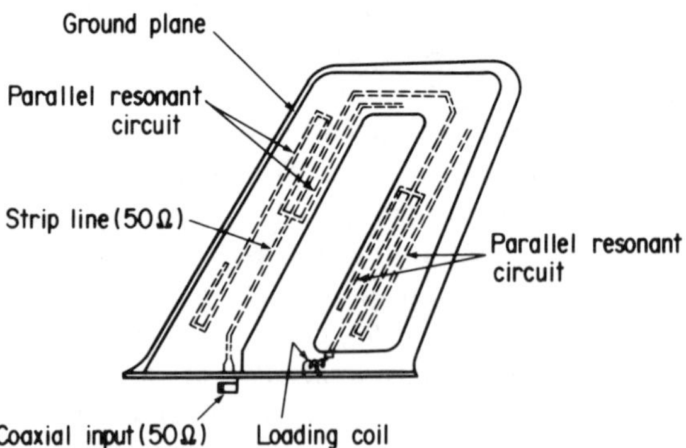

Figure 7.16 Structure of a printed type of VHF blade antenna on YS-11 aircraft. (After [17].)

The input impedance is matched to 50Ω by two parallel resonant circuits and loading coil, so that no external matching circuits are required. The height has been accommodated in less than one-eighth wavelength by employing the bent monopole structure.

Another way to minimize a drag is to use the tail fin itself as an antenna. A folded monopole or shunt antenna may be easily included in the tail fin. In this case, the tail fin itself may be regarded as an antenna. However, the radiation pattern is sensitive to any change in the environment, so successful operation may be possible only where the fin has no rudder, and its chord and height are less than a quarter and half wavelength, respectively.

A number of monopole designs have been developed to be installed on a fin, in which the top portion of the metal structure has been removed and replaced by a suitable dielectric housing. This type of antenna is the tail-cap antenna. Figure 7.17 shows three typical examples of VHF tail-cap antennas [2]. Various forms of the vertically polarized VHF/UHF radiator have been designed for tail-cap installation, because the top of the tail fin is a very good position for the communication antennas in the 100-MHz band, although a disadvantage is that the feed cable attenuation may increase as frequency increases.

Slot Antennas

A flush mounted antenna introduces no drag, and an example of this type of antenna is an annular slot antenna that is vertically polarized. This antenna can be visualized as the open end of a large-diameter coaxial line with low characteristic impedance, but it becomes an effective radiator only when the circumference of the slot approaches a wavelength. The structural difficulty of burying the antenna in the

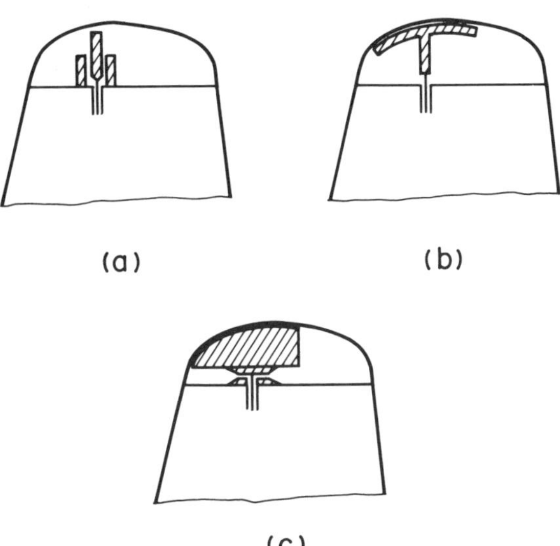

Figure 7.17 Typical structures of tail-cap antennas for VHF band (after [2]): (a) partial sleeve type; (b) pickax type; (c) mailbox type.

fuselage, dictates that the smallest possible diameter consistent with the required bandwidth be used. Figure 7.18 shows one example of such an antenna and its approximate equivalent circuit [2]. This antenna has a diameter of about 24 in (610 mm) to cover a frequency band of 225 to 400 MHz, and a VSWR of less than 2:1 has been obtained when the cavity depth is 4.5 in (114.3 mm). In practice, the flush-mounted antenna design is dominated by the construction tolerances and the airframe effect on the impedance.

Notch Antennas

The use of notch antennas is also possible for the VHF/UHF band, although it is mechanically undesirable because the mechanical strength is weakened. Figure 7.19 shows a notch antenna available for the full VHF communication band near the tip of the Herald fin [22], where the notch length is approximately a quarter

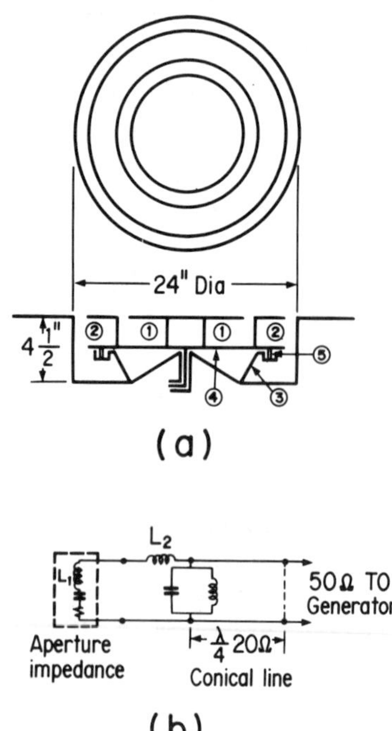

Figure 7.18 Annular slot antenna for 225- to 400-MHz band (after [2]): (a) structure; (b) equivalent circuit.

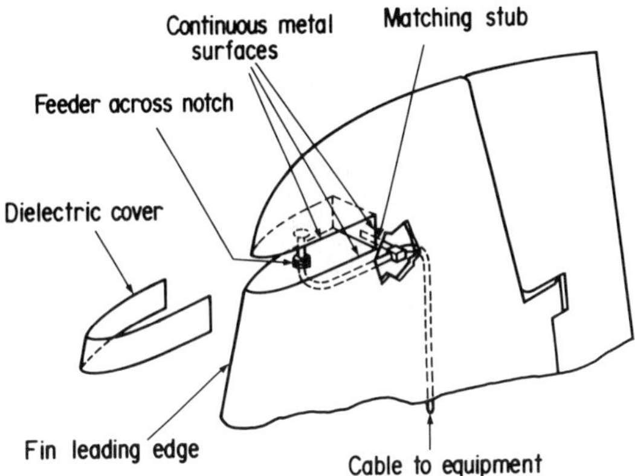

Figure 7.19 Basic configuration of VHF notch antenna in tail fin. (After [22].)

wavelength at 100 MHz and the width is 2 in. Over the 100- to 156-MHz band, a simple coaxial feed is sufficient, but a short-circuited stub is necessary at an appropriate point along the coaxial feed cable to acquire the desired impedance matching. In the case of a notch antenna, however, the notch length is undoubtedly a limiting factor. If the structural limitations prevent the use of a straight notch, then the required length may be obtained by bending or folding the notch.

Multiband and Shared Antennas

The increase in the number of antennas to be mounted on an aircraft makes it more difficult to ensure the individual electrical requirements, because the electrical and mechanical interference between them increases. The electrical interference exerts some unfavorable influence on the radio systems through coupling, radiation pattern degradation, and so on. A reduction in the number of antennas reduces the aerodynamic drag, and multiband antennas or shared antennas are a distinct advantage in this respect. A multiband antenna is defined here as an antenna having a single terminal only, and an example is shown in Figure 7.20 [23]. This is a VHF/UHF communication antenna operating in two frequency bands over 100 to 156 MHz and 225 to 400 MHz. In this sectional view, a pair of sleeves, A and B, composes a conventional sleeve monopole for the UHF band. A horizontal top element C is electrically connected only to a metallic rod D in the sleeve B, and the rod D is connected to an inner conductor of a coaxial feeder at the lower end of the sleeve B. As a result, the rod D and the inside surface of the sleeve B form a coaxial line having a high impedance at its open upper end over the UHF band,

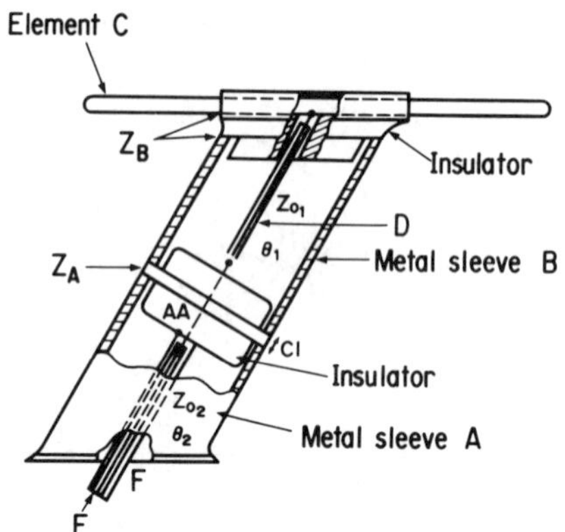

Figure 7.20 Dual-frequency sleeve antenna for VHF/UHF band. (After [23].)

so the horizontal top element C can be ignored without significant effects in the UHF band. This line acts as a series inductance in the VHF band, and the whole antenna thus operates as a top-loaded sleeve antenna when the length of the top element C is adjusted to obtain a good impedance match over this frequency band. This antenna is about 30 cm in total height and some variants have been produced for fuselage and fin cap mounting. Shared antennas are defined as antennas having multiterminals that match in an individual band. Figure 7.21 shows one example of shared antennas [21]. This is a tail-cap antenna designed to support the VHF/UHF communication system.

7.3.3.2 Other Antennas

Although a variety of communication antennas have been described in the previous section, there are also many interesting antennas other than the communication antennas in the VHF/UHF band. Some examples are now described.

Homing Antennas

The airborne homing system permits an aircraft to fly directly toward a signal source that is radiated from the ground station supporting the aeronautical mobile communications. So this system must have two patterns, which are vertically polarized and are symmetrical with respect to the line of flight. The amplitudes of signals received through the two patterns are compared by an inflight homing system, and then the homing system presents an indication regarding whether or not the aircraft

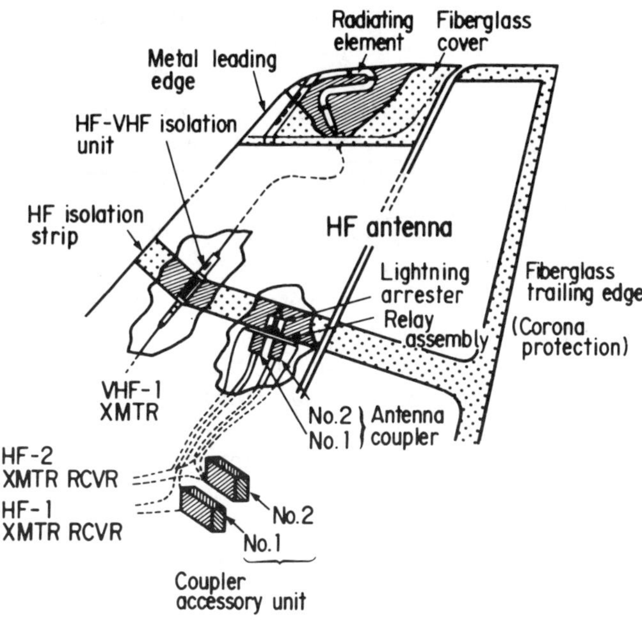

Figure 7.21 Tail-cap antenna for VHF/UHF band. (After [21].)

is flying on a homing course. In practical applications, a pair of symmetrical cardioid patterns is desired.

The desired pair of patterns can be easily synthesized by alternately feeding a symmetrical array antenna with two elements. However, it should be noted that a limitation exists in the wide-band operation of the array antenna, because the resultant pattern depends on the frequency. As a general guide, a quarter wavelength with respect to the highest frequency in the desired frequency band should be chosen as an element spacing. Furthermore, if better homing accuracy is desired, not only patterns but also impedances between two elements need to be as identical as is practical. That is, similarity with the VSWRs only is not sufficient, except for the case of a very low VSWR. A quarter-wave monopole antenna is a commonly used element, but the transmission line antenna is useful only for single-frequency applications, because the bandwidth is narrow due to the height limitation. For the 30- to 76-MHz frequency band, dipoles are mounted in front of the fuselage. In this case, the dipole lengths are selected to be a half wavelength at 76 MHz, and the low radiation resistance at 30 MHz is compensated for by loading the dipole with resistance.

A pair of symmetrical cardioid patterns is also produced by two separate antennas. A pair of short Yagi antennas is one such candidate, because a two-element or three-element Yagi antenna can produce the desired cardioid pattern by itself. In order to minimize pattern degradation by the airframe, the pair is normally mounted near the wing tips. In this case, monopole and dipole types are both available. The former is rather more sensitive to position and is then usually mounted on the wing surface. A three monopole Yagi antenna, which has a frequency bandwidth of 1.36:1 at the center frequency of 200 MHz, has been designed. The latter has been realized by mounting the dipole (or sleeve dipole) and director combinations on both sides of the fuselage, where a suitable position has been determined by trial and error based on the actual measurements; the radiation patterns are influenced by diffraction from the nose and their theoretical estimation is difficult. This example typically has a 10% bandwidth, but has the disadvantage that the gain characteristics sharply depend on the frequency. Another candidate for separate types of antennas is a pair of slot antennas mounted on both sides of the fuselage. In this case, a skewed radiation pattern can be provided by skewing the slots by 45 deg from the line of flight. The slots may be directly fed or indirectly fed by probes. The latter has the advantage that it can improve the impedance matching by using an off-center feeding technique. Although a horizontal slot in the side of a metal nose can provide a low-drag solution, the limitation is in the physical cavity dimensions. For example, the typical cavity dimensions may be 0.5λ (length), 0.22λ (height), and 0.1λ (depth) for a 10% bandwidth. So the available frequency is practically limited to more than 300 MHz for most aircraft. The terminated loop antenna is also another candidate [24]. For example, a vertical loop with a termination can synthesize a cardioid pattern by itself if the terminating impedance is suitable. The typical dimensions are about 0.3λ (height) and 0.07λ (width). The actual example is mounted vertically on a horizontal metal surface so that the height can be halved by using the image effect. Such an antenna has a practical advantage in that distortion in the loop shape is not critical and it is also possible to maintain satisfactory electrical performances over a 2.5:1 frequency range. At present, this type of antenna has been mainly used in the 30- to 75-MHz band, though its application to higher frequency bands is, of course, possible.

Marker Beacon Antennas

The marker beacon is one of three components making up the instrument landing system (ILS) and it can give the aircraft the distance information to the landing point. The ground system consists of an outer, middle, and inner marker and informs the aircraft that it is passing over each marker position. The aircraft receives the signal emitted from the marker beacon just under the aircraft and is aware of when the signal level exceeds a given threshold. The marker beacon receiver operates on a fixed frequency of 75 MHz, and the antenna is required to have a downward-looking pattern polarized parallel to the axis of the fuselage. The ICAO standard antenna is a half-wavelength dipole spaced 0.038λ below the aircraft fuselage and parallel to the axis of the fuselage. However, the radiation pattern of this standard antenna is affected by the fuselage, so it is rarely used.

Two types of antennas shown in Figures 7.22(a,b) are commonly used [19]. In these examples, however, high capacitance loading is necessary to obtain good electrical performances from the physically small elements. Figure 7.22(a) is the low-drag type, consisting of a simple vertical loop oriented in the longitudinal plane of the aircraft. The antenna elements are contained in a streamlined plastic housing. The feed line couples inductively with the loop resonated by a series capacitor. Figure 7.22(b) shows a flush-mounted type of antenna, where the antenna conductor is also series-resonated by a capacitor.

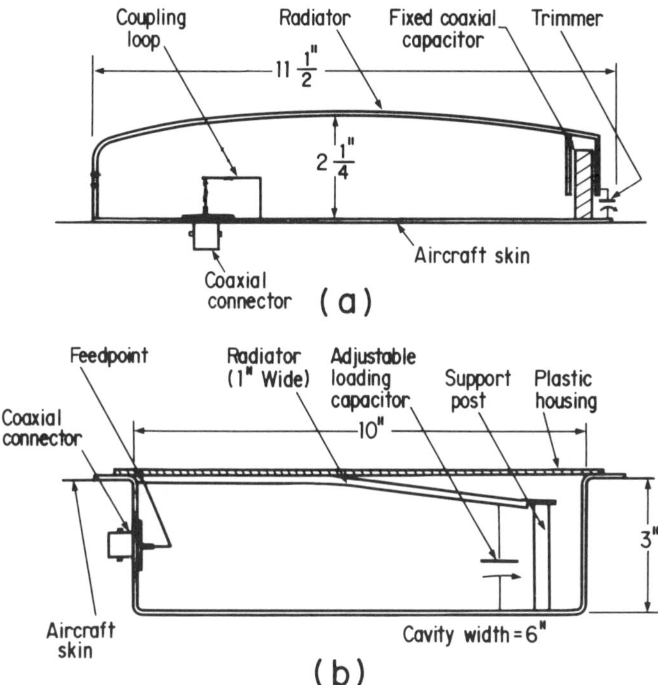

Figure 7.22 Structures of typical marker beacon antennas (after [19]): (a) for COLLINS 37X-1 marker beacon antenna; (b) for Electronic Research Inc. AT-134/ARN flush marker beacon antenna.

Localizer Antennas

The localizer system is in charge of azimuth guidance in ILS operation and uses a frequency between 108 and 112 MHz. This system informs the aircraft of a displacement in the azimuth plane landing course (or glide path) by comparing the modulation depth of two kinds of received signals, which are emitted with a horizontal polarization from the ground station placed at the edge of the runway. The accuracy of the information depends on the ability of the aircraft antennas to reject false multipath signals due to diffraction, reflection, and cross-polarization. On the other hand, mechanically moving parts such as propellers and helicopter rotors may impose further modulation on the received signal and then degrade the accuracy. So, the installation of antenna needs the greatest care. The gain requirement is modest and usually acceptable at -17 dBi in the forward direction, while the coverage requirements vary with the flight procedures; but, in general, these minima should apply up to 45 deg in azimuth from the line of flight. When an antenna is shared with the VOR antenna described later, however, an azimuthally omnidirectional coverage is required in addition to a rather higher gain because of the longer ranges involved. Typical positions for localizer antennas have been classified as shown in Figure 7.23 [24].

The ramshorn in this figure consists of a horizontal half-wave dipole bent into a V- or U-shape, and is mounted approximately 0.1λ above the aircraft fuselage, as shown in Figure 7.24, though it is rarely mounted below the fuselage. In either case, this antenna should be mounted well forward to minimize the phase center shift, which has a fairly large influence on the measurement accuracy. However, it should be noted in this installation that the maximum radiation arises in the

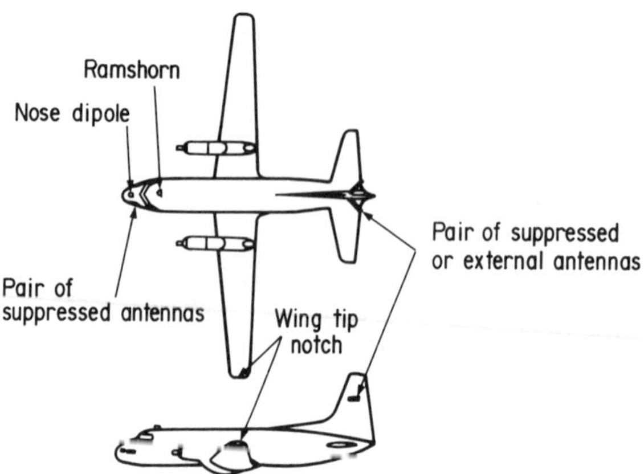

Figure 7.23 Typical positions for localizer antennas. (After [24].)

Figure 7.24 V-shape ramshorn antenna.

normal direction to the fuselage surface and the gain is low in the desired azimuth plane, because the polarization is horizontal and parallel to the fuselage surface. The radiation pattern examples for the ramshorn located on the top center line near the nose of a DC-6B aircraft are shown in Figure 7.25 [2]. The gain is seen to be low in the azimuth plane from these results.

The gain in the azimuth plane can be improved by mounting the V-dipole through the tail fin, though it is not suitable to be mounted on thick fins more than about 6 in wide at the mounting point. The V-dipole can be conveniently composed of a pair of monopoles [22], raked back by about 45 deg and fed in antiphase, respectively. Other candidates, the half loops and flush-mounted elements with backing cavities, are also available. Since these fin-mounted types suffer significantly from the influences of reflection and diffraction from the fuselage and wings, they may be unacceptable for automatic landing systems, despite their acceptability for conventional ILS use. A pair of antenna elements may be mounted on each side of the aircraft nose if their separation is not more than about 0.9m and this installation produces a good radiation pattern in the azimuth plane.

Structurally, the wing tip may often be a suitable position for a flush monopole or capacitance-loaded notch. In Figure 7.23, a single element is used as a wing tip

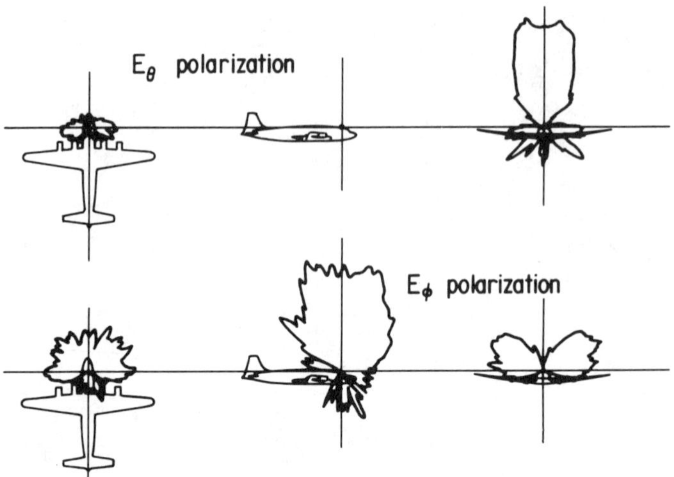

Figure 7.25 Radiation patterns of ramshorn antenna mounted on top center line near nose of a DC-6B aircraft. (After [2].)

notch. In this case, the leading edge of the wing is expected to have a strong current flow and the current may couple with the fuselage so that the antenna response is very sensitive to the aircraft attitude and the phase center shift is very marked. The above implies that this type of antenna is also unsuitable for automatic landing system use.

On the other hand, a dipole mounted in a nose radome, as shown in Figure 7.23, is naturally shielded from the remainder of the aircraft, so the dependence on aircraft attitude is minimal. If the antenna itself has an adequate gain and good radiation patterns and the installation is physically and electrically symmetrical, then such a composition may be used not only for automatic landing, but also for inspection and calibration of the ILS ground stations, though it is necessary to pay attention to vertically polarized radiation excited from lightning strips on the radome.

Glide Path Antennas

The glide path system is in charge of elevation guidance in ILS operation and uses a frequency between 328 and 336 MHz. This system informs the aircraft of a displacement in the elevation plane landing course (or glide path) by comparing the modulation depth of two kinds of horizontal polarized signals emitted from the ground station placed at the side of a runway. The criteria for localizer antennas previously described are mostly applied to the glide path antennas, except for some items. The accuracy also depends on the ability of the antenna to reject false information caused by diffraction, reflection, and cross-polarization. On the other hand, mechanically moving parts of the aircraft, such as propellers, helicopter rotors, and radar scanners may impose further modulation on the received signal and then degrade the accuracy. The gain requirement is modest and usually acceptable at −13 dBi in the forward direction. The coverage requirements are, in general, an angular sector 60 deg on either side of the nose and 20 deg above and below the horizon. Considering the above requirements, the most suitable antenna position is towards the nose of the aircraft. Also, the glide path system does not need a wide frequency band, so it is possible to use a small-sized antenna, and two kinds of variants of the loop arrangement are commonly used, which can also be regarded as a variation of a transmission line antenna [19]. The loop shape and loading are adjusted to give a maximum forward signal. In practice, the antenna is mounted on the aircraft nose or on a metal plate in the nose radome, as shown in Figure 7.26 [17]. A variation of this antenna has two connectors for dual glide path systems.

Another candidate is a half-wavelength slot with backing cavity, which is very suitable as a flush-mounted antenna. The probe feeding method can provide for two isolated feeds in the same cavity for dual glide path systems. Typical cavity dimensions are 0.44λ (length), 0.16λ (width), and 0.13λ (depth). If the aircraft nose is occupied by other equipment, a slot can be cut in the metal skin immediately aft of the radome. Ideally, the glide path antennas should be on the pitch axis of the aircraft to minimize movement of the phase center due to pitch change. For this reason, a tail fin is not a suitable antenna position for an aircraft having a large pitch change during its approach and landing phase.

When the antenna is mounted in the nose radome as in the above example, the received signal may be modulated by the movement of weather radar scanners.

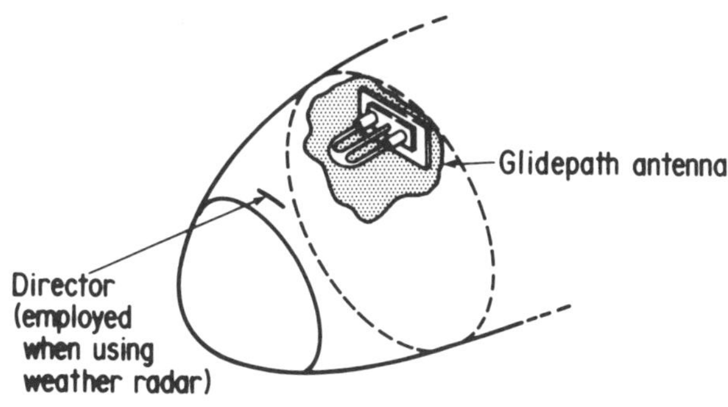

Figure 7.26 Installation example of glide path antenna. (After [17].)

Furthermore, a small or medium propeller-driven aircraft may not have any positions free from modulation for the conventional antennas. In such cases, some success has been achieved by mounting a short Yagi antenna in monopole form on the side of the aircraft nose.

VOR Antennas

The VOR beacon covers all directions at a frequency between 108 and 117.975 MHz. It can be used to inform the aircraft of the direction from the north by centering on the VOR ground station. The VOR system is composed of the $\rho - \theta$ navigation system along with the DME system to be described later. The VOR antennas can be shared with the localizer antennas, as mentioned previously, because both systems use a nearly equal frequency band. The loop, the turnstile, and the longitudinal slot in a vertical cylinder of small diameter are three basic antenna elements that yield azimuthally omnidirectional patterns with horizontal polarization. These antennas and their variants have been widely used on aircraft, but all have a basic disadvantage in that the gain is low at angles near the horizon, because they must be mounted near a horizontal conducting surface. In general, the greater the spacing from the conducting surface, the higher the horizontal gain becomes. So the locations at or near the top of the vertical fin are popular for VOR antennas. The gain requirements are comparable with those for line-of-sight communication systems, and a minimum gain of -6 dBi is adequate. Since the ramshorn has good electrical performance, it is shared in practice with the localizer antenna, as mentioned previously.

Another candidate is the E-fed cavity antenna shown in Figure 7.27(a). This is an example that has been designed into the tail fin tip of the L-1011 aircraft [19]. The antenna consists of flush-mounted dual E-slots and a stripline rat-race circuit mounted on the forward bulkhead of the cavity between two antenna halves. The antenna halves are mirror-image assemblies, consisting of 0.508-mm-thick aluminum elements bonded to the inside surface of honeycomb-fiberglass windows. In this configuration, the input VSWR is less than 2:1 over the 108- to 118-MHz band, and the cross-polarization component level is less than -18 dB. The voltage radiation patterns in the principal plane are shown in Figure 7.27(b) [19].

The VOR antenna differs from the localizer antenna in the coverage requirements and needs to have a higher grade omnidirectional azimuth pattern, so there exist a number of problems to be solved. Reflection, scattering, diffraction due to the fuselage, and undesired radiation from the discontinuities in the structure must be considered. The discontinuities work as parasitic elements, which couple with the primary antenna, causing impermissible effects on the radiation pattern in some cases.

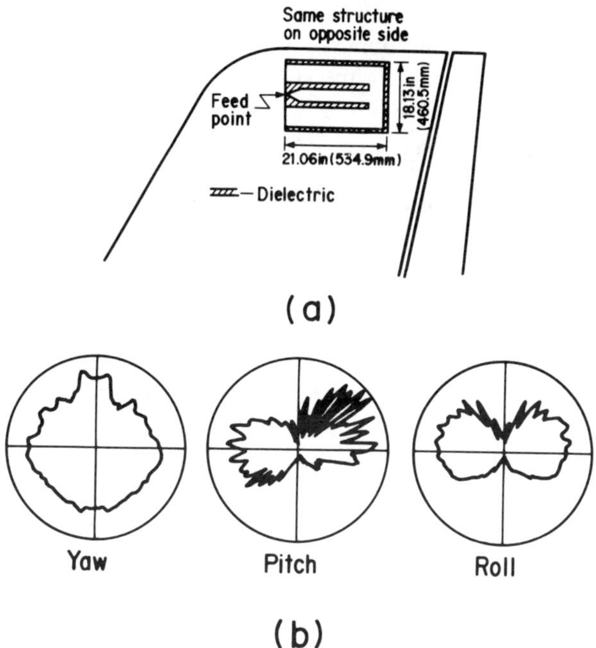

Figure 7.27 E-slot VOR antenna for the L-1011 aircraft (after [19]): (a) structure of antenna; (b) voltage radiation pattern on the L-1011 aircraft.

The VOR system employs a signal modulated at 30 Hz embodying the desired angular position information. The VOR system is particularly vulnerable to the modulation effects of, for example, a helicopter rotor, because the third harmonic of the fundamental blade rotation frequency on a typical helicopter corresponds closely to this 30-Hz modulation tone.

DME and ATC Antennas

DME is used to inform the aircraft of the distance from the DME base station and is operated using a frequency between 962 and 1,213 MHz. The ATC system is used to control aircraft traffic, and the system using the radar beacon is called an ATC radar beacon system (ATCRBS), which consists of the airborne and ground-based equipment. The former is the ATC transponder, while the latter is the secondary surveillance radar (SSR). In this system, the ATC transponder installed in the aircraft first receives the 1,030-MHz interrogation pulse signals radiated from the SSR antenna, and then automatically transmits the 1,090-MHz return pulse signals to the SSR antenna. The ground station can identify the aircraft by decoding the return pulse received by the SSR antenna.

The DME and ATC systems are operated using a nearly equal frequency, and they require the same coverage and vertical polarization so that both antennas can be shared. Figure 7.28 shows one example of an SSR/DME antenna mounted on the bottom of an aircraft. The pattern examples at 1,000 MHz, for a monopole on the bottom center line of a C-141 aircraft, are shown in Figure 7.29 [19]. In these patterns, the coverage is found to be limited to the hemisphere below the aircraft, where the deep lobing in the roll-plane pattern is due to reflections from the strongly illuminated engine cavities. A second antenna may therefore be necessary if the aircraft must install identification friend or foe (IFF), which requires some upward coverage for air-to-air interrogation. The upward coverage is satisfied by mounting the antenna on the top of the fuselage. A vertical monopole and its variants generally meet the requirements for DME and ATC systems. If there are no suitable surfaces for them, however, the horizontal slots or notches may be available.

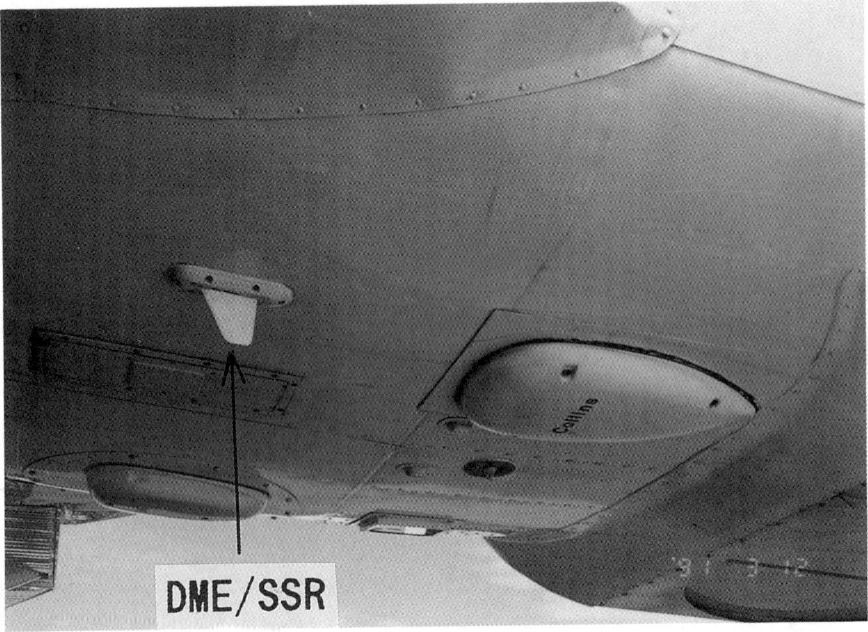

Figure 7.28 SSR/DME antenna.

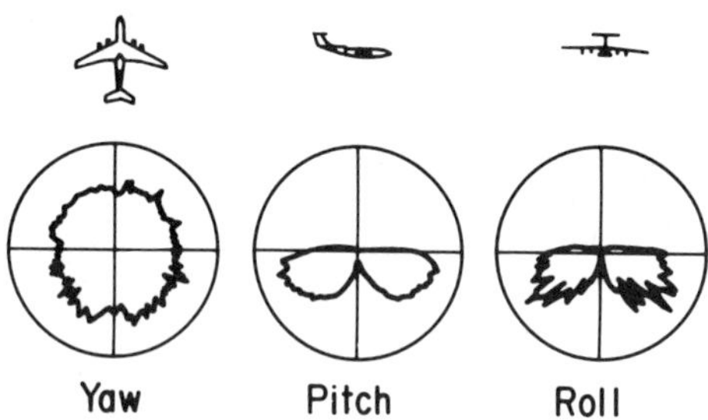

Figure 7.29 Voltage radiation patterns of monopole antenna on the bottom of a C-141 aircraft at 1,000 MHz. (After [19].)

Shared Antennas

Figure 7.30 shows one example of the shared antennas with two terminals [23] formed on a printed circuit board for ILS localizer and glide path systems. The glide path antenna consists of the forward-driven notch and a pair of parasitic notches on each side. The parasitic notches are arranged to suppress the unwanted forward radiation. The notch for the localizer antenna is cut out at the rear of the board and synthesizes a nearly omnidirectional coverage pattern in the azimuth plane.

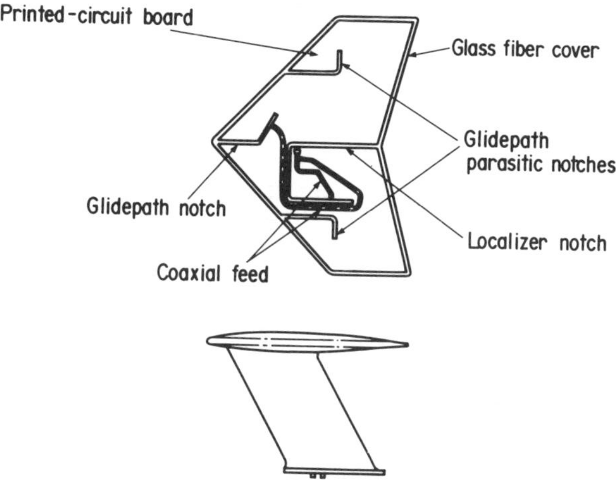

Figure 7.30 Localizer/glide path antenna in common surface. (After [23].)

7.4 ADVANCED CIRCULARLY POLARIZED ANTENNAS

Satellite communication systems require other kinds of antennas, which must radiate circularly polarized waves, and some suitable elements are as follows:

- Crossed dipole;
- Crossed slot;
- Helical;
- Conical spiral;
- Cavity-backed planar spiral;
- Quadrifilar helical;
- Microstrip patch.

These elements are the basis of circularly polarized high-gain antennas, phased-array antennas, shaped-beam antennas, and so on.

7.4.1 Crossed-Dipole Antennas

The resonance of a dipole antenna is obtained when the length is somewhat shorter than a free-space half wavelength. As the thickness is increased, the resonant length is reduced more. The bandwidth is, in general, proportional to the thickness, and the following relationship exists between the unloaded Q (Q_0) and the relative bandwidth (B_r), in which the input VSWR is less than ρ [25]

$$Q_0 B_r = (\rho^2 - 1)/2\rho \tag{7.4}$$

Circular polarization can be produced by a pair of orthogonally positioned dipoles driven in quadrature phase with equal amplitudes. However, this crossed-dipole arrangement cannot provide a good axial ratio off boresight, because the radiation patterns for the straight dipole are different in both principal planes, called the H-plane and the E-plane. This shortcoming can be improved by modifying the straight dipole to a nonstraight version, such as the V-form and U-form. These improved dipoles are called V-type and U-type dipoles. According to some measurements, the U-type provides better electrical performance than the V-type, though the V-type is simpler in mechanical structure and is less complex than the U-type. Figure 7.31(a) shows an inverted V-type crossed-dipole antenna, where the environmental metal frame is employed to widen the beamwidth still more and to broaden the pattern in the elevation plane. A quadrature phase condition with equal amplitudes to produce circular polarization is achieved by a hybrid splitter inserted at the input. Figure 7.31(b) shows one example of the radiation pattern, where the pattern was measured at 2044.25 MHz.

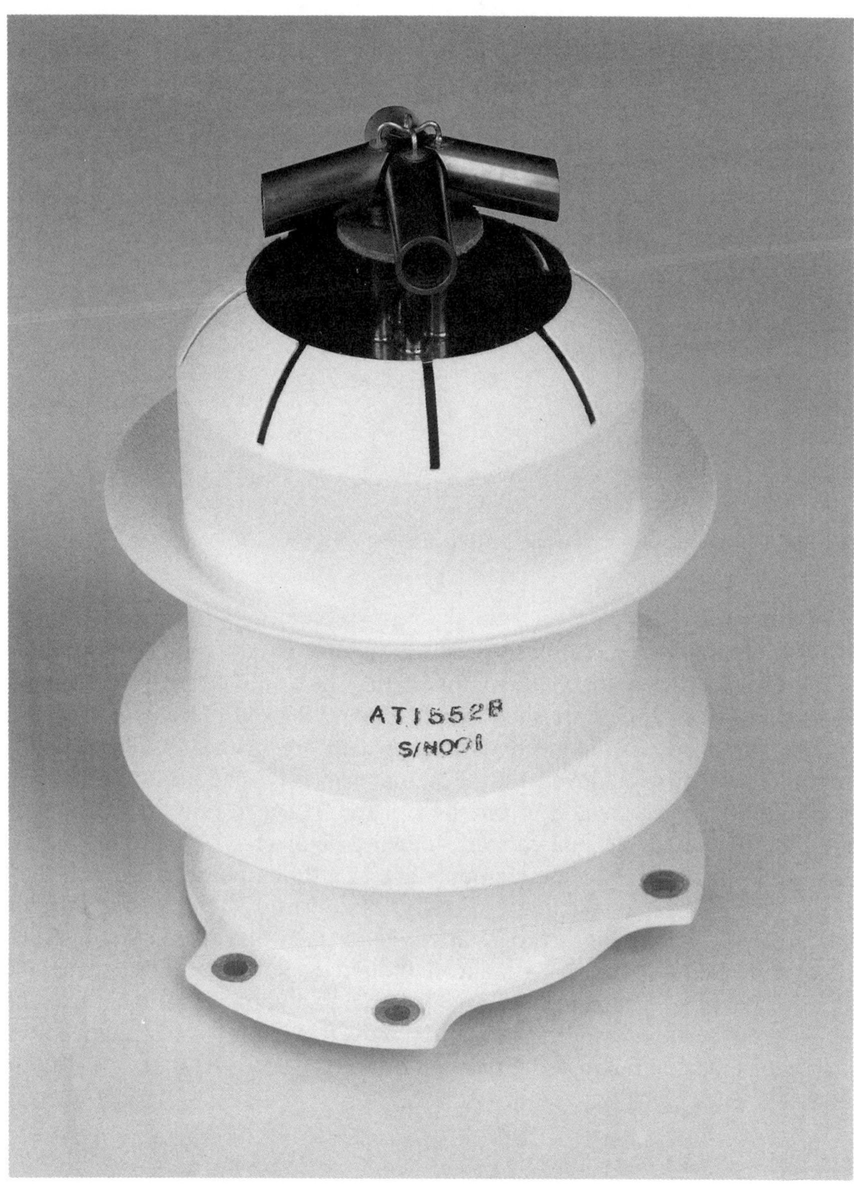

Figure 7.31a Inverted-V-type crossed dipole antenna.

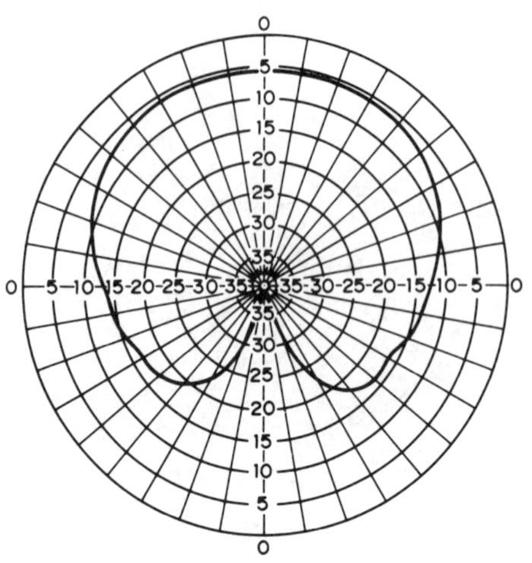

Figure 7.31b One example of radiation pattern at 2,044.25 MHz.

The crossed-dipole can also produce circular polarization without using any external circuits such as the hybrid component. The condition to excite the circularly polarized waves can be established by a balun and the self-phasing of four radiating elements. That is, the balanced outputs, which are in the antiphase with equal amplitudes, are first achieved by the balun. Next, two adjacent elements of the crossed dipole are connected to one-half of the balanced outputs from the balun. The other two adjacent elements are likewise connected to the other half of the balun. As a result, two of the elements are at a 0 deg phase angle and the other two are at a 180 deg phase angle. Finally, the desired 90 deg phase difference is obtained by designing the orthogonal elements such that one is larger relative to the desired resonant length to make it inductive, while the other is smaller, to make it capacitive. The resultant crossed-dipole antenna requires no external circuits to produce the circular polarization. This technique is also applicable to the quadrifilar helical antenna to be described later.

7.4.2 Crossed-Slot Antennas

Slot antennas are useful in many applications for high-speed aircraft because they are low-profile in structure and suitable for a flush-mount application. The slot antennas are electromagnetically complementary with the corresponding dipole antennas, so that the radiation pattern is the same as that for the complementary horizontal dipole consisting of a perfectly conducting flat strip. This holds true when both antennas are driven at the same position, but with two differences. One difference is the property that the electric and magnetic fields are interchanged. The other is that the slot electric field component normal to the perfectly conducting sheet is discontinuous from one side of the sheet to the other because the direction of the field reverses. In this case, the tangential component of the magnetic field is, likewise, discontinuous.

The crossed-slot antennas can be said to be also complementary with the corresponding crossed-dipole antenna, although the feeding method for the circular polarization is more complicated. Figure 7.32 shows an example of a crossed-slot antenna where the feeding method has been improved [26]. This antenna belongs to a cavity-backed crossed-slot antenna category and is made of copper-clad 3.3-mm-thick Teflon fiberglass with a 2.55 dielectric constant. In this example, the slot length and width are about 112 and 5 mm, respectively, and the cavity dimensions are about 80 by 80 by 20 mm. Each slot antenna, etched on the inside of the cavity, is fed by a balanced feed technique from two feed points on the microstrip feed lines etched on the opposite side of the slot. Although this antenna needs one 90 deg hybrid to produce the circular polarization, this feed technique is effective not only to suppress undesired coupling between the cross slots, but also to match the input impedance over a wider frequency band.

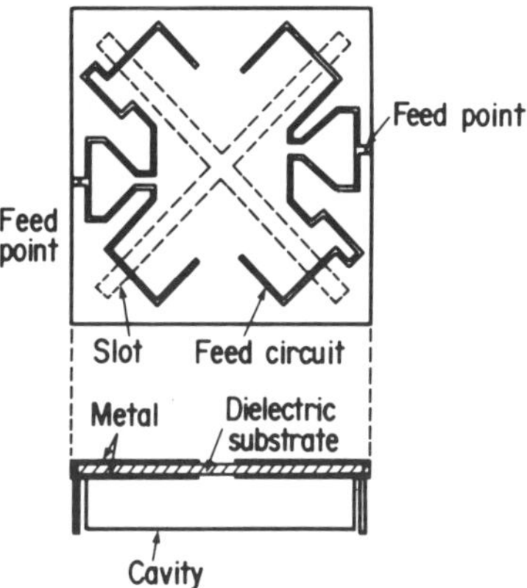

Figure 7.32 Structure of crossed-slot antenna. (After [26].)

7.4.3 Helical Antennas

A helical antenna is composed of a single conductor or multiple conductors wound into a helical shape. A monofilar type is most common, though a helix can be designed with bifilar, quadrifilar, or multifilar windings. This antenna can radiate in many modes depending on the helix diameter. When the helix circumference is on the order of one wavelength, the axial mode is excited, which is most suitable for the low-gain antennas of interest here, having maximum radiation along the helix axis.

Denoting the diameter of the helix, the length of each turn, the spacing between turns, and the helix pitch angle with the symbols D, L, S, and α, respectively, the geometrical relationship between them is illustrated by Figures 7.33(a,b), where C denotes the helix circumference. The relationship between the helix dimensions presented graphically in rectangular coordinates (S,C) and in polar coordinates (L,α) is shown in Figure 7.33(c), along with some useful design information [27]. The subscript λ signifies that the dimensions are measured in free-space wavelengths, m denotes the order of space harmonics for a wave, and the curve corresponding to $m = 1$ is of special interest for the axial mode. This figure is called a *circumference-spacing chart,* where the ordinate axis represents simple loops, the abscissa axis represents straight conductors, and the entire area between the two axes describes the general case of the helix. The frequency dependence can also be predicted by moving the corresponding coordinates along a line of constant pitch angle.

In practice, the helical conductor consists of a single wire or narrow tape, and, unless self-supporting, is wound directly on a low-permittivity dielectric cylinder. If the helical conductor is wound like a right-handed screw, it produces right-hand circular polarization and vice versa. The antenna is usually excited by a coaxial line from the back of a small ground plane, and the other end of the conductor is left as an open circuit. Although a normal mode helical antenna has a highly sensitive impedance to frequency changes, an axial mode helical antenna has a nearly pure resistance R that is approximated by the following empirical relation.

$$R = 140C/\lambda \tag{7.5}$$

where the error of this approximation is within $\pm 20\%$. The far-field pattern can be, as a first approximation, obtained by the product of the pattern of one turn and that of an end-fire array consisting of N isotropic elements with spacing S, where N is the number of turns. In the case of the axial mode, the pattern for a single turn can be approximated by $\cos(\theta)$, where θ is the angle measured from the axis of the helix, so that the total radiation pattern is calculated by

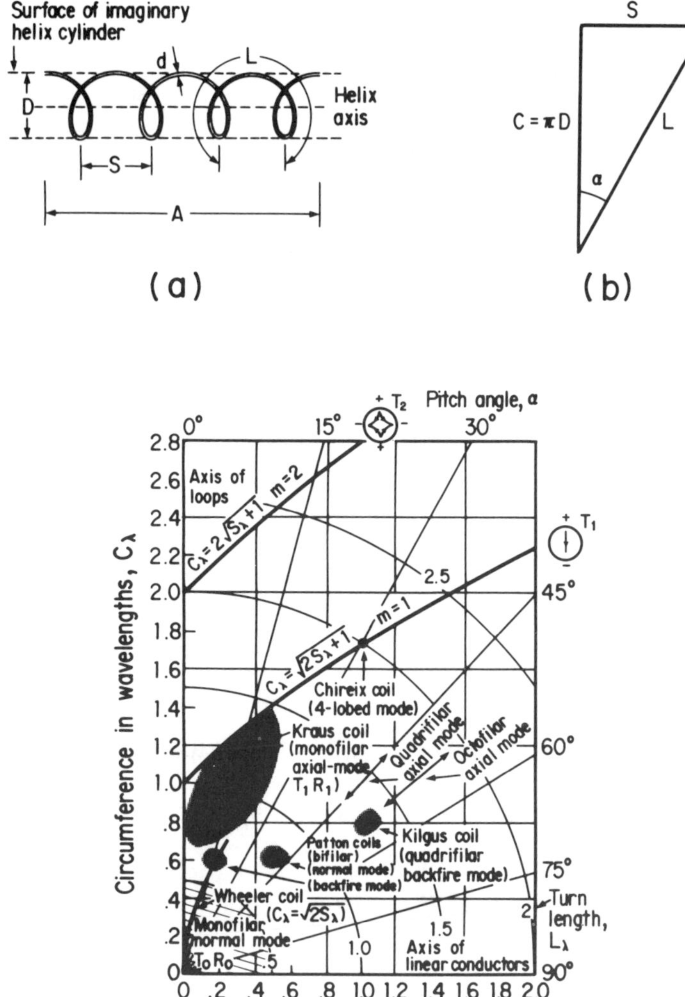

Figure 7.33 (a) Helix and design parameters; (b) relationship between circumference, spacing, turn length, and pitch angle of a helix; (c) circumference spacing chart (after [27]).

$$E(\theta) = \sin(\pi/2N) \frac{\sin(N\psi/2)}{\sin(\psi/2)} \cos\theta \tag{7.6}$$

where

$$\psi = \frac{2\pi}{\lambda} S \cos\theta - \delta \qquad \delta = \frac{2\pi}{\lambda} \frac{L}{\left(\frac{v}{c}\right)} \tag{7.7}$$

which equals progressive phase between turns, where v is the phase velocity along the helical conductor and c is the velocity of light in free space. If the directivity is significant, ψ is written as

$$\psi = \frac{2\pi}{\lambda} S(\cos\theta - 1) - \frac{\pi}{N} \tag{7.8}$$

In this case, the polarization is not purely circular and the ellipticity ratio or axial ratio (AR) is expressed by

$$\text{AR} = (2N + 1)/N \tag{7.9}$$

However, the practically observed axial ratio may be expected to deviate somewhat from the above approximate value because of various reasons. The peak gain can be empirically expressed, from the results of many experiments, as [28]

$$G_p = 8.3 \left(\frac{\pi D}{\lambda_p}\right)^{\sqrt{N+2}-1} \left(\frac{NS}{\lambda_p}\right)^{0.8} \left(\frac{\tan 12.5^\circ}{\tan\alpha}\right)^{\sqrt{N/4}} \tag{7.10}$$

where λ_p is the wavelength at a frequency where the peak gain is obtained. The reasonable value of $(\pi D/\lambda_p)$ in (7.10) is about 1.15 when the number of turns is on the order of 5, and is about 1.06 when it is on the order of 35. The gain is, of course, dependent on the frequency and has been found to vary approximately as $f^{\sqrt{N}}$ for $f < 0.962f_p$ and $f^{-3\sqrt{N}}$ for $f > 1.03f_p$, where f_p is the frequency at which the peak gain is obtained. The gain bandwidth can also be estimated by using the following frequency ratio [28].

$$\frac{f_h}{f_l} \cong 1.07 \left(\frac{0.91}{G/G_p}\right)^{4/(3\sqrt{N})} \tag{7.11}$$

where G_p is the peak gain in (7.10) and f_h and f_l are the frequency limits depending on the choice of a −3-dB or −2-dB bandwidth, respectively. The calculated bandwidths for $G/G_p = -3$ dB and −2 dB agree reasonably well with the measurement. Finally, the half power beamwidth (HPBW) is empirically expressed as [29]

$$\text{HPBW} = \frac{61.5\left(\dfrac{2N}{N+5}\right)^{0.6}}{\left(\dfrac{\pi D}{\lambda}\right)^{\sqrt{N/16}}\left(\dfrac{NS}{\lambda}\right)^{0.7}}\left(\frac{\tan\alpha}{\tan 12.5°}\right)^{\sqrt{N/16}} \tag{7.12}$$

When considering L-band satellite communication use, a typical helix diameter D and spacing S may be 6 and 4 cm, respectively. In this case, about 7.5 dBic gain is obtained from (7.10), and the HPBW is on the order of 75 deg according to (7.12) when the number of turns is three. Three may be the minimum number of turns, and very short antennas with fewer than three turns exhibit poor electrical performances. So the cylindrical helix antenna may not be, in general, suitable for wide-beamwidth applications, though the beam-width may be widened by surrounding the antenna with dielectric. If more turns are permitted, the radiation pattern and VSWR characteristics can be improved by adopting nonuniform-diameter helical structures and tapering the end section turn diameters to suppress the currents reflected from the end [30,31].

The conical helix antenna, shown in Figure 7.34, can be regarded as a low-gain development of the cylindrical helix antennas and is suitable for wide-beamwidth applications with good efficiency. With a suitable choice of the cone angle and turn spacing, for example, it is possible to achieve a beamwidth on the order of 100 deg. This type of antenna can also achieve an input VSWR of 1.5:1 or better over a 5% frequency bandwidth merely by incorporating a simple quarter-wavelength transformer. The typical size for an L-band application is on the order 15 cm in length, and the ground plane is about 20 cm in diameter. The resultant gain is approximately 4 to 7 dBic.

7.4.4 Conical Spiral Antennas

In comparison with the conical helix antenna described above, the bifilar conical spiral, shown in Figure 7.35, provides better performance and is more versatile,

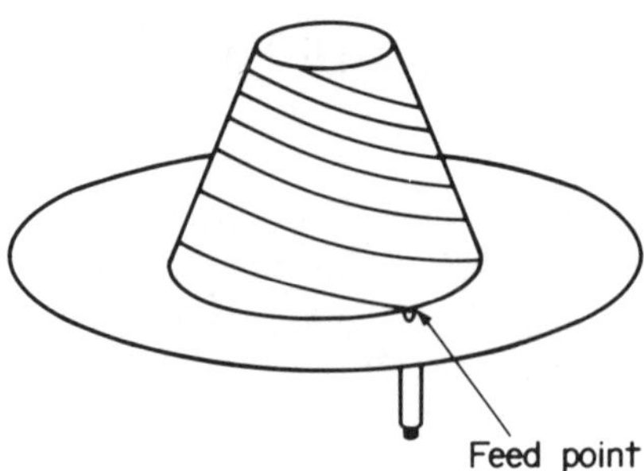

Figure 7.34 Monofilar conical helix antenna.

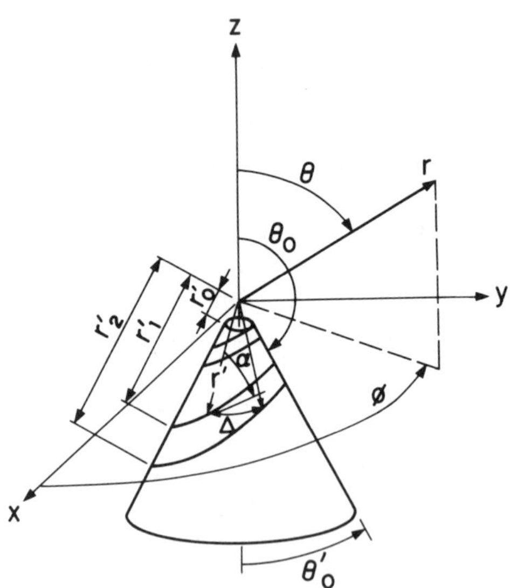

Figure 7.35 Conical spiral antenna and design parameters.

though the geometry is somewhat complex. The conical spiral antenna discussed here is the equiangular version, which can be specified by angular dimensions only. So such an antenna is independent of frequency and its geometry can be prescribed in spherical coordinates (r, θ, ϕ) as

$$r = e^{a\phi}g(\theta) \tag{7.13}$$

where a and $g(\theta)$ are an arbitrary constant and angular function, respectively. Although (7.13) predicts the geometry of an antenna with a radius extending from zero to infinity, it is practically possible to truncate it to an extent which depends on the required bandwidth. For the typical model, the angular function $g(\theta)$ needs to satisfy the following equation.

$$\mathrm{d}g(\theta)/\mathrm{d}\theta = g'(\theta) = A\delta(\theta_0 - \theta) \tag{7.14}$$

where δ is the Dirac delta function. Combining (7.13) and (7.14) gives

$$r_1' = r_0' e^{a\phi \sin\theta_0} \tag{7.15}$$

$$r_2' = r_0' e^{a(\phi - \Delta)\sin\theta_0} \tag{7.16}$$

where r_1' and r_2' define both edges of one spiral arm in Figure 7.35; r_0', θ_0, and Δ are also defined in Figure 7.35. The dimensions for another arm of a bifilar conical spiral can be obtained by rotating the above angular functions 180 deg with respect to the ϕ axis. Thus,

$$r_3' = r_0' e^{a(\phi \pm \pi)\sin\theta_0} \tag{7.17}$$

$$r_4' = r_0' e^{a(\phi - \Delta \pm \pi)\sin\theta_0} \tag{7.18}$$

In this structure, the spiral (pitch) angle of the cone is given by

$$\alpha = \tan^{-1}(1/a) \tag{7.19}$$

The radiation mechanism for this antenna can be understood by regarding the two spirals as a transmission line. When two conductor arms are fed in antiphase at the cone apex, waves travel out from the feed point and propagate along the spirals without radiating until a resonant length has been traversed. Strong radiation occurs at that point, and very little energy is reflected by the outer limits of the spiral. Broadband characteristics are achieved by this mechanism, and Figure 7.36 shows a set of the typical pattern variations with respect to cone angle θ_0', spiral angle α, and angular arm width Δ [32]. These results are presented for very narrow arm structures ($\Delta = 16$ deg), for very wide arm structures ($\Delta = 164$ deg), and for self-complementary arm structures ($\Delta = 90$ deg). For the latter structure, the geometries of the arms and the spaces between them are identical except for a

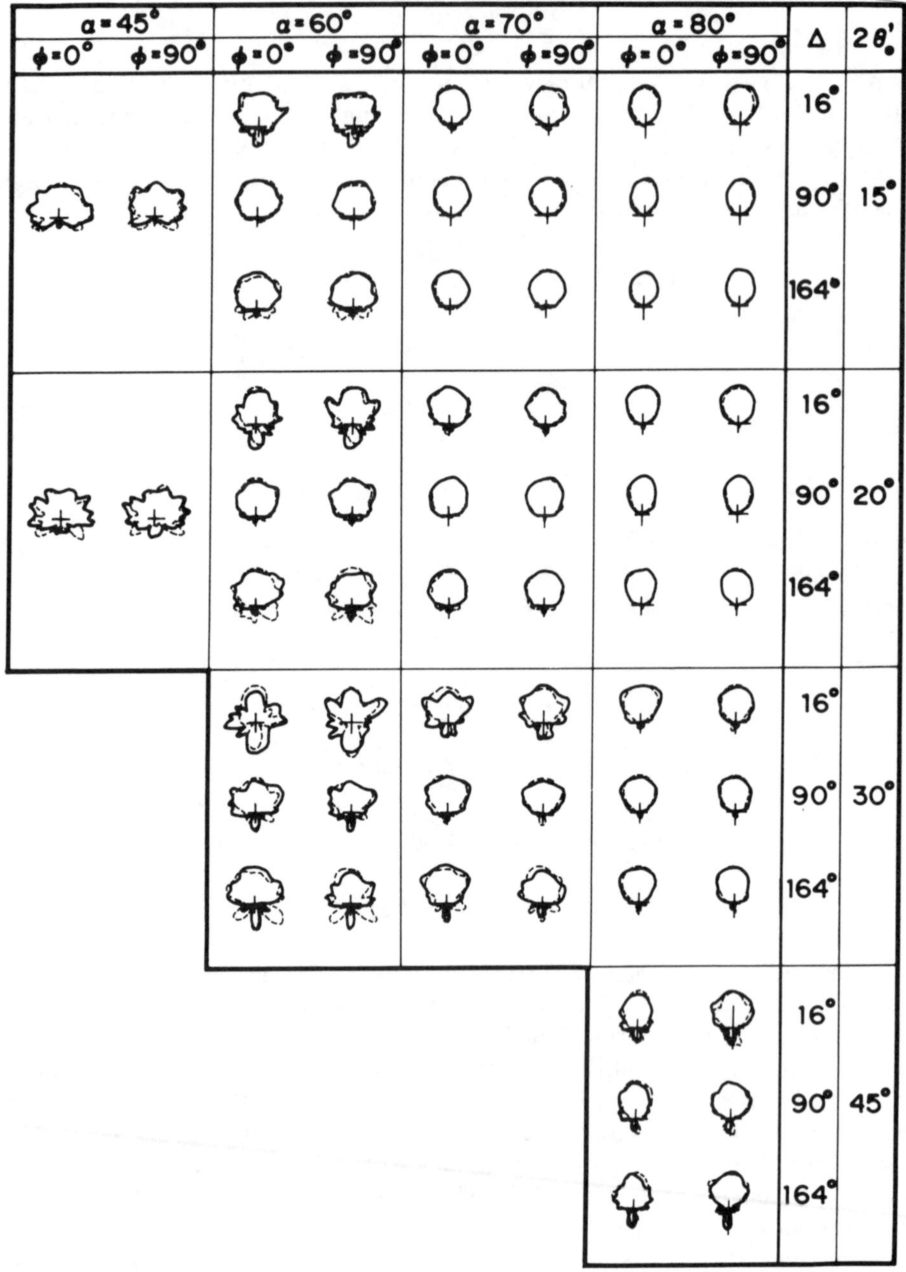

Figure 7.36 Typical radiation patterns indicating general change in shape with cone angle $2\theta_0'$ spiral angle α, and angular arm width δ. (After [32].)

rotation of 90 deg around the axis of the antenna. Figure 7.37(a) is an example of this kind of antenna, with $\alpha = 49$ deg, $\Delta = 90$ deg, and $2\theta_0' = 15$ deg, where a quarter-wavelength coaxial balun is employed. The typical radiation pattern, measured with a rotating linearly polarized source at 1,537.5 MHz, is shown in Figure 7.37(b).

Conveniently, two conductor arms can also be fed directly at the centerpoint or apex from a coaxial cable bonded to one of the spiral arms without any external baluns, because the spiral arm can itself act as a balun [33]. In this case, a dummy cable may be bonded to another arm to maintain the symmetrical performance. If

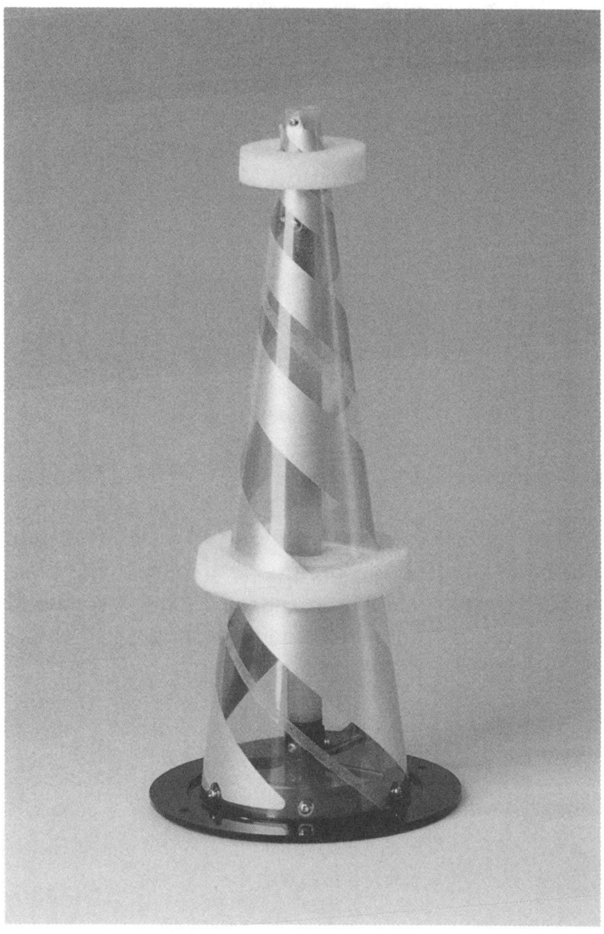

Figure 7.37a Two-arm balanced conical spiral antenna using $\lambda/4$ balun.

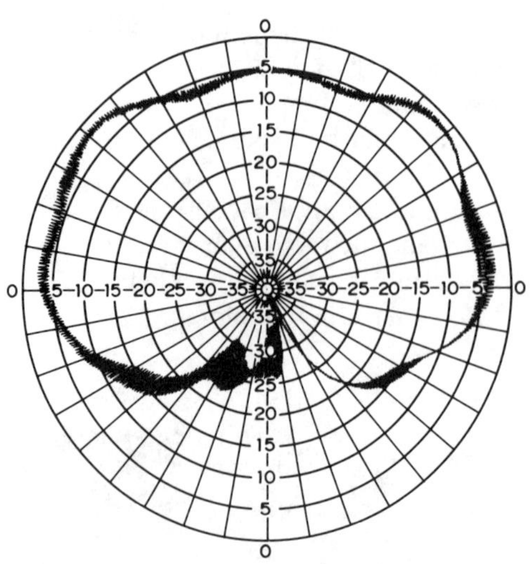

Figure 7.37b Radiation pattern at 1,537.5 MHz.

the width of arms is decreased to a narrow constant value, the arms can be formed by the cable alone.

7.4.5 Cavity-Backed Planar Spiral Antennas

The planar spiral antennas are commonly divided into three categories: an equiangular spiral, logarithmic spiral, and Archimedean spiral. The planar equiangular spiral antenna has the geometry shown in Figure 7.38 and corresponds to the special case of the conical spiral antennas mentioned previously. The geometry can be obtained by substituting a $\pi/2$ into θ_0 in (7.15) to (7.18) to give

$$r_1 = r_0 e^{a\phi} \tag{7.20}$$

$$r_2 = r_0 e^{a(\phi - \Delta)} \tag{7.21}$$

$$r_3 = r_0 e^{a(\phi \pm \pi)} \tag{7.22}$$

$$r_4 = r_0 e^{a(\phi - \Delta \pm \pi)} \tag{7.23}$$

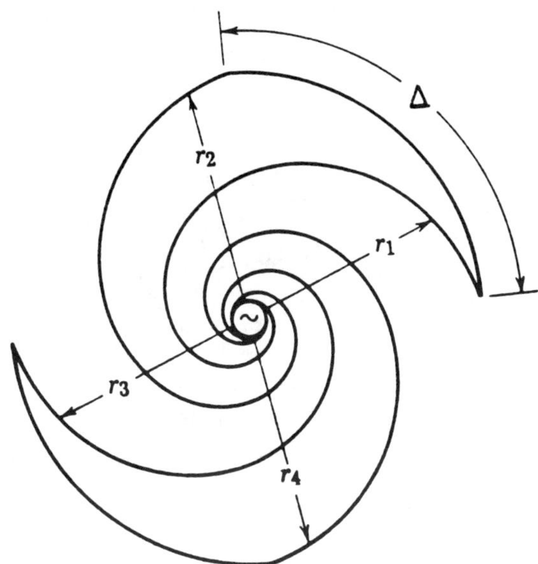

Figure 7.38 Two-arm equiangular spiral antenna and design parameters.

This antenna needs no external hybrid circuits to produce a circular polarization, and the example shown in Figure 7.38 can radiate left-hand circularly polarized waves outward from the page and right-hand circularly polarized waves into the page when the pair of spirals is excited in antiphase at the center. According to the experimental measurements, the axial ratio is near unity and the HPBW is on the order of 90 deg over a decade bandwidth or more. As for the input impedance, the resistive part depends on the thickness of the antenna elements, and thin elements lead to high impedance values. This implies that the impedance depends on the arm width when the structure is planar. If the angular extent Δ in Figure 7.38 is chosen to be 90 deg, the geometries of the arm and the space between arms are identical, except for a rotation of 90 deg around an axis. This structure is defined as self-complementary, just like the case of the previous conical spiral antenna, but it should be noted that the planar spiral antenna has a constant impedance of $60\pi\ \Omega$ for the two arm configurations.

The logarithmic spiral, shown in Figure 7.39, can be prescribed by

$$r_1 = a^{\phi} \tag{7.24}$$

$$r_2 = a^{(\phi - \Delta)} \tag{7.25}$$

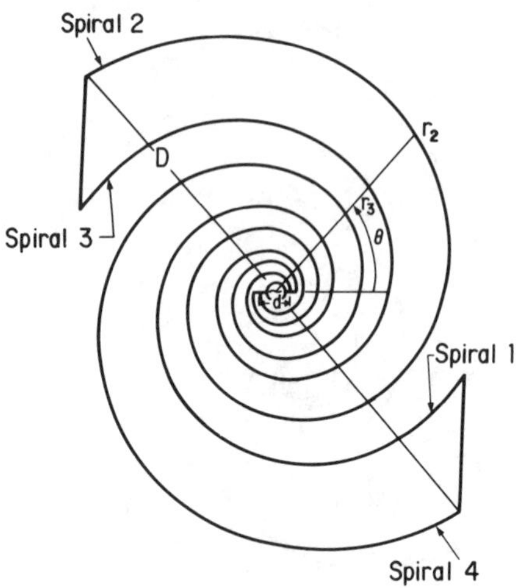

Figure 7.39 Logarithmic spiral antenna and design parameters.

$$r_3 = a^{(\phi \pm \pi)} \tag{7.26}$$

$$r_4 = a^{(\phi - \Delta \pm \pi)} \tag{7.27}$$

This antenna can radiate right-hand circularly polarized waves outward from the page and left-hand circularly polarized waves into the page without any external hybrid circuits if a pair of spirals is excited with antiphase at the center.

In general, the standard planar spiral antenna has been fed by using the external balun, but it can also be fed at the center point or apex from a coaxial cable bonded to one of the arms without any external baluns [34], like the conical spiral antenna. This antenna can, furthermore, be composed of thin wire arms.

The Archimedean spiral, shown in Figure 7.40, is another geometry of the planar spiral [35]. This antenna has superior bandwidth properties when fully optimized, and typically consists of a pair of thin wire arms, of which the geometry can be prescribed by

$$r_1 = r_0\phi \tag{7.28}$$

$$r_2 = r_0(\phi - \pi) \tag{7.29}$$

This antenna also needs no external hybrid circuits to produce circular polarization, and the example shown in Figure 7.40 can radiate the right-hand circularly polarized

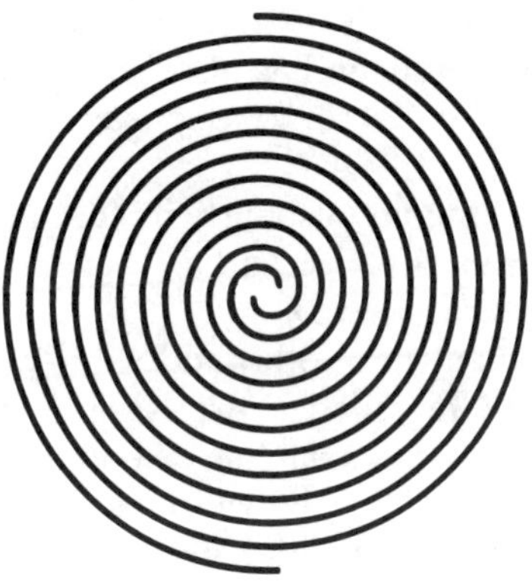

Figure 7.40 Archimedean spiral antenna.

waves outward from the page and the left-hand circularly polarized waves into the page if the pair of thin wire arms is excited in antiphase at the center. The Archimedean spiral is a broadband antenna and has properties similar to the standard planar spiral antennas, although it is not theoretically a frequency-independent structure.

When placed in a quarter-wave cavity, the Archimedean spiral antenna can achieve near-octave bandwidths even when the cavity consists of a metal-based cylinder without any absorber. If an absorber-loaded cylinder is employed in the cavity, a greater-than-decade bandwidth may be achieved, although about half the power is dissipated into heat by the absorber. A typical Archimedean spiral antenna has an octave bandwidth for a VSWR of less than 2, and axial ratio of less than 2 dB, and beamwidth of about 70 deg, while a gain of 7 to 8 dBic is achieved without any absorber. The structure has several mechanical advantages: it is compact and fairly simple to construct, and the spiral arms can be easily fed using a suitable impedance-transforming balun.

7.4.6 Quadrifilar Helical Antennas

The quadrifilar helical antenna consists of four helices equally wound circumferentially on a cylinder, and the four terminals are excited with equal amplitude but with relative phases of 0, 90, 180, and 270 deg. It is classified into two categories that are a resonant and nonresonant type. The former is very useful for an aircraft antenna, because of its small size, lack of a ground plane, and insensitivity to nearby metal structures, although the impedance bandwidth is narrow (e.g., the bandwidth for a VSWR less than 2 is typically 3% to 5%). A wider bandwidth can be achieved by using larger diameter wire as a helical arm, as in the case of the dipole antenna. This helical antenna has also been used for various applications requiring a wide beamwidth of 120 to 180 deg over a relatively narrow frequency range.

In order to make the antenna resonate, its four arms must be equal to $m\lambda/4$ in length, where $m = 1, 2, 3, 4, \ldots$, and are generally wound to form a small-diameter helix with an $n/4$ turn, where $n = 1, 2, 3, 4, \ldots$. The ends of the helices must be open-circuited, when m = odd, and short-circuited, when m = even. These structures may also be regarded as two bifilar helices. Each bifilar helix can be balun-fed at the top through a coaxial line extending to the top along the central axis. In order to produce a circular polarization, some external circuit is required: for example, a 90 deg hybrid is conventionally employed. The radiation pattern is basically characterized by the radius r_0 and the pitch distance for each element measured along the axis of helix p, defined in Figure 7.41(a) [36]. If these helical parameters are appropriately chosen, the antenna can produce shaped-conical beams having 90 to 240 deg beamwidth, with excellent circular polarization, over the wide angular region. Excellent circular polarization naturally produces high front-to-back ratios. Many measurements indicate that the shaped-conical radiation patterns with full-cone angles from 100 to 180 deg and center minimums down 3 to 20 dB can be realized by an appropriate choice of helical parameters, according to the design curves in Figure 7.41(b) [36].

Figure 7.42(a) shows an example of the resonant quadrifilar helical antennas with $\lambda/4$ coaxial balun, which is designed for use at a frequency in the L-band. A typical radiation pattern, measured with a rotating linearly polarized source at 1,537.5 MHz, is also shown in Figure 7.42(b).

7.4.7 Microstrip Patch Antennas

The typical microstrip patch antenna consists of a metallic ground plane, a thin dielectric or air sheet, and a strip conductor patch on top of the dielectric substrate. It can be made conformal to a metallic surface and can be produced at low cost with photo-etch techniques. When low profile, lightweight, small size, and low cost are particularly required, the microstrip patch antennas are very important. In practical applications, typical shapes of the patch radiator are circular and rectangular. Their dimensions are approximately determined by

$$a = a_e\left[1 + \frac{2t}{\pi a_e \epsilon_r}\left\{\ln\left(\frac{\pi a_e}{2t}\right) + 1.7726\right\}\right]^{-1/2} \tag{7.30}$$

$$\because a_e = \frac{1.8412\lambda_0}{2\pi\sqrt{\epsilon_r}} \tag{7.31}$$

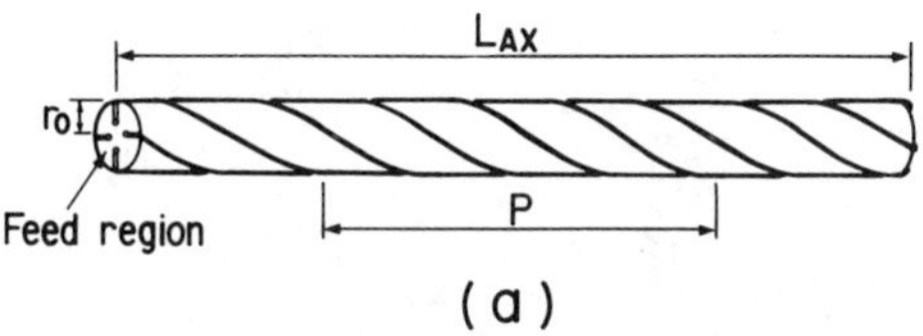

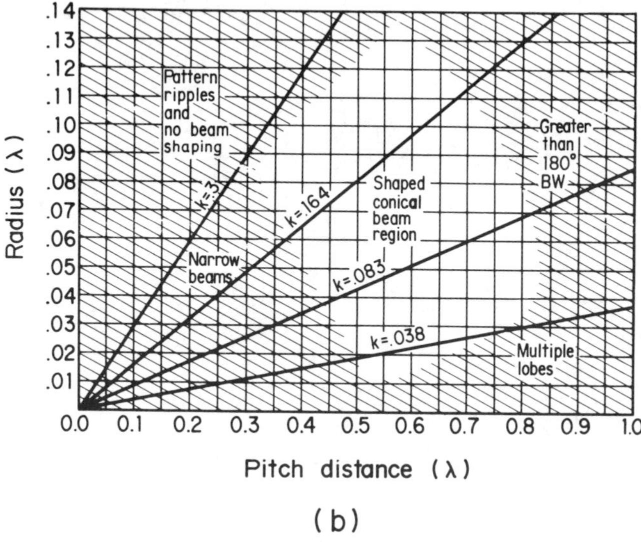

Figure 7.41 (a) Quadrifilar helical antenna and design parameter; (b) region of shaped conical beam performances (after [36]).

for circular patch antennas with the TM_{110} mode excited, and

$$L = \frac{\lambda_0}{2\sqrt{\epsilon_e}} - 2\Delta\ell \tag{7.32}$$

$$\because \epsilon_e = \frac{\epsilon_r + 1}{2} + \frac{\epsilon_r - 1}{2}[1 + 12(t/W)]^{-1/2} \tag{7.33}$$

$$\because \Delta\ell = 0.412t\frac{(\epsilon_e + 0.3)[(W/t) + 0.264]}{(\epsilon_e - 0.258)[(W/t) + 0.8]} \tag{7.34}$$

$$\because W = \frac{\lambda_0}{2}\left(\frac{\epsilon_r + 1}{2}\right)^{-1/2} \tag{7.35}$$

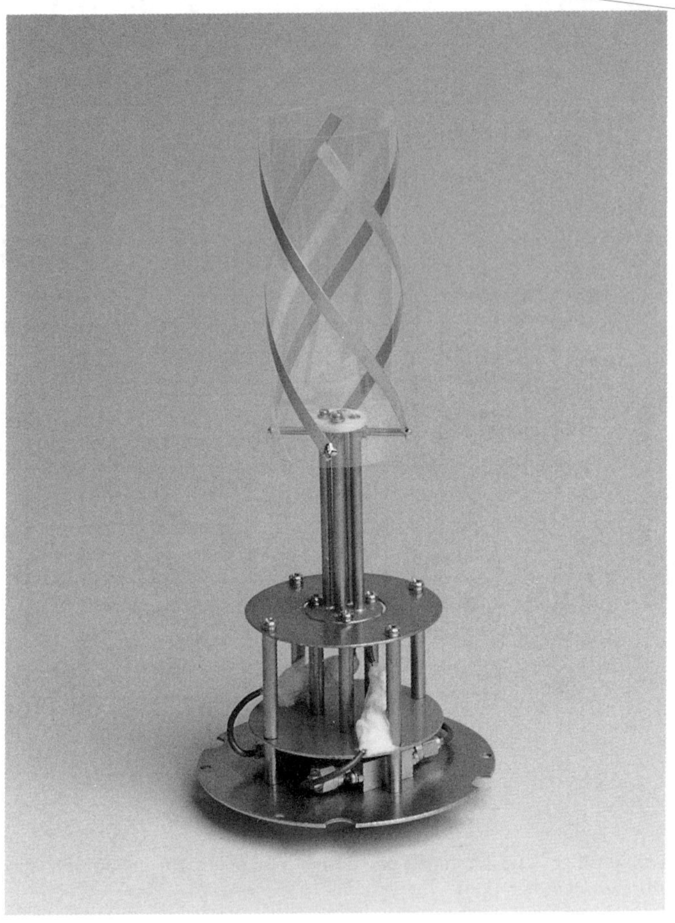

Figure 7.42a One-half-turn quadrifilar helical antenna.

for rectangular patch antennas with the TM_{010} or TM_{100} mode excited; a is the radius of circular patch, L and W are the length and width of the rectangular patch, respectively, t and ϵ_r are the thickness and dielectric constant of the dielectric sheet (or substrate), respectively, and λ_0 is the resonant wavelength.

However, the microstrip antennas have the serious disadvantage that the frequency bandwidth is narrow and only on the order of 2%. Various impedance matching networks have been investigated [37], but the feed network may become complex and lossy. There are two other techniques for bandwidth extension, as follows. The first technique is to increase the thickness of the substrate and decrease

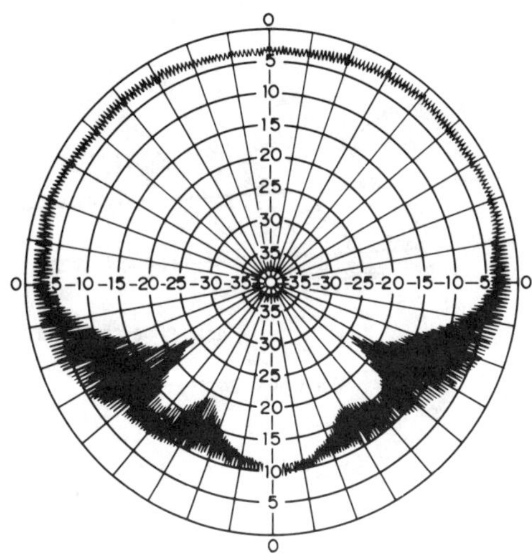

Figure 7.42b Radiation pattern at 1,537.5 MHz.

its dielectric constant. Figure 7.43(a) shows an example of this kind of antenna [25], consisting of a paper honeycomb substrate sandwiched between epoxy fiberglass skins. Figure 7.43(b) shows the return loss characteristics, and an approximate 8.4% bandwidth is obtained for a VSWR less than 2 [30]. When circular polarization is required, however, this technique may need some countermeasures [38] against the degradation in the axial ratio, due to the generation of higher order modes.

The second technique is achieved by using the stacked patches, which electromagnetically couple together. One example is shown in Figure 7.44(a), where the lower patch, which does not appear in this view, is excited with coaxial probes using the four-point feeding technique to improve the axial ratio. The electromagnetic energy from the lower patch couples to the upper parasitic patch that appears in this view. The external circuit to produce the circular polarization consists of a stripline circuit and is attached on the back of the ground plane attached to the lower patch antenna. The radiation patterns, measured by 1,540 and 1,600 MHz, are shown in Figure 7.44(b) and Figure 7.44(c), where the diameters of lower and upper patches are 67.88 and 73.13 mm, respectively, the spacing between them is 20.2 mm, and the substrate thickness and dielectric constant are 3.2 mm and a nominal 2.55, respectively.

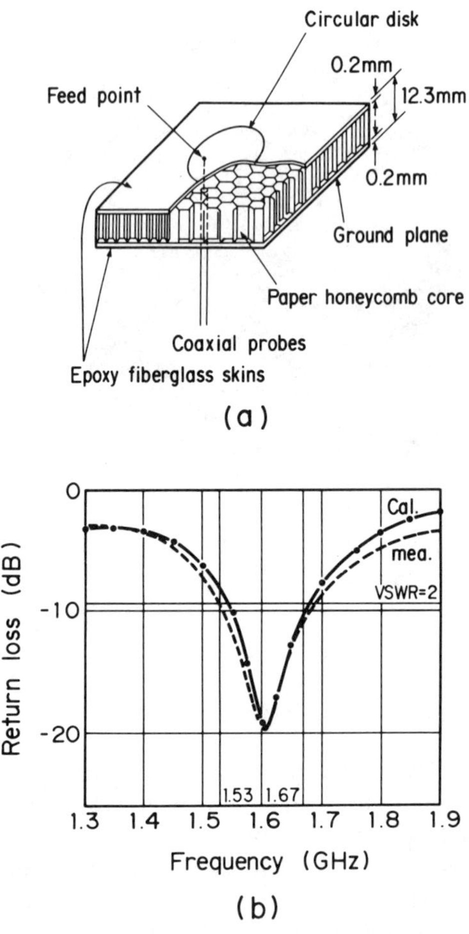

Figure 7.43 (a) Configuration of circular microstrip antenna consisting of epoxy fiberglass skins and paper honeycomb core (after [25]); (b) one example of return loss characteristics (after [25]), where radius = 40.5 mm and ϵ_r = 1.21.

When the receive and transmit frequency bands are completely separate, a dual-frequency technique is also available. This technique assigns the separate frequency bands to each patch antenna so that the antenna system does not need an external duplexer. Figure 7.45 shows an example of a two-layer dual-frequency microstrip patch antenna configuration [26], where the antenna is composed of a 3.2-mm-thick glass-microfiber-reinforced PTFE substrate, with a dielectric constant of 2.3. The upper patch is for transmission, while the lower patch deals with

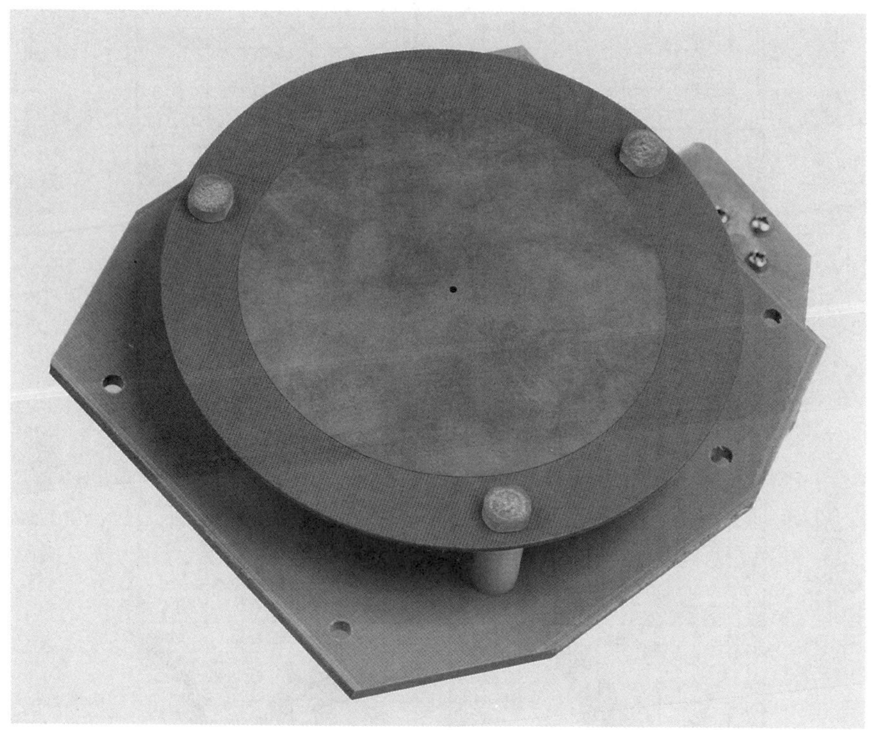

Figure 7.44 (a) Stacked microstrip antenna.

reception. Each patch is individually fed at two points with a phase difference of 90 deg to produce the circular polarization. The distances from the center to the feed points are about 10 and 20 mm for the upper and lower patch, respectively. In this configuration, the upper layer is a conventional circular microstrip patch antenna with a 66-mm diameter, while the lower patch, with a diameter of 84 mm, functions as a circular microstrip antenna with an electric shielding ring. The electric shielding ring is necessary to avoid any electrical interferences between the various feed lines and consists of 12 short-circuited pins arranged in a circle with a 27-mm diameter.

The above examples all needed some external circuit to produce the degenerate modes for circular polarization. This can, however, be achieved by giving an appropriate perturbation to the patch dimensions and selecting a suitable feed

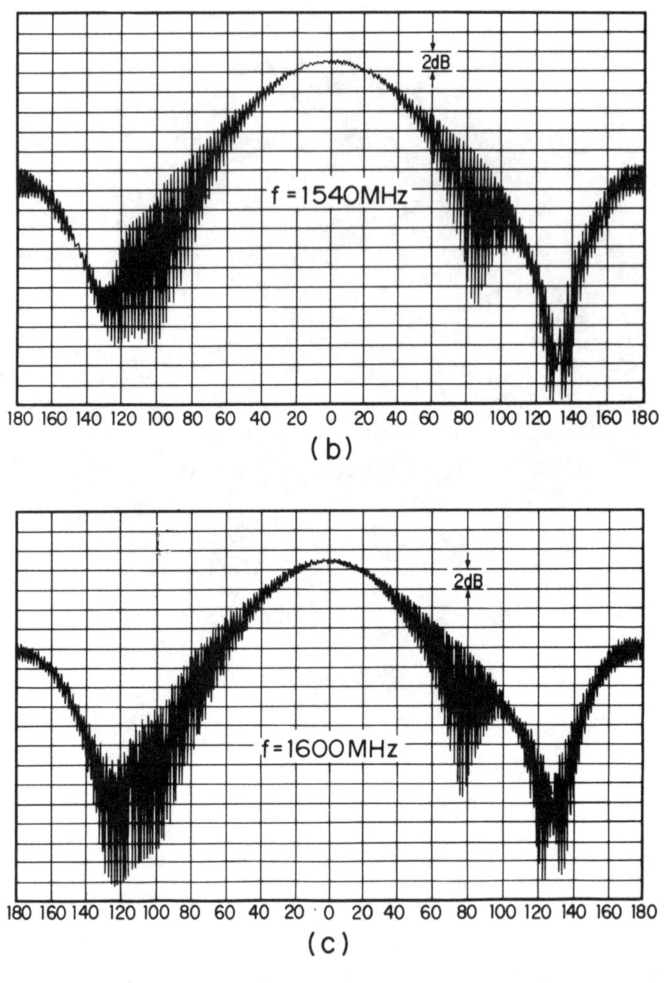

Figure 7.44 (b) Radiation pattern at 1,540 MHz; (c) radiation pattern at 1,600 MHz; where diameter of lower patch = 67.88 mm, diameter of upper patch = 73.13 mm, spacing between them = 20.2 mm, thickness of substrate = 3.2 mm, and ϵ_r = 2.55.

point. Such an antenna can simultaneously radiate a pair of orthogonally polarized waves at some frequency between the resonant frequencies of the two orthogonal modes, because the degenerate modes are, in general, vectorially orthogonal. Figure 7.46 shows typical examples of these circularly polarized microstrip antennas [39], which radiate circularly polarized waves when fed at a point on the chain-dotted lines. In this figure, the conditions, which should be imposed on each patch

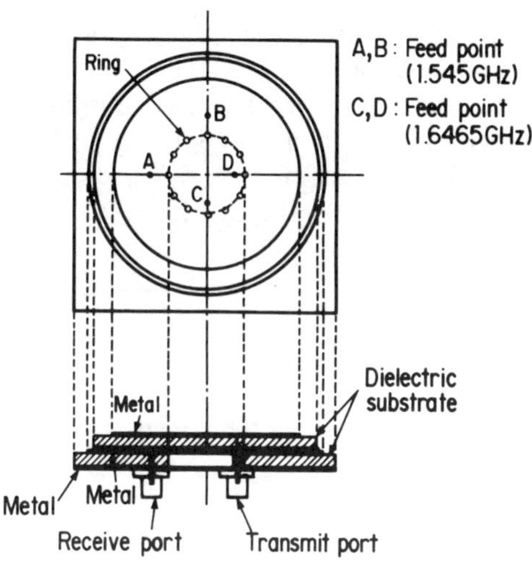

Figure 7.45 Structure of two-layer dual-frequency microstrip antenna. (After [26].)

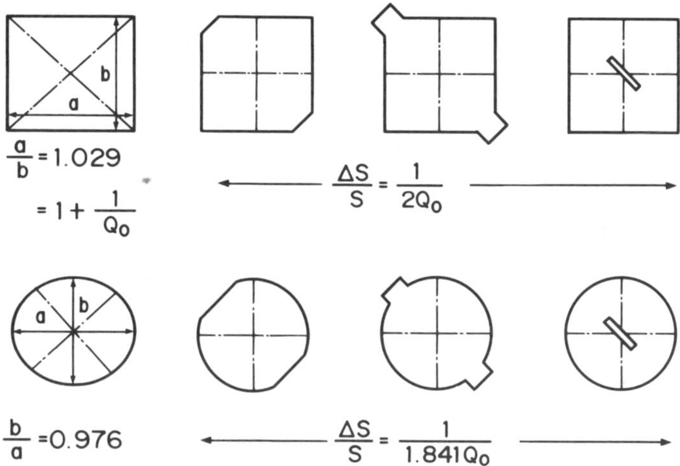

Figure 7.46 Current examples of singly fed circularly polarized microstrip antenna. (After [39].)

geometry to produce the circular polarization, are also presented, where ΔS is the perturbation given to the area of the original patch radiator of area S and Q_0 is the unloaded Q for the original patch antenna. In addition to the above examples, there are various other shapes, including the pentagon proposed by Weinschel [40]. Figure 7.47 shows some of these variants [41], where a and c are defined in Figure 7.47(b) and $b/a = 1.0603$ here. The solid line drawn in each pentagon shows the feed location loci for exciting the left-hand circularly polarized waves, while the broken line shows the feed location loci for exciting the right-hand circularly polarized waves. The symbols f_{c1} and f_{c2} are the two frequencies at which the circularly polarized waves can be excited when $a = 100$ mm, $t = 3.2$ mm, $\epsilon_r = 2.55$, and $\tan \delta = 0.0018$.

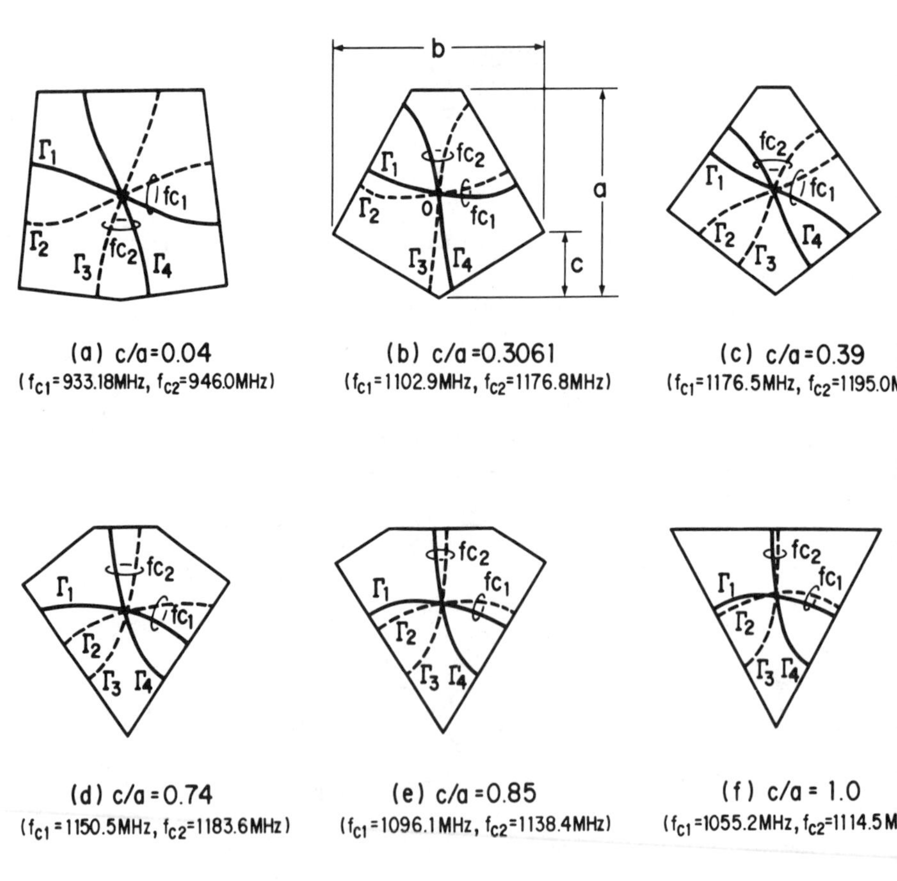

(a) $c/a = 0.04$ ($f_{c1} = 933.18$ MHz, $f_{c2} = 946.0$ MHz)

(b) $c/a = 0.3061$ ($f_{c1} = 1102.9$ MHz, $f_{c2} = 1176.8$ MHz)

(c) $c/a = 0.39$ ($f_{c1} = 1176.5$ MHz, $f_{c2} = 1195.0$ M

(d) $c/a = 0.74$ ($f_{c1} = 1150.5$ MHz, $f_{c2} = 1183.6$ MHz)

(e) $c/a = 0.85$ ($f_{c1} = 1096.1$ MHz, $f_{c2} = 1138.4$ MHz)

(f) $c/a = 1.0$ ($f_{c1} = 1055.2$ MHz, $f_{c2} = 1114.5$ MH

Figure 7.47 Variations of singly fed circularly polarized pentagonal microstrip antenna. (After [41].)

REFERENCES

[1] INMARSAT RFP, "Aeronautical High Gain Antenna Subsystem," Request for Proposal (RFP) No. 082, HK/86-2063M/pt, March 1986.
[2] Granger, J.V.N., and J. T. Bolljahn, "Aircraft Antennas," *Proc. IRE,* Vol. 43, May 1955, pp. 533–550.
[3] Harrington, R. F., *Field Computation by Moment Methods,* New York: Macmillan, 1968.
[4] Burke, G. L., and A. J. Poggio, "Numerical Electromagnetic Code (NEC)—Method of Moments, Part 1: Program Description-Theory," Tech. Doc. 116, Naval Electronic Systems Command (ELEX 3041), July 1977.
[5] Richmond, J. H., "A Wire-Grid Model for Scattering by Conducting Bodies," *IEEE Trans. Ant. Propag.,* Vol. AP-14, Nov. 1976, pp. 782–786.
[6] Wang, J. J., and C. E. Ryan, Jr., "Application of Wire-Grid Modelling to the Design of a Low-Profile Aircraft Antenna," *IEEE AP-S Int. Symp. Digest,* June 1977, pp. 222–225.
[7] Knepp, D. L., and J. Goldhirsh, "Numerical Analysis of Electromagnetic Radiation Properties of Smooth Conducting Bodies of Arbitrary Shape," *IEEE Trans. Ant. Propag.,* Vol. AP-20, May 1972, pp. 383–388.
[8] Keller, J. B., "Geometrical Theory of Diffraction," *J. Optic. Soc. Am.,* Vol. 52, Feb. 1962, pp. 116–130.
[9] Kouyoumjian, R. G., and P. H. Pathak, "A Uniform Geometrical Theory of Diffraction for an Edge in a Perfectly Conducting Surface," *Proc. IEEE,* Vol. 62, Nov. 1974, pp. 1448–1460.
[10] Burnside, W. D., M. C. Gilreath, R. J. Marhefka, and C. L. Yu, "A Study of KC-135 Aircraft Antenna Patterns," *IEEE Trans. Ant. Propag.,* Vol. AP-23, May 1975, pp. 309–316.
[11] Yu, C. L., W. D. Burnside, and M. C. Gilreath, "Volumetric Pattern Analysis of Airborne Antennas," *IEEE Trans. Ant. Propag.,* Vol. AP-26, Sept. 1978, pp. 636–641.
[12] Cooke, W. P., and C. E. Ryan, Jr., "A GTD Computer Algorithm for Computing the Radiation Patterns of Aircraft-Mounted Antennas," *IEEE AP-S Int. Symp. Digest,* June 1980, pp. 631–634.
[13] Ryan, C. E., Jr., "Analysis of Antennas on Finite Circular Cylinders With Conical or Disk End Caps," *IEEE Trans. Ant. Propag.,* Vol. AP-20, July 1972, pp. 474–476.
[14] Balanis, C. A., and Y. B. Cheng, "Antenna Radiation Modeling for Microwave Landing System," *IEEE Trans. Ant. Propag.,* Vol. AP-24, July 1976, pp. 490–497.
[15] Marin, L., J. P. Castillo, and K.S.H. Lee, "Broad-Band Analysis of VLF/LF Aircraft Wire Antennas," *IEEE Trans. Ant. Propag.,* Vol. AP-26, Jan. 1978, pp. 141–145.
[16] Kossey, P. A., E. A. Lewis, and E. C. Field, "Relative Characteristics of TE/TM Waves Excited by Airborne VLF/LF Transmitters," *AGARD Conf. Proc.,* Vol. 305, 1982, p. 19–1.
[17] IECE Japan, *Antenna Engineering Handbook,* (Japanese), Tokyo, Ohm-Sha Co., 1980.
[18] Fujii, K. "Loop Antenna for Aircrafts," Japanese Utility model 914431, 1970.
[19] Jasik, H., ed., *Antenna Engineering Handbook,* New York: McGraw-Hill, 1984.
[20] Blackband, W. T., "A Comparison of the HF Aerials Fitted to Comet XV814," *IEE Int. Conf. Antennas for Aircraft and Spacecraft,* June 1975, pp. 171–177.
[21] Hirano, M., ed., *New Radio Wave Technologies,* Part 2 (Japanese), Tokyo Japan: RATESU Co., 1973.
[22] Burberry, R. A., "Progress in Aircraft Aerials," *Proc. IEE,* Vol. 109, Pt. B, Nov. 1962, pp. 431–444.
[23] Burberry, R. A., "The Rationalization of Aircraft Antennas," *IEE Int. Conf. Antennas for Aircraft and Spacecraft,* June 1975, pp. 204–209.
[24] Rudge, A. W., K. Milne, A. D. Oliver, and P. Knight, *The Handbook of Antenna Design,* Vol. 2, London: Peter Peregrinus, 1983.

[25] Suzuki, Y., and T. Chiba, "Designing Method of Microstrip Antenna Considering the Bandwidth," *IECE Japan Trans.,* Vol. E67, Sept. 1984, pp. 488–493.

[26] Yasunaga, Y., F. Watanabe, T. Shiokawa, and H. Yamada, "Phased Array Antennas for Aeronautical Satellite Communications," *IEE ICAP 87,* Part 1, March 1987, pp. 47–50.

[27] Kraus, J., "The Helical Antenna," *Proc. IRE,* Vol. 37, March 1949, pp. 263–272.

[28] King, H. E., and J. L. Wong, "Characteristics of 1 to 8 Wavelength Uniform Helical Antennas," *IEEE Trans. Ant. Propag.*, Vol. AP-28, March 1980, pp. 291–296.

[29] Wong, J. L., and H. E. King, "Empirical Helix Antenna Design," *IEEE AP-S Int. Symp. Digest,* May 1982, pp. 366–369.

[30] Nakano, H., J. Yamauchi, and H. Mimaki, "Tapered Balanced Helices Radiating in the Axial Mode," *IEEE AP-S Int. Symp. Digest,* June 1980, pp. 700–703.

[31] Angelakos, D. J., and D. Kajfez, "Modifications on the Axial-Mode Helical Antenna," *IEEE Proc.,* Vol. 55, April 1967, pp. 558–559.

[32] Dyson, J. D., "The Characteristics and Design of the Conical Log-Spiral Antenna," *IEEE Trans. Ant. Propag.,* Vol. AP-13, July 1965, pp. 488–499.

[33] Dyson, J. D., "The Unidirectional Equiangular Spiral Antenna," *IRE Trans. Ant. Propag.,* Vol. AP-7, Oct. 1959, pp. 329–334.

[34] Dyson, J. D., "The Equiangular Spiral Antenna," *IRE Trans. Ant. Propag.,* Vol. AP-7, April 1959, pp. 181–188.

[35] Kaiser, J. A., "The Archimedean Two-Wire Spiral Antenna," *IRE Trans. Ant. Propag.,* Vol. AP-8, May 1960, pp. 312–323.

[36] Kilgus, C. C., "Shaped-Conical Radiation Pattern Performance of the Backfire Quadrifilar Helix," *IEEE Trans. Ant. Propag.,* Vol. AP-23, May 1975, pp. 392–397.

[37] Pues, H. F., and A. R. Van De Capelle, "An Impedance-Matching Technique for Increasing the Bandwidth of Microstrip Antennas," *IEEE Trans. Ant. Propag.,* Vol. AP-37, Nov. 1989, pp. 1345–1354.

[38] Chiba, T., Y. Suzuki, N. Miyano, S. Miura, and S. Ohmori, "A Phased Array Antenna Using Microstrip Patch Antennas," *Proc. 12th European Micro. Conf.,* Sept. 1982, pp. 472–477.

[39] Hirasawa, K., and M. Haneishi, ed., *Introductory Analysis Design and Measurement of Small or Low-Profile Antennas,* Dedham, MA: Artech House, 1992.

[40] Weinschel, H. D., "A Cylindrical Array of Circularly Polarized Microstrip Antennas," *IEEE AP-S Int. Symp. Digest,* June 1975, pp. 177–180.

[41] James, J. R., and P. S. Hall, ed., *Handbook of Microstrip Antennas,* IEE Electromagnetic Waves Series 28, London: Peter Peregrinus, 1989.

Appendix
Glossary
K. Fujimoto and J.R. James

The demand for mobile antenna systems has brought with it an ever increasing variety of antennas, as discussed in Chapter 1 and amply illustrated in subsequent chapters. To the uninitiated, the various antennas often appear as distinct types having different fundamental behavior, but this is not necessarily so. In fact, there are surprisingly few generic forms, and even experienced antenna designers may ponder as to the purpose of some seemingly small modification to an otherwise conventional antenna configuration. For instance, does a spiral top-loaded monopole perform very differently from a circular plate top-loaded monopole? The fundamental electromagnetic properties that underlie this perpetual creation of antenna variants has been elaborated on elsewhere [1], where reasons are given explaining why such variety exists. In this glossary, we aim to help the antenna designer collate the likely types of antennas that might fit a given mobile requirement, and, furthermore, to narrow down the choice of antennas. We have kept in mind many special features of mobile antenna systems, which are typically electrically small because of size and shape constraints; furthermore, mobile antennas commonly have to operate on small, ill-defined ground planes and they may also embody additional system functions, such as diversity, polarization selectivity, and beam scanning. As far as possible, the catalog of types in Section A.1 attempts to identify generic forms such as dipoles and loops, but in many cases it is obvious that a simple classification is likely to be ambiguous and serves no useful purpose. The catalog also includes some mention of related issues such as baluns, ground plane effects, and array techniques. Antennas with little or no relevance to mobile systems are excluded, and this includes large parabolic reflectors, large phased arrays, and large broadcast transmitting antennas.

The tables in Sections A.2 to A.5 relate the antenna types to the nature of the mobile terminals, while Section A.6 gives a breakdown in terms of applications and frequency bands. The overlap of information between these sections is intentional, as are the alternative presentations, but we leave it to the designer to explore how best to sift the data when making an initial choice of antenna for a given mobile system requirement.

A.1 CATALOG OF ANTENNA TYPES

For ease of presentation, diagrammatic sketches are given, together with a literature reference which may be of historical or general interest, or perhaps, may be a more recent research reference. Note that the number in the text corresponds to both the number and the literature reference number up to reference [74].* A few additional references are included and annotated with a, b, c and so on (for example, [7a] is an additional figure reference to the spiral-loaded antenna [7]).

A.1.1 Dipole Derivatives

The wire dipole [2] and its unbalanced counterpart, the monopole [3], are universally known generic antennas from which have evolved a multitude of variants offering special features such as a more compact size, greater bandwidth, circular polarization, a desired impedance level, or particular physical characteristics. When each arm of the dipole or monopole is a quarter wavelength long, resonant action occurs and the resonant resistance is compatible with conventional transmission line feeders. When the arms are much less than a quarter of a wavelength, matching and efficiency problems are severe, and the feed region radiation can corrupt the overall pattern characteristics. Compactness of size can therefore only be obtained in exchange for some curtailment in performance. When physical size is not a constraint, many desirable characteristics can be achieved by altering the antenna geometry; that is, replacing the dipole or monopole wire with a flat metal plane or metal cone increases the impedance bandwidth, but the cross-polarization is also increased. A selection of dipole derivatives is as follows. The inverted-L antenna [4] readily gives a reduced height h at the expense of accommodating the horizontal sections of length l and decreasing the radiation resistance, bandwidth, and some change in the radiation patterns. The antenna is resonant when $l + h$ is near a quarter wavelength. A more symmetrical form of top loading is [5], while [6] is loaded with a circular plate. The spiral [7] top loading gives good control of resonance, but is a less rigid structure. The spiral can also be mounted on a planar substrate [7a,7b]. The inverted antenna [4] is often referred to as a transmission line antenna when an impedance Z_L loads the line [8]. This loading enables a shorter horizontal line to be deployed, hence saving on space. If Z_L is a short circuit, the structure is rigidly supported, but the horizontal section is then half a

*With the exception of reference [1]. The figures will appear on pages 552–565.

wavelength long. The inverted-F antenna [9] is a version of the inverted-L antenna with the additional freedom to tap the input along the horizontal wire to achieve some control of the input impedance [9a]. A parasitic inverted-L section can be coupled to the inverted-F antenna to increase bandwith [10]. In another version of the transmission line antenna, a boxlike structure has been derived [11]. The inverted-V antenna [12] offers another way of top loading, and when several wires are used, it is called the *umbrella antenna*. The hula-hoop antenna [13] is essentially an inverted-L antenna bent in a circular arc, and as such is more compact, but efficiency is very low. When the thickness of the wire is increased in all the above antennas, we can expect some increase in bandwidth. The Goubau antenna [14] invokes coupled circuits to give a large-impedance bandwidth, and four top-loaded monopoles are used with various coupling arrangements. The input impedance of a monopole can be isolated from the effects of a small ground plane by raising it on a sleeved section [15], but the structure is now a half wavelength in height. Short monopoles commonly contain a series loading coil to achieve resonance [16]. In the extreme, the loading coil itself can act as the radiator, and this is then the helical antenna [17], which gives essentially monopole radiation patterns when dimensioned to function in the normal mode. Antennas composed of compact inductors admit many variations, such as a counterwound version [18], giving wide bandwidth at low efficiency, a means of tapping into the helix to increase the input impedance [19], and a more compact form of transmission line antenna [20]. The quadrifilar helical antenna [21] consists of orthogonal windings excited in quadrature to generate circular polarization. A helical antenna designed to operate in the backfire mode [22] has a conical beam suitable for satellite communication reception on a vehicle. Lumped and continuous impedance loadings [23] have also been designed to modify radiation pattern characteristics. The mutual coupling between two small, wide monopoles can be used to create a wide impedance bandwidth by connecting the radiators by a transmission line Z_0 and magic T circuit [24]. Further freedom in design can be achieved by incorporating active devices in the antenna, assuming that noise and intermodulation constraints can be tolerated [25]. Diodes incorporated in a helical antenna [26] give wide-band electronic tuning at HF and VHF. Orthogonal dipoles [27] [27a] have useful polarization diversity characteristics, and considerable progress has been made in designing car window heaters with the capability to function also as antennas [28]. Leaky-cable radiators are used extensively in tunnel communication systems [29] and train telephone systems [29a].

A.1.2 Loop and Slot Derivatives

The small dipole antenna [2] is commonly referred to as an *electric dipole* in recognition of the fact that its fields are similar to those of the Hertzian electrical dipole. In the same way, a small loop [30] has the field characteristics of a Hertzian magnetic dipole. The electric and magnetic Hertzian dipoles exhibit duality in their field components, and this property is retained to a high degree in small dipoles and loops. The latter have, however, differing radiation resistances and Q factors for a given occupancy of space. The loop also behaves differently from a dipole in the presence of a reflecting object, and these properties are extremely useful in the design of antennas for pagers [30a], which have to function on or near the human body. The loop can also function in a monopolelike mode [31] in the presence of a ground plane. The loop yields a resonant input impedance when its circumference is one wavelength, but it is more typical to use a much smaller loop and tune it with a capacitor [32]. The loop need not be circular; for instance, a rectangular version of [31] and [32] is [33], which is identical to the tuned inverted-L or transmission line antenna [8]. In fact, it could be argued that [7–9, 16–20, 25] are derivatives of the loop rather than the dipole antenna. Multiturn loops embody mutual inductive coupling [34], but coupled isolated loops [35] offer better control of bandwidth which can be extended somewhat. A resonant slot in a ground plane is a useful radiator [36,37], and the excitation is applied across the slot at a and b. The polarity of the dipolelike fields can be deduced with Babinet's Principle [37a], showing that [36] and [37] are, respectively, dipole and loop radiators with E and H fields interchanged, and satisfying the conducting boundary conditions. A bent slot in a small ground plane [38] gives useful cross-polarization characteristics in multipath conditions, while the combination of a slot and patch antenna gives a compact diversity arrangement [39].

A.1.3 Material Loading

It is well known that most antennas can be detuned to a lower frequency when dielectric material is placed around the radiator. Appreciable size reduction in a monopole [40] and Yagi array [41] can be achieved, but the additional volume occupied by the material and the considerable increase in weight make this technique less attractive in many applications. Partial cladding, as in the skeleton discone antenna [42], is a useful compromise. Ferrite cladding of wire antennas [42a] gives enhanced performance compared to dielectric material, but suitable ferrites are only available at low frequencies. The common ferrite coil antenna [43] has a very low efficiency, but functions well for broadcast reception where the receiver is externally noise-limited. Slot antennas [44] can be size-reduced when fed by a material-loaded cavity, and ceramic material has been used in a compact top-loaded monopole [45]. The use of superconducting materials to fabricate electronically small dipoles and loops and their matching components drastically improves efficiency and noise performance at the expense of bandwidth and signal handling capacity [45a]. The use of pyroelectric polymer films to make sensitive probes at microwave frequency has been reported [46].

A.1.4 Printed Elements

The rectangular or circular patch microstrip antenna [47] is the basic printed element. In a sense, it is an extreme case of material loading whereby wave trapping occurs between the conducting patch and the ground plane; cavity action takes place with a leakage of radiation from the patch edges. For the basic rectangular patch, l is about a half wavelength long in the material, while h is typically a few millimeters. The radiation pattern is essentially that of a small loop antenna, but is modified by the contribution of the surface currents on the patch. Higher permittivity substrates give more compact radiators with less efficiency and bandwidth. The conducting patch can take numerous shapes, such as annular, elliptical, and triangular [48]. Higher cavity modes give multilobe patterns. Patch elements can be back-fed with a probe, excited by a coplanar line, excited by capacitive coupling, and so on. Some surface waves are generated in the surrounding substrate. An electric wall across the central region of a rectangular patch can be created by metal rivets [49] and enables the length l to be reduced to a quarter wavelength. This is particularly useful in portable equipment. Given appropriate excitation, patches can be angular segments of disks and annular rings [50]. Circular polarization can be created by generating both orthogonal modes in a symmetric patch, but only one feed is required in patches with appropriate perturbation notches [51,52], at the expense of a narrower bandwidth. Several techniques are available to increase the impedance bandwidth of printed antennas apart from using a low-permittivity substrate. Essentially, the techniques rely on coupled-circuit methods whereby parasitic patches are coupled to the main patch antenna by stacking [53] or planar proximity [54]. Clearly, the additional impedance bandwidth is achieved by an increase in space occupancy. Furthermore, the bandwidth of the radiation characteristics may be a constraint. The positioning of conducting posts in a patch [55] gives further control of the resonant frequency, and slots have similar advantages [55a]. Patches and ground planes can be composed of wire grids, offering many possibilities for multiband operation [56]. Hemispherical coverage with circular polarization has been reported for a circular element [57], and a bent patch on a small ground plane gives adequate cross-polarization for multipath operation [58]. Orthogonally placed quarter-wave elements are used for polarization diversity [59].

A.1.5 Balun Requirements and Imperfect Ground Planes

Coaxial cables are the most common means of interconnecting equipment units, and these unbalanced feeders can be connected directly to any antenna mounted on a ground plane [60]. Balanced antennas without a ground plane, such as a dipole and loop, require a balanced feed. A balun [61] is a device that converts a coaxial cable into a balanced feeder, and it takes a variety of forms. At lower frequencies, ferrite-loaded transformers and various component arrangements [62] can be utilized with tolerable losses, while bifilar winding techniques give tight coupling over a wide bandwidth [63]. At microwave frequencies, resonant transmission line lengths in baluns act as wave traps [64] and corporate feed phase inverters [65]. In practice, a ground plane may not be perfectly conducting or flat, and certainly not of infinite extent [66]. Radiation patterns and input impedances are modified by these effects depending on the extent to which the ground plane departs from ideal conditions. This is particularly important for antennas mounted on mobile units that are comparable in size to the antennas. Analytical solutions for a monopole on a sphere or cylinder [66a] indicate the nature of the perturbations in performance, while computational methods [66a] give information about specific geometries. An obstacle in close proximity to an antenna [67] will perturb the radiation patterns and input impedance characteristics. In some cases, conducting materials near an antenna can be used to enhance or improve the antenna performance. Analytical and computational methods [66a] are again a useful indication of practical situations.

A.1.6 Arrays and Diversity Systems

Circularly polarized patch antennas are clustered to form a conical system [68]. A circular patch operating in the TM_{210} mode gives similar characteristics [69], while circular elements with coupled parasitics are used in a 3 × 3 array [70] with beam control. Flat-plate direct broadcast satellite antennas [71] have been exhaustively addressed worldwide and have reached a high degree of engineering sophistication. Smaller printed arrays are important in airborne satellite receivers [72]. Beam squint techniques have been exploited for tracking systems [73]. Coupled stacked patches can operate in different modes in a novel beam scan arrangement [74], and strip-slot foam-inverted patch elements offer wider bandwidths [74a].

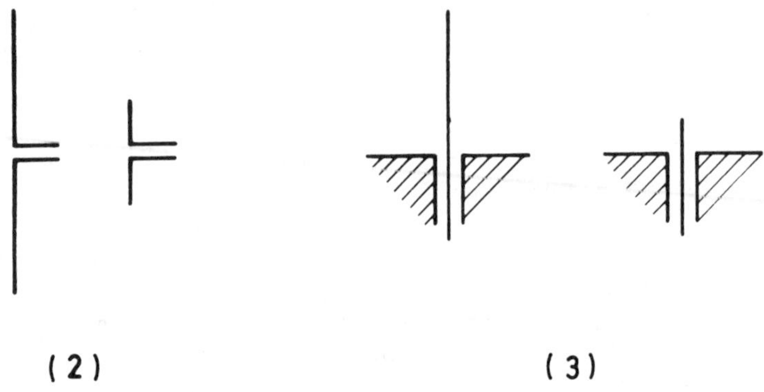

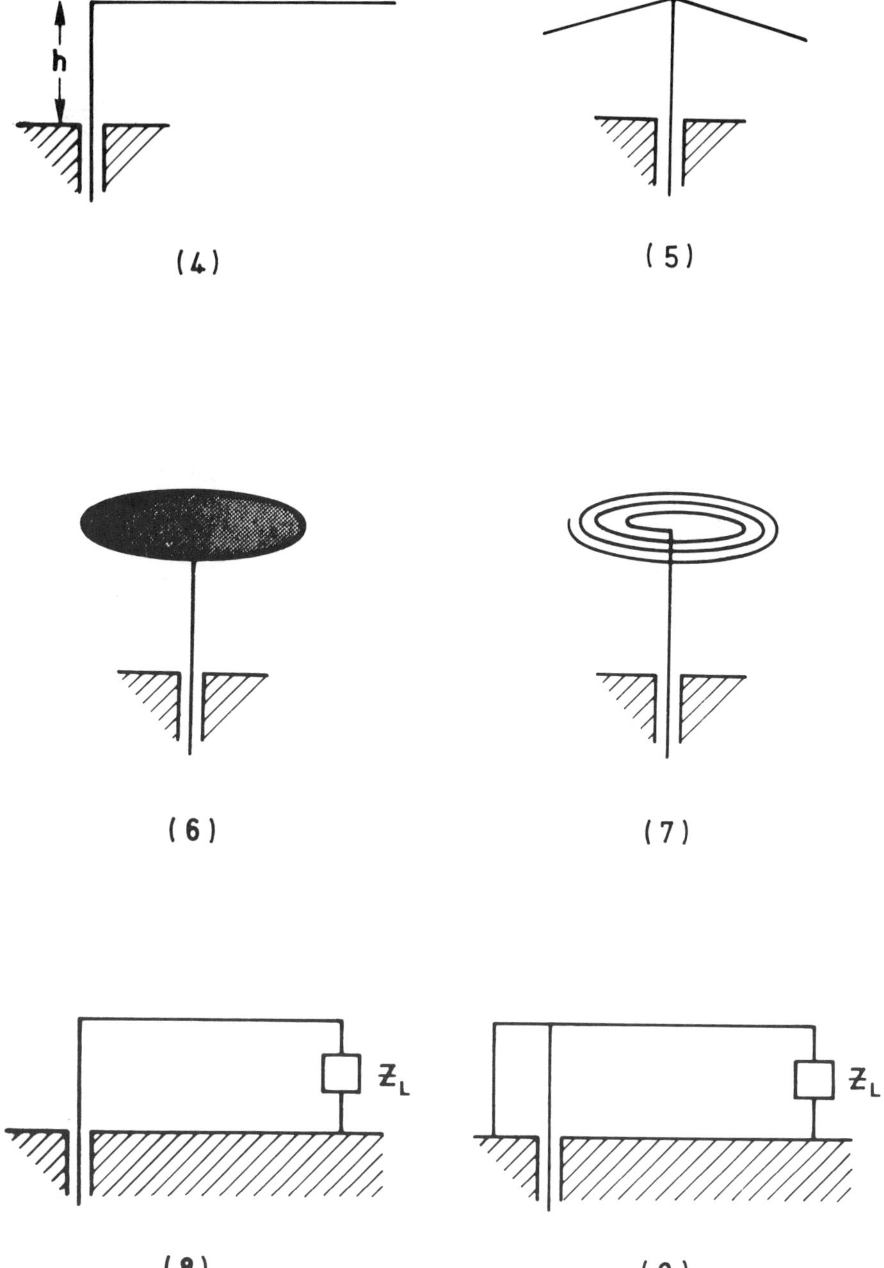
l
h
(4)
(5)
(6)
(7)
Z_L
(8)
Z_L
(9)

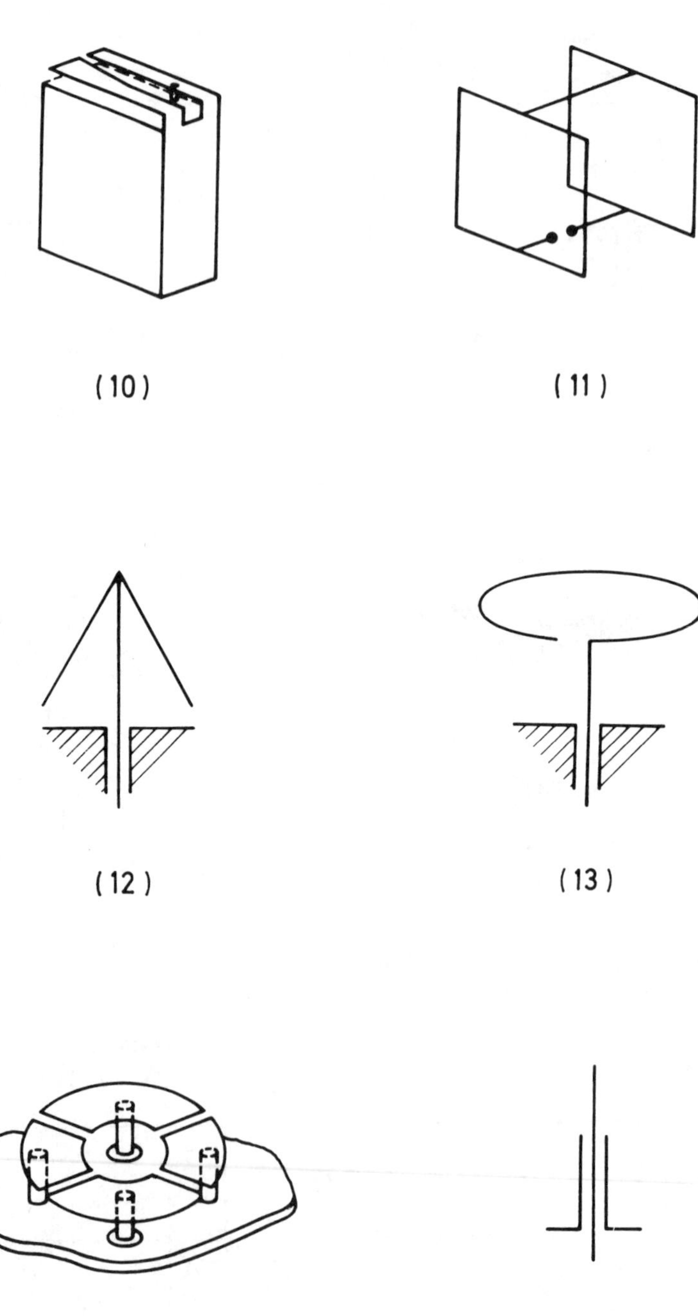
(10)
(11)
(12)
(13)
(14)
(15)

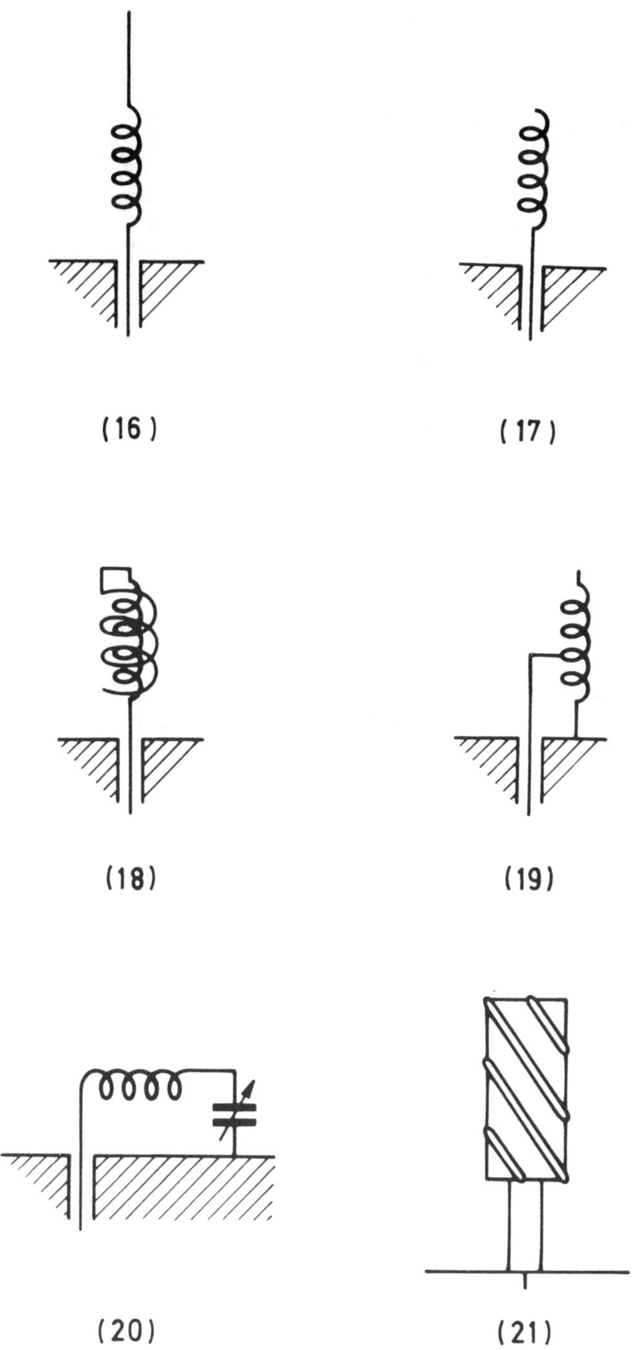
(16)
(17)
(18)
(19)
(20)
(21)

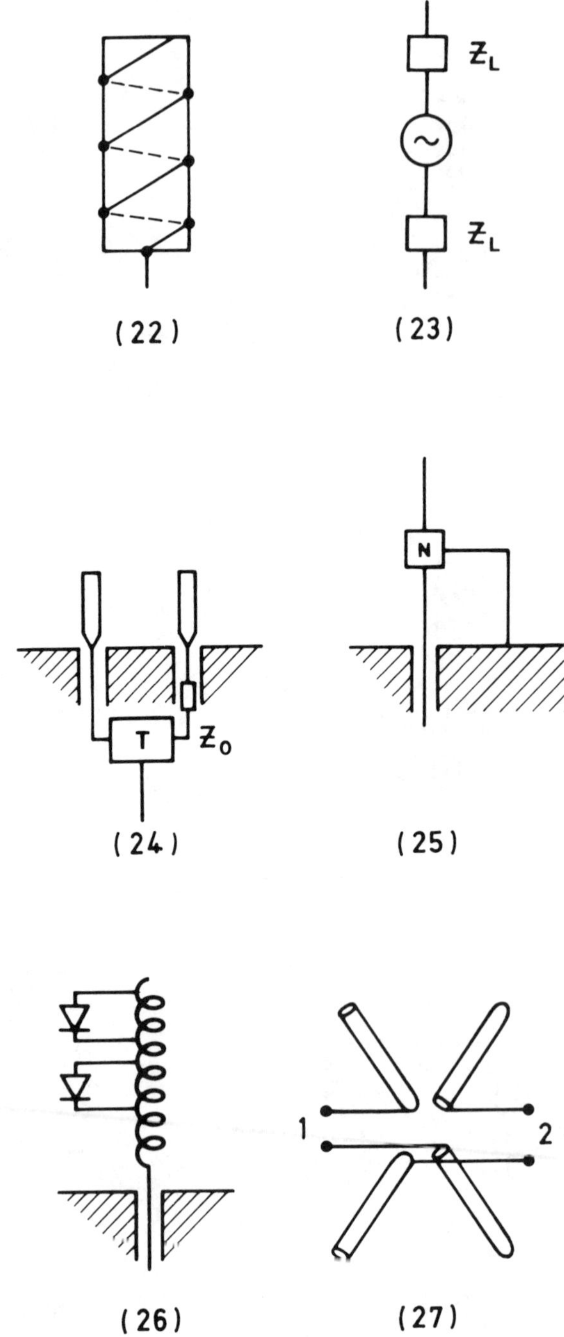

(22) (23)

(24) (25)

(26) (27)

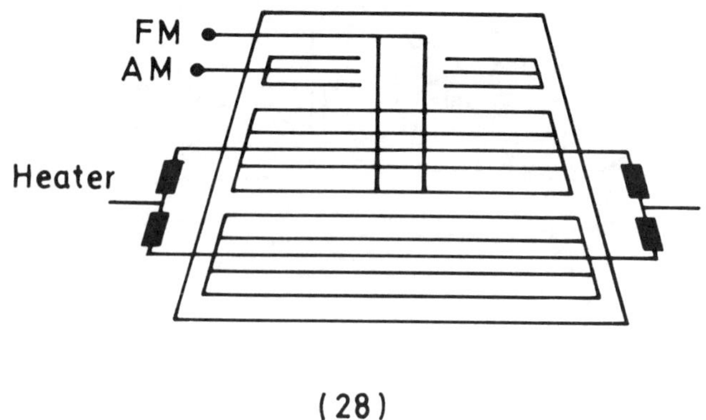

(28)

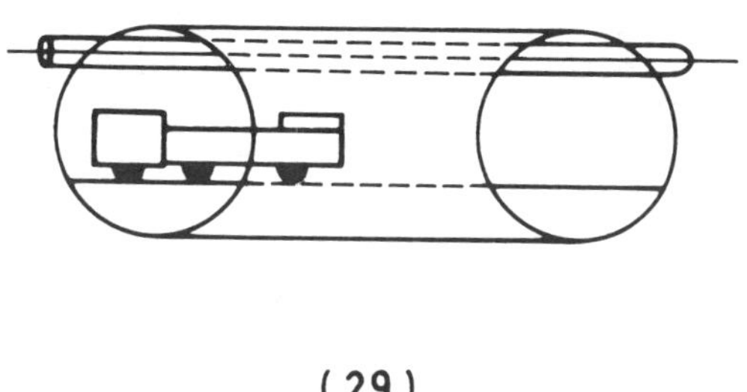

(29)

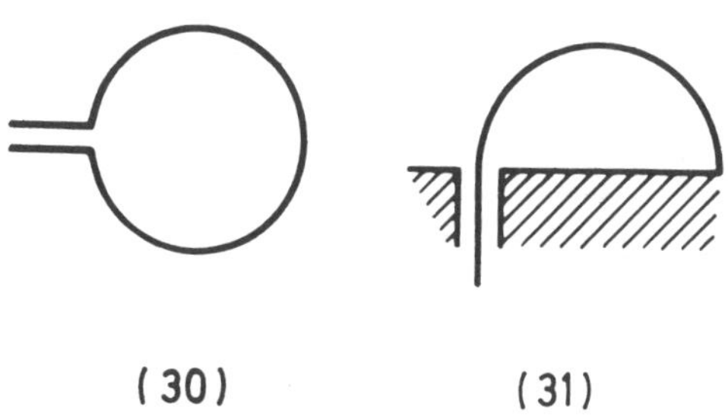

(30) (31)

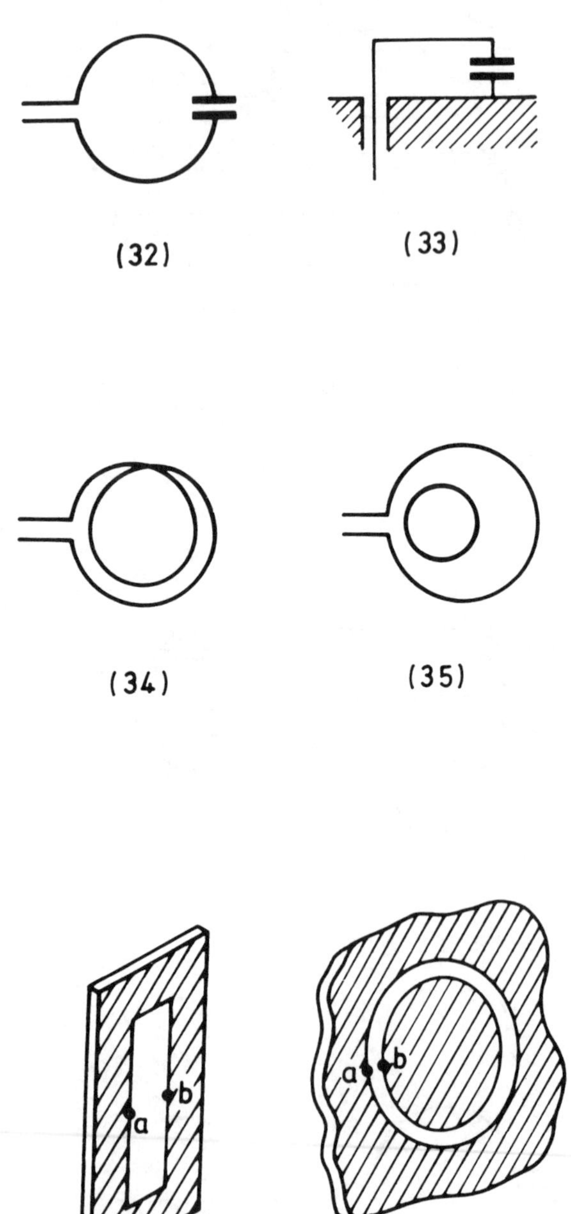
(32)
(33)
(34)
(35)
a
b
a
b
(36)
(37)

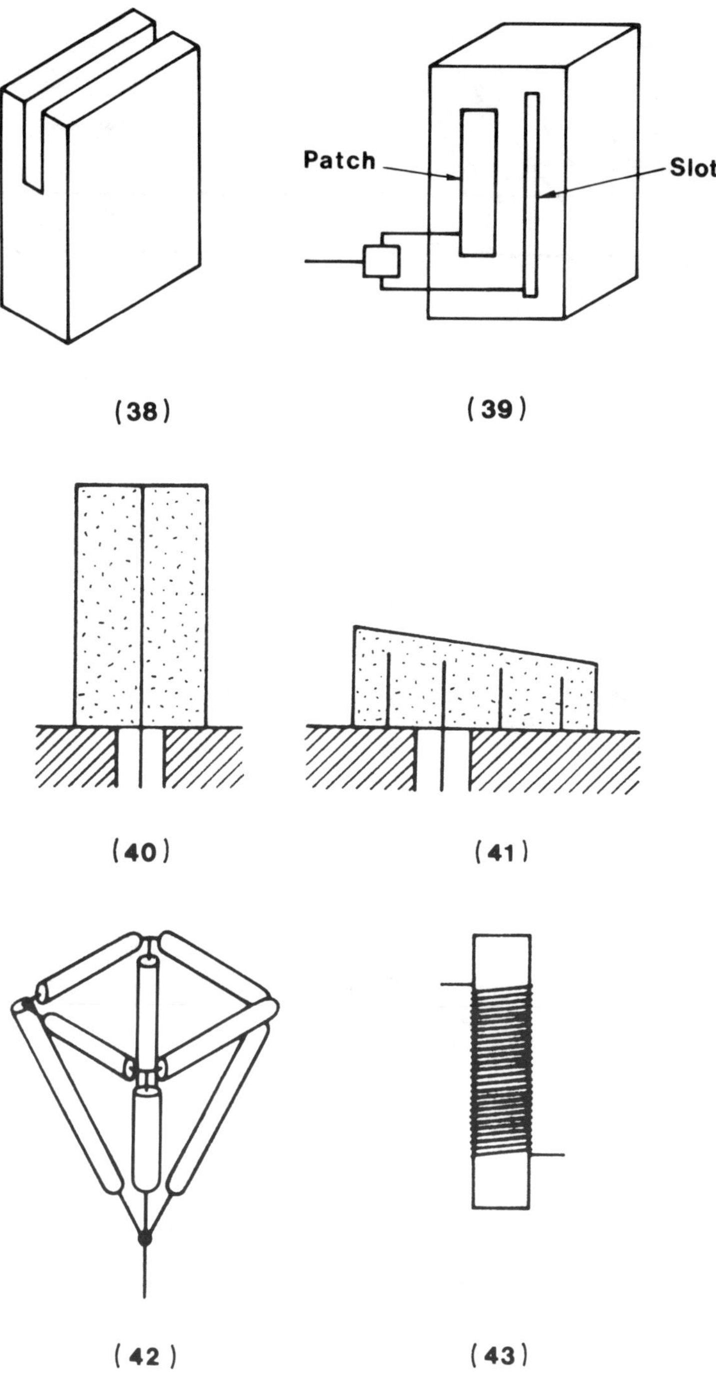
Patch
Slot
(38)
(39)
(40)
(41)
(42)
(43)

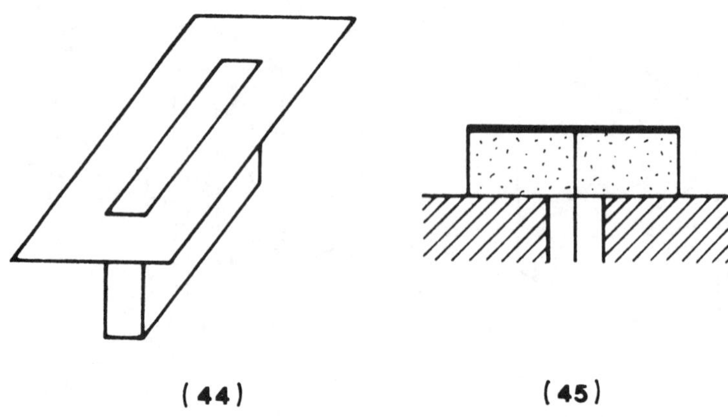
(44) (45)

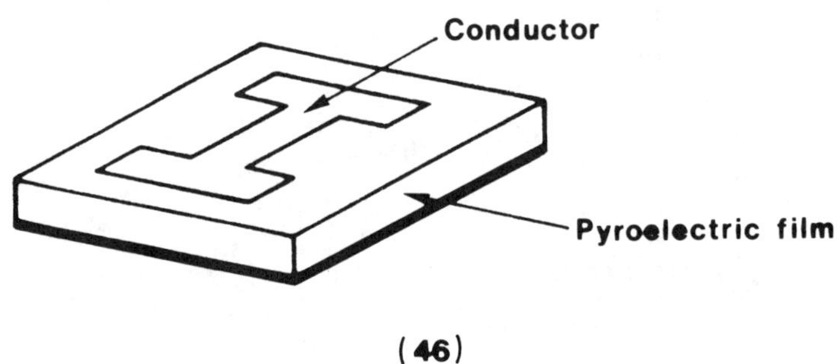

(46)

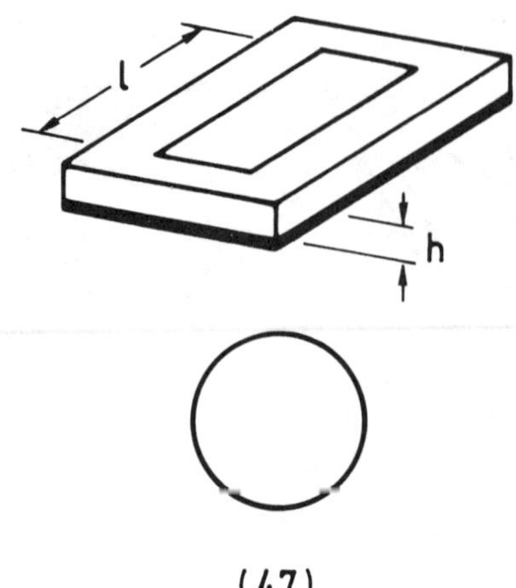

(47)

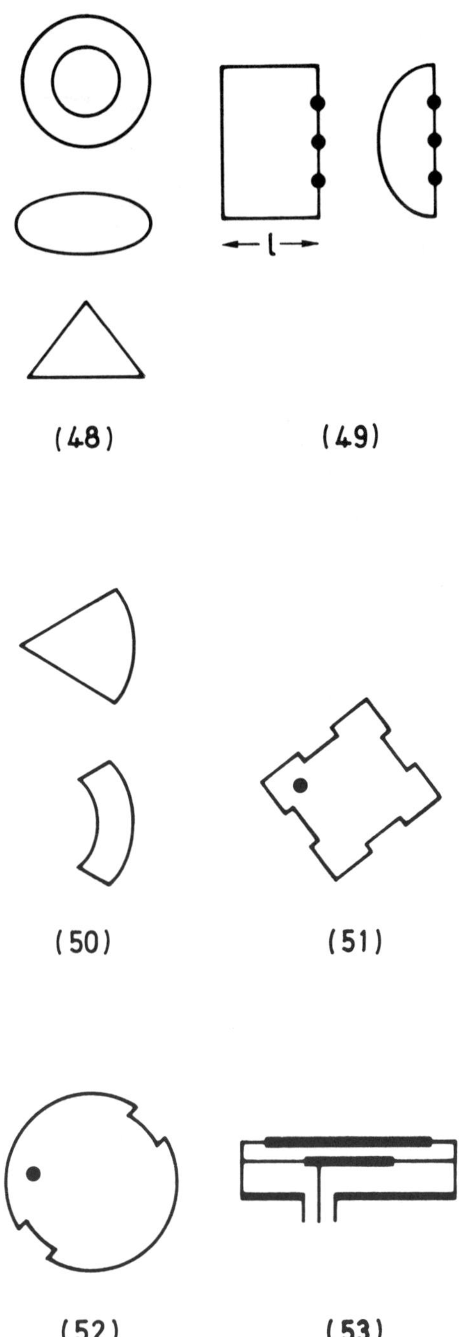
l
(48)
(49)
(50)
(51)
(52)
(53)

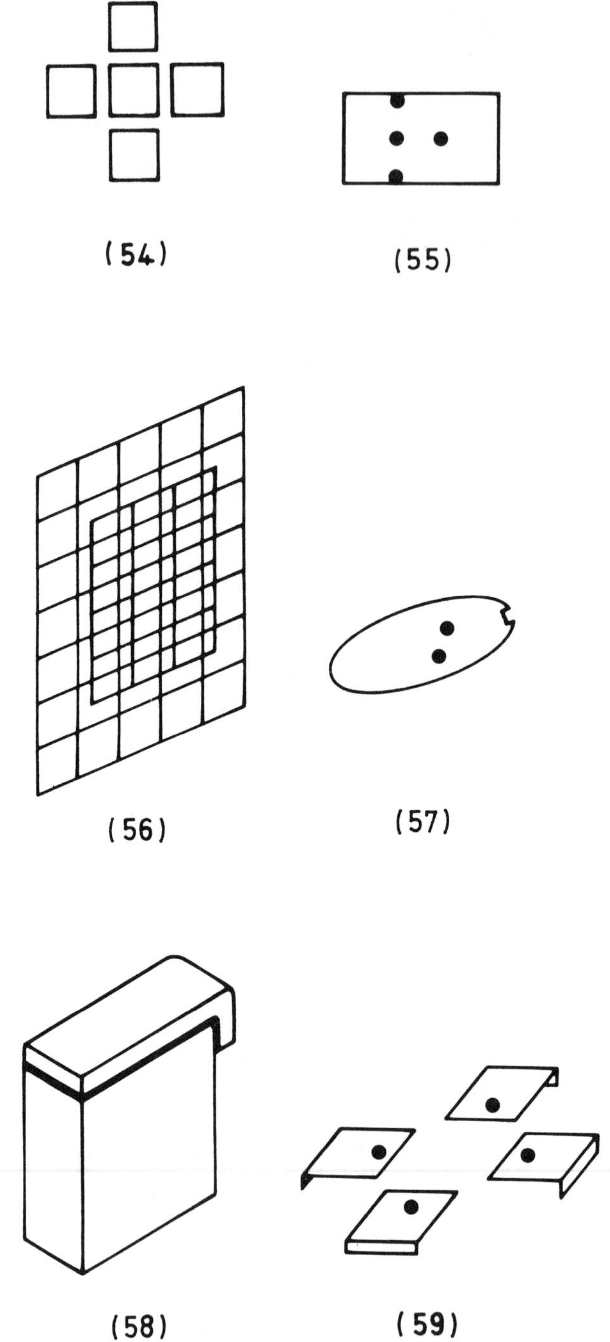

(54) (55)

(56) (57)

(58) (59)

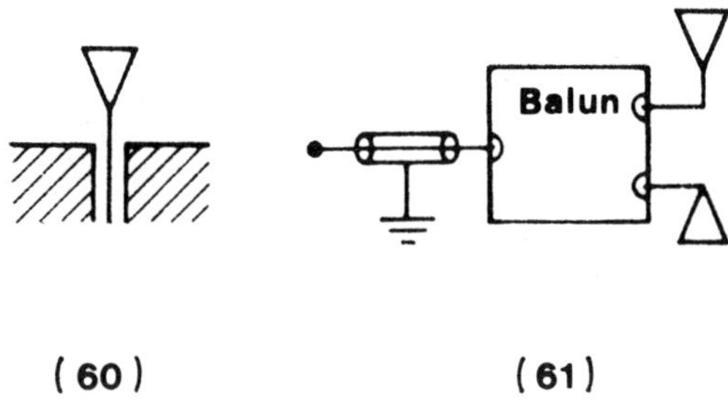

(60) (61)

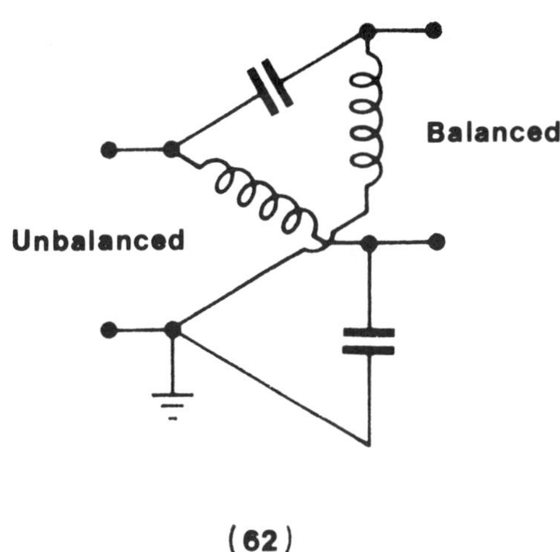

(62)

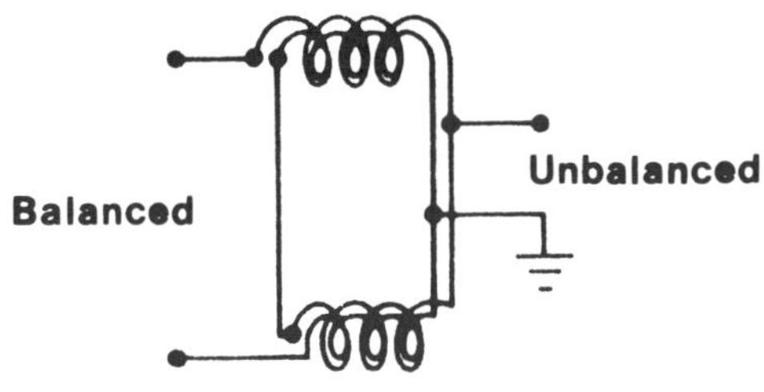

(63)

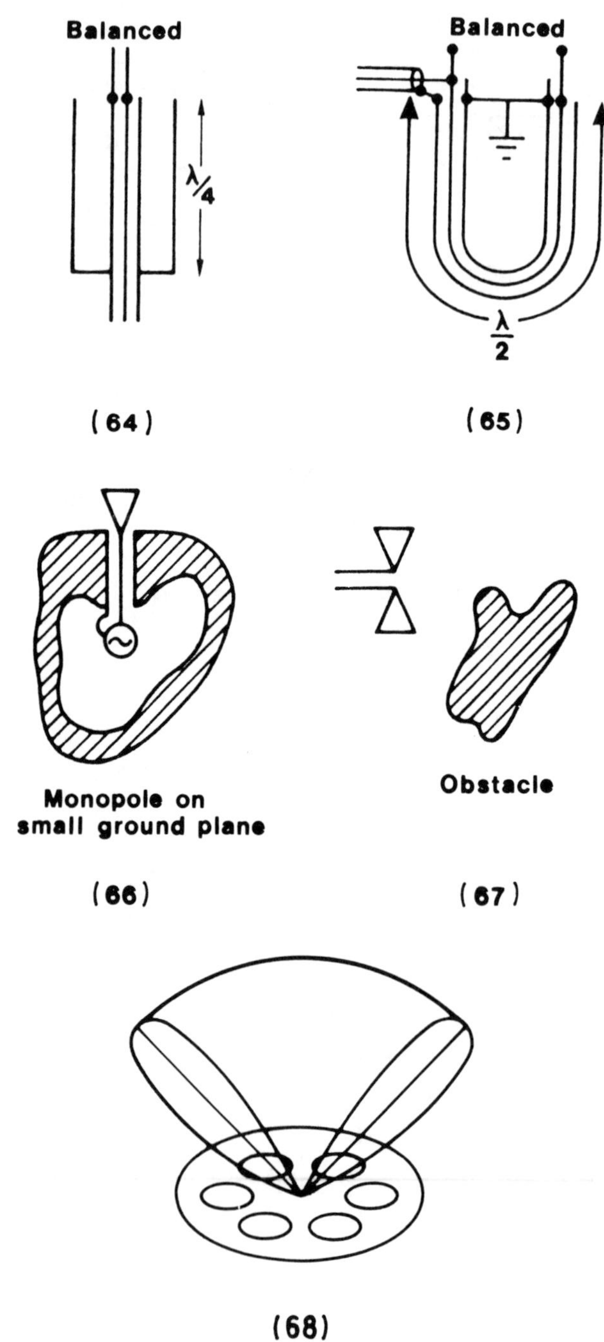
Balanced
λ/4
(64)
Balanced
λ
2
(65)
Monopole on
small ground plane
(66)
Obstacle
(67)
(68)

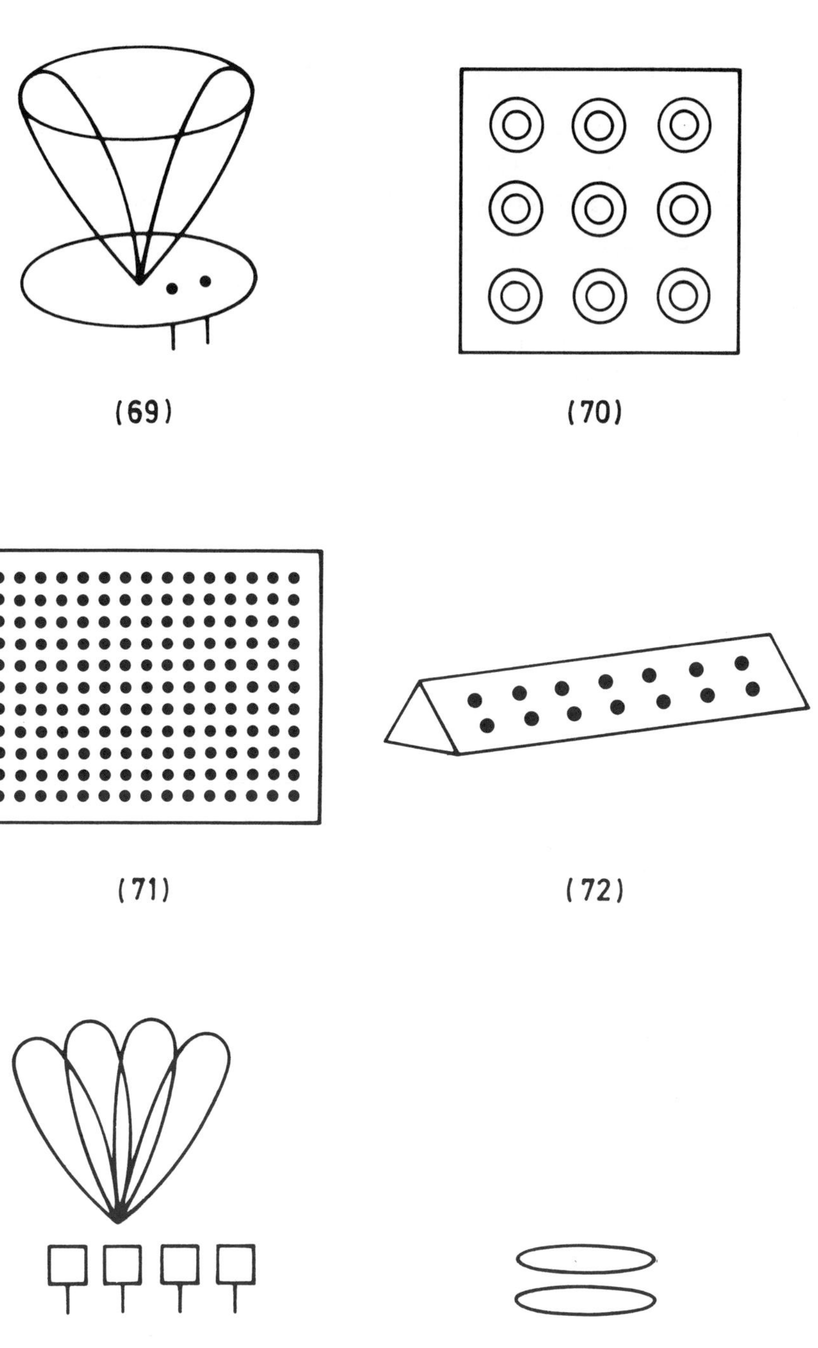
(69)
(70)
(71)
(72)
(73)
(74)

A.1.7 Key to Symbols and Acronyms Used in Sections A.2 to A.6

For compactness the following cataloging symbols are used to denote the type of mobile system and antenna. The first letter denotes the type of terminal, and the second symbol denotes the antenna type followed by the sequence number.

Examples

C-Mo-3	Automobile monopole antenna number 3
P-S-2	Portable slot antenna number 2
A-A-1	Aeronautical array number 1

Type of Mobile Terminals/Base Station

Automobiles and buses	C	Maritime	M
Portable equipment	P	Aeronautical	A
Train	T	Base station	B
Satellite mobile	S		

Types of Antennas

Monopole	Mo	Printed on glass	G
Dipole	D	Array	A
Inverted-L and F	I	Composite	C
Loop	Lo	Diversity and beam shaping	R
Planar	P	Slot	S
Microstrip	MS	Wire	W
Helix	H	Spiral	Sp
Ferrite	F	Reflector	Re

Antenna Acronyms

Short backfire	SBF	Inverted-L antenna	ILA
Normal-mode helix antenna	NMHA	Inverted-F antenna	IFA
Microstrip antenna	MSA	Singly fed circularly polarized	SFCP
Planar inverted-F antenna	PIFA		

Below, we provide a list of antennas, organized according to mobile terminal.

Brief View	*Type and Location*	*Technical Data*	*Cataloging Symbols and Relevant References*

A.2 LAND MOBILE SYSTEMS

A.2.1 Automobiles

Monopole

Brief View	Type and Location	Technical Data	Cataloging Symbols and Relevant References
① ②	Short monopole Chapters 3, 5	A single element (about 1/300λ for MF and 1/3λ for VHF), used commonly for AM/FM broadcast reception, matching circuits with very high inductive impedance necessary for both AM and FM receivers mounted on the front fender or trunk lid of a car	C-Mo-1 [3,75]
	Short monopole Chapters 3, 5	An element mounted along the pillar, AM/FM broadcast reception, telescopic element, simple element, better performance than other type	C-Mo-2 [75]
	Short monopole Chapters 3, 5	An element mounted on the bumper, inductive loading at the feeding point for the matching, mainly HF and VHF bands	C-Mo-3 [76–79]
	1/4λ monopole Chapters 3, 5	An element mounted on the roof, detachable, mainly used in VHF and UHF bands, almost omnidirectional horizontal pattern	C-Mo-4 [3,75,76,80]
	On-glass antenna Chapters 3, 5	Simple structure, EM coupling through window glass, directivity about 3 dBd, for car radios and mobile phones in 800 MHz	C-Mo-5 [81]

Brief View	Type and Location	Technical Data	Cataloging Symbols and Relevant References
Dipoles			
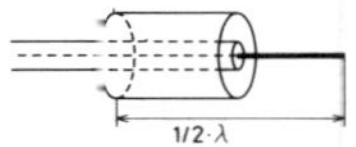 1/2·λ	λ/2 with a sleeve Chapters 3, 4, 5	A λ/4 element with a λ/4 sleeve, balun not necessary, used generally in VHF/UHF bands	C-D-1 [2,82,83]
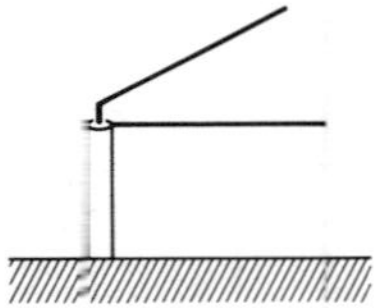	V-shaped Chapters, 5, 7	Horizontal polarization, directional pattern, low-drag structure, VHF/UHF bands, TV reception in cars	C-D-2 [84]
Inverted-L and Inverted-F			
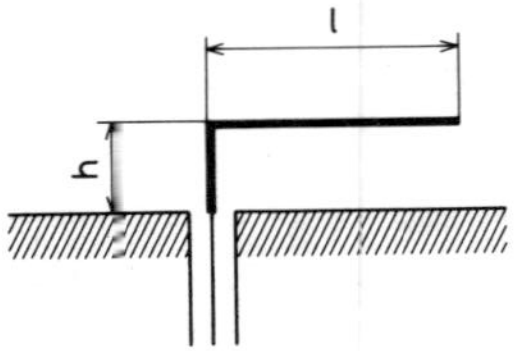 l, h	Short inverted-L $(h + \ell \ll \lambda/4)$ Chapters 3, 4, 5, 7	Low-profile, mainly HF bands, narrow bandwidth, matching circuit necessary, vertical and horizontal polarizations	C-I-1 [85,86]
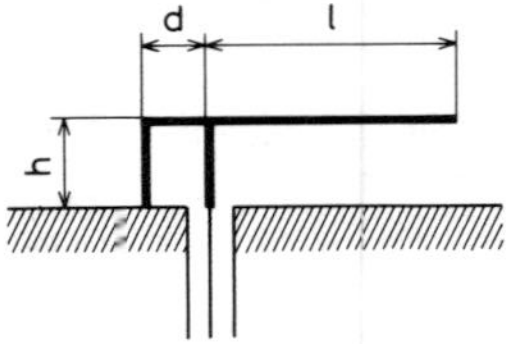 d, l, h	λ/4 inverted-F $(h + 1 \doteqdot \lambda/4)$ $d \ll \lambda$ Chapters 3, 4, 5, 7	Low-profile, VHF/UHF bands, narrow bandwidth, matching impedance obtained by adjusting d and h, both vertical and horizontal polarizations, used on cars, portable units, missiles, and so on.	C-I-2 [85]

Wires printed on glass

Type	Application	Ref.
Printed wire Chapter 5	AM/FM radios, printed on the rear window	C-G-1 [87]
Printed wire + monopole element Chapter 5	AM/FM radios, TV reception, diversity by a monopole and printed element on the rear window	C-G-2 [87]
Printed wire Chapter 5	AM/FM radios, diversity by two wire elements printed on the quarter-window and the rear window	C-G-3 [88]
Printed wires Chapter 5	AM/FM radios, TV reception, diversity by two wire elements printed on both side quarter windows (from [10, Ch. 5]. Reprinted with permission)	C-G-4 [88]

Loops

Type	Application	Ref.
Small loop Chapters 3, 5	A loop built in side mirror box, MF band, used for door remote control	C-Lo-1 [89]

Brief View	Type and Location	Technical Data	Cataloging Symbols and Relevant References
Planar			
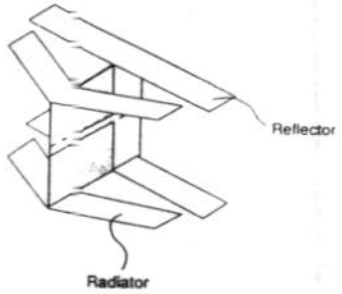	Printed dipole Chapters 3, 5	Thin, lightweight elements, VHF/UHF mobile telephones, dashboard mount, more than 8% bandwidth, semicircular horizontal radiation pattern (after [59, Ch. 3])	C-P-1 [90]
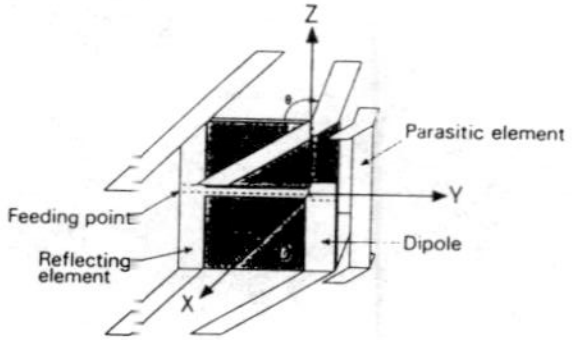	Dual-frequency-band folded printed dipole antenna Chapters 3, 5	Two frequency bands: 900-MHz and 1.5-GHz bands, thin, lightweight compact structure, semicircular horizontal radiation pattern, directivity (4.5 dBi) (from [62, Ch. 3], reprinted with permission)	C-P-2
Diversity			
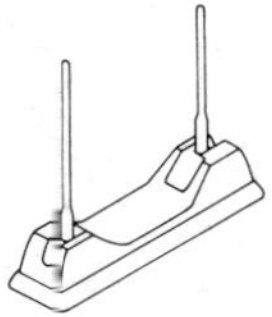	λ/4 monopoles Chapters 3, 5	Two vertical monopoles, mobile telephones in UHF bands, space diversity (horizontal), mounted on the roof or trunk or a car, detachable	C-R-1 [91,92]
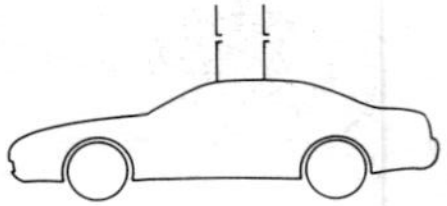	λ/2 dipoles Chapters 3, 5	Two vertical dipoles, UHF bands, space diversity (horizontal), mounted on the roof or trunk of a car	C-R-2 [93]

λ/2 collinear dipoles Chapters 3, 5	Two sleeve dipoles, VHF bands, space diversity (vertical), mounted on the bumper of a car	C-R-3 [94]
λ/2 collinear sleeve dipoles Chapters 3, 5	Higher gain than a dipole, UHF bands, space diversity (vertical), self-resonant (from [60, Ch. 3], reprinted with permission)	C-R-4 [76,95]
4-λ/2 elements with reflector Chapters 3, 5	Corner reflector to divide a zone into four, UHF bands, angle diversity	C-R-5
Dipole loop Chapters 3, 5	Combination of a monopole and a loop element, UHF bands, polarization diversity	C-R-6 [96]
Cross-dipole Chapters 3, 5	λ/2 dipole elements arrayed perpendicularly, polarization diversity	C-R-7
Dual-frequency-band antenna Chapters 3, 5	Sleeve dipoles with parasitic elements, two frequency bands: 900-MHz and 1.5-GHz bands, vertical diversity, trunk lid mount (after [61, Ch. 3])	C-R-8

Brief View	Type and Location	Technical Data	Cataloging Symbols and Relevant References
Microstrip patch			
Dielectric ε_1 ε_2 Ground plane Coaxial cable	Circular patch Chapters 6, 7	Circular polarization, low-profile, narrow bandwidth, semisphere coverage pattern, UHF bands, GPS	C-MS-1 [47,97]
	Circular patch with a parasitic element Chapters 6, 7	Mobile telephone, 800-MHz bands low-profile, vertical and horizontal polarizations, self-resonant, broadband	C-MS-2 [98,99]
Helix			
	Bifilar Chapters 6, 7	Satellite mobile telephone, conical-beam radiation pattern, balanced feed, backfire type, UHF bands	C-H-1 [100]
Composite			
	Land mobile/satellite mobile–compatible antenna Chapter 3	Frequency bands: 870–940 MHz/2.5/2.6 GHz, sleeve antenna and two-wire helical antenna combined, each antenna has different radiation pattern (after [63, Ch. 3])	C-C-1

Ferrite

Antenna	Description	Ref.
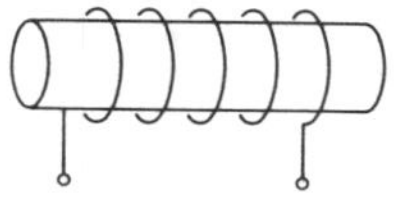 Cylinder Chapter 5	Very small, narrow bandwidth, LF band, inductive communication, for receiving and transmitting use, bus-roadside communications	C-F-1

A.2.2 Portable Equipment

Monopole

Antenna	Description	Ref.
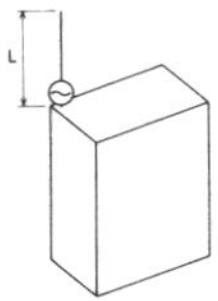 $\lambda/4$ monopole ($L = \lambda/4$) Chapter 4	A $\lambda/4$ monopole mounted on the body of a unit, simple, popular in VHF and UHF portable units, the body of the portable unit included in the antenna system as a part of radiator	P-Mo-1 [9,101]
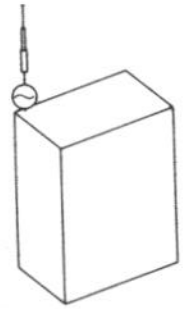 $\lambda/4$ (telescopic) Chapter 4	A telescopic $\lambda/4$ monopole, convenient for handling, VHF/UHF bands, cordless and portable phones	P-Mo-2
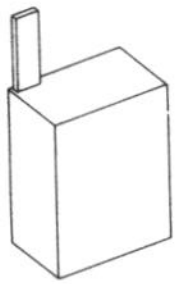 $\lambda/4$ (ribbon) Chapter 4	A ribbon-type element mounted on a portable unit, VHF/UHF bands, portable telephones, push-in pull-out type, flexible and thin, convenient for handling	P-Mo-3
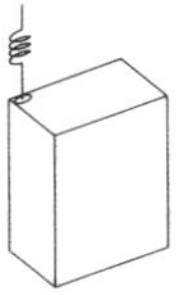 Less than $\lambda/4$ (inductance loaded) Chapter 4	A monopole element mounted on a portable unit, VHF bands, inductance-loaded, self-resonant, narrower bandwidth than a monopole	P-Mo-4 [16,78,79]

Brief View	Type and Location	Technical Data	Cataloging Symbols and Relevant References
See P-Mo-1, above	3/8λ, 5/8λ, 7/8λ Chapter 4	A monopole, not a λ/4 length, higher gain than λ/4 monopole, input resistance higher than 50Ω, inductive input reactance, matching circuit necessary, upward radiation pattern, portable telephones, VHF/UHF bands	P-Mo-5 [81]
Dipole			
	Sleeve (λ/2) Chapter 4	A λ/2 sleeve dipole mounted on a portable unit, Balun not necessary, portable telephones, two-frequency operation in UHF bands possible, less body effect than a λ/4 monopole	P-D-1 [2,101]
IFA			
	PIFA Chapter 4	A PIFA mounted on or built-into a portable unit, vertical and horizontal polarizations, portable telephones, relay transmitter, UHF bands	P-I-1 [102–104]
Helix			
	Normal-mode helical antenna Chapter 4	NMHA mounted on a portable unit, shorter than a λ/4 monopole, self-resonant, narrow bandwidth, flexible structure, portable telephone, UHF/VHF	P-H-1 [17,105]

Slot			
	Bent slot Chapter 4	Slot mounted on the top and sides of a portable unit, the length about $\lambda/2$, bandwidth 3% ~ 8%, vertical and horizontal polarization, self-resonant, portable telephones, semidiversity performance	P-S-1 [38]
	Bent slot Chapter 4	A bent slot mounted on the side of a portable unit, the length about $\lambda/2$, narrow bandwidth, self-resonant, portable telephones, UHF bands, vertical and horizontal polarization	P-S-2 [38]
Planar			
	Square patch Chapter 4	A square patch with one end shorted, length about $\lambda/4$, UHF bands, narrow bandwidth, vertical polarization, self resonant, portable telephones, pagers	P-MS-1 [97]
Microstrip Patch			
	Square patch Chapter 4	A square patch with one end shorted, length about $\lambda/4$, UHF bands, narrow bandwidth, vertical polarization, self resonant, portable telephones, pagers	P-Ms-1 [97]
Composite			
	Monopole PIFA Chapter 4	Diversity by a monopole and PIFA, mobile telephones, UHF bands, element spacing of 0.1λ possible with low correlation factor	P-C-1 [109]

Brief View	*Type and Location*	*Technical Data*	*Cataloging Symbols and Relevant References*
Loop			
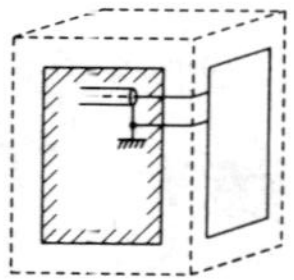	Small rectangular Chapter 4	Electrically small loop, self-resonant, VHF/UHF bands, narrow bandwidth, composite mode: loop plus dipole modes, omnidirectional receiving pattern, sensitivity improved operated near a human body, small volume and lightweight, pager	P-Lo-1 [110]
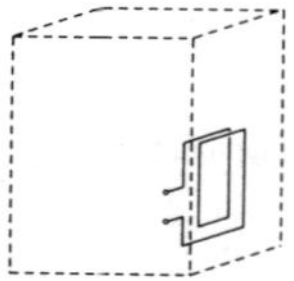	Small rectangular (2 turns) Chapter 4	Multiturn small loops, VHF/UHF bands, pagers, almost same performance as the above	P-Lo-2 [111]
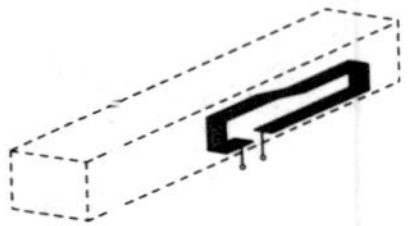	Small rectangular (modified) Chapter 4	A ribbon type element, VHF/UHF, pagers (pencil type), almost same performance as the above	P-Lo-3 [110]
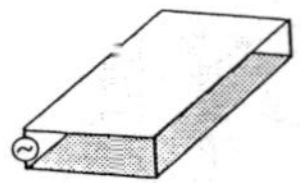	Parallel plate (magnetic loop) Chapter 4	Aperture between two plates acts as a magnetic loop, 280 MHz, receiving pattern changes according to position of the short point, pager (card type)	P-Lo-4 [112]
Ferrite			
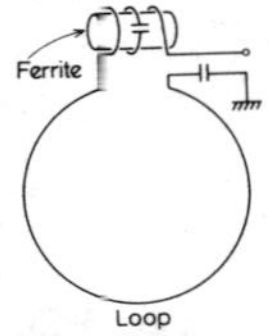	Chapter 4	Wrist-carried pager, a loop antenna connected in series with a ferrite-cored loop antenna, VHF bands	P-F-1 [113]

Ferrite	Chapter 4	MF bands, very small size, very high permeability, very low efficiency, portable radio	P-F-2 [111,114]

A.2.3 Trains

	Inverted-L Chapter 5	An inverted element mounted on the roof of the train, 150-MHz bands, communication with nearby stations only	T-I-1 [85,115]

A.2.4 Base Stations

	Dipole Chapter 3	A printed dipole with feeding circuit integrated, 180° beamwidth with corner reflector, UHF bands, telephone, mass production feasible	B-D-1 [97,116]
	Array Chapter 3	Three-dipole element triangle array, pattern shaping, diversity, UHF bands, telephone	B-R-1 [117]
printed dipole	Corner reflector array antenna with printed dipole Chapter 3	16 printed dipole elements, beam tilt, frequency band: 870~940 MHz, beamwidth controllable by adjusting corner angle	B-R-2
microstrip	Microstrip patch array antenna Chapter 3	MSA array, electrical beam tilt downward, frequency band 810 ~ 940 MHz, narrow beamwidth	B-R-3

Brief View	*Type and Location*	*Technical Data*	*Cataloging Symbols and Relevant References*
	Dual-frequency band corner reflector Chapter 3	A dipole with a parasitic element and corner reflector, dual frequency bands, narrow beamwidth from 60° ~ 150° by adjusting corner length L (after [23, Ch. 3])	B-R-4
	Pattern diversity antenna with omnidirectional pattern Chapter 3	Four dipole elements combined to perform pattern diversity, each element with omnidirectional pattern, element spacing can be small, mobile telephones	B-R-5
	Polarization diversity array antenna Chapter 3	MSA patch array, polarization diversity, thin and compact configuration, beamwidth controllable by adjusting element spacing, mobile telephones (after [32, Ch. 3])	B-R-6

A.3 MARITIME SYSTEMS

Brief View	*Type and Location*	*Technical Data*	*Cataloging Symbols and Relevant References*
	Crossed-drooping dipole Chapter 6	Two crossed-drooping elements, omnidirectional pattern, low gain (4 dBi), L-band, INMARSAT-C	M-D-1
	Quadrifilar helix Chapter 6	Long-structured helix, omnidirectional pattern, low gain (3 dBi), L-band, INMARSAT-C	M-H-1

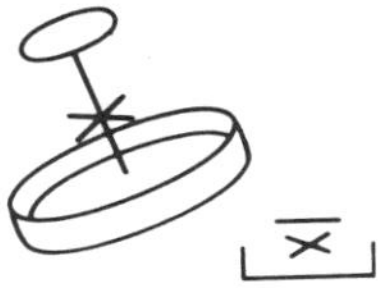	SBF Chapter 6	Backfire antenna with disk reflectors, high efficiency, compact, low sidelobe, INMARSAT-M, ETS-V, PROSAT	M-Re-1
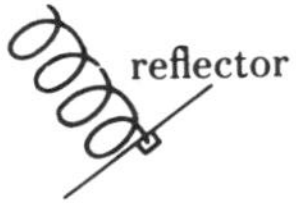 reflector	Helical antenna Chapter 6	Axial mode, wide-band, circular polarization, gain: 8 ~ 15 dBi	M-H-2
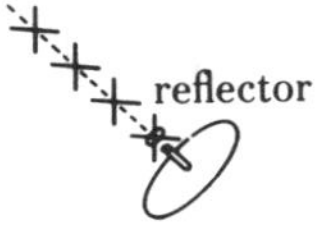 reflector	Yagi-Uda array of cross-dipoles Chapter 6	Endfire array, circular polarization, mediun gain: (8 ~ 15 dBi)	M-A-1
	Improved SBF Chapter 6	SBF with two small reflectors, high aperture efficiency, low sidelobe pattern, wide-band, VSWR	M-Re-2
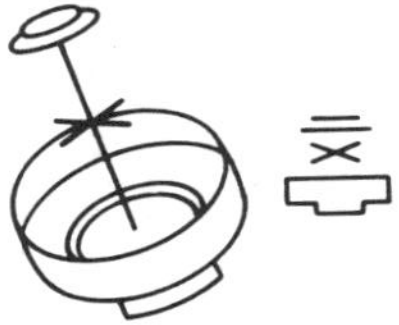	Modified SBF Chapter 6	SBF with two small reflectors, high aperture efficiency, low-sidelobe pattern	M-Re-3
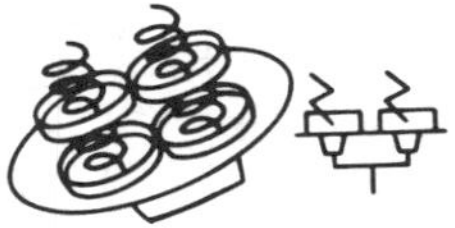	Quad-helix array Chapter 6	Array of four-turn helices, high aperture efficiency, circular polarization	M-A-2

Brief View	Type and Location	Technical Data	Cataloging Symbols and Relevant References
	Four-element SBF array Chapter 6	High aperture efficiency, circular polarization, gain: 18 ~ 20 dBi	M-A-3
	Cross-dipole array Chapter 6	Array of 16 cross-dipoles, circular polarization, medium gain: ~15 dBi	M-A-4

A.4 AERONAUTICAL SYSTEMS

Dipole

Brief View	Type and Location	Technical Data	Cataloging Symbols and Relevant References
	Cross dipole Chapters 6, 7	Inverted V-form with thick elements, Ku-bands, circular polarization with good axial ratio over wide angle, satellite communications	A-D-1
	Ramshorn Chapter 7	Horizontal V- or U-form, dragless structure, VHF bands, horizontal polarization, VOR, localizer	A-D-2 [118]

Monopole

Type	Description	Ref.
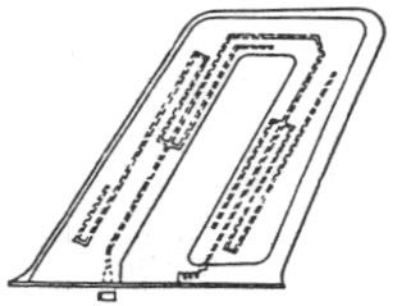 Blade Chapter 7	A flush-mounted antenna of printed stripline elements, vertical polarization, VHF bands, communications (after [17, Ch. 7])	A-Mo-1 [119]
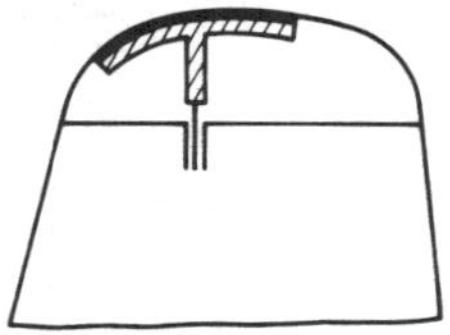 Pickax Chapter 7	Tail-cap installation, vertical polarization, VHF bands, communications	A-Mo-2 [122]
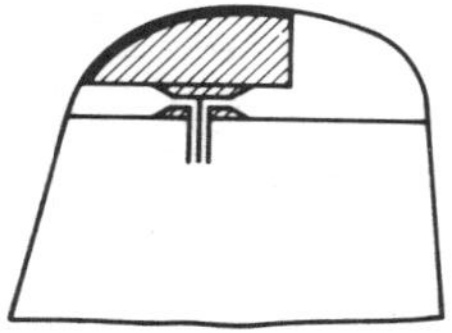 Mailbox Chapter 7	Tail-cap installation, vertical polarization, VHF bands, communications	A-Mo-3 [122]
 Loaded wire Chapter 7	A type of transmission line antenna, suitable for lower speed aircraft, horizontal polarization, HF bands, communications	A-W-1
 Trailing wire Chapter 7	VLF/LF bands, suitable for lower speed and large-size aircraft, mainly horizontal polarization, communications	A-W-2 [124,125]
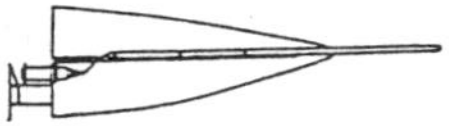 Probe Chapter 7	A type of isolated-cap antenna, dragless structure, characteristics depend on entire airframe, HF bands, communications	A-Mo-4 [126]

Brief View	*Type and Location*	*Technical Data*	*Cataloging Symbols and Relevant References*
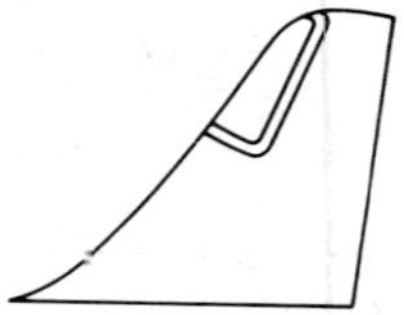	Tail-cap antenna Chapter 7	Flush-mount, dragless structure, characteristics depend on entire airframe, HF bands, communications	A-S-1 [118]
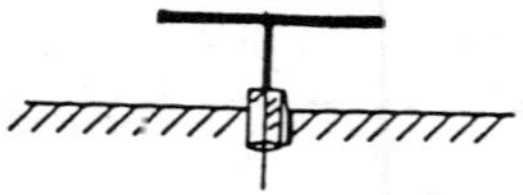	Top-loaded monopole Chapter 7	A short monopole with a disk loaded, low-profile structure, narrow bandwidth, LF/MF bands, ADF sense antenna	A-Mo-5 [119]
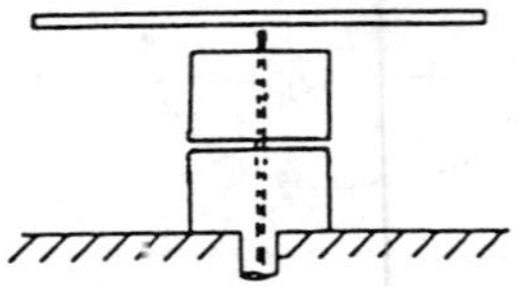	Multiband sleeve Chapter 7	A top-loaded sleeve with a sleeve in a fin cap, multifrequency operation in VHF/UHF bands, communications	A-Mo-6 [120]
	Bent sleeve Chapter 7	An inverted-L-shaped sleeve, dragless structure, vertical polarization, VHF bands, communications	A-Mo-7 [121]
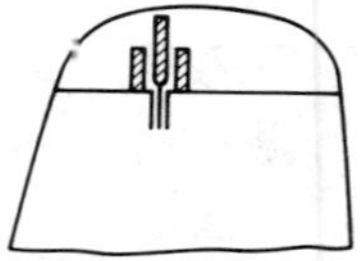	Partial sleeve Chapter 7	A monopole partially sleeved, tail-cap installation, vertical polarization, VHF bands, communications	A-Mo-8 [122]

Loop

Type	Description	Reference
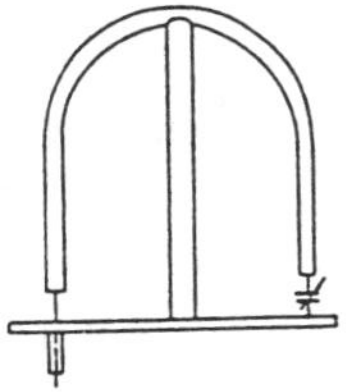 Half-loop loaded with a reactance Chapter 7	Small series-resonant half loop, horizontal polarization, UHF bands, glide path	A-Lo-1 [123]
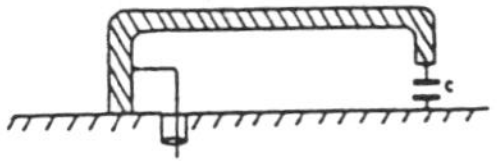 Vertical loop Chapter 7	A half loop loaded with a reactance, low-profile structure, polarization parallel to the axis of fuselage, VHF bands, marker beacon	A-Lo-2 [123]
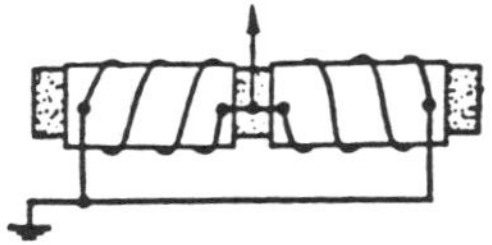 Loop wound on a ferrite core Chapter 7	Small loop antenna, narrow bandwidth, LF/MF bands, ADF	A-F-1 [119,127]
 Terminated loop Chapter 7	A loop terminated with a resistance, small dimensions, cardioid pattern, VHF/UHF bands, homing	A-Lo-3 [128]

Slot

Type	Description	Reference
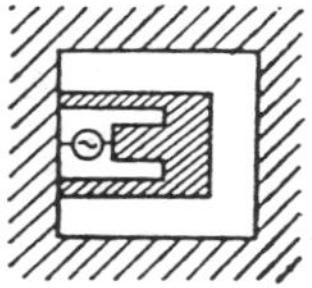 E-slot Chapter 7	Buried structure, cavity backed, horizontal polarization, VHF/UHF bands, VOR, localizer	A-S-2 [118]

Brief View	Type and Location	Technical Data	Cataloging Symbols and Relevant References
	Crossed slot Chapters 6, 7	Flush-mounted, cavity backed, circular polarization, L-bands, satellite communications	A-S-3 [129]
	Annular slot Chapter 7	One-wavelength circumference loop, vertical polarization, flush-mounted, cavity-backed, communications	A-S-4 [122,130]
Notch			
Notch	Standard notch Chapter 7	Buried structure, normally inductive impedance and small radiation resistance, narrow bandwidth, HF/VHF/UHF bands, communications	A-S-5 [121,131]
	Small gap Chapter 7	Body mount (a gap is driven), vertical polarization, LF/MF bands, loran	A-S-6 [123]
Helix			
	Cylindrical helix Chapter 6	Monofilar or multifilar, a normal, axial, or conical mode, circular polarization without hybrid, L-bands, satellite communications	A-H-1 [132–136]

		Conical helix Chapter 6	HPBW on the order of 100-deg, circular polarization without hybrid gain (4 to 7 dBi), L-band, satellite communications	A-H-2
		Quadrifilar helix Chapter 6	Two bifilar helices, high front-to-back ratio without ground plane, circular polarization with hybrid, narrow band (3% to 5% for VSWR $\leqq 2$), L-band, satellite communications	A-H-3 [137]
Spiral				
		Conical spiral Chapters 6, 7	Bifilar with self-complementary, circular polarization without hybrid, frequency-independent, L- to Ku-band, satellite communications	A-Sp-1 [138,139]
		Equiangular planar spiral Chapters 6, 7	Bifilar with logarithmic period, cavity-backed, flush-mounted, constant impedance of $60\pi\ \Omega$, circular polarization without hybrid, L- to Ku-bands, satellite communications	A-Sp-2 [140]
		Logarithmic planar spiral Chapters 6, 7	Bifilar with logarithmic period, cavity-backed, flush-mounted, circular polarization without hybrid, L- to Ku-band, satellite communications	A-Sp-3
		Archimedean Chapters 6, 7	Thin-wire bifilar, cavity-backed, flush-mounted, broadband, circular polarization without hybrid, L- to Ku-band, satellite communications	A-Sp-4 [141]

Brief View	*Type and Location*	*Technical Data*	*Cataloging Symbols and Relevant References*
Microstrip Patch			
	Standard patch Chapters 6, 7	A square or circular patch, conformal structure, self-resonant, circular polarization with hybrid, L- to Ku-band, satellite communications	A-MS-1 [142, 143]
	Thick patch Chapters 6, 7	A patch on a thick substrate with low dielectric constant, conformal structure, higher mode suppression using balanced feeding technique, circular polarization with hybrid, L-bands, satellite communications	A-MS-2 [142–146]
	Stacked patch Chapters 6, 7	A patch stacked with a parasitic element, dual resonance, higher mode suppression using balanced feeding technique, circular polarization with hybrid, L-bands, satellite communications	A-MS-3 [143]
	Stacked-patch dual-frequency operation Chapters 6, 7	Separate stacked configuration with individual feeding, dual frequency, self-duplexing, circular polarization with hybrid, L-bands, satellite communications	A-MS-4 [129]
	SFCP patch Chapters 6, 7	Circular polarization produced by orthogonal modes degenerated, with single feeding, narrow bandwidth, L- and Ku-band, satellite communications	A-MS-5 [142,143,145]

Reflector

	Parabola Chapter 7	Noncircular periphery mounted on the bottom of fuselage with radome, horizontal polarization, X-bands, weather radar	A-Re-1
	Cassegrain Chapter 7	Dual dishes, mounted on the nose, with radome, either normal pencil beam or modified cosec-square beam, also either circular or horizontal polarization, X-bands, weather radar	A-Re-2

Quadrifilar

See M–H–1 in A.3	Quadrifilar Chapters 6, 7	Omnidirectional radiation pattern, low gain, low air drag INMARSAT-Aero, GPS	A-H-4

Phased Array

	Phased array Chapters 6, 7	MSA arrays on two planes, mounted on the top of fuselage, gain: 12 dBi, airplane telephones, INMARSAT-Aero, ETS-V	A-A-1

A.5 SATELLITE SYSTEMS

Brief View	*Type and Location*	*Technical Data*	*Cataloging Symbols and Relevant References*
Land Mobile Systems			
	Quadrifilar Chapters 6, 7	Two one-and-half turn helices, circular polarization, UHF bands, mobile telephones, satellite mobile systems	S-H-1
	Conical spiral Chapters 6, 7	Spiral elements on a cone, circular polarization, wide bandwidth in 1.5-GHz bands, GPS	S-Sp-1
Maritime Mobile Systems			
See M–H–1 in A.3	Quadrifilar Chapter 6	Omnidirectional pattern, low gain, INMARSAT-C, GPS, AMSC, MSAT, Mobilesat ETS-V, MSAT-X	S-H-2
	Array mechanical tracking Chapter 6	Spiral element array, thin structure, medium gain: 15 dBi, directional pattern, INMARSAT-M, Mobilesat, ETS-V, MSAT-X	S-A-1

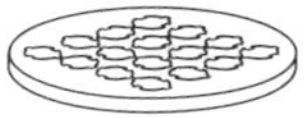	Phased-array electronic tracking Chapter 6	MSA patch phased array, thin structure, directional pattern, medium gain 15 dBi, ETS-V, MSAT-X	S-A-2
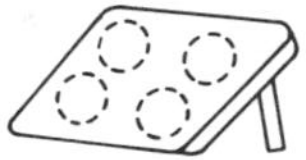	Patch array Chapters 6, 7	Four-element patch array, transportable antenna, INMARSAT-C	S-A-3
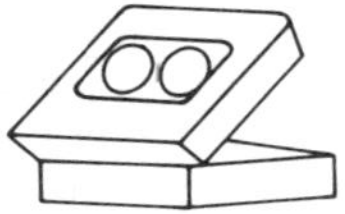	Patch array Chapter 6	Two-MSA patch array mounted on the lid of briefcase terminal, low-speed data transmission, ETS-V	S-A-4
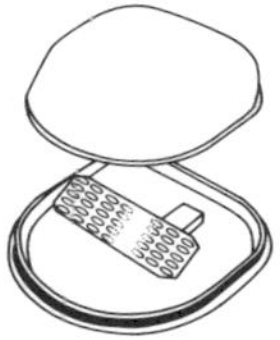	Planar-array mechanical tracking Chapter 6	MSA patch array, high gain (40 dBi), Ku-bands (12 GHz), TV and BS reception on a vehicle	S-A-5
	Parabolic mechanical tracking Chapter 6	A parabolic antenna, high gain (40 dBi), Ku-band (12 GHz), TV and BS reception	S-R-1
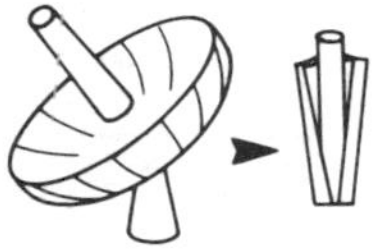	Umbrella-type antenna Chapter 6	A deployable parabola antenna, transportable, compact, lightweight	S-R-2

A.6 TYPICAL ANTENNA TYPES AND THEIR APPLICATIONS

Antenna Type	*Applications*					
	LF (to *0.3 MHZ*)	*MF* (to *3 MHZ*)	*HF* (to *30 MHZ*)	*VHF* (to *300 MHZ*)	*UHF* (to *3,000 MHZ*)	*SHF*
ILA (long wire)	Ship, airplane					
(small, rigid)				Train, airplane		
Monopole	Car (radio, AM)		Car (FM, TV)		Land vehicle, ship	
(1/4λ)			Portable, car, ship, airplane			
(3/8λ)				Car	Portable	
(5/8λ)						
Backfire					Ship (INMARSAT)	
IFA			Vehicle		Missile	
Helx (bifilar)					Satellite mobile	
(NMHA)				Portable		
(quadrifilar)					INMARSAT	
Loop	Car (remote control)			Pager		
(small square)				Pager, portable, telephone		
(multiturn)						
Ferrite	Portable radio (AM)					
				Pager		

PIFA	Portable telephone Missile			
MSA		Pager	Airplane telephone	Portable (data) terminal
Array (MSA) (Wire)	Car (diversity)		Mobile phone (base)	
Composite (diversity)			Portable telephone Mobile phone (base)	
Reflector		Telephone (base) Pager (base)		Airplane (radar)
Spiral			GPS, INMARSAT	
Slot		Portable Telephone		Airplane (radar)

REFERENCES

[1] James, J. R., "What's New in Antennas," *Proc. IEE Int. Conf. Ant. Propag.*, ICAP '89, Warwick, England, April 1989, pp. 84–97.

[2] Johnson, R. C., and H. Jasik, Ch. 4 in *Antenna Engineering Handbook*, 2nd edition, New York: McGraw-Hill, 1984.

[3] Ibid.

[4] Prasad, S., and R.W.P. King, "Experimental Study of Inverted L, T and Related Transmission-Line Antennas," *J. Research*, NBS, 65D, Sept./Oct. 1961.

[5] Simpson, T. L., "The Theory of Top-Loaded Antennas: Integral Equations for the Currents," *IEEE Trans.*, Vol. AP-19, 1971.

[6] Weeks, W. L., *Antenna Engineering*, New York: McGraw-Hill, 1968.

[7] Bhojwani, H. R., and L. W. Zelby, "Spiral Top-Loaded Antenna: Characterization and Design," *IEEE Trans.*, Vol. AP-21, 1973, pp. 293–298.

[7a] Fenwick, R. C., "A New Class of Electrically Small Antennas," *IEEE Trans.*, Vol. AP-13, 1965, pp. 379–383.

[7b] Nakano, H. "Research on Spiral and Helical Antennas," *IEEE AP-S Newsletter*, June 1988, p. 20.

[8] King, R.W.P., et al., "Transmission-Line Missile Antennas," *IRE Trans.*, Vol. AP-8, 1960, pp. 88–90.

[9] Hirasawa, K., and K. Fujimoto, "Characteristics of Wire Antennas on a Rectangular Conducting Body," *Trans. IECE Japan*, Vol. J65-B, 1982, pp. 1113–1139.

[9a] Fujimoto, K., A. Henderson, K. Hirasaura, and J. R. James, "Small Antennas," England: Research Studies Press, 1987, distributed by Wiley & Sons, p. 117–122.

[10] Rasinger, J., et al., "A New Enhanced Bandwidth Internal Antenna for Portable Communication Systems," *40th IEEE Vehicular Technology Conf.*, Orlando, May 1990, pp. 7–12.

[11] Tsukiji, R., and Y. Kumon, "On a Modified Transmission Line Type Antenna," *Proc. IEE Int. Conf. Ant. Prop.*, ICAP '91, York, England, April 1991, pp. 38–41.

[12] Gangi, A. F., et al., "The Characteristics of Electrically Short, Umbrella Top-Loaded Antennas," *IEEE Trans.*, Vol. AP-13, 1965, pp. 864–871.

[13] Burton, R. W., and R.W.P. King, "Theoretical Considerations and Experimental Results for the Hula-Hoop Antenna," *Microwave J.*, Nov. 1963, pp. 89–90.

[14] Goubau, G., "Multi-Element Monopole Antennas," *Proc. ECOM-ARO Workshop on Electrically Small Antennas*, G. Goubau and F. Schwering, ed., 6–7 May 1976, Fort Monmouth, pp. 63–67.

[15] Johnson, R. C., and H. Jasik, Ch. 4 in *Antenna Engineering Handbook*, 2nd edition, New York: McGraw-Hill, 1984.

[16] Hansen, R. C., "Optimum Inductive Loading of Short Whip Antennas," *IEEE Trans.*, Vol. VT-24, 1975, pp. 21–29.

[17] Hansen, L. H., "A New Helical Ground-Plane Antenna for 30 to 50 MHz," *IRE Trans.*, Vol. VC-10, 1961, pp. 36–39.

[18] Fujimoto, K., A. Henderson, K. Hirasaura, and J. R. James, "Small Antennas," England: Research Studies Press, 1987, distributed by Wiley & Sons, pp. 236–238.

[19] Ramsdale, P. A., and T.S.M. Maclean, "Active Loop-Dipole Aerials," *Proc. IEE*, Vol. 118, 1971, pp. 1697–1710.

[20] Smith, G. S., "Radiation Efficiency of Electrically Small Multi-Turn Loop Antennas," *IEEE Trans.*, Vol. AP-20, 1972, pp. 656–657.

[21] Kilgus, C. C., "Resonant Quadrifiler Helix Design," *Microwave J.*, Vol. 13, 12 Dec. 1970, pp. 49–54.

[22] Nakano, H., et al., "Generation of a Circularly Polarized Conical Beam From Backfire Helical Antennas," *Proc. Int. Conf. Ant. Propag.* ICAP '91, York, England, April 1991, pp. 42–45.

[23] Hirasawa, K., and K. Fujimoto, "On Electronically-Beam-Controllable-Dipole Antenna," *IEEE AP-S Int. Symp. Digest*, 1980, pp. 692–695.

[24] Schroeder, K. G., and K. M. Soo Hoo, "Electrically Small Complementary Pair (ESCP) With Interelement Coupling," *IEEE Trans.*, Vol. AP-24, 1976, pp. 411–418.

[25] Hiroi, Y., and K. Fujimoto, "Active Inverted-L Antenna," paper of technical grop TGAP 70-29, *IECE Japan*, 1970, pp. 1–12 (in Japanese). See also [19].

[26] Ploussios, G., "An Electronically Tuned HF/VHF Helical Antenna," *Microwave J.*, May 1991, pp. 223, 224, 227, 228, 231, 234, 237, 239, 240.

[27] Taga, T. et al., "Correlation Properties of Antenna Diversity in Indoor Mobile Communication Environments," *IEEE Vehicular Technology Conf.* 1989, pp. 446–451.

[27a] Gatti, M. S., and D. J. Nybakken, "A Circular Polarized Crossed Drooping Dipole Antenna," *IEEE AP-S Int. Symp. Digest*, Dallas, May 1990.

[28] Lindenmeier, H. K. et al., "Multiple FM Window Antenna System for Scanning Diversity With an Integrated Processor," *IEEE Vehicular Technology Conf.*, Orlando, May 1990, pp. 1–6.

[29] Delogne, P., *Leaky Feeders and Subsurface Radio Communications*, London: Peter Peregrinus, IEE, Vol. 14, 1982.

[29a] Kishimoto, T., and S. Sakai, "LCX Communication Systems," *IECEJ*, Tokyo, 1982.

[30] Johnson, R. C., and H. Jasik, Ch. 5 in *Antenna Engineering Handbook*, 2nd edition, New York: McGraw-Hill, 1984.

[30a] Fujimoto, K., A. Henderson, K. Hirasaura, and J. R. James, "Small Antennas," England: Research Studies Press, 1987, distributed by Wiley & Sons, pp. 100–107.

[31] Johnson, R. C., and H. Jasik, Ch. 5 in *Antenna Engineering Handbook*, 2nd edition, New York: McGraw-Hill, 1984.

[32] Fujimoto, K., A. Henderson, K. Hirasaura, and J. R. James, "Small Antennas," England: Research Studies Press, 1987, distributed by Wiley & Sons, pp. 86–89.

[33] King, R.W.P., et al., "Transmission-Line Missile Antennas," *IRE Trans.*, Vol. AP-8, 1960, pp. 88–90. Same generic form.

[34] Smith, G. S., "Radiation Efficiency of Electrically Small Multiturn Loop Antennas," *IEEE Trans.*, Vol. AP-20, 1972, pp. 656–657.

[35] Technique used by radio amateurs.

[36] Johnson, R. C., and H. Jasik, Ch. 8 in *Antenna Engineering Handbook*, 2nd edition, New York: McGraw-Hill, 1984.

[37] Ibid.

[37a] Silver, S., "Microwave Antenna Theory and Design," McGraw-Hill, 1949, pp. 167–168.

[38] Kuboyama, H., K. Fujimoto, and K. Hirasawa, "UHF Bent-Slot Antenna System for Portable Equipment-I," *IEEE Trans.*, Vol. VT-36, No. 2, May 1987, p. 78.

[39] Ito, K., and S. Sasaki, "A Small Printed Antenna Composed of Slot and Wide Strip for Indoor Communication Systems" *IEEE Int. AP-S Digest*, 1988, pp. 716–719.

[40] James, J. R., and A. Henderson, "Electrically Short Monopole Antennas With Dielectric or Ferrite Coatings," *Proc. IEE*, Vol. 125, No. 9, 1978, pp. 793–803.

[41] James, J. R., "Reduction of Antenna Dimensions by Dielectric Loading," *Electronics Letters*, Vol. 10, 1974, pp. 263–265.

[42] Woodman, K. F., "Dielectric-Clad Discone," *Electronics Letters*, Vol. 13, 1977, pp. 264–265.

[42a] James, J. R., and A. Henderson, "Investigation of Electrically Small VHF and HF Cavity-Type Antennas," *Proc. Int. Conf. Ant. Propag.*, London, 1978, pp. 322–326.

[43] Johnson, R. C., and H. Jasik, Ch. 5 in *Antenna Engineering Handbook*, 2nd edition, New York: McGraw-Hill, 1984.

[44] Johnson, R. C., and H. Jasik, Ch. 8 in *Antenna Engineering Handbook*, 2nd edition, New York: McGraw-Hill, 1984.

[45] Brunner, J. E., and G. Seward, "Low Profile VHF Antenna for Armor," *Proc. ECOM-ARO Workshop on Electrically Small Antennas*, G. Goubau, and F. Schwering, ed., 6–7 May 1976, Fort Monmouth, pp. 153–157.

[45a] Hansen, R. C., "Superconducting Antennas," *IEEE Trans.*, AES, Vol. 26 No. 2, 1990, pp. 345–355.

[46] Lee, T. M., et al., "Microwave Field-Detecting Element Based on Pyroelectric Effect in PVDF," Electronic Letters, Vol. 22, No. 4, 1986, pp. 200–202.

[47] James, J. R., and P. S. Hall, eds., *Handbook of Microstrip Antennas*, Vols. 1 and 2, London: Peter Peregrinus, IEE, 1989.

[48] James, J. R., and P. S. Hall, eds., *Handbook of Microstrip Antennas*, Vols. 1 and 2, London: Peter Peregrinus, IEE, 1989, pp. 25–26.

[49] Penard, E., and J. P. Daniel, "Open and Hybrid Microstrip Antennas," *IEE Proc.*, Vol. 131, pt. H, 1984.

[50] Richards, W. F., et al. "Theoretical and Experimental Investigation of Annular, Annular Sector and Circular Sector Microstrip Antennas," *IEEE Trans.*, Vol. AP-12, 1984, pp. 864–866.

[51] Oswald, L. T., and C. W. Garvin, "Microstrip Command and Telemetry Antennas for Communications and Technology Satellites," *IEE Int. Conf. on Antennas for Aircraft Spacecraft*, London, 1975, pp. 217–222.

[52] Haneishi, M., et al., "Broadband Microstrip Array Composed of Single Feed Type Circularly Polarized Microstrip Element," *IEEE AP-S Int. Symp. Digest*, May 1982, pp. 160–163.

[53] Sabban, A., "New Broadband Stacked Two Layer Microstrip Antenna," *IEEE AP-S Int. Symp. Digest*, Houston, 1983, pp. 63–66.

[54] Kumar, G., and K. C. Gupta, "Non-Radiating Edge and Four Edges Gap Coupled Multiple Resonator Broadband Microstrip Antenna," *IEEE Trans.*, Vol. AP-33, 1985, pp. 173–177.

[55] James, J. R., and P. S. Hall, eds., *Handbook of Microstrip Antennas*, Vols. 1 and 2, London: Peter Peregrinus, IEE, 1989, pp. 1096–1099.

[55a] James, J. R., and P. S. Hall, eds., *Handbook of Microstrip Antennas*, Vols. 1 and 2, London: Peter Peregrinus, IEE, 1989, pp. 1099–1102.

[56] James, J. R., and G. Andrasic, "Superimposed Dichroic Microstrip Antenna Arrays," *Proc. IEE.*, Vol. 135H, No. 5, 1988, pp. 304–312.

[57] James, J. R., and P. S. Hall, eds., *Handbook of Microstrip Antennas*, Vols. 1 and 2, London: Peter Peregrinus, IEE, 1989, pp. 1124–1125.

[58] Microstrip used as bent slot. See [38].

[59] Arai, H., "Flat 4—Direction Antenna for Antenna Pattern Diversity Reception," *Proc. Int. Conf. Ant. Prop.*, ICAP'91, York, England, April 1991, pp. 46–49.

[60] Johnson, R. C., and H. Jasik, Ch. 4 in *Antenna Engineering Handbook*, 2nd edition, New York: McGraw-Hill, 1984.

[61] Johnson, R. C., and H. Jasik, Ch. 43 in *Antenna Engineering Handbook*, 2nd edition, New York: McGraw-Hill, 1984.

[62] Johnson, R. C., and H. Jasik, Ch. 43 in *Antenna Engineering Handbook*, 2nd edition, New York: McGraw-Hill, 1984.

[63] Johnson, R. C., and H. Jasik, Ch. 43 in *Antenna Engineering Handbook*, 2nd edition, New York: McGraw-Hill, 1984.

[64] Johnson, R. C., and H. Jasik, Ch. 43 in *Antenna Engineering Handbook*, 2nd edition, New York: McGraw-Hill, 1984.

[65] Johnson, R. C., and H. Jasik, Ch. 43 in *Antenna Engineering Handbook*, 2nd edition, New York: McGraw-Hill, 1984.

[66] Parhami, P., et al., "Technique for Calculating the Radiation and Scattering Characteristics of Antennas Mounted on a Finite Ground Plane," *Proc. IEE*, Vol. 124, 1977, pp. 1009–1016.

[66a] Fujimoto, K., A. Henderson, K. Hirasaura, and J. R. James, Ch. 4 in "Small Antennas," England: Research Studies Press, 1987, distributed by Wiley & Sons.

[67] Hansen, J. E. ed., *Spherical Near-Field Antenna Measurements*, London: Peter Peregrinus, IEE, 1988, Chapter 2.

[68] Hori, T., et al., "Circularly Polarized Microstrip Array Antenna With Conical Beam," *Natl. Conv. Rec. IECE Japan*, Vol. 655, 1982, (in Japanese).

[69] Hori, T., et al., "Circularly Polarized Broadband Microstrip Antenna Radiating Conical Beam," *Natl. Conv. Rec. IECE Japan*, Vol. 637, 1986, (in Japanese).

[70] Yasunaga, M., et al., "Phased Array Antennas for Aeronautical Satellite Communications," *IEE Int. Conf. Ant. Prop.*, ICAP '87, York, pp. 47–50.

[71] James, J. R., and P. S. Hall, eds., *Handbook of Microstrip Antennas*, Vols. 1 and 2, London: Peter Peregrinus, IEE, 1989, pp. 1112–1121.

[72] James, J. R., and P. S. Hall, eds., *Handbook of Microstrip Antennas*, Vols. 1 and 2, London: Peter Peregrinus, IEE, 1989, pp. 1142–1145.

[73] Hawkins, G. J., et al., "Electronic Beam Squint Tracking for Land Mobile Satellite Terminals," *IEEE Vehicular Technology Conf.*, 1989, pp. 735–741.

[74] Shafai, L., "MSAT Vehicular Antennas With Self Scanning Array Elements," *Proc. Int. Mobile Satellite Conf.*, IMSC '90, Ottawa, June 1990, pp. 523–528.

[74a] Sanford, J., et al., "Shaped Beam Patch Arrays for Mobile Communication Base Stations," *Microwave Engineering Europe*, June/July 1991, pp. 31–33.

[75] Rudge, A. W., et al., eds., *The Handbook of Antenna Design*, London: Peter Peregrinus, 1986, pp. 1470–1488.

[76]

[77] Rudge, A. W., et al., eds., *The Handbook of Antenna Design*, London: Peter Peregrinus, 1986, pp. 1470–1488.

[78] Hansen, R. C., "Optimum Inductive Loading of Short Whip Antennas," *IEEE Trans.*, Vol. VT-24, 1975, p. 21.

[79] Fournier, M., and A. Pomerleau, "Experimental Study of Inductively Loaded Short Monopole," *IEEE Trans. Vehicular Tech.*, Vol. VT-27, 1978, p. 1.

[80] Nishikawa, K., and Y. Asano, "Vertical Radiation Patterns of Mobile Antenna in UHF Band," *IEEE Trans. Vehicular Tech.*, Vol. VT-35, May 1986, pp. 57–62.

[81] Horn, D. W., "Cellular 'On-Glass' Antenna Technology," *Proc. 34th IEEE Vehicular Technology Conf.*, Pittsburgh, 21–23 May, 1984, pp. 65–68.

[82] Johnson, R. C., and H. Jasik, *Antenna Engineering Handbook*, 2nd edition, New York: McGraw-Hill, 1984, pp. 4-18–4-20.

[83] Rudge, A. W., et al., eds., *The Handbook of Antenna Design*, London: Peter Peregrinus, 1986, pp. 1420–1436.

[84] Jasik, H., ed., Ch. 22 in *Antenna Engineering Handbook*, 1st edition, New York: McGraw-Hill, 1961, pp. 24-4–24-5.

[85] Fujimoto, K., A. Henderson, K. Hirasaura, and J. R. James, "Small Antennas," England: Research Studies Press, 1987, distributed by Wiley & Sons, pp. 116–127.

[86] Johnson, R. C., and H. Jasik, *Antenna Engineering Handbook*, 2nd edition, New York: McGraw-Hill, 1984, pp. 27-18–27-21.

[87] Toriyama, H., et al., "Development of Printed-On Glass TV Antenna System for Car," *IEEE Vehicular Tech. Conf.*, 1987, pp. 334–342.

[88] Fukumura, H., "Space Diversity FM Radio System for Automobiles," IECE Tech. G, SANE 82-20, 1982, pp. 9–13.

[89] Hirano, M., et al., "Keyless Entry System With Radio Card Transponder," *IEEE Trans. Industrial Electronics*, Vol. IE-35, No. 2 1988, pp. 208–216.
[90] Ebine, Y., and M. Karikomi, "Printed Dipole Antenna With a Parasitic Element," *IEICE Tech. G.*, Vol. AP89-12, 1989. (in Japanese).
[91] "Vehicular Antenna," Japan Design Patent 793435, Aug. 1990.
[92] "Vehicular Antenna," Japan Design Patent 793436, Aug. 1990.
[93] Lee, W.C.Y., *Mobile Communication Engineering*, New York: Wiley & Sons, 1982, p. 174.
[94] Ibid., p. 175.
[95] Sakitani, A., and S. Egashira, "Analysis of Coaxial Collinear Antenna: Recurrence Formula of Voltages and Admittances at Connections," *IEEE Trans. Ant. Propag.*, Vol. AP-39, No. 1, Jan. 1991, pp. 15–20.
[96] Lee, W.C.Y., *Mobile Communication Engineering*, New York: Wiley & Sons, 1982, p. 165.
[97] Fujimoto, K., et al., "Application in Mobile and Satellite Systems," Chapter 19 in *Handbook of Microstrip Antennas*, J. R. James and P. S. Hall, eds., London: Peter Peregrinus, 1989.
[98] Mishima, H., and T. Taga, "Mobile Antennas and Duplexer for 800 MHz Band Mobile Telephone System," *IEEE AP-S Int. Symp. Digest.*, Quebec, June 1980, pp. 508–511.
[99] Mishima, H., and T. Taga, "Antenna and Dupelexer for New Mobile Radio Unit," *Review of the ECL, NTT*, Vol. 30, No. 2, March 1982, pp. 359–370.
[100] Kumar, A., "Fixed and Mobile Terminal Antennas," Norwood, MA: Artech House, 1991, pp. 116–169.
[101] Fujimoto, K., A. Henderson, K. Hirasaura, and J. R. James, Ch. 4 in "Small Antennas," England: Research Studies Press, 1987, distributed by Wiley & Sons.
[102] Sato, K., et al., "Characteristics of a Planar Invented-F Antenna on a Rectangular Conducting Body," *Electronics and Communications in Japan*, Scripta Publishing, Part 1, Vol. 72, October 1989, pp. 43–51.
[103] Taga, T., and K. Tsunekawa, "Performance Analysis of a Built-In Planar Inverted-F Antenna for 800 MHz Band Portable Radio Units," *IEEE Trans.*, Selected Areas in Communication, Vol. SAC-5, June 1987, pp. 921–929.
[104] Taga, T., "Analysis of Planar Inverted-F Antennas and Antenna Design for Portable Radio Equipment," Chapter 5 in *Analysis Design and Measurement of Small and Low-Profile Antennas*, K. Hirasawa and M. Haneishi, eds., Norwood, MA: Artech House, 1991.
[105] Fujimoto, K., A. Henderson, K. Hirasaura, and J. R. James, "Small Antennas," England: Research Studies Press, 1987, distributed by Wiley & Sons, pp. 59–75.
[106] Kobayashi, K., S. Nishiki, T. Taga, and A. Sasaki, "Detachable Mobile Radio Units for the 800 MHz Land Mobile Radio System," *Proc. 34th IEEE Vehicular Technology Conf.*, Pittsburgh, 21–23 May 1984, pp. 6–11.
[107] Shimizu, I., S. Urabe, K. Hirasawa, K. Nagata, and S. Yuki, "A New Pocket-Size Cellular Telephone for NTT High-Capacity Land Mobile Communication System," *Proc. 41st IEEE Vehicular Technology Conf.*, St. Louis, 19–22 May 1991, pp. 114–119.
[108] Taga, T., K. Tsunekawa, and A. Sasaki, "Antennas for Detachable Mobile Radio Units," *Review of the ECL, NTT*, Japan, Vol. 35, No. 1, Jan. 1987, pp. 59–65.
[109] Tsunekawa, K., "Diversity Antennas for Portable Telephones," *Proc. 39th IEEE Vehicular Tech. Conf.*, San Francisco, 1989, pp. 50–56.
[110] Fujimoto, K., A. Henderson, K. Hirasaura, and J. R. James, "Small Antennas," England: Research Studies Press, 1987, distributed by Wiley & Sons, pp. 89–100.
[111] Wheeler, H. A., "Small Antennas," *IEEE Trans. Ant. Propag.*, Vol. AP 23, July 1975, pp. 462–469.
[112] Ishii, N., K. Itoh, "A Consideration on the Numerical Method for a Card-Sized Thin Planar Antenna," Paper of Technical Group, *IEICE Japan*, Vol. AP91-36 (in Japanese), 1991, pp. 9–14.

[113] U.S. Patent 4873527, Oct. 1989.

[114] Sato, S., and Y. Naito, "Dipole Antenna Covered With Ferrite Sleeve," *Trans. IECE Japan*, Vol. J58-B, June 1975, pp. 285–292 (in Japanese).

[115] Watanabe, H., "Electronic Control and Communication System of Shinkansen," *Korona-sha*, Tokyo, 1982, p. 104 (in Japanese).

[116] Nakajima, N., et al., "A Major Angle Corner Reflector Antenna With 180° Beam Width," *Natl. Conv. Rec. IECE Japan*, Vol. 752, 1985, (in Japanese).

[117] Lee, W.C.Y., *Mobile Communication Engineering*, New York: Wiley & Sons, 1982, p. 306.

[118] Johnson, R. C., and H. Jasik, *Antenna Engineering Handbook*, 2nd edition, 1984, pp. 37-18–37-33.

[119] IECE Japan, *Antenna Engineering Handbook*, Tokyo Japan: Ohm-Sha Co., 1980, pp. 319–388 (in Japanese).

[120] Burberry, R. A., "The Rationalization of Aircraft Antennas," *IEE Int. Conf. Antennas for Aircraft and Spacecraft*, June 1975, pp. 204–209.

[121] Burberry, R. A., "Progress in Aircraft Aerials," *Proc. IEE*. Vol. 109, pt. B, Nov. 1962, pp. 431–444.

[122] Granger, J.V.N., and J.T. Bolljahn, "Aircraft Antennas," *Proc. IRE*, Vol. 43, May 1955, pp. 533–550.

[123] Jasik, H., ed., Ch. 22 in *Antenna Engineering Handbook*, 1st edition, New York: McGraw-Hill, 1961. pp. 27-8–27-40.

[124] Marin, L., J. P. Castillo, and K.S.H. Lee, "Broad-Band Analysis of VLF/LF Aircraft Wire Antennas," *IEE Trans. Ant. Propagat.*, Vol. AP-26, Jan. 1978, pp. 141–145.

[125] Kossey, P. A., E. A. Lewis, and E. C. Field, "Relative Characteristics of TE/TM Waves Excited by Airborne VLF/LF Transmitters," *AGARD Conf. Prog.*, Vol. 305, 1982, p. 19–1.

[126] Hirano, M., ed., *New Radio Wave Technologies*, Part 2, Tokyo Japan: RATESU Co., 1973, pp. 276–281 (in Japanese).

[127] Fujii, K., "Loop Antenna for Aircrafts," Japanese Utility Model 914431, 1970.

[128] Rudge, A. W., et al., eds., *The Handbook of Antenna Design*, London: Peter Peregrinus, 1986, pp. 736–820.

[129] Yasunaga, Y., F. Watanabe, T. Shiokawa, and M. Yamada, "Phased Array Antennas for Aeronautical Satellite Communications," *IEE ICAP '87*, Part 1, March 1987, pp. 47–50.

[130] Cumming, W. A., and M. Cormier, "Design Data for Small Annular Slot Antennas," *IRE Trans. Ant. Propagat.*, Vol. AP-6, April 1958, pp. 210–211.

[131] Blackband, W. T., "A Comparison of the HF Aerials Fitted to XV814," *IEE Int. Conf. Antennas for Aircraft and Spacecraft*, June 1975, pp. 171–177.

[132] Kraus, J., "The Helical Antenna," *Proc. IRE*, Vol. 37, March 1949, pp. 263–272.

[133] King, H. E., and J. L. Wong, "Characteristics of 1 to 8 Wavelength Uniform Helical Antennas," *IEEE Trans. Ant. Propag.*, Vol. AP-28, March 1980, pp. 291–296.

[134] Wong, J. L., and H. E. King, "Empirical Helix Antenna Design," *IEEE AP-S Int. Symp. Digest*, May 1982, pp. 366–369.

[135] Nakano, H., J. Yamauchi, and H. Mikami, "Tapered Balanced Helices Radiating in the Axial Mode," *IEEE AP-S Int. Symp. Digest*, June 1980, pp. 700–703.

[136] Angelakos, D. J., and D. Kajfez, "Modifications on the Axial-Mode Helical Antenna," *IEEE Proc.*, Vol. 55, April 1967, pp. 558–559.

[137] Kilgus, C. C., "Shaped-Conical Radiation Pattern Performance of the Backfire Quadrifilar Helix," *IEEE Trans. Ant. Propag*. Vol. AP-23, May 1975, pp. 392–397.

[138] Dyson, J. D., "The Characteristics and Design of the Conical Log-Spiral Antenna," *IEEE Trans. Ant. Propag.*, Vol. AP-13, July 1965, pp. 488–499.

[139] Dyson, J. D., "The Unidirectional Equiangular Spiral Antenna," *IRE Trans. Ant. Propag.*, Vol. AP-7, Oct. 1959, pp. 329–334.

[140] Dyson, J. D., "The Equiangular Spiral Antenna," *IRE Trans. Ant. Propag.*, Vol. AP-7, April 1959, pp. 181–188.

[141] Kaiser, J. A., "The Archimedean Two-Wire Spiral Antenna," *IRE Trans. Ant. Propag.*, Vol. AP-8, May 1960, pp. 312–323.

[142] Hirasawa, K., and M. Haneishi, ed., *Introductory Analysis Design and Measurement of Small or Low-Profile Antennas*. Dedham, MA: Artech House, 1992, pp. 83–159.

[143] James, J. R., and P. S. Hall, eds., *Handbook of Microstrip Antennas*, Vols. 1 and 2, London: Peter Peregrinus, IEE, 1989, pp. 1057–1191.

[144] Chiba, T., Y. Suzuki, N. Miyano, S. Miura, and S. Ohmori, "A Phased Array Antenna Using Microstrip Patch Antennas," *Proc. 12th European Micro.*, Sept. 1982, pp. 472–477.

[145] Weinschel, H. D., "A Cylindrical Array of Circularly Polarized Microstrip Antennas," *IEEE AP-S Int. Symp. Digest*, June 1975, pp. 177–180.

[146] Suzuki, Y., and T. Chiba, "Designing Method of Microstrip Antenna Considering the Bandwidth," *IECE Trans.*, Vol. E67, Sept. 1984, pp. 488–493.

Acronyms and Abbreviations

A/D	analog-to-digital
ADF	automatic direction finder
AF	audio frequencies
AGC	automatic gain control
AM	amplitude modulation
AMPS	Advanced Mobile Phone System
AMSC	American Mobile Satellite Corporation
ANSI	American National Standards Institute
AR	axial ratio
ATC	air traffic control
ATCRBS	ATC radar beacon system
ATU	automatic tuning unit
BER	bit error rate
BS	broadcast satellite
C/A	coarse acquisition
C/I	carrier-to-interference ratio
C/M	average power ratio of carrier to multipath wave
CAD	computer-aided design
CAT	computerized axial tomography
CBOCS	Comprehensive Bus Operaton Control System
CCIR	International Committee on Radio
CFM	compounded frequency modulation
CIR	carrier-to-interference ratio
CNR	carrier-to-noise ratio
CONUS	continental United States
CPU	central processing unit
CRC	Communications Research Center
CRL	Communications Research Laboratory
CW	continuous wave
D/U	desired-to-undesired signal strength ratio

DBS	direct broadcast satellite
DME	distance measuring equipment
EGC	equal-gain combining
EIRP	equivalent isotropically radiated power
EM	electromagnetic
EMF	electromagnetic field
ENRI	Electronic and Navigation Research Institute
ERP	effective radiated power
ESA	European Space Agency
ESTEC	European Space Research and Technology Center
ETS/V	Engineering Test Satellite-V
EW	east-west
FDTD	finite difference time domain
FEC	forward error correction
FET	field-effect transistor
FM	frequency modulation
FPLMTS	Future Public Land Mobile Telecommunication System
FSK	frequency shift keying
FSPL	free-space path loss
G/T	ratio of system gain to system noise temperature
GEO	geostationary earth orbiting
GLONAS	Global Orbiting Navigation Satellite System
GMDSS	Global Maritime Distress and Safety System
GPS	Global Positioning Satellite
GSM	Groupe Special Mobile
GTD	geometrical theory of diffraction
HP	horizontally polarized
HPBW	half-power beamwidth
IDO	Nippon Idou Tshushin Corporation
IF	intermediate frequency
IFA	inverted-F antenna
IFF	identification friend or foe
ILA	inverted-L antenna
ILS	instrument landing system
INMARSAT	International Maritime Satellite Organization
ISDN	integrated services digital network
JPL	Jet Propulsion Laboratory
KDD	Kokusai Denshin Denwa Co.
LCX	leaky coaxial cable
LEO	low earth orbit
LNA	low-noise amplifier
loran	long-range navigation

LSI	large-scale integration
MCA	multichannel access
MEG	mean effective gain
MES	mobile earth station
MIC	microwave integrated circuit
MOM	method of moments
MRC	maximal ratio combining
MRI	magnetic resonance imaging
MSA	microstrip antenna
MSS	mobile satellite system
NASA	National Aeronautics and Space Administration
NAVSTAR	Navigation System with Time and Ranging/Global Positioning System
NDB	nondirectional beacon
NMHA	normal-mode helical antennas
NNSS	Navy Navigation Satellite System
NS	north-south
NTT	Nippon Telegraph and Telephone Corporation
PAG	pattern averaging gain
PCN	personal communication network
PCS	personal communication system
PHP	Personal Handy Phone
PIFA	planar IFA
PIM	passive intermodulation
PSM	polarization shaping method
PSTN	public service telephone network
PTFE	polytetrafluoroethylene
QHA	quadrifilar helix antenna
R	receiving
RAM	random access memory
RF	radio frequency
RFM	random field measurement
RHCP	right hand polarization
RT	roadside transceiver
SA	specific absorption
SAR	specific absorption rate
SAW	surface acoustic wave
SBF	short backfire antenna
SDM	System Definition Manual
SEC	selective combining
SES	ship earth station
SFCP	singly fed circularly polarized

SMR	special mobile radio
SSB	single-sideband
SSR	secondary surveillance radar
SWC	switch combining
TDD	time division duplex
TDMA	time division multiple access
TEM	transverse electromagnetic
TLM	transmission line method
TR	transmitting and receiving
UHF	ultrahigh frequency
VCO	voltage-controlled oscillator
VHF	very high frequency
VLSI	very-large-scale integration
VOR	VHF omnidirectional range
VP	vertically polarized
VSWR	voltage standing wave ratio
VT	vehicle transceiver
WARC	World Administrative Radio Committee
XPR	cross-polarization ratio
XSA	cross-slot antenna

About the Authors

Dr. Quirino Balzano
Vice President of Technical Staff
Motorola Inc.
Fort Lauderdale
FL 33322
USA

Professor Kyohei Fujimoto
(Professor Emeritus, University of Tsukuba)
Department of Information Engineering
Niigata University
Japan 950-21

Associate Professor Kazuhiro Hirasawa
Institute of Applied Physics
University of Tsukuba
Tsukuba
Japan 305

Dr. Jochen Hopf
Institute for High Frequency Systems
University of the Bundeswehr
Munich
D8014 Neubiberg
Germany

Professor Jim James
Director, Wolfson RF Engineering Centre
Royal Military College of Science
Shrivenham
Swindon
SN6 8LA
England

Dr. Kenichi Kagoshima
Antenna Group Leader
Nippon Telegraph and Telephone Corporation
Kanagawa 238-03
Japan

Dr. William C. Y. Lee
Vice President
Pac Tel Corporation
MS600 Walnut Creek
California
USA

Professor Heinz K. Lindenmeier
Institute for High Frequency Systems
University of the Bundeswehr
Munich
D8014 Neubiberg
Germany

Mr. Richard Mumford
Group Head of Development
Multitone Electronics PLC
Basingstoke
RG22 4AD
England

Dr. Shingo Ohmori
Manager, Planning Section
Communications Research Laboratory
Ministry of Posts and Telecommunications
Koganei-Shi
Tokyo 184
Japan

Dr. Leopold Reiter
Institute for High Frequency Systems
University of the Bundeswehr
Munich
D8014 Neubiberg
Germany

Dr. Takayasu Shiokawa
Senior Research Engineer
KDD R and D Laboratories
2-1-15 Ohara Kamifukuoka-shi Saitama
356 Japan

Dr. Yasuo Suzuki
Senior Research Scientist
Research and Development Center
Toshiba Corporation
Kawasaki
210 Japan

Dr. Tokio Taga
Senior Researcher, Supervisor
Radio Communications Systems Laboratories
Nippon Telegraph and Telephone Corporation
Yokosuka 238-03
Japan

Dr. Tasuku Teshirogi
Director, Electromagnetic Technology Division
Communications Laboratory
Ministry of Posts and Telecommunications
Koganei-Shi
Tokyo 184
Japan

Index

The Artech House Telecommunications Library

Vinton G. Cerf, Series Editor

Advanced Technology for Road Transport: IVHS and ATT, Ian Catling, editor

Advances in Computer Communications and Networking, Wesley W. Chu, editor

Advances in Computer Systems Security, Rein Turn, editor

Analysis and Synthesis of Logic Systems, Daniel Mange

Asynchronous Transfer Mode Networks: Performance Issues, Raif O. Onvural

A Bibliography of Telecommunications and Socio-Economic Development, Heather E. Hudson

Broadband: Business Services, Technologies, and Strategic Impact, David Wright

Broadband Network Analysis and Design, Daniel Minoli

Broadband Telecommunications Technology, Byeong Lee, Minho Kang, and Jonghee Lee

Cellular Radio: Analog and Digital Systems, Asha Mehrotra

Cellular Radio Systems, D. M. Balston and R. C. V. Macario, editors

Client/Server Computing: Architecture, Applications, and Distributed Systems Management, Bruce Elbert and Bobby Martyna

Codes for Error Control and Synchronization, Djimitri Wiggert

Communication Satellites in the Geostationary Orbit, Donald M. Jansky and Michel C. Jeruchim

Communications Directory, Manus Egan, editor

The Complete Guide to Buying a Telephone System, Paul Daubitz

Computer Telephone Integration, Rob Walters

The Corporate Cabling Guide, Mark W. McElroy

Corporate Networks: The Strategic Use of Telecommunications, Thomas Valovic

Current Advances in LANs, MANs, and ISDN, B. G. Kim, editor

Design and Prospects for the ISDN, G. Dicenet

Digital Cellular Radio, George Calhoun

Digital Hardware Testing: Transistor-Level Fault Modeling and Testing, Rochit Rajsuman, editor

Digital Signal Processing, Murat Kunt

Digital Switching Control Architectures, Giuseppe Fantauzzi

Digital Transmission Design and Jitter Analysis, Yoshitaka Takasaki

Distributed Multimedia Through Broadband Communications Services, Daniel Minoli and Robert Keinath

Distributed Processing Systems, Volume I, Wesley W. Chu, editor

Disaster Recovery Planning for Telecommunications, Leo A. Wrobel

Document Imaging Systems: Technology and Applications, Nathan J. Muller

EDI Security, Control, and Audit, Albert J. Marcella and Sally Chen

Enterprise Networking: Fractional T1 to SONET, Frame Relay to BISDN, Daniel Minoli

Expert Systems Applications in Integrated Network Management, E. C. Ericson, L. T. Ericson, and D. Minoli, editors

FAX: Digital Facsimile Technology and Applications, Second Edition, Dennis Bodson, Kenneth McConnell, and Richard Schaphorst

FDDI and FDDI-II: Architecture, Protocols, and Performance, Bernhard Albert and Anura P. Jayasumana

Fiber Network Service Survivability, Tsong-Ho Wu

Fiber Optics and CATV Business Strategy, Robert K. Yates *et al.*

A Guide to Fractional T1, J. E. Trulove

A Guide to the TCP/IP Protocol Suite, Floyd Wilder

Implementing EDI, Mike Hendry

Implementing X.400 and X.500: The PP and QUIPU Systems, Steve Kille

Inbound Call Centers: Design, Implementation, and Management, Robert A. Gable

Information Superhighways: The Economics of Advanced Public Communication Networks, Bruce Egan

Integrated Broadband Networks, Amit Bhargava

Integrated Services Digital Networks, Anthony M. Rutkowski

International Telecommunications Management, Bruce R. Elbert

International Telecommunication Standards Organizations, Andrew Macpherson

Internetworking LANs: Operation, Design, and Management, Robert Davidson and Nathan Muller

Introduction to Satellite Communication, Bruce R. Elbert

Introduction to T1/T3 Networking, Regis J. (Bud) Bates

Introduction to Telecommunication Electronics, A. Michael Noll

Introduction to Telephones and Telephone Systems, Second Edition, A. Michael Noll

An Introduction to U.S. Telecommunications Law, Charles H. Kennedy

Introduction to X.400, Cemil Betanov

The ITU in a Changing World, George A. Codding, Jr. and Anthony M. Rutkowski

Jitter in Digital Transmission Systems, Patrick R. Trischitta and Eve L. Varma

Land-Mobile Radio System Engineering, Garry C. Hess

LAN/WAN Optimization Techniques, Harrell Van Norman

LANs to WANs: Network Management in the 1990s, Nathan J. Muller and Robert P. Davidson

The Law and Regulation of International Space Communication, Harold M. White, Jr. and Rita Lauria White

Long Distance Services: A Buyer's Guide, Daniel D. Briere

Measurement of Optical Fibers and Devices, G. Cancellieri and U. Ravaioli

Meteor Burst Communication, Jacob Z. Schanker

Minimum Risk Strategy for Acquiring Communications Equipment and Services, Nathan J. Muller

Mobile Antenna Systems Handbook, K. Fujimoto, J. R. James

Mobile Communications in the U.S. and Europe: Regulation, Technology, and Markets, Michael Paetsch

Mobile Information Systems, John Walker

Narrowband Land-Mobile Radio Networks, Jean-Paul Linnartz

Networking Strategies for Information Technology, Bruce Elbert

Numerical Analysis of Linear Networks and Systems, Hermann Kremer *et al.*

Optimization of Digital Transmission Systems, K. Trondle and Gunter Soder

Packet Switching Evolution from Narrowband to Broadband ISDN, M. Smouts

Packet Video: Modeling and Signal Processing, Naohisa Ohta

The PP and QUIPU Implementation of X.400 and X.500, Stephen Kille

Principles of Secure Communication Systems, Second Edition, Don J. Torrieri

Principles of Signals and Systems: Deterministic Signals, B. Picinbono

Private Telecommunication Networks, Bruce Elbert

Radiodetermination Satellite Services and Standards, Martin Rothblatt

Residential Fiber Optic Networks: An Engineering and Economic Analysis, David Reed

Secure Data Networking, Michael Purser

Setting Global Telecommunication Standards: The Stakes, The Players, and The Process, Gerd Wallenstein

Smart Cards, José Luis Zoreda and José Manuel Otón

Secure Data Networking, Michael Purser

The Telecommunications Deregulation Sourcebook, Stuart N. Brotman, editor

Television Technology: Fundamentals and Future Prospects, A. Michael Noll

Telecommunications Technology Handbook, Daniel Minoli

Telephone Company and Cable Television Competition, Stuart N. Brotman

Teletraffic Technologies in ATM Networks, Hiroshi Saito

Terrestrial Digital Microwave Communciations, Ferdo Ivanek, editor

Transmission Networking: SONET and the SDH, Mike Sexton and Andy Reid

Transmission Performance of Evolving Telecommunications Networks, John Gruber and Godfrey Williams

Troposcatter Radio Links, G. Roda

UNIX Internetworking, Uday O. Pabrai

Virtual Networks: A Buyer's Guide, Daniel D. Briere

Voice Processing, Second Edition, Walt Tetschner

Voice Teletraffic System Engineering, James R. Boucher

Wireless Access and the Local Telephone Network, George Calhoun

Wireless LAN Systems, A. Santamaría and F. J. Lopez-Hernandez

Writing Disaster Recovery Plans for Telecommunications Networks and LANs, Leo A. Wrobel

X Window System User's Guide, Uday O. Pabrai

For further information on these and other Artech House titles, contact:

Artech House
685 Canton Street
Norwood, MA 01602
617-769-9750
Fax: 617-762-9230
Telex: 951-659
email: artech@world.std.com

Artech House
Portland House, Stag Place
London SW1E 5XA England
+44 (0) 71-973-8077
Fax: +44 (0) 71-630-0166
Telex: 951-659